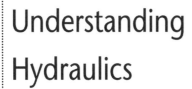

Understanding

Hydraulics

Understanding Hydraulics

SECOND EDITION

LES HAMILL
School of Civil and Structural Engineering
University of Plymouth

First published 1995
Reprinted three times
Second edition 2001
Published by
PALGRAVE MACMILLAN
Houndmills, Basingstoke, Hampshire RG21 6XS and
175 Fifth Avenue, New York, N. Y. 10010
Companies and representatives throughout the world

PALGRAVE MACMILLAN is the new global academic imprint of St. Martin's Press LLC Scholarly and Reference Division and Palgrave Publishers Ltd (formerly Macmillan Press Ltd).

ISBN 0–333–77906–1
ISBN 13978–0–333–77906–4

This book is printed on paper suitable for recycling and made from fully managed and sustained forest sources.

A catalogue record for this book is available from the British Library.

10   9   8   7
15   14   13   12   11   10   09   08   07

Printed by Cromwell Press, Trowbridge

Contents

Preface to first edition

Many conventional text books on hydraulics or fluid mechanics appear to be written for the benefit of people who already understand the subject, so their principal function is as a reference or memory aid. Since the material covered on the first and second years of an honours degree course or HNC/HND is the easiest, this usually gets a brief treatment in a couple of introductory chapters with little in the way of explanation. Thus anyone new to the subject often finds it difficult.

This text book is an ideal introduction to hydraulics for students on degree courses, BTEC HNC or HND, GNVQ levels 3–5 or the equivalent. It assumes very little in the way of prior knowledge, and can be used as a student workbook for student centred learning. It offers advice and guidance on how to approach the subject, avoid making mistakes and solve problems. There are numerous examples, Self Test Questions (with guide solutions in Appendix 2) and Revision Questions (a Solutions Manual for lecturers is available from the Publishers). It can be used in conjunction with other more conventional texts, if required.

The book tries to foster an ***understanding*** of hydraulics, so some of the chapters contain simple experiments that readers can perform using, for example, a ping-pong ball and a funnel. Thus it approaches hydraulics in a more light hearted way than some other texts. As a means of highlighting key points or questions there is sometimes a dialogue between a student representative (Spike, who appears on the left below) and the Prof, who has the answers.

66 This is Spike.
Spike is trying to understand the basic principles of hydraulics.
He has a lot of questions he wants to ask. **99**

66 This is the Prof.
He will guide you through the text. **99**

Spike and the Prof are a device that has been designed to help readers to learn, that is as a means of asking questions on your behalf and receiving answers. The fact that they are included does not diminish the academic treatment of the subject, but provides a more user-friendly means of presenting it. How many people, for instance, can honestly say that they have completely read an academic text book, or that they found parts of it interesting or enjoyable? For this reason some of the more mathematical or longer derivations of equations are presented in Appendix 1, while the text concentrates on explaining what happens and why.

Start by reading the Introduction. Then move on to the beginning of Chapter 1 which deals with some fundamentals of the subject. Please try to work through the Self Test Questions for yourself, and avoid looking at the guide solutions in Appendix 2. Knowing how someone else solved the problem is not the same as being able to solve it yourself.

Good luck with your studies.

L. Hamill

 ## Acknowledgements

The author wishes to thank all those who have contributed in any way to the preparation of this book. In particular, the following organisations are thanked for specific permission to reproduce text and figures from other published works.

BSI Standards – for permission to reproduce, as Figures 5.10 and 9.29, material extracted from BS 3680: Part 4A 1981 and Part 4C: 1981. Complete copies of BS 3680 may be obtained by post from BSI Sales, Linford Wood, Milton Keynes, MK14 6LE.

Chapman & Hall Ltd – for permission to reproduce the following figures taken from *Fluid Mechanics for Civil Engineers*, by N. B. Webber, SI Edition, 1971 (original figure number and page shown first). Figure 5.7, p. 92 – Figure 6.14; Figure 8.7, p. 178 (modified) – Figure 8.24; Figure 8.8b, p. 179 (modified) – Figure 8.24; Figure 8.9, p. 180 (modified) – Figure 8.25; Figure 10.1, p. 255 – Figure 11.19; Figure 10.3, p. 258 – Figure 11.15; Figure 10.5, p. 261 – Figure 11.16; Figure 10.13, p. 270 – Figure 11.11; Figure 10.21(b), p. 285 – Figure 11.14; Figure 10.22(b), p. 286 – Figure 11.14; Figure 10.23, p. 287 – Figure 11.14; Figure 10.27, p. 296 – Figure 11.20.

HR, Wallingford – for Figure 10.2.

Hodder & Stoughton Ltd – for permission to reproduce in modified form as Figure 11.22, Figure 12.8 on p. 384 of *Water Supply*, by A. C. Twort, R. C. Hoather and F. M. Law, 1974 (1st edn).

New Civil Engineer – for Figure 11.13.

Sulzer-Escher Wyss Ltd, Zurich – for Figures 11.3, 11.8 and 11.12.

United States Department of the Interior, Bureau of Reclamation, Denver, USA – for permission to use Figure 9.42(a) Type II Basin Dimension, p. 395 from *Design of Small Dams*, 3rd edition, 1987 – as Figure 8.26.

Every effort has been made to obtain copyright permission where necessary. Any omissions notified will be rectified at the earliest opportunity.

Preface to second edition

The new edition of *Understanding Hydraulics* has 13 chapters, instead of the original 10. A new Chapter 9 covers hydraulic structures, such as dams, sluice gates, bridges and culverts. It also now includes broad crested weirs, which were in Chapter 8. That chapter, which is concerned with flow in open channels, now additionally covers compound channels, non-uniform gradually varied flow and the determination of the water surface profile. Chapters 12 and 13 have been added to provide an introduction to engineering hydrology. At first, hydrology may not seem to be an obvious topic to include in a text on hydraulics, but a moment's reflection soon reveals that there are obvious overlaps. For example, before sewers, open channels and flood alleviation works can be designed, it is necessary to calculate the flow they have to carry. The new Flood Estimation Handbook produced by the Institute of Hydrology arrived just in time for some of the concepts to be included in Chapter 13. Other topics such as the effects of global warming, water supply reservoirs and groundwater are also covered.

It is natural for readers to ask why this or that topic was not included. To some extent the answer is 'length'. There is a limit to the number of topics that can be included, while simultaneously trying to maintain the clarity and simplicity of explanation that were the reason for writing the first edition. Perhaps another time. Nevertheless, the new material extends the book from a first or second year undergraduate text into the later years of such courses, and into professional practice.

The book also recognises changes in the way students study and learn. In a few places the addresses of websites are given so that some topics can be explored in more detail. It is not certain what the longevity of Internet sites will be but, since several of them are Government sites, perhaps they stand a better chance than many of still being there in 5 or 10 years' time.

I would like to thank everyone who has helped to supply information for the second edition, including Felicity Sanderson, Terry Marsh and Duncan Reed at the Centre for Ecology and Hydrology (CEH). Until recently the CEH was known as the Institute of Hydrology (IOH), and Chapters 12 and 13 include acknowledgements to both. The maps of rainfall, evapotranspiration, soil moisture deficit and runoff were derived by the CEH from Meterological Office data, while river flow is recorded by the Environment Agency. My thanks are extended to them also. Tim Wood at the Environment Agency in Exeter has been helpful on many occasions over the years. Simon Woods at TecQuipment Ltd has provided several of the illustrations, and I would like to express my gratitude to him and his colleagues. When appropriate, acknowledgements are attached to all of the illustrations and tables. I apologise if I have accidently omitted anyone; once notified this will be rectified at the earliest opportunity.

I hope you find the new edition useful. It may also be a curious thing to say of a text book, but I hope you also find parts of it interesting!

October 2000   Les Hamill

Acknowledgements

The author and Publisher wish to thank all those who have contributed in any way to the preparation of this book. In particular, the following:

Pearson Education – for permission to reproduce *page 811, Appendix 2* from *Fluid Mechanics, 3rd edition*, by J F Douglas, J M Gasiorek and J A Swaffield. Figure 4.30 (page 116).

Water Authorities Association. Figure 8.14 (page 248).

Routledge – for permission to reproduce Figure 8.13 from *Fluid Mechanics for Civil Engineers* by N B Webber. Figure 8.29 (page 275).

The McGraw-Hill Companies – for permission to reproduce Table 8.2 from *Water Resources Engineering, 3rd edition (1979)* by Linsley Franzini, Freyberg and Tchobanoglous. Table 9.1 (page 303).

Technomic Publishing Co., Inc for permission to reprint Figures 17 and 19 of Chapter 15, Scour at Bridge Sites by B W Melville from *Civil Engineering Practice 2 (Hydraulics/Mechanics) 1988* by P N Cherememisinoff, N P Cherememisinoff and S L Gheng (Eds). Figures 9.17 and 9.18 (pages 322 and 324).

United States Department of the Interior, Bureau of Reclamation, Denver, USA for permission to include Table 6.3 *Representative Friction Factors for Foundation Material from Design of Small Dams, USBR, 1960*, and from *Hydraulic Engineering* (ISBN 0471124664) by Robertson published by John Wiley & Sons Inc. Table 9.2 (page 307). And for supplying photographs of the Monticello Dam (page 304).

University of Toronto Press, Ontario, Canada for permission to reprint Tables 4.2 and 4.3 on pages 95 and 96 of *Guide to Bridge Hydraulics 1973* (ISBN 0–8020–1961–7) by C R Neill (Ed.). Table 9.4 (page 323).

American Society of Civil Engineers, Reston, USA for permission to reproduce Fig. 35 of *Diffusion of Submerged Jets* by H R Henry (Discussion of paper by Albertson, Dai, Jenson and Rouse) from the *ASCE Transactions*, December 1948. Figure 9.13 (page 317).

John Paul Photography for permission to reproduce a photograph of the Inverness Railway Viaduct Collapse, 1989. Figure 9.15 (page 320).

ITPS Ltd, Andover, Hampshire on behalf of Routledge for permission to reproduce Figure 5.1 of *Bridge Hydraulics* by L Hamill, 1999 [ISBN 0–419–20570–5]. Figure 9.20 (page 326).

ITPS Ltd, Andover, Hampshire on behalf of Routledge for permission to reproduce data in Table 1.1.5, Chapter 1, The World Hydrological Cycle, Author R G Barry from *Water, Earth, and Man, 1969* [ISBN 0–416–12030–X] by R J Chorley (Ed.). Table 12.2 (page 437).

National Water Archive, Wallingford, Oxford for permission to quote the following data: *Commissioned maps of 6190 average rainfall, potential evapotranspiration, runoff and soil moisture deficit* by Centre for Ecology and Hydrology; *Observed hydrograph and rainfall event 4083 and flow data for River Warleggan at Trengoffe* by Centre for Ecology and Hydrology; *Annual maximum flood peak data from the CD-ROM for River Warleggan, Tamar and St Neot* by Institute of Hydrology, 1999. Figures 12.4, 12.5, 12.6 and 12.7 (pages 446, 447, 448 and 449).

Centre for Ecology and Hydrology, Wallingford, Oxford for permission to quote the following data: Figure 3.1 from *Flood Estimation Handbook, Volume 2, 1999* by Institute of Hydrology. Figure 12.9 (page 455).

Thomas Telford Publishing, London for permission to reproduce information in Table 1 from *Floods and Reservoir Safety, 3rd Edition*, by Institution of Civil Engineers, published by Institution of Civil Engineers, London, 1996.

Centre for Ecology and Hydrology, Wallingford, Oxford for permission to include Part of Figure 4 (MORECS square 174), Figure 10 and part of Figure 11 (Dial Farm and Compton House) from *Hydrological Data United Kingdom, 1995 Yearbook, 1996* [ISBN 0 948540 78 8]. Figures 13.1 and 13.20 (pages 479 and 526).

HR Wallingford, Wallingford, Oxford for permission to quote data in table on page 6 of *Design and analysis of urban storm drainage, The Wallingford Procedure, Volume 4, The Modified Rational method, 1981* by The Standing Technical Committee on Sewers and Water Mains [ISBN 0–901090]. Table 13.10 (page 517).

Western Morning News Co. Ltd, Plymouth for permission to reproduce a photograph of the aftermath of the 1952 Lynmouth flood. Figure 13.11 (page 508).

Butterworth Heinemann Publishers, a division of Reed Educational & Professional Publishing Ltd and Hodder Headline Group for permission to include Figure 3.6 from *Water Supply, Second Edition* by Twort, Hoather and Law. Figure 13.12 (page 510).

Environment Agency, Exeter for permission to use aerial photograph of Exwick flood relief channel. Figure 13.14 (page 512).

F Walters for photography showing flooding in Teignmouth. Figure 13.17 (page 516).

Principal notation

a	acceleration, area
a_J	area of jet (at vena contracta)
A	area (e.g. of pipe, cross-section, catchment)
AE	actual evaporation
AET	actual evapotranspiration
A_P	area of wetted perimeter
A_{WS}	area of water surface
b	width or breadth (e.g. of weir)
B	width (e.g. of channel)
B_S	water surface width (e.g. in a channel)
c	velocity of sound
C	Chezy coefficient (e.g. channel roughness)
C_C	coefficient of contraction (e.g. orifice)
C_D	coefficient of discharge (e.g. orifice, weir)
C_{DR}	coefficient of drag
C_L	coefficient of lift
C_V	coefficient of velocity (e.g. orifice)
d	diameter, depth
d_J	diameter of jet
D	diameter, depth of flow
D_M	hydraulic mean depth
E	energy
E	evaporation
ET	evapotranspiration
f	infiltration rate
$f( )$	function of thing in brackets
F	force
F	Froude number
F_R	resultant force having two components (e.g. F_H and F_V or F_{RX} and F_{RY})
g	gravitational acceleration ($9.81 \, \mathrm{m/s^2}$)
h	head, depth below water surface
h_F	head loss due to friction (e.g. in a pipe)
h_L	minor head loss in a pipe (e.g. exit loss)
H	head, depth below water surface

i	rainfall intensity
I_G	second moment of area about centroid (m^4)
I_M	moment of inertia about centre of mass ($kg\,m^2$)
I_{WS}	second moment of area in plane of water surface (m^4)
k	roughness (e.g. of a pipe surface)
K	permeability or hydraulic conductivity
L	length
M	mass
Ma	Mach number
n	Manning roughness coefficient
N	rotational speed (e.g. of a pump or turbine)
N_S	specific speed of a pump or turbine
P	probability ($= 1/T$)
P	pressure, pressure intensity
P	wetted perimeter
PE	potential evaporation
PET	potential evapotranspiration
Pow	power (e.g. output of a turbine)
q	discharge per unit width (m^3/s per m)
Q	discharge = volumetric flow rate
Q_T	flood discharge of return period T
QMED	median annual maximum flood
r	radius (e.g. of pipe)
R	radius, hydraulic radius
Re	Reynolds number
S	coefficient of storage
S	slope (e.g. sides of an open channel)
S_O	longitudinal bed slope of a channel
S_F	friction slope = slope of energy line
SMD	soil moisture deficit
t	time
T	time
T	coefficient of transmissivity
T	return period (years)
u	velocity
U	velocity (e.g. of Pelton wheel bucket)
v	velocity
V	velocity (usually mean velocity = Q/A)
V_X	component of velocity in x direction
V_Y	component of velocity in y direction
V_Z	component of velocity in z direction
Vol	volume
w	weight density
W	weight
We	Weber number
z	potential head
Z	elevation (e.g. water surface in reservoir above datum level)
α	angle
α	energy (velocity distribution) coefficient

β	momentum coefficient
δ	difference, increment
Δ	change in
ε_P	overall efficiency of a pump
ε_T	overall efficiency of a turbine
ϕ	angle (e.g. of resultant force)
η	efficiency, proportion of original velocity
λ	pipe friction factor (Darcy equation)
μ	coefficient of dynamic viscosity
ν	coefficient of kinematic viscosity
θ	angle
ρ	mass density
τ	shear stress

Introduction (. . . or read this first!)

66 I am having trouble understanding hydraulics.
Is there anything I can do to make it easier? **99**

66 Firstly, you are not alone. Many, if not most, students find hydraulics difficult when they first meet it. I will make some suggestions that, if you follow them, will help you to understand the subject. You may not always appreciate the significance of what I am telling you at first, but in time you will. To start with, always try to do these three things: **99**

THINK LOGICALLY

ASK QUESTIONS

TRY TO UNDERSTAND WHAT IS HAPPENING

66 OK, that sounds clever but how do I do it?
What does think logically mean exactly? **99**

66 All right, let me give you an example. I will ask you some questions (Q) and you give me the answers (A). **99**

Q: How many piano tuners are there in the city of Plymouth?

A: How could I possibly know that?

Q: You will have to make an estimate. In hydraulics, and indeed engineering as a whole, you often have to estimate the value of things from what you already know. So do that.

A: How? I do not know anything about piano tuners in Plymouth.

Q: You know more than you think, you are just not aware of it. In hydraulics you will often have an intuitive understanding and some basic knowledge without realising it. The trick is to learn to use it, like this. What is the population of Plymouth: 27 000 or 270 000 or 2.70 million?

A: Well you said it was a city so 27000 is too small, and the population of Britain is only about 55 million so it cannot be 2.70 million. It must be 270000.

Q: Good, very logical. Now what is the size of the average household? One, two, three, four, or five people?

A: About three I think.

Q: Near enough, and it makes the maths easy. So that means something like 90000 households. Now how many families or households own a piano: 1 in 1, 1 in 10, 1 in 50, 1 in 100, 1 in 500, or 1 in 1000?

A: Well, I know lots of people who had piano lessons when they were kids, so it's not 1 in 1000. But obviously not everyone owns a piano. Many modern houses are quite small, and there are lots of small electronic keyboards around, so how about 1 in 50?

Q: That gives us a figure of 1800 pianos in Plymouth. Now would there be one piano tuner for each piano, or 1 for 10, 1 for 100, 1 for 200, 1 for 500, or 1 for each 1000 pianos?

A: Difficult. I guess some people like musicians, theatres with orchestras and so on have their piano tuned regularly, but most people do not. How about 1 for 200 pianos?

Q: OK, that gives 9 piano tuners in Plymouth. This may not be totally accurate, but it is usually better to have a rough estimate of a figure than none at all. How can you check the accuracy of the estimate? As an engineer or scientist it is always a good idea to check that your answers are sensible.

A: How about looking in 'Yellow Pages', the 'phone directory?

Q: Yes, that is possible. In fact there are about 7 or 8 listed, although it's not clear if some piano tuners operating in Plymouth live outside the city, and some advertisers may have more than one tuner. You could also try working out how many pianos a piano tuner can tune in one week, then a year, and then how long it would take to tune all the pianos in Plymouth. On average, most people have their piano tuned very infrequently, so this again would confirm that your answer was in the right field. Got the idea?

66 So what you are saying is break down a large problem to which you do not know the answer into smaller steps to which you can either estimate or guess an answer. **99**

That is correct. This book will provide you with many examples of how to think your way through a problem. However, it is also important that you ask yourself questions, not just as we did above, but questions about a particular hydraulic phenomenon. The questions you should always ask yourself are:

WHAT happens?
WHERE does it happen?
WHEN does it happen?
WHY does it happen?
HOW does it happen?

If you can answer these questions (when appropriate) then you are well on your way to understanding hydraulics. This subject becomes much easier when you understand and can visualise what is happening. Incidentally, the word 'hydraulic' if used correctly only applies to water (being derived from the Greek word for water). However, we always refer to hydraulic jacks and hydraulic excavators which use oil. Thus hydraulics usually means the study of the properties and movement of all liquids. Fluid mechanics is the study of gases in addition to liquids, although hydraulics may sometimes also be so defined. This book deals mainly with liquids, hence its title, but some principles apply equally to fluids like air. Indeed, we will sometimes use air to illustrate what happens in a liquid.

❝ OK, any other suggestions?
What other mistakes am I likely to make that I could easily avoid? ❞

Well, one thing that is very important, and which students never give enough attention to, is the question of units. A numerical answer is not correct unless it has the units written after it. For example, if you tell me that a distance is 42, what does this mean? It could be 42 mm, 42 cm, 42 inches, 42 feet, 42 m, 42 km or 42 miles. Furthermore, if it was 42 miles then I might guess the value was the distance between two places. If it was 42 mm it might be the distance between two points on a piece of paper. On the other hand, if something has a value of $42 \, N/m^2$ then I know that we are talking about a pressure. Thus the units can convey a meaning that is not clear just from a numerical value.

Another thing, you must always work in one consistent set of units. If you had £10 and $15 you would not add them together and say that you had £25, you would have to convert the dollars to pounds first then add them up. So when it comes to calculations, you must make sure everything is in the same units: you cannot have one value in mm and another in metres, because the answer will be wrong if you do. To avoid making this type of mistake in your calculations, remember this.

◀ Remember

Always work in metres (m), kilogrammes (kg), seconds (s) and Newtons (N). In hydraulics, never work in anything else.

Although you may be tempted to use other units, do not. There are traps for those who try to be clever. For example, a common mistake is to try to work in mm, while forgetting that the value of gravity has been taken as $9.81 \, m/s^2$. Another is to work in mm while taking the density of water as 1000, which is only correct when the units are kg/m^3. In this book only m, kg, s and N are used. *Get into this habit as well, and it will cut out a lot of mistakes.* Yes, I know that you might be using other units like N/mm^2 in structures and megaNewtons (MN) in geotechnics, but do not try to use them in hydraulics. It will only lead to mistakes.

❝ Throughout the book you will find 'Remember Boxes' like the one above. They contain some key information, or a summary, or a procedure that you should remember. There is also a 'Remember' symbol that is used to flag important pieces of information. One is shown below. ❞

The book contains some Self Test Questions for you to work through. Use these to confirm that you understand the text. Brief guide solutions are given in Appendix 2, but try to obtain the correct answers for yourself. It is very important that you develop the habit of working logically and systematically through a problem. If you always get the wrong answer to numerical problems as a student, how are you suddenly going to start getting the right answers as an engineer in industry where mistakes may cost thousands or millions of pounds?

It was stated above that you would not add pounds to dollars, and that everything must be in the same units. This also applies to equations. All the terms of an equation must have the same units otherwise the equation will be meaningless. Later on in the book you will meet the energy or Bernoulli equation, which can be written like this:

$$V^2/2g + P/w + z = \text{constant}$$

Now let us think of the units. V is a velocity in m/s, g is the acceleration due to gravity in m/s^2, P is a pressure in N/m^2, w is the weight density of the liquid in N/m^3 and z is an elevation in m. Now if we substitute these units into the Bernoulli equation we get:

$$\frac{(\text{m/s})^2}{\text{m/s}^2} + \frac{\text{N/m}^2}{\text{N/m}^3} + \text{m} = \text{constant}$$

Cancelling similar units gives:

$$\text{m} + \text{m} + \text{m} = \text{constant}$$

In other words, all three terms of the equation have the same units, which is metres, and it follows that the constant must also be measured in metres. In fact these terms are often called 'heads' because they represent a head or the height of a column of water measured in metres.

◀ Dimensional homogeneity

For dimensional homogeneity, both sides of an equation must have the same units. Similarly, all the terms of an equation must have the same units, otherwise they could not be added together. This simple fact can be useful sometimes. For example if you cannot remember if P in the above equation should be in m or N/m^2, then by considering the requirements of dimensional homogeneity you should be able to determine that it must be in N/m^2. A careful consideration of the dimensions can help you to check that you have remembered an equation correctly, (REMEMBER!!!) and that you have conducted a valid analysis.

❝ The reason this is called 'dimensional' homogeneity is that the principle is usually applied using dimensions not units. There are only three dimensions: length (L), mass (M) and time (T). Thus the metre is a unit of length, and length is a dimension. The three dimensions can be grouped together so that, for example, pressure becomes $ML^{-1}T^{-2}$. However, on the basis that it is better to learn to walk before trying to run, stick to thinking in terms of units initially, with pressure in N/m^2. The ideas of dimensions, dimensional homogeneity and dimensional analysis are discussed in detail in Chapter 10. ❞

❝ Another thing you should be aware of is the difference between accuracy and precision. Accuracy is how near an answer is to the true value of the variable. Precision is how many significant figures or decimal places are given. You should never use a greater degree of precision than the accuracy of the variable in question justifies. ❞

Let me give you an example. If you press the π key on your calculator you will get a value of 3.141592654. This has nine figures after the decimal point, which is many more than you actually need in most circumstances. If you take π as 3.142, which is the value people remember, the percentage error caused by this approximation is only 0.013%, which is trivial in most calculations. So you could say:

3.141592654 is a very accurate and very precise value of π.

3.142 is an accurate but less precise value.

3 is an inaccurate (4.5% error) and imprecise value.

2.568295684023 is a totally inaccurate but very precise value.

You should note from the last figure (which I made up, of course) that quoting a lot of decimal places does not make a value correct. In fact it looks stupid quoting so many decimal places for a value which is so obviously wrong. You would be surprised how many people get drawn into this mistake, particularly when conducting practical or laboratory work. Let me give you a simple example to try for yourself.

REMEMBER!!!

Draw a circle on a piece of paper.

Now use a ruler to measure its radius, r.

Work out the length of the circumference of the circle = $2\pi r$ using the π key of your calculator.

Now write down the answer on the paper.

For instance, if $r = 34$ mm, or 0.034 m, then the length of the circumference is $2 \times \pi \times 0.034$. Using a calculator and its π key the answer displayed is 0.2136283 m. Now how many decimal figures should be written down? Well since the radius could only be measured to the nearest mm, it would be logical to quote the length of the circumference to the nearest mm also, that is 0.214 m or 214 mm. It would be silly quoting the answer to the nearest 10 000th of a mm since there is no way this could be measured with a ruler, and this degree of precision is not required. *How many figures did you write down after the decimal point?*

❝ So what you are saying is do not copy all the figures off the calculator display, only those that match the level of accuracy of the input data. ❞

❝ OK. We have covered a few general introductory points, now work your way through the book, remembering the things that we have been talking about. They can all help you to avoid mistakes. ❞

Ask questions

Think logically

Try to understand what is happening

Use only m, kg, s *and* N *in your calculations*

Whenever possible, try to check your answer by another means

Get used to thinking in terms of units and dimensional homogeneity

Remember the sort of accuracy you are working to and do not be over-precise

▶And finally, consider this . . .

As long ago as 1637, René Descartes, the French philosopher, published '*Discourse on Method*'. In this book he gave four rules for scientific enquiry which are just as valid today and which underline much of what has been said above.

1. **Never accept as true anything which cannot be clearly seen as such** (or question the accuracy of your input data).

2. **Divide difficulties into as many parts as possible** (or break down a problem into smaller components).

3. **Seek solutions of the simplest problems first and proceed step by step to the most difficult.**

4. **Review all conclusions to make sure there are no omissions** (or check your answers).

Hydrostatics

This chapter introduces some of the fundamental quantities involved in hydraulics, such as pressure, weight, force, mass density and relative density. It then considers the variation of pressure intensity with depth below the surface of a static liquid, and shows how the force on a submerged surface or body can be calculated. The principles outlined are used to calculate the hydrostatic forces on dams and lock gates, for example. These same principles are applied in Chapter 2 in connection with pressure measurement using piezometers and manometers, and in Chapter 3 to the analysis of floating bodies. Thus the sort of questions that are answered in this chapter are:

What is meant by pressure?

What is the difference between force and weight?

What is the difference between mass and weight?

How and why does pressure intensity vary with depth in a liquid?

How can we calculate the pressure intensity at any depth?

How can we calculate the force on a flat immersed surface, such as the face of a dam?

How can the hydrostatic force be calculated when the immersed surface is curved?

Does hydrostatic pressure act equally in all directions, and if it does – why?

How can the buoyancy force on a body be calculated?

What do we do if the liquid is stratified with layers of different density?

1.1 Fundamentals

1.1.1 Understanding pressure and force

❝ Have you ever asked yourself why a trainer will not damage a soft wooden floor, but a stiletto heel will? ❞

The answer is because the average pressure, P_{AV}, exerted on the floor is determined by the weight of the person, W, and the area of contact, A, between the sole of the shoe and the floor. Thus:

$$P_{AV} = W/A \qquad (1.1)$$

So, because a trainer has a flat sole with a large area of contact, it exerts a relatively small pressure on the floor (Fig. 1.1). On the other hand, the sharp point of a stiletto means that much of the weight is transmitted to the floor over a small area, giving a large pressure. Similarly a drawing pin (or a 'thumb tack' in American) creates a large, penetrative pressure by concentrating a small applied force at a sharp point.

❝ I understand that, but can you now tell me what is the difference between weight and force? ❞

The answer is basically 'none'. Weight is simply one particular type of force, namely that resulting from gravitational attraction. So equation (1.1) can also be written as $P_{AV} = F/A$, where F is the force. This can be rearranged to give:

$$F = P_{AV}A \qquad (1.2)$$

The unit of force is the **Newton** (N), named after Sir Isaac Newton, so pressure has the units N/m^2. A Newton is defined as the force required to give a mass of 1 kg an acceleration of 1 m/s². Hence:

Figure 1.1 Illustration of the pressure exerted on a floor by two types of shoe. The stiletto is the more damaging because the weight is distributed over a small area, so giving a relatively large pressure

Force = mass × acceleration

$$F = Ma \qquad (1.3)$$

where M represents mass and a is the acceleration. For weight, W, which is the force caused by the acceleration due to gravity, g, this becomes:

Weight = mass × gravity

$$W = Mg \qquad (1.4)$$

On Earth, gravity, g, is usually taken as $9.81\,\text{m/s}^2$.

1.1.2 Understanding the difference between mass and weight

 66 OK, so what is the essential difference between mass and weight, and why is it important? 99

It is important to have a clear understanding of the difference between mass and weight, because without it you will make mistakes in your calculations. The essential difference is that mass represents the amount of matter in a body, which is constant, so mass stays the same everywhere in the universe, while weight varies according to the local value of gravity since $W = Mg$ (equation (1.4) and Fig. 1.2).

 66 So what is mass density and weight density
What is meant by relative density? And how heavy is water? 99

Density, ρ, is the relationship between the mass, M, of a substance and its volume, V. Thus:

$$\rho = M/V \qquad (1.5)$$

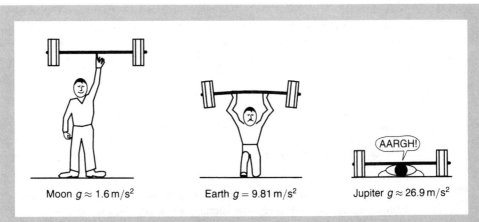

Figure 1.2 The concept of weight, which varies according to the local value of gravity

---

Box 1.1 ▶ **Remember**

It is important to realise that water is heavy! Each cubic metre of water weighs 9.81 $\times 10^3$ N, that is **one tonne**. Thus every cubic metre weighs about the same as a large car.

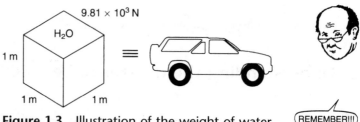

Figure 1.3 Illustration of the weight of water

The density of fresh water (ρ) is 1000 kg/m³. This can be thought of as the **mass density** of the water, since it gives the mass per unit volume. Alternatively, the weight (W) per unit volume may be quoted, which is the **weight density**, w (also called the specific weight). Using equations (1.4) and (1.5), weight density can be expressed in several ways:

$$w = W/V \quad \text{or} \quad w = Mg/V \quad \text{or} \quad w = \rho g \tag{1.6}$$

Thus the weight density of fresh water is 1000×9.81 N/m³.

Another term you may come across is the **relative density** (or specific gravity) of a liquid, s. This is the ratio of the density of a substance, ρ_S, to the density of fresh water, ρ. Of course, the same value can be obtained by using the ratio of the weight densities (equation (1.6)), since g is the same for both substances. Thus:

$$s = \rho_S/\rho \quad \text{or} \quad s = w_S/w \tag{1.7}$$

where w_S is the weight density of the substance. Since s represents a ratio of the mass or weight of equal volumes of the two substances, it has no dimensions. For example, water has a relative density of 1.0 while mercury has a relative density of 13.6.

Box 1.2 ▶ **Using relative density**

It is important to remember that s usually has to be multiplied by the density of water before it can be used in your calculations, otherwise the answer you obtain will be wrong, both numerically and dimensionally. For example, the density of mercury (ρ_M) is 13.6×1000 kg/m³. Quoting the relative density as 13.6 is just a shorter and more convenient way of writing this.

1.1.3 An application of what you have learned so far – the hydraulic jack

❝ You may not realise it, but you now have a sufficient understanding of hydrostatics to understand how a hydraulic jack works. ❞

The hydraulic jack uses two cylinders (Fig. 1.4), one with a large cross-sectional area (CSA), A, and one with a small area, a. By using a handle, or something similar, a small force, f, is applied to the piston in the small cylinder. From equation (1.2), it can be seen that this generates a pressure in the liquid of $P_{AV} = f/a$. Now one of the properties of a liquid is that it transmits pressure equally in all directions (more of this later), so this means that the same pressure P_{AV} acts over the whole cross-sectional area (A) of the large piston. As a result, the force exerted on the large piston is $F = P_{AV}A$ (equation (1.2)). Because $A > a$, the output force $F > f$, even though the pressure of the liquid is the same. Thus the jack acts as a kind of hydraulic amplifier. This simple but extremely useful effect can be used to lift weights of many tonnes while applying only a relatively small force to the input end of the jack.

◢◣ 1.2 Hydrostatic pressure and force

❝ Now let us try to determine how we can work out the hydrostatic force, F, on a dam, or on a lock gate, or on the flap gate at the end of a sewer. ❞

The term '**hydrostatic**' means, of course, that the liquid is not moving. Consequently there are no viscous or frictional resistance forces to worry about (see section 4.1). Also, in a stationary liquid there can be no shear forces, since this would imply movement. The water pressure must act at right angles to all surfaces with which the liquid comes into contact. If the pressure acted at any other angle to the surface, then there would

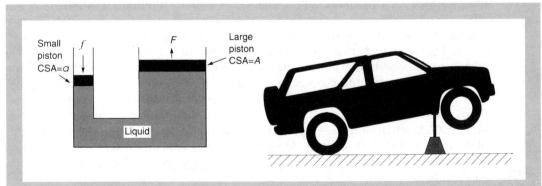

Figure 1.4 A hydraulic jack. The hydraulic pressure that results from applying a small force to the small piston is transmitted to the large piston, so enabling a relatively heavy load to be raised

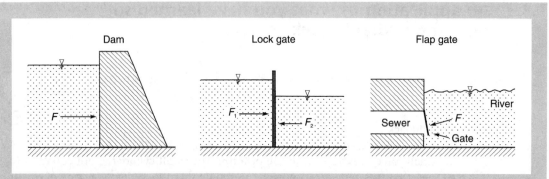

Figure 1.5 Typical examples of situations where the hydrostatic force may have to be calculated

be a component of force along it which would cause the liquid to move. However, this component is zero when the pressure is normal to the surface since $\cos 90° = 0$. Hence in a static liquid the pressure acts at right angles to any surface. This fact comes in useful later.

(REMEMBER!!!)

❝ OK, so the pressure acts at 90° to the surface.
Please can you now explain why a submarine can only dive to a certain depth, as in all those old war movies? ❞

The answer is quite simple. The pressure intensity increases with depth. Beyond a certain depth the water pressure would crush the hull of the submarine.

❝ But what causes the pressure, and how can you calculate what it is? After all, if you were in the submarine you would want to know, right? ❞

The weight of the water above the submarine causes the pressure. Remember, every cubic metre of fresh water equals 1 tonne, which is 9810N (that is ρg N with $\rho = 1000\,\text{kg/m}^3$ and $g = 9.81\,\text{m/s}^2$). This makes it quite easy to calculate the pressure. Try thinking of it like this.

Imagine a large body of fresh water. Then consider a column of the liquid with a plan area of $1\,\text{m}^2$ extending from the surface all the way to the bottom, as in Fig. 1.6. Now, suppose we draw horizontal lines at one metre intervals from the surface, so that the column is effectively separated into cubes with a volume of $1\,\text{m}^3$. Every cube weighs $9.81 \times 10^3\,\text{N}$. Since the pressure on the base of each of the cubes is equal to the weight of all the cubes above it divided by $1\,\text{m}^2$ ($P_{AV} = W/A$), it can be seen that the pressure increases uniformly with depth. Similarly, if the column of liquid has a total depth, d, then the total weight of all the cubes is $9.81 \times 10^3 \times d\,\text{N}$. Dividing this by $1\,\text{m}^2$ to obtain the pressure on the base of the column gives $9.81 \times 10^3 \times d\,\text{N/m}^2$. Therefore, at any depth, h, below the water surface the pressure is:

$$P = \rho g h \text{ N}/\text{m}^2 \tag{1.8}$$

Equation (1.8) shows that there is a linear relationship between pressure, or pressure intensity, and depth. This pressure–depth relationship can be drawn graphically to obtain

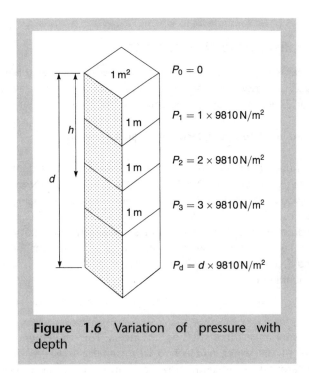

Figure 1.6 Variation of pressure with depth

a **pressure intensity diagram** like that in Fig. 1.7. This diagram shows the pressure intensity on a vertical surface that is immersed in a static liquid and which has the same height, h, as the depth of water. The arrows can be thought of as vectors: they are drawn at 90° to the surface indicating the direction in which the pressure acts, while the length of the arrow indicates the relative magnitude of the pressure intensity. When analysing a problem, a pressure intensity diagram is used to help visualise what is happening, while equation (1.8) provides the means to calculate the pressure intensity.

The relationship described by equation (1.8) is very useful; it can be used to calculate the pressure at any known depth, or alternatively, to calculate the depth from a known pressure. The fact there is a precise relationship between pressure and depth forms the basis of many instruments that can be used to measure pressure, such as manometers, which are described in Chapter 2.

Now one important point. Figure 1.7 only shows the pressure caused by the weight of the water. This is called the **gauge pressure**, and is

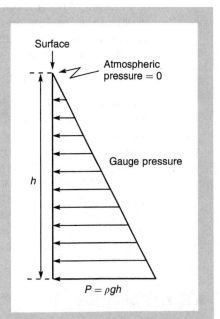

Figure 1.7 A pressure intensity diagram corresponding to Fig. 1.6

Box 1.3	▶ **Visualising the size of units**

You can easily visualise a metre, because it is just over three feet in length, and, of course, you know how long a second is. You may also be aware that a kilogramme is about 2.2 lb, that is about the equivalent of a bag of sugar.

But do you know how large or small a Newton is?

If you use equation (1.8) to work out the pressure at a depth of 0.3 m of fresh water you get $P = 1000 \times 9.81 \times 0.3 = 2943\,N/m^2$. So every (REMEMBER!!!) time you have a bath at home, parts of your body are being subjected to almost 3000 N/m^2. It does not cause any discomfort, in fact you do not even notice. So you may deduce that a Newton is a relatively small unit of force. For this reason it is frequently not worthwhile quoting a value to less than a Newton (the exception being if you are dealing with very, very small values where accuracy may be affected by rounding off).

---

the pressure most often used by engineers. For convenience, gauge pressure measures the pressure of the water **relative to atmospheric pressure**, that is it takes the pressure of the air around us as zero. Now in reality, the atmosphere exerts a pressure of about $101 \times 10^3\,N/m^2$ on everything at sea level (this is equivalent to the pressure at the bottom of a column of water about 10.3 m high, that is a 'head' of 10.3 m of water). So if we want to obtain the **absolute pressure** measured relative to an absolute vacuum, that is the total pressure exerted by both the water and the atmosphere, we have to add atmospheric pressure, P_{ATM}, to the gauge pressure (Fig. 1.8). Thus the absolute pressure, P_{ABS}, is:

$$P_{ABS} = \rho gh + P_{ATM}\ \mathbf{N/m^2}$$

(1.9)

A good way to think of this is that you can measure the height of a table top either from the floor, which is the most convenient way, or above sea level (ordnance datum). Similarly, it is more convenient to measure temperature above the freezing point of water than above absolute zero. Consequently in this book we will always use gauge pressures (unless stated otherwise). For future reference, note that under some circumstances, such as in pipelines, a pressure less than atmospheric may occur (Fig. 1.8). This is a negative gauge pressure, $-\rho gh$, but equation (1.9) is still valid. Note also that if absolute pressure is used then the gauge pressure intensity diagram shown in Fig. 1.7 will have to have P_{ATM} added to it, as shown in Fig. 1.9.

Now try Self Test Question 1.1. A short guide solution is given in Appendix 2, if you need it.

SELF TEST QUESTION 1.1

Oil with a weight density, w_O, of 7850 N/m^3 is contained in a vertically sided, rectangular tank which is 2.0 m long and 1.0 m wide. The depth of oil in the tank is 0.6 m.

(a) What is the gauge pressure on the bottom of the tank in N/m^2?

(b) What is the weight of the oil in the tank?

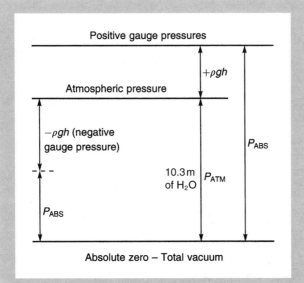

Figure 1.8 Relationship between gauge pressure and absolute pressure

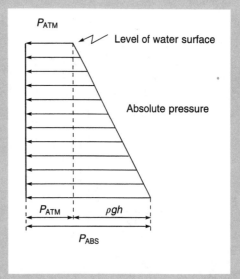

Figure 1.9 Pressure intensity diagram including atmospheric pressure

(c) If the bottom of the tank is resting not flat on the ground but on two pieces of timber running the width of the tank, so that each piece of timber has an area of contact with the tank of 1.0 m × 0.1 m, what is the pressure on the timber?

1.3 Force on a plane (flat), vertical immersed surface

    66 How do you work out the force on something as a result of the hydrostatic pressure? Say, something like a rectangular gate at the end of a sewer or culvert? **99**

OK, there are two thing to remember. First of all, equation (1.2) tells us that $F = P_{AV}A$, so a force is a pressure multiplied by an area. However, the second thing we have to remember is that the pressure varies with depth. So, on a vertical surface such as the gate in Fig. 1.10, the pressure at the top of the gate is ρgh_1. At the bottom of the gate the pressure is ρgh_2. Hence the **average pressure** on the gate is $P_{AV} = (\rho gh_1 + \rho gh_2)/2$. Now if we multiply this by the area of the gate in contact with the water, A, we get the force, F:

$$F = \rho g[(h_1 + h_2)/2]A \tag{1.10}$$

For a rectangle, $(h_1 + h_2)/2$ is the depth to the centre of the area, that is the vertical depth to the centroid, G, of the immersed surface. This depth is represented by h_G, so the expression for the resultant hydrostatic force, F, becomes:

$$F = \rho gh_GA \tag{1.11}$$

This equation can be applied to surfaces of any shape. For geometrical shapes other than a rectangle, the depth to the centroid can be found from Table 1.1. For the full derivation of equation (1.11), see Proof 1.1 in Appendix 1.

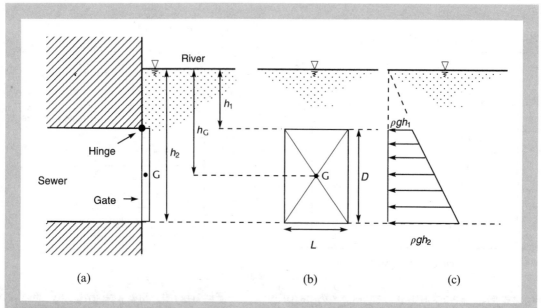

Figure 1.10 A vertical gate at the end of a sewer which discharges to a river. The gate hangs from a hinge at the top: (a) side view, (b) front view, (c) pressure intensity diagram. Note that only the part of the pressure intensity diagram at the same depth as the gate contributes to the hydrostatic force acting on it

Table 1.1 Geometrical properties of some simple figures

Shape	Dimensions	Location of the centroid, G	Second moment of area, I_G
Rectangle	breadth L height D	$D/2$ from base	$LD^3/12$
Triangle	base length L height D	$D/3$ from base	$LD^3/36$
Circle	radius R	centre of the circle	$\pi R^4/4$
Semicircle	radius R	$4R/3\pi$ from base	$0.1102R^4$

The next paragraph can be helpful in some circumstances, since it reconciles what can appear to be different ways to solve a particular problem. However, you may omit it the first time you read the chapter, or if it confuses you.

From equation (1.10), the resultant force, F = average pressure intensity × area of the immersed surface (A). For simple, flat surfaces like that in Fig. 1.10, the average pressure intensity is $(\rho g h_1 + \rho g h_2)/2$. If $A = DL$, then equation (1.10) can be written as $F = \rho g[(h_1 + h_2)/2]DL$. The same expression can be obtained by calculating the area of the trapezoidal pressure intensity diagram in contact with the gate, $\rho g[(h_1 + h_2)/2]D$ and multiplying by the length of the gate, L. This can sometimes provide a useful check that what you are doing is correct, or a means of remembering the equation. However, your best approach initially is usually to go straight to equation (1.11).

Box 1.4 ▶ **Remember**

Whenever you are faced with calculating the horizontal hydrostatic force on a plane, vertical immersed surface, the equation $F = \rho g h_G A$ is the one to use. This simple equation can solve a lot of problems. We will also use it later on when we progress to the force on inclined and curved immersed surfaces. *Remember that A is the area of the immersed surface in contact with the liquid.*

1.4 ▶ Location of the resultant force on a vertical surface

❝ How do you know where the resultant force, *F*, acts? I assume that there must be some way of working it out? ❞

Yes, there is a way of calculating where the resultant force acts, and normally you would work this out at the same time as the magnitude of the force itself. However, the proof is a bit complicated, so I have put it in Appendix 1 (the second half of Proof 1.1). You can go through it later if you want to. For the time being, though, let us try to deduce something about where the force must act.

Consider the dam in Fig. 1.11. In this case the pressure intensity diagram is triangular, since the gauge pressure varies from zero (atmospheric pressure) at the surface to $\rho g h$ at the bottom. The average pressure intensity on the dam is therefore $(0 + \rho g h)/2$ or $\rho g h/2$. This pressure occurs at G, half way between the water surface and the bottom of the dam.

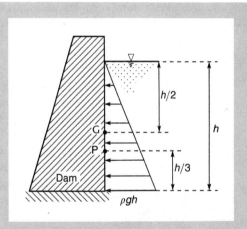

Figure 1.11 Pressure intensity on a dam. G is the centroid of the wetted area, P is the centre of pressure where the resultant force acts

But where would the resultant force act?

At G, half way down? Above? Below?

Can you deduce where it would be?

Think of it this way. The resultant force on the dam is the result of the average pressure intensity acting over the area of the dam face in contact with the water. The longer the arrows of the pressure intensity diagram, the greater the pressure. The larger the area of the pressure intensity diagram, the greater the force.

Box 1.5 Note that the centre of pressure, P, is always below the centroid, G, of the surface in contact with the water. In many problems it is not obvious where P is located, so this has to be calculated using equation (1.12). However, as the depth of immersion of the surface increases, P moves closer to G. This is apparent from equation (1.12): the distance between P and G is ($h_P - h_G$). If A and I_G have constant values, then the equation can be rearranged as ($h_P - h_G$) = C/h_G where C represents the value of the constants. Thus ($h_P - h_G$) decreases as h_G increases.

REMEMBER!!!

Look at the triangular area that forms the top half of the pressure intensity diagram, and compare it with the area of the trapezoidal bottom half. The area of the bottom part of the diagram is much larger, indicating that the resultant force would act *below* half depth. In fact, the resultant force acts horizontally through the centroid of the pressure intensity diagram. For the **triangular** pressure intensity diagram in Fig. 1.11, this is located at $h/3$ from the base (**but note that this is only the case when the pressure intensity diagram is triangular**). The point, P, at which the resultant force acts is called the **centre of pressure** (Fig. 1.11).

With more complex problems, like that in Fig. 1.10, there is no simple rule to give the location of P, but if h_P is the vertical depth to the centre of pressure then this can be calculated from:

$$h_P = (I_G/Ah_G) + h_G \qquad (1.12)$$

where the value in the brackets gives the vertical distance of P below the vertical depth to the centroid of the surface, h_G. The appropriate expression for the second moment of area calculated about an axis through the centroid, I_G, can be found from Table 1.1. For a rectangle $I_G = LD^3/12$, where L is the length of the body and D its height. A is the surface area of the body. The derivation of equation (1.12) can be found in Appendix 1.

Examples 1.1 and 1.2 show how equations (1.11) and (1.12) are used to solve a couple of typical problems, one involving the flap gate at the end of a sewer and the other a lock gate. Study these carefully and then try Self Test Question 1.2 (a short solution is given in Appendix 2).

SELF TEST QUESTION 1.2

A rectangular culvert (a large pipe) 1.8 m wide by 1.0 m high discharges to a river. At the end of the culvert is a rectangular gate which seals off the culvert when the river is in flood (as in Fig. 1.10). The gate hangs vertically from hinges at the top. If the flood level in the river rises to 1.9 m above the top of the gate, calculate the magnitude and location of the resultant hydrostatic force on the gate caused by the water in the river.

EXAMPLE 1.1

A rectangular gate is 2 m wide and 3 m high. It hangs vertically with its top edge 1 m below the water surface. (a) Calculate the pressure at the bottom of the gate. (b) Calculate the resultant hydrostatic force on the gate. (c) Determine the depth at which the resultant force acts.

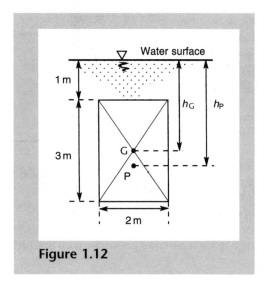

Figure 1.12

(a) From equation (1.8), $P = \rho g h$

Therefore $P = 1000 \times 9.81 \times (3+1)$
$$= 39.24 \times 10^3 \, \text{N/m}^2$$

(b) From equation (1.11), $F = \rho g h_G A$

Now $h_G = 1 + (3/2) = 2.50 \, \text{m}$
$$A = 2 \times 3 = 6 \, \text{m}^2$$
Thus $F = 1000 \times 9.81 \times 2.50 \times 6$
$$= 147.15 \times 10^3 \, \text{N}$$

(c) From equation (1.12)
$$h_P = (I_G / A h_G) + h_G$$
where $I_G = LD^3/12 = 2 \times 3^3/12 = 4.50 \, \text{m}^4$
A and h_G are as above
so $h_P = (4.50/6 \times 2.50) + 2.50$
$$= 2.80 \, \text{m}$$

EXAMPLE 1.2

A lock on a canal is sealed by a gate that is 3.0 m wide. The gate is perpendicular to the sides of the lock. When the lock is used there is water on one side of the gate to a depth of 3.5 m, and 2.0 m on the other side. (a) What is the hydrostatic force of the two sides of the gate? (b) At what height from the bed do the two forces act? (c) What is the magnitude of the overall resultant hydrostatic force on the gate and at what height does it act?

(a) Using $F = \rho g h_G A$

$F_1 = 1000 \times 9.81 \times (3.5/2) \times (3.5 \times 3.0)$
$$= 180.26 \times 10^3 \, \text{N}$$
$F_2 = 1000 \times 9.81 \times (2.0/2) \times (2.0 \times 3.0)$
$$= 58.86 \times 10^3 \, \text{N}$$

(b) Since both pressure intensity diagrams are triangular, both forces act at one-third depth from the bed:

$Y_1 = 3.5/3 = 1.17 \, \text{m}$
$Y_2 = 2.0/3 = 0.67 \, \text{m}$

(c) Overall resultant force $F_R = F_1 - F_2$

$F_R = 121.40 \times 10^3 \, \text{N}$

Figure 1.13

Taking moments about O to find the height, Y_R, of the resultant:

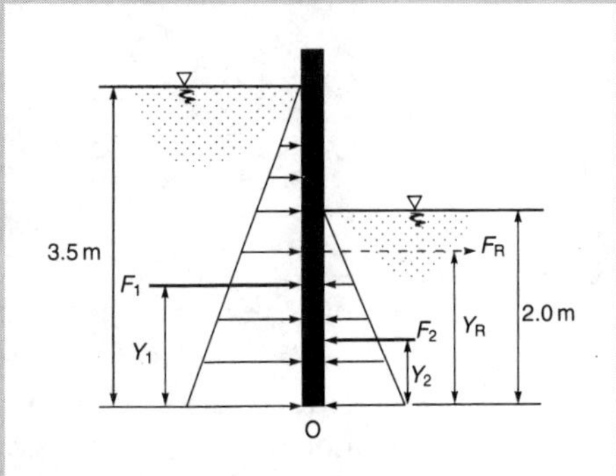

$121.40 \times 10^3 \times Y_R = 180.26 \times 10^3 \times 1.17 - 58.86 \times 10^3 \times 0.67$
$$Y_R = 1.41 \, \text{m above the bed.}$$

The value of Y_R obtained in part (c) of the above example may have surprised you. Possibly you expected Y_R to be somewhere between 0.67 m and 1.17 m, whereas it is actually 1.41 m. This is a situation where the pressure intensity diagrams (which are not really needed to conduct the calculations) can be used to visualise what is happening. In Fig. 1.13 the slope of the two pressure intensity triangles is the same, since the water has the same density on both sides of the gate. Thus if the triangle on the right is subtracted from the triangle on the left, the result is as in Fig. 1.14. This is the net pressure intensity on the gate. The diagram is more rectangular than either of the triangles so, employing a similar argument to that used with Fig. 1.11, this indicates that Y_R would be higher above the base than either Y_1 or Y_2.

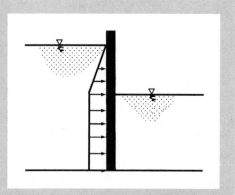

Figure 1.14 Net pressure intensity diagram for Example 1.2

Figure 1.15 The dam on the bottom left of the photograph is holding back a considerable quantity of water. The force exerted by the water on the structure must be calculated before the dam can be designed. Many lay people believe, incorrectly, that the greater the *volume* of water stored behind the dam, the larger the force on the structure. This is not the case. Equation (1.8) indicates that the pressure on the dam is related to the *depth* of water, while the force is the product of the average pressure and the area of the dam in contact with the water (equation (1.2))

1.5 Force on a plane, inclined immersed surface

66 I understand how to work out the force on a flat vertical surface, but how about one that is inclined at an angle to the water surface? Surely this is much more difficult? 99

The answer is 'no'. The calculations are still very simple and almost identical to those above. There are three things that you should remember when analysing these situations:

(1) The resultant force acts at right angles to the immersed surface.

(2) The hydrostatic pressure on the inclined surface is still caused only by the weight of water above it, so $P = \rho g h$.

(3) When calculating the location of the resultant force on an inclined surface, always use equation (1.13) (never equation (1.12), see below).

To illustrate simply that the resultant force can be calculated in the same way as for a vertical surface, consider this. The pressure at the top of the rectangular, inclined surface in Fig. 1.16a is $\rho g h_1$ while that at the bottom is $\rho g h_2$. Thus the average pressure intensity on the surface is $\rho g (h_1 + h_2)/2$, or $\rho g h_G$ since $h_G = (h_1 + h_2)/2$. The resultant force is the average pressure intensity multiplied by the area of the surface, and since the pressure acts at right angles to the inclined surface the actual area, A, should be used. Thus $F = \rho g h_G A$, as in equation (1.11). Note that the inclination of the surface is automatically taken into account by the value of h_G. For example, if h_1 in Fig. 1.16a is fixed, and the surface rotated upwards about its top edge, then h_G will decrease so that $h_G = h_1$ when it is horizontal. Similarly, the maximum possible value of h_G would be obtained when the surface is vertical.

One other important point, the resultant force on the inclined surface, F, has components in both the vertical and horizontal directions. These can be calculated separately, as in section 1.6 and Example 1.4, but the procedure outlined above is quicker for flat (plane) surfaces.

To calculate the location of the resultant force, the following equation should be used:

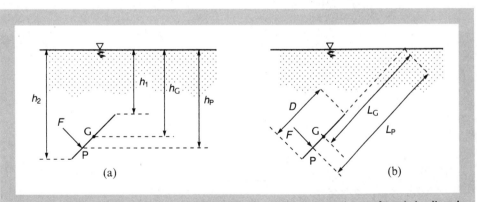

Figure 1.16 (a) Force on an inclined surface. (b) When the surface is inclined always use the dimensions L_G and L_P with equation (1.13) (never the vertical dimensions h_G and h_P with equation (1.12))

$$L_P = (I_G/AL_G) + L_G \qquad\qquad (1.13)$$

This is similar to equation (1.12), but the inclined lengths, L_P and L_G, are used to denote the location of the centre of pressure and centroid of surface (Fig. 1.16b), not the vertical depths.

EXAMPLE 1.3

A sewer discharges to a river. At the end of the sewer is a circular gate with a diameter (D) of 0.6 m. The gate is inclined at an angle of 45° to the water surface. The top edge of the gate is 1.0 m below the surface. Calculate (a) the resultant force on the gate caused by the water in the river, (b) the vertical depth from the water surface to the centre of pressure.

(a) Vertical height of gate $= 0.6 \sin 45° = 0.424\,\text{m}$

Vertical depth to G $= h_G = 1.000 + 0.424/2 = 1.212\,\text{m}$

Area of gate, $A = \pi D^2/4 = \pi 0.6^2/4 = 0.283\,\text{m}^2$

$F = \rho g h_G A = 1000 \times 9.81 \times 1.212 \times 0.283 = 3365\,\text{N}$

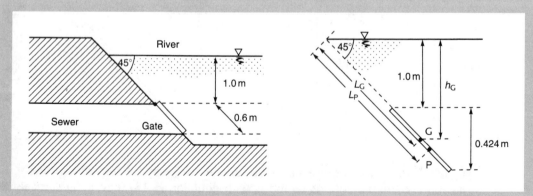

Figure 1.17 An inclined, circular gate at the end of a sewer

(b) Slope length to G, $L_G = 1.212/\sin 45° = 1.714\,\text{m}$

For a circle (Table 1.1) $I_G = \pi R^4/4 = \pi(0.3)^4/4 = 0.0064\,\text{m}^4$

$L_P = (I_G/AL_G) + L_G = (0.0064/0.283 \times 1.714) + 1.714 = 1.727\,\text{m}$

Vertical depth to P, $h_P = L_P \sin 45° = 1.727 \sin 45° = 1.221\,\text{m}$

Figure 1.18 This vertical lift gate on the Old Bedford River provides another example of where the engineer may be required to calculate the resultant hydrostatic force. If the horizontal force is large it may be difficult for a vertical lift gate to slide up and down, the gate being pushed hard against the guide channels. In the Fens of East Anglia much of the drainage is controlled by man, using pumps and sluice gates like the one above

1.6 Force on a curved immersed surface

❝ I suppose that you are now going to tell me that working out the force on a curved surface is just as easy as calculating the force on a flat or inclined surface? ❞

Well, the calculations are perhaps a little longer, but no more difficult. Let me clarify this by breaking the analysis of the force on an immersed curved surface down into steps.

(1) The resultant force (F) acts at right angles to the curved surface. This force can be thought of as having both a horizontal (F_H) and a vertical (F_V) component (Fig. 1.19).

(2) To calculate the horizontal component of the resultant force (F_H), project the curved surface onto a vertical plane, as in Fig. 1.20. This effectively is what you would see if you looked at the curved surface from the front. Calculate the force on this projected vertical surface as you would any other vertical surface using $F_H = \rho g h_G A$, where *A is the area of the projected vertical surface* (not the area of the actual curved surface).

(3) Calculate the vertical component of the resultant force (Fig. 1.21) by evaluating the weight of the volume (V) of water above the curved surface, that is:

$$F_V = \rho g V \tag{1.14}$$

(4) The resultant force, F, is given by:

$$F = \left(F_H^2 + F_V^2\right)^{1/2} \tag{1.15}$$

(5) The direction of the resultant force (Fig. 1.22) can be found from:

$$\tan \phi = F_V / F_H \tag{1.16}$$

This gives the angle, ϕ, of the resultant to the horizontal.

Remember, the resultant also acts at 90° to the curved surface, so it passes through the centre of curvature (for example, the centre of the circle of which the surface is a part).

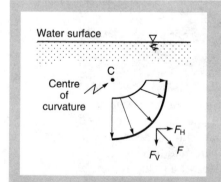

Figure 1.19 Pressure intensity on a curved surface. *F* passes through the centre of curvature, C

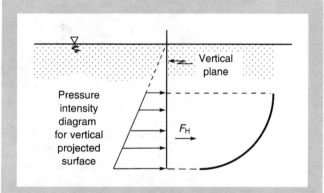

Figure 1.20 Projection of the curved surface onto a vertical plane

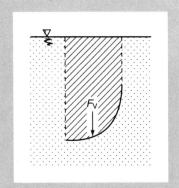

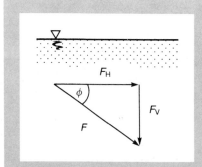

Figure 1.21 The vertical component of force, F_V, caused by the weight of water above the surface

Figure 1.22 The direction of the resultant force, F, which must also pass through C

(6) The above steps enable the resultant force on the upper side of the surface to be calculated. Always remember that there is an equal and opposite force acting on the other side of the surface. This fact comes in useful later, because it is always easier to calculate the force on the upper surface, even if this is not the surface in contact with the water.

EXAMPLE 1.4

A surface consists of a quarter of a circle of radius 2.0 m (Fig. 1.22). It is located with its top edge 1.5 m below the water surface. Calculate the magnitude and direction of the resultant force on the upper surface.

Step 1 Project the curved surface onto a vertical plane and calculate F_H
$F_H = \rho g h_G A$ where A is the area of the projected vertical surface.
Since the length of the gate is not given, calculate the force per metre length with $L = 1.0$ m.
Thus $A = 2 \times 1.0 = 2.0\,\text{m}^2$ per metre length
The value of h_G is that for the projected vertical surface: $h_G = 1.5 + (2.0/2) = 2.5$ m.

$$F_H = 1000 \times 9.81 \times 2.5 \times 2.0 = 49.05 \times 10^3\,\text{N/m}$$

Step 2 Calculate F_V from the weight of water above the surface
$F_V = \rho g V$ where V is the volume of water above the curved surface. Again using a 1 m length:

$$V = (1/4 \times \pi 2.0^2 \times 1.0) + (2.0 \times 1.5 \times 1.0) = 6.14\,\text{m}^3\ \text{per metre length}$$
$$F_V = 1000 \times 9.81 \times 6.14 = 60.23 \times 10^3\,\text{N/m}.$$

Step 3 Calculate the magnitude and direction of the resultant force

$$F = \left(F_H{}^2 + F_V{}^2\right)^{1/2} = 10^3 (49.05^2 + 60.23^2)^{1/2} = 77.68 \times 10^3\,\text{N/m}.$$
$$\phi = \tan^{-1}(F_V/F_H) = \tan^{-1}(60.23/49.05) = 50.8°.$$

The resultant passes through the centre of curvature, C, at an angle of 50.8°.

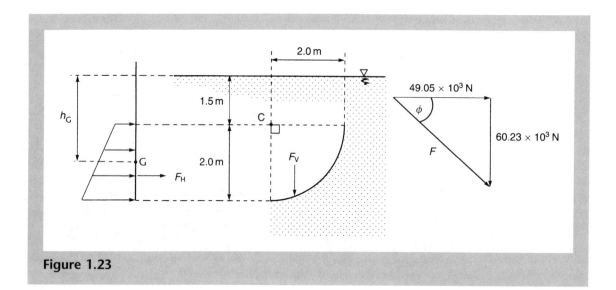

Figure 1.23

SELF TEST QUESTION 1.3

An open tank which is 4.0 m wide at the top contains oil to a depth of 3.4 m as shown in Fig. 1.24. The bottom part of the tank has curved sides which have to be bolted on. To enable the force on the bolts to be determined, calculate the magnitude of the resultant hydrostatic force (per metre length) on the curved surfaces and its angle to the horizontal. The curved sections are a quarter of a circle of 1.5 m radius, and the oil has a relative density of 0.8.

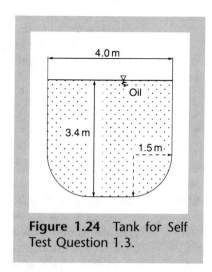

Figure 1.24 Tank for Self Test Question 1.3.

66 I understand Example 1.4, but when you described the steps used to analyse the force on a curved surface, in point 6 you said something about always analysing the upper side of the surface. You said that we should do this even if the upper side of the surface was not in contact with the water. How can this be right? No water, no hydrostatic force I would have thought. 99

I suppose this is one of the tricks you have to learn to make hydraulics easy. Think of it like this. The curved surface in Fig. 1.25 is an imaginary one, drawn in a large body of static liquid. Now it is possible to calculate the force on the upper side of this imaginary surface using the same procedure as in Example 1.4. However, the surface is only imaginary, so what resists this force? Something must because the liquid is static, that is not moving. The answer is that there is an equal and opposite force acting on the underside of the imaginary surface, so that this balances the force on the top. It does not matter which force you calculate, because they are numerically equal, but it is easier to calculate that on the upper surface. The same is true with real surfaces. Remember this when you encounter problems like Example 1.5 with air on the upper surface and water underneath.

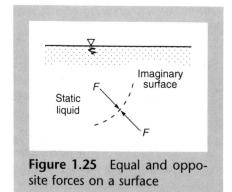

Figure 1.25 Equal and opposite forces on a surface

Something to note from Example 1.5 is that the vertical component of the resultant force acts **upwards**, which means that it is a **buoyancy** force. Sometimes there is a tendency to think of a buoyancy force as being different from the hydrostatic force, but in fact they are the same thing. The buoyancy force on a body, such as a ship, is the result of the hydrostatic pressure acting on the body. This will be explored in more detail in section 1.7.

REMEMBER!!!

EXAMPLE 1.5

A radial gate whose face is part of a circle of radius 5.0 m holds back water as shown in Fig. 1.26. The sector of the circle represented by the gate has an angle of 30° at its centre. Water stands to a depth of 2.0 m above the top of the upstream face of the gate. The other side of the gate is open to the atmosphere. Determine the magnitude and direction of the resultant hydrostatic force. The gate is 3.5 m long.

Step 1 Project the curved surface onto a vertical plane and calculate F_H
Vertical height of projection = BC = $5.0 \cos 60° = 2.5$ m.

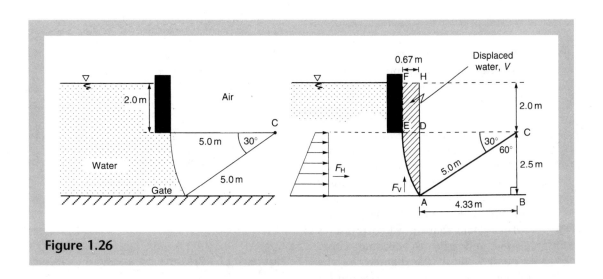

Figure 1.26

$h_G = 2.0 + (2.5/2) = 3.25\,\text{m}$, and $A = 2.5 \times 3.5 = 8.75\,\text{m}^2$.

$F_H = \rho g h_G A = 1000 \times 9.81 \times 3.25 \times 8.75 = 278.97 \times 10^3\,\text{N}$.

Step 2   Calculate the vertical component, F_V, from the weight of water above the surface
In this case calculate the weight of water that would be above the gate if it was not there, *that is the weight of the water displaced by the gate*. This is shown in the diagram as AEFH. The width of this area, DE, can be calculated as follows:

AB $= 5.0 \sin 60° = 4.33\,\text{m}$, so DE $= 5.00 - 4.33 = 0.67\,\text{m}$.

The area of ADE (and subsequently AEFH) can be found using geometry, as follows.
Area sector ACE $= (30/360)$ths of a 5.0m radius circle $= (30/360)\pi 5.0^2 = 6.54\,\text{m}^2$.
Area triangle ACD $= (1/2) \times 4.33 \times 2.5 = 5.41\,\text{m}^2$.
Area ADE $= 6.54 - 5.41 = 1.13\,\text{m}^2$.
Therefore, the total area AEFH $= 1.13 + (0.67 \times 2.00) = 2.47\,\text{m}^2$.
Volume of water displaced, $V = 2.47 \times 3.5 = 8.65\,\text{m}^3$.

$F_V = \rho g V = 1000 \times 9.81 \times 8.65 = 84.86 \times 10^3\,\text{N}$.

Step 3   Calculate the magnitude and direction of the resultant force
$$F = \left(F_H^2 + F_V^2\right)^{1/2} = 10^3(278.97^2 + 84.86^2)^{1/2} = 291.59 \times 10^3\,\text{N}.$$
$$\phi = \tan^{-1}(F_V/F_H) = \tan^{-1}(84.86/278.97) = 16.9°$$
Resultant acts at 16.9° to the horizontal passing *upwards* through the centre of curvature, C.

 1.7 **Variation of pressure with direction and buoyancy**

❝ We have already discussed the fact that hydrostatic pressure acts at right angles to any surface immersed in it, so it follows that on the underside of a horizontal surface the resultant force is acting vertically upwards. This is a buoyancy force, and it is caused simply by the hydrostatic pressure on the surface. Try thinking it through like this. ❞

Imagine a sphere some distance below the water surface as in Fig. 1.27. The hydrostatic pressure acts at 90° to the surface of the sphere. Looking at this two-dimensionally, as in the diagram, then the smallest pressure intensity is $\rho g h_1$ at the top, and the largest is $\rho g h_2$ at the bottom.

Now, consider what would happen if the diameter of the sphere gradually decreased so that the difference between h_1 and h_2 decreased. This would cause the two pressures $\rho g h_1$ and $\rho g h_2$ to become closer numerically. If the diameter of the sphere continued to decrease until it

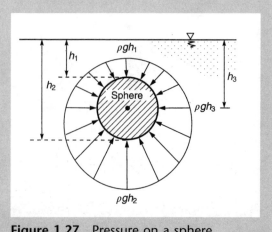

Figure 1.27   Pressure on a sphere

became infinitesimally small then the difference between h_1 and h_2 would be negligible so that $\rho gh_1 = \rho gh_2$. By the same argument, the pressure intensity in any other direction, such as ρgh_3 acting horizontally, would also have the same value (see Proof 1.2, Appendix 1).

Thus the pressure at a point in a static liquid acts equally in all directions, up, down, sideways or whatever.

❝ That's all very interesting, but does it have any practical purpose, and how can I work out the value of the buoyancy force? ❞

Yes it has a practical purpose, and working out the value of the buoyancy force is quite easy. In fact you can do so using what you have already learnt. Let me illustrate by using a similar situation to the sphere in Fig. 1.27, but this time we will make the body a cube because it simplifies the calculations. The cube is shown in Fig. 1.28. The pressure intensities on the vertical sides cancel each other out, so only the pressure acting on the top and bottom faces need be considered. Let the area of each face of the cube be A. Then:

Assuming the top and bottom faces are in a horizontal plane, then the pressure is constant over the face so:

Pressure on the top face $= \rho gh_1$
Pressure on the bottom face $= \rho gh_2$

The force on the face is equal to the pressure multiplied by the area of the face, A. So:

Force on the top face $= \rho gh_1 A$
Force on the bottom face $= \rho gh_2 A$

Since $h_2 > h_1$ there will be a net force acting vertically upwards, F. This is:

$$F = \rho gh_2 A - \rho gh_1 A = \rho g(h_2 - h_1)A$$

Now $(h_2 - h_1)A$ is the volume of the cube, V, so:

$$F = \rho gV \qquad (1.14)$$

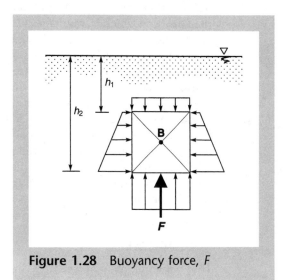

Figure 1.28 Buoyancy force, F

Box 1.7 ▶ **Remember**

The buoyancy force, F, acts vertically upwards through the centre of gravity of the displaced liquid (such as the centre of the cube). The point at which F acts is called the **centre of buoyancy**, B. The force, F, is equal to the weight of the volume of liquid displaced by the body, that is ρgV. This is known as Archimedes' Principle. Now go back and look at Step 2 of Example 1.5. You should be able to see that a buoyancy force is just the vertical force caused by hydrostatic pressure. See also Chapter 3 and Box 3.1.

66 When we analysed the buoyancy force on the cube in Fig. 1.28 we only considered the hydrostatic forces acting vertically on it. The weight of the cube was irrelevant. However, if we wanted to know whether or not the completely immersed cube would float or sink, we would have to compare the weight of the cube (W) with the buoyancy force (F), remembering that weight is a force. 99

W = weight density of cube material × volume = $\rho_s g V$ N acting vertically downwards.

F = weight of liquid displaced by the cube = $\rho g V$ N acting vertically upwards.

Since g and V are the same, it follows that if the density of the substance, ρ_s, that forms the cube is greater than the density of the liquid, ρ, then the cube would sink ($W > F$). Conversely, if $\rho_s < \rho$, then the cube would float ($F > W$). If $\rho_s = \rho$ then the cube has neutral buoyancy and would neither float nor sink, but would stay at whatever depth it was located ($F = W$).

The analysis above explains why a concrete or steel cube would sink, and a cork or polystyrene cube would float. However, this assumes that the cube is solid. If the cube was hollow, its average density would have to be used in the calculations, not the density of the material from which it was made. Submarines provide an interesting example, because they must be able to sink and, more importantly, rise to the surface again. This can be achieved by adjusting the average density of the submarine, by changing its weight by admitting or expelling water from tanks on the outside of the hull.

Floating bodies, such as ships and the pontoon in Example 1.7, are quite easy to analyse. If the depth of immersion is constant, then obviously W and F in Fig. 1.29 are exactly equal (otherwise the body would move up or down). Hence the starting point for many calculations involving floating bodies is:

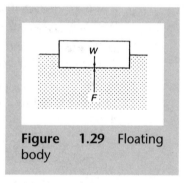

Figure 1.29 Floating body

$$W = F$$
or $$W = \rho g V \qquad (1.17)$$

Thus a floating body of weight W displaces a volume of water (V) that has a weight ($\rho g V$) equal to its own. Since $W = Mg$ this can also be written as:

$$Mg = \rho g V \qquad (1.18)$$

or $$M = \rho V \qquad (1.19)$$

Therefore it is also true to say that a floating body of mass M displaces a volume of water (V) that has a mass (ρV) equal to its own. Of course, equation (1.19) is a rearrangement of equation (1.5). Remember to use W with the weight density (ρg) and M with the mass density (ρ). Typically the body's weight or mass is known, so the relationships above allow the volume of water (V) displaced by a floating body to be calculated. Then for pontoons which are rectangular in plan and cross-section like those in Fig. 3.1:

depth of imersion = V/plan area $\qquad (1.20)$

By now it should be apparent that a solid steel cube sinks, but a ship made from steel plates floats because it is hollow and can displace a much larger volume of water (V) that has a

Figure 1.30 Lock gates provide another example of where it may be necessary to calculate hydrostatic forces. The buoyancy, depth of immersion and freeboard (the distance from the deck to the waterline) of the barge may also be the subject of an engineer's calculations

mass (or weight) equal to that of the ship. This is why we say that a ship has a displacement of 10 000 tonnes, for instance. When $W = F$ the depth of immersion is constant, but if W is increased by adding cargo the ship settles deeper in the water, increasing its displacement and consequently F, until $W = F$ again.

From 1876 onwards, British ships have had a Plimsoll line painted on their hull to indicate the maximum safe loading limit. Since the density of water changes according to temperature and salt content, the Plimsoll line includes marks for sea or fresh water, winter or summer, in tropical or northern waters.

Box 1.8 ▶ **Try this – amaze your friends**

Get an empty fizzy drink bottle, fill it completely with water and put a sachet of ketchup in it (Fig. 1.31a). You need one that just floats, so you may have to try a few different types until you find one that works. Now challenge your friends to concentrate their minds and use the power of thought to make the sachet sink. Unless they really do have telekinetic powers, they won't be able to do it of course. Now here's the trick. When it is your turn, make sure you have your hands around the bottle, and gently squeeze it. Try to disguise the fact you are doing this. If you squeeze hard enough the sachet will sink, and you can claim to have a better brain than all of your friends combined.

The reason the sachet sinks is as follows. A body in water has two forces acting on it: its weight (W) acting vertically down and the buoyancy force ($F = \rho g V$) acting vertically up. The weight of the sachet cannot change, so W is constant. However, F depends upon the volume (V) of water displaced by the sachet. When you squeeze the bottle you are exerting pressure on the water inside. The water is incompressible, but the air in the sachet can be compressed. So by compressing the air, V is reduced and so is F. When $W > F$ the sachet sinks. When you stop squeezing $F > W$ so the sachet rises.

Human divers can control their buoyancy and move up and down like this, either by inflating or deflating their dry suits or by controlling the amount of air in their lungs. Usually a Cartesian diver consists of a small length of open ended glass tubing with a bubble at one end (Fig. 1.31b). It can be used instead of the sachet and works in the same way.

(a)

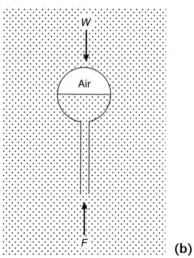

(b)

Figure 1.31 (a) Alternative Cartesian diver using a sachet of sauce. Squeezing and releasing the bottle makes the diver sink and then rise. (b) Conventional glass diver

EXAMPLE 1.6

A pipe which will carry natural gas is to be laid across an estuary which is open to the sea. The weight of the pipe is 2360 N per metre length and its outside diameter is 1.0 m. The weight of the gas can be ignored. The density of sea water is 1025 kg/m³. Determine whether the pipe will remain on the sea bed or float. If it does float, what force would be required to hold the pipe on the sea bed?

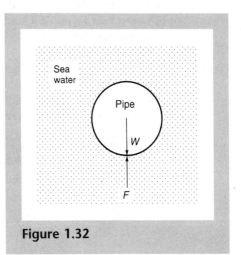

The maximum buoyancy force occurs when the pipe is fully submerged. Thus:

Buoyancy force, $F = \rho g V$.

$V = 1 \times \pi(1.0)^2/4 = 0.785\,\text{m}^3/\text{m length}$.

$F = 1025 \times 9.81 \times 0.785 = 7893\,\text{N/m}$.

Weight of pipe, $W = 2360\,\text{N/m}$.

Therefore, net force on pipe $= (7893 - 2360)$
$$= 5533\,\text{N/m}$$

The net force acts upwards since $F > W$. The pipe would float and a force of at least 5533 N/m would be required to hold it down.

Figure 1.32

EXAMPLE 1.7

A pontoon which is being used to conduct some construction work on a pier built into the sea has a mass of 50 tonnes (1 tonne = 1000 kg). The pontoon is rectangular in plan and cross-section. Its length is 10 m, its width 5 m, and its sides are 2 m high. The density of sea water (ρ_{SW}) is 1025 kg/m³. (a) Determine the volume of water displaced by the pontoon. (b) Determine the depth of immersion and the freeboard of the pontoon. (c) Determine the buoyancy force on the pontoon.

(a) A floating body displaces its own mass (or weight) of water, so from equation (1.19), volume displaced = mass of pontoon/ρ_{SW} = 50 000/1025 = 48.78 m³.

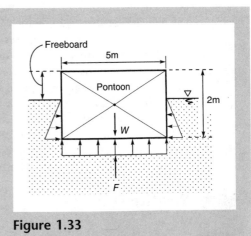

(b) Depth of immersion (the depth in the water) = volume displaced/plan area = 48.78/(10 × 5) = 0.98 m.

Freeboard = (2.00 − 0.98) = 1.02 m.

(c) Buoyancy force = weight of water displaced = weight of pontoon = 50 000 × 9.81 = 490.5 × 10³ N.

(Alternatively $F = \rho g V = 1025 \times 9.81 \times 48.78 = 490.5 \times 10^3\,\text{N}$)

Figure 1.33

1.8 ▶ The hydrostatic equation

The hydrostatic equation is really a statement of what, by now, should be obvious to you. Nevertheless, it can be useful, so the meaning of the equation is considered below while the derivation of the equation can be found in Appendix 1.

Basically, the hydrostatic equation states that the change in pressure intensity between two levels of a homogeneous (uniform) liquid is proportional to the vertical distance between them.

Consider points 1 and 2 at some distance below the surface as in Fig. 1.34. This time let us measure the depth of the points from the **bottom** (not from the surface) and let these distances be denoted by z_1 and z_2.

Pressure at point 1, $P_1 = \rho g h_1 = \rho g(d - z_1)$

Pressure at point 2, $P_2 = \rho g h_2 = \rho g(d - z_2)$

The difference in pressure between the two points is $(P_2 - P_1)$ where:

$$(P_2 - P_1) = \rho g(d - z_2) - \rho g(d - z_1)$$
$$= \rho g(d - z_2 - d + z_1)$$
$$= \rho g(-z_2 + z_1)$$
$$(P_2 - P_1) = -\rho g(z_2 - z_1) \quad (1.21)$$

Equation (1.21) shows that the difference in pressure between two points is equal to the vertical distance $(z_2 - z_1)$ between them. However, the equation is more useful when rearranged so:

$$(P_2/\rho g) + z_2 = (P_1/\rho g) + z_1 \quad (1.22)$$

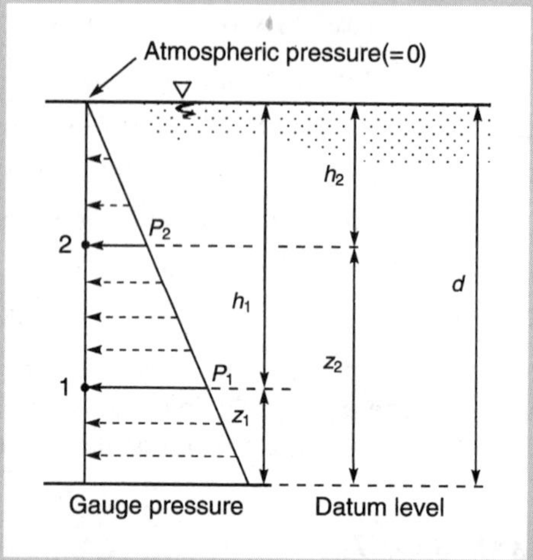

Figure 1.34 Pressure intensity at two points

This equation contains four of the six terms of the Bernoulli (or energy) equation that will be discussed in Chapter 4. The two terms that are missing from the Bernoulli equation are the velocity heads $(V^2/2g)$, which is logical since the velocity (V) is zero in a static liquid.

When we start considering pressure measurement using manometers in the next chapter we will be using equations (1.21) and (1.22), or at least the meaning of the equations if not the actual equations themselves. Perhaps you can see from the equations that if you know the pressure (say P_1) at some point in a static liquid, then you can calculate the pressure (P_2) at any other point so long as you can measure the vertical distance between them. Manometers are designed to enable the pressure difference $(P_2 - P_1)$ to be determined from the difference in the height of two columns of liquid $(z_2 - z_1)$, knowing the weight density of the liquid ρg. Do not worry if you do not fully understand this at the moment, since manometers are explained in the next chapter.

1.9 ▶ Stratified fluids

❝ How do you calculate the pressure if you have two liquids of different density? Does this make things more difficult? ❞

Box 1.9 # The equal level, equal pressure principle

One final thing about the hydrostatic equation, which again is a statement of the obvious, is that at a constant depth (or height z in the case of Fig. 1.35) the pressure is constant. It has to be since $P = \rho gh$. However, this gives rise to the '*equal level, equal pressure*' principle. This simply states that if you draw a horizontal line in a continuous body of static, uniform fluid then the pressure is the same anywhere on that line. The meaning and significance of this will be clearer if you look at Fig. 1.35. Again, this principle is used with manometers and will be used in the next chapter. However, remember the liquid must have a uniform density (otherwise see section 1.9).

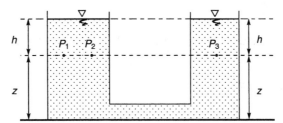

Figure 1.35 Equal level, equal pressure principle. The broken line is horizontal and the liquid has a constant density, so the hydrostatic pressure is constant along the line and $P_1 = P_2 = P_3$

Well, again this is nothing new. If you look back to section 1.2 you will see that we discussed the fact that hydrostatic pressure is caused simply by the weight of the liquid above the point (or surface) that we are considering. This is still true when you have two or more liquids of different densities. All you have to do is work out the weights (or pressures) of the liquids one at a time then add them together. Let us analyse the situation in Fig. 1.36.

Say that the column of liquid has a plan area of A m². The weight of the upper block of liquid is given by:

W_1 = weight density × volume

$\quad = \rho_1 gh_1 A$

Similarly, the weight of the lower block is:

$W_2 = \rho_2 gh_2 A$

Total weight $W_T = W_1 + W_2$

$\quad\quad = \rho_1 gh_1 A + \rho_2 gh_2 A$

Now equation (1.1) told us that:

pressure = weight/area

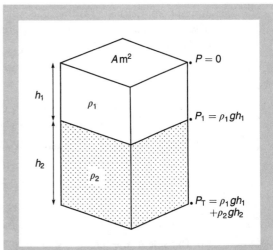

Figure 1.36 A stratified liquid with layers of density ρ_1 and ρ_2

Remember

Because the liquid is stratified and has two different densities, the pressure intensity diagram does not have the same gradient over the whole depth (as it did in Fig. 1.7, for instance). Instead, there is a change in gradient at the interface between the two liquids. However, within a particular liquid the gradient is uniform. Figure 1.37b in Example 1.8 provides an illustration of this.

so the total pressure, P_T, at the base of the column of liquid is:

$$P_T = W_T/A$$
$$P_T = \rho_1 g h_1 + \rho_2 g h_2 \tag{1.23}$$

66 We have analysed many different situations in this chapter. To help you to remember how to approach the different types of problem, I have provided a summary for you at the end of the chapter. This may prove useful when revising or when tackling the revision questions. 99

EXAMPLE 1.8

A tank with vertical sides contains both oil and water. The oil has a depth of 1.5 m and a relative density of 0.8. It floats on top of the water, with which it does not mix. The water has a depth of 2.0 m and a relative density of 1.0. The tank is 3.0 m by 1.8 m in plan and open to the atmosphere. Calculate (a) the total weight of the contents of the tank; (b) the pressure on the base of the tank; (c) the variation of pressure intensity with depth; (d) the force on the side of the tank.

(a) $W_T = (\rho_1 g h_1 + \rho_2 g h_2)A$

Plan area $A = 3.0 \times 1.8 = 5.4 \, m^2$

$W_T = (0.8 \times 1000 \times 9.81 \times 1.5 + 1.0 \times 1000 \times 9.81 \times 2.0)5.4$
    $= (11\,772 + 19\,620)5.4 = 169\,517 \, N$

(b) Total pressure at base of tank = $W_T/A = 169\,517/5.4 = 31\,392 \, N/m^2$

(c) Pressure at the surface = atmospheric = 0

Pressure at the bottom of the oil = $\rho_1 g h_1 = 11\,772 \, N/m^2$

Total pressure at the bottom of the tank = $31\,392 \, N/m^2$

The pressure intensity diagram is shown in Fig. 1.37b.

(d) The side of the tank is 3.0 m long. The force on the side of the tank can be obtained from equation (1.2) by multiplying the area of the tank in contact with each of the liquids by the average pressure intensity of the particular liquid.

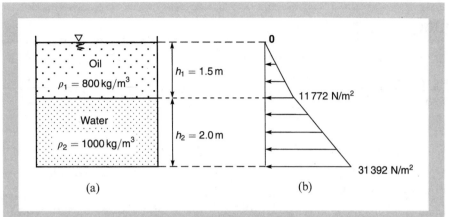

Figure 1.37 (a) Tank containing a stratified liquid, and (b) the corresponding pressure intensity diagram

Average pressure of the oil = $(0 + 11\,772)/2 = 5886\,\text{N/m}^2$

Force due to the oil = $3.0 \times 1.5 \times 5886 = 26\,487\,\text{N}$

Average pressure of the water = $(11\,772 + 31\,392)/2 = 21\,582\,\text{N/m}^2$

Force due to the water = $3.0 \times 2.0 \times 21\,582 = 129\,492\,\text{N}$

Total force on the side = $26\,487 + 129\,492 = 155\,979\,\text{N}$

Summary

· ·

PLANE VERTICAL SURFACE EXTENDING TO WATER SURFACE

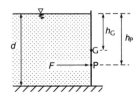

$F = \rho g h_G A$
h_G = depth to centroid of vertical surface
A = area of vertical surface
Depth to centre of pressure, $h_P = (2/3)d$

PLANE VERTICAL SURFACE **NOT** EXTENDING TO WATER SURFACE

$F = \rho g h_G A$
h_G = depth to centroid of vertical surface
A = area of vertical surface
$h_P = (I_G / A h_G) + h_G$
I_G = 2nd moment of area about G in vertical plane

PLANE INCLINED SURFACE

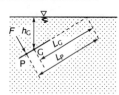

$F = \rho g h_G A$
h_G = depth to centroid of inclined surface
A = actual area of inclined surface
$L_P = (I_G / A L_G) + L_G$
L_P = inclined length to centre of pressure, P
L_G = inclined length to centroid of surface, G
I_G = 2nd moment of area calculated about G in the plane of the inclined surface

CURVED SURFACE

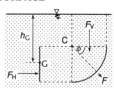

HORIZONTAL COMPONENT OF RESULTANT
$F_H = \rho g h_G A$
h_G = depth to centroid of projected vertical surface
A = area of projected vertical surface
VERTICAL COMPONENT OF RESULTANT
$F_V = \rho g V$
V = volume of water above curved surface
RESULTANT $F = (F_H^2 + F_V^2)^{1/2}$
Angle to horizontal $\phi = \tan^{-1}(F_V/F_H)$

BUOYANCY FORCE

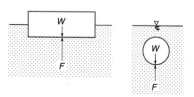

$F = \rho g V$
V = volume of water displaced
F acts through centre of displaced water
Net buoyancy force = $F - W$

STRATIFIED FLUID

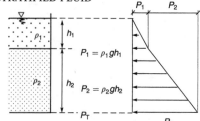

Total pressure $P_T = P_1 + P_2 = \rho_1 g h_1 + \rho_2 g h_2$

Revision questions

1.1 Define clearly what is meant by the following, and give the appropriate units in each case: (a) pressure; (b) force; (c) weight; (d) gravity; (e) mass; (f) mass density; (g) weight density; (h) relative density; (i) hydrostatic pressure; (j) buoyancy force.

1.2 (a) Explain what is meant by gauge pressure and absolute pressure. (b) What is the approximate numerical value of atmospheric pressure expressed in N/m^2 and as a head of water? (c) Calculate atmospheric pressure expressed as a head of mercury (the relative density of mercury is 13.6)

[(c) 0.76 m]

1.3 A rectangular tank is 1.0 m long and 0.7 m wide and contains fresh water to a depth of 0.5 m. (a) What is the gauge pressure at the bottom of the tank in N/m^2? (b) What is the absolute pressure at the bottom of the tank?

[4905 N/m^2; 105 905 N/m^2]

1.4 For the tank in question 1.3, using gauge pressure, calculate (a) the mean pressure intensity on the 0.7 m wide end of the tank; (b) the mean pressure intensity on the 1.0 m long side of the tank; (c) the force on the end of the tank; and (d) the force on the side.

[2453 N/m^2; 2453 N/m^2; 858 N; 1226 N]

1.5 A dam that retains fresh water has a vertical face. Over a one metre length of the face at the centre of the valley the water has a depth of 38 m. (a) Calculate the resultant force on this unit length of the face. (b) At what depth from the surface does the resultant force act?

[7083 × 10^3 N; 25.33 m]

1.6 (a) A rectangular culvert 2.1 m wide by 1.8 m high discharges to a river channel as in Fig. 1.10. At the end of the culvert is a vertical flap gate which is hinged along its top edge, the gate having the same dimensions as the culvert. During a flood the river rises to 3.5 m above the hinge. What is the force exerted by the floodwater on the gate, and at what depth from the surface does it act? (b) A circular gate, also hinged at the top, hangs vertically at the end of a pipe discharging to

the river. The gate has a radius of 0.5 m, and during a flood the hinge is 3.5 m below the water surface. What is the force exerted by the floodwater on the gate, and at what depth from the surface does it act?

[(a) 163.16 × 10^3 N at 4.461 m;
(b) 30.82 × 10^3 N at 4.015 m]

1.7 A gate at the end of a sewer measures 0.8 m by 1.2 m wide. It is hinged along its top edge and hangs at an angle of 30° to the vertical, this being the angle of the banks of a trapezoidal river channel. (a) Calculate the hydrostatic force on the gate and the vertical distance between the centroid of the gate, G, and the centre of pressure, P, when the river level is 0.1 m above the top of the hinge. (b) If the river level increases to 2.0 m above the hinge, what is the force and the distance GP now? (c) Has the value of GP increased or decreased, and why has it changed in this manner?

[(a) 4.21 × 10^3 N, 0.090 m;
(b) 22.10 × 10^3 N, 0.017 m]

1.8 A circular gate of 0.5 m radius is hinged so that it rotates about its horizontal diameter, that is it rotates about a horizontal line passing through the centroid of the gate. The gate is at the end of a pipe discharging to a river. Measured above the centroid of the gate, the head in the pipe is 6.0 m while the head in the river is 2.0 m. Assuming that the gate is initially vertical: (a) calculate the force exerted by the water in the pipe on the gate, and the distance GP between the centre of the gate, G, and the centre of pressure, P; (b) calculate the force exerted by the river water on the gate, and the distance GP; (c) by taking moments about the hinge, using the results from above, determine the net turning moment on the gate caused by the two forces acting at their respective centres of pressure on opposite sides of the gate. Explain your answer.

[(a) 46 228.5 N at 0.0104 m;
(b) 15 409.5 N at 0.0312 m;
(c) 0 exactly, allowing for rounding errors]

1.9 A gate which is a quarter of a circle of radius 4.0 m holds back 2.0 m of fresh water as shown in the diagram.

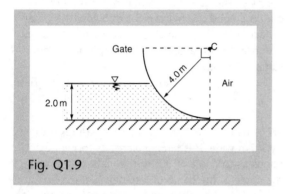

Fig. Q1.9

Calculate the magnitude and direction of the resultant hydrostatic force on a unit length of the gate.

[52.05 × 10^3 N/m at 67.9° to the horizontal, acting upwards through the centre of curvature, C]

1.10 The dam in Fig. Q1.10 has a curved face, being part of a 40 m radius circle. The dam holds back water to a depth of 35 m. Calculate the magnitude and direction of the resultant hydrostatic force per metre length.

[7840.6 × 10^3 N/m at 40° to the horizontal, acting downwards through the centre of curvature, C]

1.11 A 7500 tonne reinforced concrete lock structure has been constructed in a dry dock. The lock is 60 m long by 30 m wide in plan and is shaped like an open shoe box. The side walls are 8 m high. (a) Will the lock structure float in sea water of density 1025 kg/m^3, and if so, what is its draught and free-

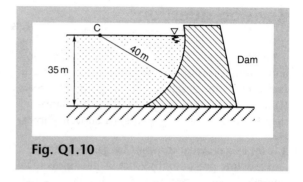

Fig. Q1.10

board? (b) What additional weight will be required to sink the structure onto the sea bed if the depth of water is 5.3 m, assuming the structure is water-tight? (c) If the additional weight is to be provided by a blanket of sand (density 2600 kg/m^3), how thick must the layer of sand be? (1 tonne = 1000 kg).

[(a) yes, 4.07 m, 3.93 m;
(b) >22 352 × 10^3 N;
(c) >0.5 m]

1.12 (a) Explain what is meant by a stratified fluid. (b) A pressure transducer is used to measure the hydrostatic pressure on the sea bed in a tidal estuary. The water in the estuary is stratified at the point where the measurement is taken, with fresh water (1000 kg/m^3) overlying saline water (1025 kg/m^3). Water sampling shows that the fresh water extends from the water surface to a depth of 2.7 m. If the transducer indicates a gauge pressure of 69.73 × 10^3 N/m^2, how thick is the layer of saline water?

[4.3 m]

2

Pressure measurement

It may be necessary to measure the pressure of a liquid for operational reasons, such as to monitor the distribution and supply of water, or to enable the discharge in a pipeline to be calculated. Whatever the reason, piezometers and manometers can be used for this purpose. The basis of these devices is the pressure–depth relationship that exists in a static liquid, and the principles described in the last chapter. The type of questions that are answered in this chapter include:

Why do we want to measure the pressure of a liquid?

What is a piezometer? What is a manometer?

How does a piezometer work?

How can we measure pressure with a mercury U-tube manometer?

How do we calculate a pressure difference with a differential U-tube manometer?

How do we analyse the results from an inverted U-tube manometer?

What affects the sensitivity of a manometer? How can we improve its accuracy?

What is a Bourdon gauge and how does it work?

2.1 Fundamentals

66 The fundamentals of pressure measurement were covered in the last chapter. All you have to do is apply them in the correct way. You will find it easier to do this if you remember the four points in Box 2.1. If you do not understand them, read the relevant parts of the previous chapter again. 99

Remember the basics

1. For a liquid of uniform density there is a clear relationship between pressure and depth, that is $P = \rho g h$ (equation (1.8)).

2. Pressure is due to the weight of the liquid above the point in question, so when calculating a pressure start at the point (that is at the bottom) and work out the pressure caused by the weight of the liquid above the point using equation (1.8).

3. At the same elevation in a continuous liquid of uniform density the pressure is constant, which is the equal level, equal pressure principle (see section 1.8).

4. With stratified liquids of different density, the pressure caused by each of the liquids can be worked out separately (equation (1.8)) and then added together to get the total (section 1.9).

REMEMBER!!!

❝ Just a minute, you are going too fast. What is a piezometer, what is a manometer, and how do you measure pressure in a liquid? Why do you want to measure pressure in a liquid anyway? ❞

A common reason for measuring the pressure of a liquid is to investigate the hydraulic characteristics of the flow through a pipeline. For example, it is important to know how the pressure varies along a pipeline distributing water to a town or city. If the pressure is too high (>70 m say) this may cause excessive leakage through the joints in the pipework and may also prevent valves from working efficiently. If the pressure is too low (say <30 m), then the supply of water to houses and fire hydrants may be unsatisfactory.

The discharge through a pipeline can be measured with an orifice plate or a Venturi meter (see Chapter 5). Both of these devices use a constriction in the pipe, that is a reduced cross-sectional area of flow, to induce a change of pressure. By measuring the pressure difference across the meter the discharge can be calculated (this is an oversimplification but gives the general idea). Other uses for piezometers and manometers will become apparent later in the book, but we will start by considering what they are, how their readings are converted into pressure, and their sensitivity. The Bourdon gauge, which is a simpler though less accurate means of measuring the pressure of a liquid, will then be described. Note that alternative electronic methods of measuring pressure, such as pressure transducers, are not considered.

2.2 ▶ Piezometers

One of the simplest ways of measuring the pressure in a pipe is to drill a hole in it and connect a tube of small diameter to the hole (Fig. 2.1a). The diameter of the tapping should be about 3 or 4 mm, and the connection should be flush with the inside surface of the pipe. The diameter must be small to prevent any disturbance to the flow in the pipe, and to ensure that the kinetic energy of the flow does not contribute to the level of the liquid in the tube. The tube is called a **piezometer**. Assuming the pipe is flowing full and that the liquid is under pressure, the liquid will be forced up the piezometer tube to some height, h, measured from the centreline of the pipe, where h is known as the **piezometric height** or the **piezometric head**. The elevation to which the liquid rises is the **piezometric level**. The principle of the piezometer is that the liquid rises up the piezometer tube until atmospheric pressure (P_{ATM}) and the weight of the column of liquid in the tube (ρgh) generate a pressure which equals the pressure in the pipeline, P.

It is very easy to calculate the pressure at the centreline of the pipe below the piezometer tapping. Referring to notes 1 and 2 in Box 2.1, it should be apparent that the gauge pressure is:

$$P = \rho gh \tag{2.1}$$

where ρ is the density of the liquid in the pipe, and thus in the piezometer as well. Gauge pressure is the pressure measured relative to atmospheric pressure. If the absolute pressure is needed then P_{ATM} which acts on the water surface in the piezometer tube, should be added to ρgh (see section 1.2 and equation (1.9)).

One problem with piezometers is that a very long tube may be required because, as mentioned in section 2.1, the pressure in watermains is usually somewhere between 30 m and 70 m of water. Clearly a 70 m piezometer tube is out of the question. There are many solutions to this problem; one would be to use a manometer, as described later. However, if two piezometers are being used to measure a difference in pressure between two points then it is possible to connect the tops of the piezometers to a manifold containing air which is not at atmospheric pressure, but some much higher pressure (Fig. 2.1b). This can be achieved by fitting a valve to the manifold and pressurising it with a pump, such as a bicycle pump. The actual air pressure, P_A, does not matter since it acts equally on the water surface in both piezometers. If the pipeline is horizontal then the liquid pressure at the two points on the pipe centreline, P_1 and P_2, are:

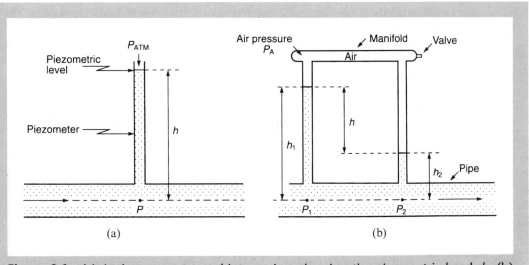

(a)   (b)

Figure 2.1 (a) A piezometer tapped into a pipe, showing the piezometric head, h. (b) A piezometer for measuring the difference in pressure ($P_1 - P_2$) between two points in a pipeline. The manifold at the top may be pressurised by pumping in air. This ensures that h_1 and h_2 have convenient values

$$P_1 = \rho g h_1 + P_A \qquad (1)$$
$$P_2 = \rho g h_2 + P_A \qquad (2)$$

Subtracting equation (2) from equation (1) gives:

$$(P_1 - P_2) = \rho g h_1 + P_A - [\rho g h_2 + P_A]$$
$$(P_1 - P_2) = \rho g h_1 - \rho g h_2$$
$$\boldsymbol{(P_1 - P_2) = \rho g h} \qquad\qquad (2.2)$$

The important thing to realise here is that it is the difference in the two piezometric levels, h, that appears in the equation. The air pressure in the manifold does not matter since P_A does not appear in equation (2.2). So in practice you can simply pump air into the manifold (or release it) until the values of h_1 and h_2 are within the range that you want, say somewhere between 0 and 2 m. If the differential head exceeds 2 m then it may be better to use a differential manometer instead.

2.3 A simple U-tube manometer

One of the simplest forms of manometer is the U-tube manometer, which commonly contains mercury in the bottom of the U and the pipe liquid in the upper part (Fig. 2.2). Again the tapping should be of a small diameter and flush with the inside of the pipe. The pressure in the pipe forces liquid out into the left-hand limb of the manometer, but the weight of the mercury in the right-hand limb balances the pressure in the pipe and prevents the pipe liquid flowing around the U and out of the end of the right-hand limb. Remember that the density of mercury is 13.6 times that of water. It is often used because its weight means that only a short U-tube is needed.

When thinking about how a mercury U-tube manometer works, remember that the differential head, h_M, will be zero only when the pipe and left-hand manometer limb contain air. In other words, when the manometer contains only the mercury. As soon as the left limb is filled with the pipe liquid (even if the pipe itself is empty) then the weight of this column of liquid will push the mercury around the U and create a differential head, h_M. Similarly, as the pressure in the pipe increases, so h_M increases.

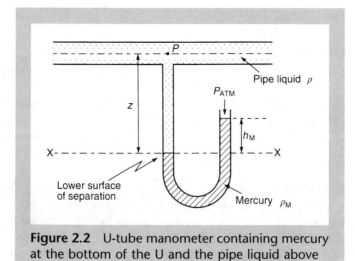

Figure 2.2 U-tube manometer containing mercury at the bottom of the U and the pipe liquid above

❝ I understand that, but how do I work out the pressure of the liquid from the mercury levels? ❞

The first step in the procedure to obtain an expression for the gauge pressure, P, is to draw a line through the lower surface of separation between the pipe liquid and the mercury. This is marked X–X in Fig. 2.2. Let the density of the liquid in the pipeline be ρ and the density of the mercury ρ_M. Using notes 1 and 2 in Box 2.1:

Pressure at level X–X in left limb $= \rho gz + P$ (1)

Pressure at level X–X in right limb $= \rho_M g h_M$ (2)

where h_M is the difference in the elevation of the surface of the mercury in the two limbs. Now apply note 3 in Box 2.1. Because the mercury occupies all of the lower part of the U-tube below X–X, at the elevation X–X the pressure in both limbs must be the same (this is the equal level, equal pressure principle). Therefore equations (1) and (2) above must be equal, and:

$$\rho gz + P = \rho_M g h_M$$
$$P = \rho_M g h_M - \rho gz \qquad (2.3)$$

This gives the gauge pressure which is measured using atmospheric pressure as a datum. Note that if $h_M = 0$ then there is a negative gauge pressure, $-\rho gz$, that is a pressure which is less than atmospheric (see Fig. 1.8). If the absolute pressure is required, atmospheric pressure (P_{ATM}) must be added to the right-hand side of equation (2) above because atmospheric pressure is acting on the surface of the mercury in the right-hand limb in Fig. 2.2. Thus equation (2.3) becomes:

$$P_{ABS} = \rho_M g h_M + P_{ATM} - \rho gz \qquad (2.4)$$

Now study Example 2.1 and then try Self Test Question 2.1.

EXAMPLE 2.1

The lower part of the U-tube manometer in Fig. 2.3 contains mercury ($\rho_M = 13\,600\,\text{kg/m}^3$). The pipe contains water ($\rho = 1000\,\text{kg/m}^3$). Determine the gauge pressure, P, at the centre of the pipe if the manometer readings are as shown in the diagram.

Step 1 Draw a horizontal line X–X through the lower surface of separation.

Step 2 Working upwards from X–X:

Pressure at level X–X in left limb $= \rho gz + P$

Pressure at level X–X in right limb $= \rho_M g h_M$

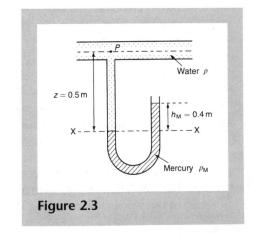

Figure 2.3

Step 3   Equate the pressures in the two limbs at X–X

$$\rho gz + P = \rho_M gh_M$$

$$P = \rho_M gh_M - \rho gz$$

Now $\rho_M = 13600 \text{kg/m}^3$, $h_M = 0.4\text{m}$,

$\rho = 1000 \text{kg/m}^3$ and $z = 0.5\text{m}$ so:

$P = 13600 \times 9.81 \times 0.4 - 1000 \times 9.81 \times 0.5 = 48.5 \times 10^3 \text{ N/m}^2$

SELF TEST QUESTION 2.1

The U-tube manometer shown in Fig. 2.4 is used to measure the pressure in a pipeline as it passes over the crest of a hill. The pipeline carries fresh water ($\rho = 1000 \text{kg/m}^3$). The bottom of the U-tube contains mercury ($\rho_M = 13\,600 \text{kg/m}^3$). Taking atmospheric pressure as the equivalent of 10.3 m of water, with the readings shown in the diagram calculate the absolute pressure at the centre of the pipeline.

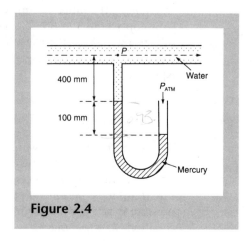

Figure 2.4

2.4 ▶ A differential U-tube manometer

If the difference in pressure between two points in a pipe is of interest (rather than the pressure at a single location) then an alternative to two piezometers (Fig. 2.1b) is a differential U-tube manometer (Fig. 2.5). In this case the free end of the U-tube in Fig. 2.2 is now connected to the pipe where the second pressure has to be measured. The liquid in the bottom of the manometer is often mercury, since this is the heaviest liquid and gives the smallest manometer. However, if the pressure difference is small, a lighter liquid can be used to increase the accuracy.

❝ To obtain an equation for the differential pressure we now have to apply all *four* notes listed in Box 2.1. Up to now we have only used the first three, but in this case we effectively have a stratified liquid. However, the way in which the problem is approached is virtually the same as for the U-tube manometer considered in the last section. Try to get used to following this procedure. It makes the problems easy and avoids mistakes. ❞

As before, the first step is to draw a line through the lower surface of separation between the pipe liquid and the mercury. This is marked X–X in Fig. 2.5. Let the density of the liquid in the pipeline be ρ and the density of the mercury ρ_M then:

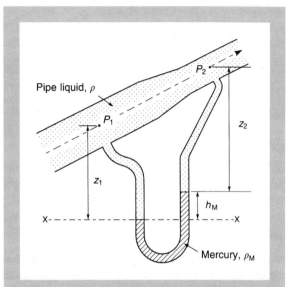

Figure 2.5 A differential U-tube mano-meter installed on an inclined pipeline of changing diameter. The manometer liquid in the bottom of the U-tube is mercury, the liquid above is that in the pipe

Pressure at level X–X in left limb = $\rho g z_1 + P_1$ (1)

Pressure at level X–X in right limb = $\rho_M g h_M + \rho g z_2 + P_2$ (2)

P_1 and P_2 are the pressures at the two points in the pipeline, and h_M is the difference in the elevation of the mercury in the two limbs. You should be able to see that the stratified liquid in the right limb presents no problem at all, the two pressures obtained from equation (1.8) [$P = \rho g h$] are simply added together to get the total (this was note 4 in Box 2.1). Now apply note 3, the equal level, equal pressure principle, which states that at the same elevation in a uniform, continuous liquid the pressure must be the same. Hence at the level X–X the pressures in both limbs must be the same, and equations (1) and (2) above must be equal, so:

$$\rho g z_1 + P_1 = \rho_M g h_M + \rho g z_2 + P_2$$
$$(P_1 - P_2) = \rho_M g h_M + \rho g z_2 - \rho g z_1$$
$$\boldsymbol{(P_1 - P_2) = \rho_M g h_M + \rho g (z_2 - z_1)} \tag{2.5}$$

In this case since a ***difference*** in pressure is being measured the question of gauge pressure and absolute pressure does not enter into the argument, because the difference would be the same in either case. Equation (2.5) applies to the general case of a sloping pipeline, but if the pipeline is horizontal then $z_1 = z_2 + h_M$ and substituting for z_1 in equation (2.5) gives:

$$(P_1 - P_2) = \rho_M g h_M + \rho g (z_2 - [z_2 + h_M])$$
$$\boldsymbol{(P_1 - P_2) = g h_M (\rho_M - \rho)} \tag{2.6}$$

Box 2.2 ▶ **For U-tube manometers**

To solve problems involving U-tube manometers all you have to do is:

Step 1. Draw a line through the lower surface of separation, X–X.

Step 2. Work out the pressure at the level X–X in each limb of the manometer. Do this by starting at the bottom and working upwards, adding the pressures together as you go.

Step 3. Equate the pressures in the two limbs and solve for the unknown pressure.

66 Box 2.2 summarises how to solve simple U-tube manometer problems. Before moving on to the next section, study Example 2.2 and then try Self Test Question 2.2. Make sure that you get into the habit of using a logical, step-by-step procedure. **99**

EXAMPLE 2.2

A differential U-tube manometer is used to measure the change in pressure between two points in a pipeline which carries oil of relative density 0.8. The lower part of the U-tube contains mercury of relative density 13.6. There is an increase in elevation between the two points of 0.26 m. If $z_1 = 0.60$ m, $z_2 = 0.73$ m and $h_M = 0.13$ m, calculate the difference in pressure.

Density oil, $\rho_O = 0.8 \times 1000 = 800\,\text{kg/m}^3$
Density mercury, $\rho_M = 13.6 \times 1000 = 13600\,\text{kg/m}^3$

Step 1 Draw a horizontal line X–X through the lower surface of separation.

Step 2 Working upwards from X–X:

Pressure at level X–X in left limb

$$P_X = \rho_O g z_1 + P_1$$
$$= 800 \times 9.81 \times 0.60 + P_1$$
$$= 4709 + P_1$$

Pressure at level X–X in right limb

$$P_X = \rho_M g h_M + \rho_O g z_2 + P_2$$
$$= 13600 \times 9.81 \times 0.13 + 800 \times 9.81$$
$$\times 0.73 + P_2$$
$$= 23073 + P_2$$

Step 3 Equate the pressures in the two limbs at level X–X.

$$4709 + P_1 = 23073 + P_2$$

$$(P_1 - P_2) = 23073 - 4709 = 18364\,\text{N/m}^2$$

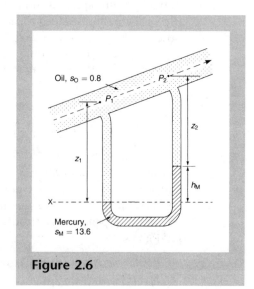

Oil, $s_O = 0.8$
P_2
P_1
z_2
z_1
h_M
X–
Mercury, $s_M = 13.6$

Figure 2.6

SELF TEST QUESTION 2.2

The differential U-tube manometer in Fig. 2.7 is used to measure the difference in pressure between two points in an expanding pipeline. The pipe liquid is water and the gauge liquid is mercury (1.0 and 13.6 relative density, respectively). For the dimensions shown in the diagram, calculate the value of ($P_1 - P_2$).

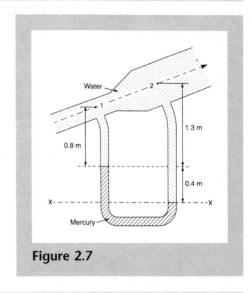

Figure 2.7

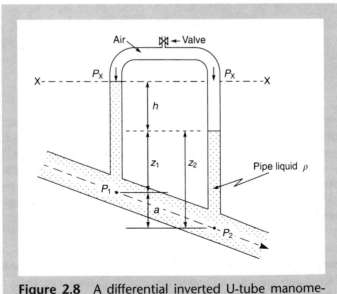

◈2.5 The inverted U-tube differential manometer

This is another type of manometer used to measure pressure differences. As the name suggests, it looks like a U-tube manometer, but turned upside down (Fig. 2.8). The pipe liquid (density ρ) fills the lower part of the manometer limbs with air in the upper part. The air

Figure 2.8 A differential inverted U-tube manometer with air above the pipe liquid

---

Box 2.3 ▶ **Inverted U-tube manometers**

To solve problems involving inverted U-tube manometers all you have to do is:

Step 1. Draw a horizontal line X–X through the **upper** surface of separation.

Step 2. Write an equation for the pressure at each of the two points in the pipe by starting at the **bottom** of each limb and equating the pressure in the pipe to the sum of the pressures caused by the weight of the various fluids in the limb up to level X–X.

Step 3. Subtract one of the equations from the other and solve for the differential pressure.

---

can be regarded as weightless, although it may not be at atmospheric pressure. As with the differential piezometer (Fig. 2.1b) air can be pumped into the top of the manometer (or released) through a valve in order to control the level of the liquid in the limbs. The diagram shows a general case where the pipeline is not horizontal.

These manometers are analysed in almost exactly the same way as the U-tube manometers in the last section, but with one logical modification. With the U-tubes in Figs 2.2 to 2.7 we drew a line X–X through the lower surface of separation. However, we have turned the ∪ upside down to obtain a ∩. So now the line X–X must pass through the **upper** surface of separation between the two fluids because it is only above X–X that we have a continuous, uniform fluid that allows us to apply the equal level–equal pressure principle. This is the one modification that we have to make to the procedure that we have been using.

Referring to Fig. 2.8, let the pressures at two points on the centreline of the pipe be P_1 and P_2. The line X–X has been drawn through the upper surface of separation. Starting at the centre of the pipe at the bottom of the left-hand limb, adding the pressures up to level X–X gives:

$$P_1 = \rho g z_1 + \rho g h + P_X \qquad (1)$$

Note that the equation has been deliberately written so that it contains the differential head, h, instead of taking z_1 as the distance from the pipe centreline to X–X. This is because the differential head is the only one that needs to be measured, as will become apparent later. P_X is the pressure of the air at level X–X. Remember that P_X does not have to be equal to atmospheric pressure, it can be used to control the height of the liquid in the limbs.

Now starting at the centre of the pipe at the bottom of the right-hand limb, working upwards to X–X adding the pressures as we go gives:

$$P_2 = \rho g z_2 + \rho_A g h + P_X$$

where $\rho_A g h$ is due to the weight of the air in the manometer limb. It was stated above that the air can be regarded as weightless so this term can be omitted from the equation. However, air actually has a density of about $1.2 \, \text{kg/m}^3$, and later on the air will be replaced with a liquid, so the term has been included above as a reminder. If it is omitted then:

$$P_2 = \rho g z_2 + P_X \qquad (2)$$

The air pressure at the level X–X is the same in both limbs of the manometer, so P_X appears in both equations (1) and (2) above. This is still the case even if the air is replaced by a liquid because of the equal level, equal pressure principle. Subtracting equation (2) from equation (1) gives:

$$(P_1 - P_2) = \rho g z_1 + \rho g h + P_X - [\rho g z_2 + P_X]$$
$$(P_1 - P_2) = \rho g z_1 + \rho g h - \rho g z_2$$
$$(P_1 - P_2) = \rho g [h + z_1 - z_2]$$

Now from Fig. 2.8 it is apparent that the difference in elevation between points 1 and 2 is a, and that $a = z_2 - z_1$ so $-a = z_1 - z_2$. Substituting $-a$ for $z_1 - z_2$ above gives:

$$(P_1 - P_2) = \rho g [h - a] \tag{2.7}$$

If the pipeline is horizontal then $a = 0$ so:

$$(P_1 - P_2) = \rho g h \tag{2.8}$$

The fact that the difference in pressure is directly proportional to the differential head, h, is the reason why h was included in equation (1) initially. Equation (2.7) also illustrates that the actual values of z_1, z_2 and P_X are not important. Indeed, the air pressure can be used to control the level of the liquid in the manometer limbs and to obtain levels that are convenient for the observer to read without affecting the result.

66 Note that equation (2.8) is the same as equation (2.2), and that provided the pipeline is horizontal the inverted U-tube manometer is the same as the two piezometers shown in Fig. 2.1b. The advantage of the inverted U-tube manometer is that instead of having air in the limbs above the pipe liquid, other liquids can be used provided that they are immiscible (do not mix with each other) as shown in Fig. 2.9. The density of the second liquid can be used to control the sensitivity and accuracy of the gauge. **99**

Let the density of the pipe liquid in Fig. 2.9 be ρ and the density of the gauge liquid, which could be a light oil, be ρ_0. The upper liquid can flow into or out of the manometer from a small header reservoir. Starting at the bottom of the left and right limbs and working up to X–X gives:

$$P_1 = \rho g z_1 + \rho g h + P_X \qquad (1)$$

$$P_2 = \rho g z_2 + \rho_0 g h + P_X \qquad (2)$$

P_X is the pressure due to the gauge liquid above level X–X. P_X is the same in both limbs as a result of the equal level, equal pressure principle. Subtracting equation (2) from equation (1) gives:

$$(P_1 - P_2) = \rho g z_1 + \rho g h + P_X - [\rho g z_2 + \rho_0 g h + P_X]$$
$$(P_1 - P_2) = \rho g z_1 + \rho g h - \rho g z_2 - \rho_0 g h$$
$$(P_1 - P_2) = \rho g [h + z_1 - z_2] - \rho_0 g h$$

It is apparent from Fig. 2.9 that $a = z_2 - z_1$, so $-a = z_1 - z_2$. Substituting for $z_1 - z_2$ in the previous equation:

$$(P_1 - P_2) = \rho g [h - a] - \rho_0 g h \tag{2.9}$$

If the pipeline is horizontal, $a = 0$ so the equation becomes:

$$(P_1 - P_2) = g h [\rho - \rho_0] \tag{2.10}$$

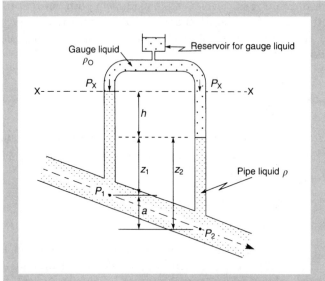

Figure 2.9 Inverted U-tube manometer utilising two liquids, the pipe liquid in the lower part of the manometer limbs with a lighter gauge liquid above

❝ In this chapter we have derived all of the equations from first principles. It is very important that you continue to do so and that you understand what you are doing. One of the most common mistakes made by students is to use equations blindly in the belief that they always apply to all situations. To become a good engineer or scientist, you must always be willing to question the applicability of an equation to a particular situation, going back to first principles if necessary. Example 2.3 shows you how easy it is to fall into the trap that waits for the unwary and to obtain the wrong answer! ❞

SELF TEST QUESTION 2.3

Everything is exactly as in Fig. 2.9 except that the pipeline slopes *upwards* from left to right. Redraw Fig. 2.9, and rederive the equation for $(P_1 - P_2)$. Does the equation change? If the relative densities of the two liquids in the manometer are 1.0 (pipe) and 0.8 (gauge), $a = 0.15\,\text{m}$ and the surfaces of separation in the left and right limbs are respectively 0.78 m and 0.45 m above the pipe centreline, what is the value of $(P_1 - P_2)$?

EXAMPLE 2.3

A manometer is used to measure the difference in pressure between two points in a pipe. The pipe carries water (density 1000 kg/m³). Above the water in the manometer is air, which is pressurised. The manometer readings are $z_1 = 100\,\text{mm}$, $z_2 = 250\,\text{mm}$ and $h = 90\,\text{mm}$. What is the value of $(P_1 - P_2)$ in N/m²?

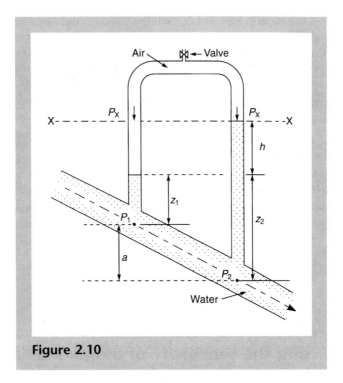

Figure 2.10

❝ First we will do the question the correct way. ❞

Step 1   Draw a line X–X through the upper surface of separation.

Step 2   Working upwards from the pipe to X–X:

$$P_1 = \rho g z_1 + P_X \qquad (1)$$

$$P_2 = \rho g z_2 + \rho g h + P_X \qquad (2)$$

Step 3   Subtract equation (2) from equation (1) and solve.

$$(P_1 - P_2) = \rho g z_1 + P_X - [\rho g z_2 + \rho g h + P_X]$$
$$(P_1 - P_2) = \rho g z_1 - \rho g z_2 - \rho g h$$
$$(P_1 - P_2) = \rho g[z_1 - z_2 - h] \qquad (3)$$
$$(P_1 - P_2) = 1000 \times 9.81[0.10 - 0.25 - 0.09]$$
$$(P_1 - P_2) = -2.35 \times 10^3 \, \text{N/m}^2$$

❝ Now this is how you get the wrong answer. You start off by saying that this is a problem involving an inverted U-tube manometer using water and air as the gauge fluids. You then turn to a text book, see an expression like equation (2.7) which is apparently applicable to this problem, put the numbers in and solve. Thus: ❞

$$(P_1 - P_2) = \rho g[h - a] \qquad (2.7)$$
$$\text{Now } a = z_2 - z_1 = 0.25 - 0.10 = 0.15 \text{m}$$
$$(P_1 - P_2) = 1000 \times 9.81[0.09 - 0.15]$$
$$(P_1 - P_2) = -0.59 \times 10^3 \, \text{N/m}^2$$

❝ Well the second method is certainly quicker, but unfortunately it is wrong! Can you see why? ❞

The reason is that the liquid levels in Fig. 2.8 to which equation (2.7) applies are not identical to those in Fig. 2.10 above. In this example the water level in the right limb is the higher, whereas in Fig. 2.8 it is higher in the left limb. Look at equation (3) above. If we say that $a = z_2 - z_1$ as before, then $-a = z_1 - z_2$. Making this substitution in equation (3) gives:

$$(P_1 - P_2) = \rho g[-a - h] \qquad (4)$$

❝ Clearly equations (2.7) and (4) are not the same. So the important thing that you should learn from this is that you should never just take an equation straight from a text book without being absolutely sure that it is applicable to the problem that you are trying to solve. The equations derived earlier should be thought of as illustrating the principles involved, not as equations that can be applied to all situations. In other words *think for yourself!* ❞ (REMEMBER!!!)

2.6 ▸ Adjusting the sensitivity of a manometer

From equation (1.8) it is apparent that $P = \rho g h$, so if a very large pressure, P, has to be measured then a heavy liquid with a large density, ρ, should be used to keep the head, h, to a sensible and convenient size. For instance, if P has a value of $2000\,\text{N/m}^2$ then h is equivalent to $0.204\,\text{m}$ of water, but only $0.015\,\text{m}$ of mercury. The ratio between the two is $0.204/0.015$ which is 13.6, the relative density of mercury. This logic applies to more complicated manometers as well. Look at the equations below, which were derived earlier in the chapter.

$$(P_1 - P_2) = g h_M (\rho_M - \rho) \qquad (2.6)$$

$$(P_1 - P_2) = \rho g h \qquad (2.8)$$

$$(P_1 - P_2) = g h (\rho - \rho_0) \qquad (2.10)$$

You should be able to see that if $(P_1 - P_2)$ is constant, then the value of the head (h or h_M) depends upon the difference in the densities of the two fluids. Or, put another way, it is possible to obtain a head which is a convenient size through the judicious choice of manometer fluids. This enables very small pressures to be measured accurately, as well as allowing large pressures to be assessed without the need for gigantic manometers. As an illustration, let us use the equations to examine the relationship between $(P_1 - P_2)$, head and density more closely. Suppose that $(P_1 - P_2)$ has a constant value of $2000\,\text{N/m}^2$.

For a differential U-tube manometer containing water and mercury (Fig. 2.5), equation (2.6) gives $h_M = 0.016\,\text{m}$ of mercury. This is too small for accuracy. An error of only $1\,\text{mm}$ when measuring h_M would be very significant.

For an inverted U-tube manometer with air above water (Fig. 2.8), equation (2.8) gives $h = 0.204\,\text{m}$ of water. This would give reasonable accuracy, and a gauge of practical size.

For an inverted U-tube manometer like that in Fig. 2.9 with a liquid of density $800\,\text{kg/m}^3$ (this is the density of paraffin oil) above water, $h = 1.019\,\text{m}$. This would allow very accurate measurements but would require quite a large manometer. If something like benzene with a density of 879 kg/m³ was used above the water then h increases to 1.685 m. It should be obvious from this result that the closer the density of the two gauge liquids, the larger the corresponding value of h. This is also apparent by inspecting the right-hand side of equation (2.10) which can be rewritten as:

$$(P_1 - P_2) = \rho g h (1 - s) \tag{2.11}$$

so $h = (P_1 - P_2)/[\rho g (1 - s)]$

where s is the relative density of the upper gauge liquid (see equation (1.7)). By definition, the relative density of water is 1.0, so mercury has a value of 13.6, paraffin 0.8, and so on. In the examples above $(P_1 - P_2)$, ρ and g are constant so h is inversely proportional to $(1 - s)$, that is:

$$h \propto 1/(1 - s) \tag{2.12}$$

The term $1/(1 - s)$ is sometimes called the **gauge coefficient**, and is a measure of how many times more sensitive a manometer is than a simple water gauge. For instance:

$s = 0.8$ gauge coefficient = $1/(1 - 0.8) = 5$ times
(check from above, 1.019 m/0.204 m = 5)

$s = 0.879$ gauge coefficient = $1/(1 - 0.879) = 8.26$ times
(check from above, 1.685/0.204 m = 8.26)

$s = 13.6$ gauge coefficient = $1/(1 - 13.6) = 0.08$ times
(check from above, 0.016 m/0.204 m = 0.08)

Example 2.4 provides an illustration of how the choice of different fluids can be used to improve the sensitivity of a manometer. Another way of increasing the sensitivity is to incline the limbs of the manometer at an angle to the vertical. Thus a given pressure appears as a longer length of fluid, while the longer distance is also easier to measure without incurring a significant error. Figure 2.11 shows a cross-section of a pipe with a simple vertical gauge on one side and an inclined gauge on the other. The pressure in the pipe is $P = \rho g z$, where z is the height of the liquid in the vertical gauge. If y is the length of the liquid in the inclined limb, then by geometry $z = y \cos \theta$. Since the pressure in the pipe is the same:

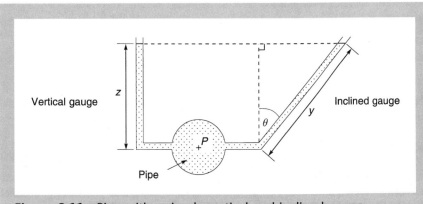

Figure 2.11 Pipe with a simple vertical and inclined gauge

$$P = \rho g y \cos \theta \qquad (2.13)$$

Thus the pressure can be determined by measuring the inclined distance, y, and then using equation (2.13). If greater sensitivity is required, this can be achieved by increasing the angle θ so that the tube is closer to the horizontal.

EXAMPLE 2.4

A differential pressure is measured using a manometer with the same geometry and readings as that in Example 2.3. The pipe liquid is again water, but now the there is oil ($\rho_O = 800\,kg/m^3$) in the upper part of the gauge (Fig. 2.12). What is the new value of ($P_1 - P_2$) and how has the sensitivity of the gauge been affected?

Step 1 Draw X–X through the upper surface of separation.

Step 2 Working upwards from the pipe to X–X:

$$P_1 = \rho g z_1 + \rho_O g h + P_X \qquad (1)$$

$$P_2 = \rho g z_2 + \rho g h + P_X \qquad (2)$$

Step 3 Subtract equation (2) from equation (1)

$$(P_1 - P_2) = \rho g z_1 + \rho_O g h + P_X - [\rho g z_2 + \rho g h + P_X]$$
$$(P_1 - P_2) = \rho g z_1 + \rho_O g h - \rho g z_2 - \rho g h$$
$$(P_1 - P_2) = \rho g [z_1 - z_2 - h] + \rho_O g h$$
$$(P_1 - P_2) = 1000 \times 9.81 [0.10 - 0.25 - 0.09] + 800 \times 9.81 \times 0.09$$
$$(P_1 - P_2) = -1.65 \times 10^3 \, N/m^2$$

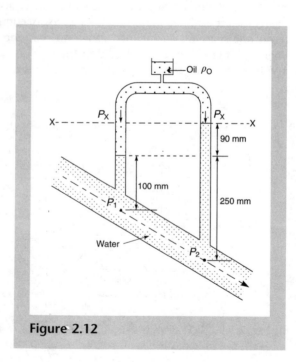

Figure 2.12

The use of oil above the water has produced a more sensitive gauge since a pressure difference of $1.65 \times 10^3 \text{N/m}^2$ now produces a differential head, h, of 90 mm whereas in Example 2.3 with a water/air gauge a pressure of $2.35 \times 10^3 \text{N/m}^2$ was required to produce the same head difference. Note that these problems involve sloping (not horizontal) pipelines, so the difference is **not** a factor of five, as the very simple gauge coefficient may incorrectly lead you to believe.

2.7 The Bourdon gauge

The Bourdon gauge provides a quick, direct means of reading pressure, although accuracy may be lost in return for ease of use. A calibration uncertainty of 0.1% of the reading is sometimes quoted, which is good enough for many applications. Once the gauge has been calibrated, accuracy can be maintained by frequently recalibrating the gauge against a manometer or some other device. Gauges can be obtained covering a wide range of pressure.

The most common form of Bourdon gauge (Fig. 2.13) measures **gauge pressure**, that is the pressure relative to the atmosphere. The stem of the gauge is tapped into the pipe (or whatever). In the diagram, an oval tube that forms about 270° of a circle can be seen branch-

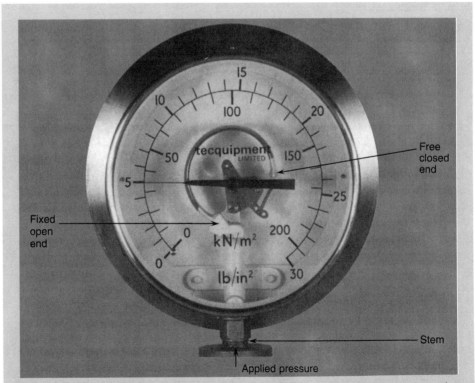

Figure 2.13 Bourdon gauge for measuring gauge pressure relative to the atmosphere

ing from the stem. This tube is fixed at the stem, and is hollow so that it contains the liquid whose pressure is to be measured. As the pressure of the liquid increases, the oval tube tends to become more circular and tries to straighten out. This results in a movement of the free, closed end of the tube. The movement can be measured by means of a mechanical linkage that causes the rotation of a pointer over a calibrated scale. The pressure inside the case of the gauge is atmospheric, so the gauge pressure is measured. It is also possible to obtain Bourdon gauges that measure absolute pressure. These incorporate a reference tube which is sealed at zero absolute pressure.

2.7.1 Calibration of a Bourdon gauge

Gauges may be recalibrated at intervals to ensure accuracy, or they may be specially calibrated prior to performing some crucial task. Calibration can be undertaken simply and easily using a dead weight tester of the type shown in Fig. 2.14. This consists of a vertical cylinder and closely fitting piston, beneath which there is a suitable liquid (oil or water). By adding weights to the piston a known, actual pressure is generated in the liquid, which is transmitted to the Bourdon gauge by a short tube. If this is the first time the gauge has been calibrated then the actual pressure can then be marked on the face of the gauge to produce a scale. Alternatively, if such a scale already exists, as in Figs 2.13 and 2.14, the actual and indicated pressures can be compared to assess the gauge's accuracy.

The actual pressure, P_{ACT}, on the gauge can be calculated from equation (1.1), which here becomes:

Figure 2.14 A dead weight tester used to calibrate a Bourdon gauge [*reproduced by permission of TecQuipment Ltd*]

$$P_{ACT} = W/A \qquad\qquad (2.14)$$

where W is the total weight added (including that of the piston itself) and A is the cross-sectional area of the piston. Usually readings are taken as the weights are both added and removed. At any time the difference between the actual and indicated pressure is the **gauge error**. This can be conveniently illustrated by drawing a graph of the indicated pressure against the actual pressure or, alternatively, the gauge error against the actual pressure.

Bourdon gauges frequently exhibit two types of error. The first is **hysteresis**: as a result of friction and backlash, the indicated pressure is smaller when the pressure is increasing and larger when it is decreasing. That is why readings are taken as weights are both added and removed. Typically this type of error is around $1\,kN/m^2$ over the entire range, which is usually acceptably small. The second type of error is **graduation error**, which is caused by inaccuracies when marking the scale on the face of the gauge. For the gauge shown, this increases to a maximum of about 2.5% of the full-scale reading. This is acceptably small for many engineering purposes, although gauges with a smaller error are commercially available.

2.8 ▶ Surface tension

Surface tension is the property of a liquid that enables insects to walk on the surface of water, which gives the surface a membrane-like quality, and which enables drops of water to form.

This phenomenon is due to **cohesion** between the molecules of a liquid where it has an interface with a gas or another immiscible liquid (see Vardy, 1990). Basically cohesion can be thought of as a force which makes the liquid contract inwards. Surface tension is often apparent when using piezometers and manometers. If a glass tube of small diameter is dipped into a body of water, the water level inside the tube will rise slightly above the water surface due to **capillarity** while the liquid in the tube will have a curved surface, a **meniscus** (Fig. 2.15a). The water surface is concave upwards, indicating that the **adhesion** between

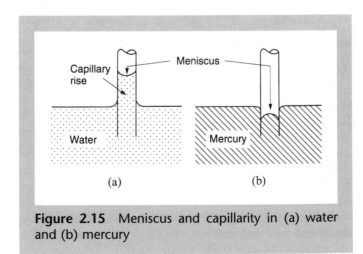

Figure 2.15 Meniscus and capillarity in (a) water and (b) mercury

the glass and water is stronger than the cohesive force, so the water is drawn up slightly at the sides. The capillary rise in the tube is due to the 'bridging' effect of surface tension. The opposite is the case with mercury, cohesion is stronger than adhesion so the liquid surface curves the opposite way to water (Fig. 2.15b). Capillarity effects disappear when the tube is larger than about 10 mm diameter, although there will still be a slight meniscus.

The force due to surface tension acts at right angles to a free surface and is proportional to its length. For water at 20°C the force is about 0.074 N/m. This is so small that it is usually ignored. It does not affect the behaviour of large-scale hydraulic systems, although it may be of significance in the laboratory. For instance, surface tension may affect the behaviour of small-scale hydraulic models (see Chapter 10).

Summary

1. For piezometers, basically $P = \rho g h$ where h is the head of water above the centreline of the pipe, or sometimes a differential head.

2. For manometers, start at the bottom of a column of fluid and calculate the total pressure (P_T) arising from each liquid in turn using equation (1.23): $P_T = \rho_1 g h_1 + \rho_2 g h_2 +$ etc. If the pipe is above the point at which P_T is being calculated, then add the pressure in the pipe P when appropriate.

3. With U shaped manometers located below the pipe, draw a line (X–X) through the lower surface of separation, use note 2 above to calculate the pressures in the two limbs at the level X–X, and then equate them (that is apply the equal level–equal pressure principle) to obtain P or $(P_1 - P_2)$ etc.

4. With ∩ shaped manometers above the pipe, draw a line (X–X) through the upper surface of separation. The pressure (say P_1) in the pipe beneath one limb is obtained by applying note 2 up to the level X–X. Remember to add any nominal air or liquid pressure above this (i.e. P_X). Repeat for the other limb to obtain P_2, then subtract to obtain $(P_1 - P_2)$ etc. as required.

5. The accuracy and sensitivity of a manometer can be adjusted by using liquids of different density: less dense liquids give a large value of h for a given P, dense liquids (e.g. mercury) a small value of h.

6. Because it is possible to have so many different manometer configurations it is always advisable to derive the equation for the pressure or differential pressure from first principles. Example 2.3 illustrated how easy it is to get the wrong answer by misapplying a 'standard' equation.

7. The advantages and disadvantages of the various pressure measuring devices are as follows. Piezometers are simple, cheap and easy to read but not suitable for very small pressures (not accurate enough) or very large pressures (too long). Manometers can employ different liquids to measure relatively small to moderate pressures accurately, but are more complex so calculation may be needed. Bourdon gauges can be read quickly and conveniently, and can be used to measure very large pressures; they are not extremely accurate and recalibration is needed.

Revision questions

2.1 Describe what is meant by (a) a piezometer; (b) piezometric head; (c) piezometric level; (d) a manometer; (e) the equal level, equal pressure principle; (f) the gauge coefficient; (g) surface tension; and (h) a meniscus.

2.2 A piezometer measures the pressure in a pipeline carrying water ($1000\,\text{kg/m}^3$). The piezometer reading is 253 mm measured from the centreline of the pipe. At this point, what is the gauge pressure and the absolute pressure in N/m^2? (Take atmospheric pressure as the equivalent of 10.3 m of water.)

$$[2.48 \times 10^3\,\text{N/m}^2;\ 103.52 \times 10^3\,\text{N/m}^2]$$

2.3 A vertical tube contains 200 mm of water on top of 270 mm of mercury (relative density 13.6). What is the pressure at the bottom of the tube?

$$[37.98 \times 10^3\,\text{N/m}^2]$$

2.4 A U-tube manometer like that in Fig. 2.2 is used to measure the pressure in a pipe which carries oil of relative density 0.88. The lower manometer liquid is mercury of relative density 13.6. The surface of separation between the oil and mercury is 0.93 m below the centreline of the pipe. (a) If the differential head, h_M, is zero, what is the gauge and absolute pressure at the centreline of the pipe? (b) If z in Fig. 2.2 remains at 0.93 m but h_M is now 37 mm of mercury, what is the gauge pressure in the pipe? Take atmospheric pressure as $P_{ATM} = 101\,040\,\text{N/m}^2$.

$$[(a)\ -8030\,\text{N/m}^2,\ 93\,010\,\text{N/m}^2;$$
$$(b)\ -3090\,\text{N/m}^2]$$

2.5 Calculate the value of $(P_1 - P_2)$ if the manometer in Fig. 2.6 (Example 2.2) has exactly the same readings but the pipeline now carries (a) oil of relative density 0.88; (b) fresh water.

$$[(a)\ 18.47 \times 10^3\,\text{N/m}^2;\ (b)\ 18.62 \times 10^3\,\text{N/m}^2]$$

2.6 A U-tube manometer has the readings shown in Fig. Q2.6. The pipe liquid is water, and the manometer liquid is mercury. Calculate $(P_1 - P_2)$.

$$[-72.71 \times 10^3\,\text{N/m}^2]$$

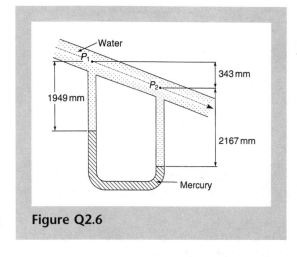

Figure Q2.6

2.7 (a) Figure Q2.7 shows an inverted U-tube manometer with oil of density $800\,\text{kg/m}^3$ above the pipe liquid, which is water. The pipeline is horizontal. What is the value of $(P_1 - P_2)$?

(b) If the manometer readings are the same but the oil is replaced by air, what is $(P_1 - P_2)$ now?

$$[(a)\ -294\,\text{N/m}^2;\ (b)\ -1472\,\text{N/m}^2]$$

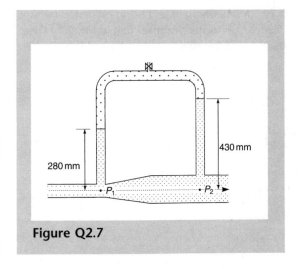

Figure Q2.7

2.8 An inverted U-tube manometer is connected to a pipeline which slopes upward, as shown in Fig. Q2.8. The pipeline carries water. Calculate $(P_1 - P_2)$ when (a) the upper part of the manometer is filled

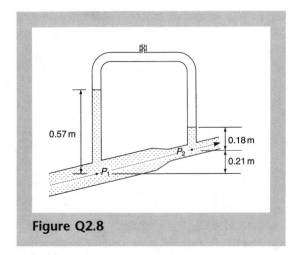

Figure Q2.8

with air, and (b) when oil of relative density 0.80 is introduced above the water.

[(a) $3.83 \times 10^3\,\text{N/m}^2$; (b) $2.41 \times 10^3\,\text{N/m}^2$]

2.9 An inclined piezometer measures the pressure in a pipeline carrying water. The piezometer is inclined at an angle of 30° to the horizontal and has an inclined reading of 330 mm. What is the gauge pressure in the pipe?

[1619 N/m²]

2.10 A Bourdon gauge is being calibrated using a dead weight tester similar to that in Fig. 2.14. The diameter of the piston is 20.5 mm and it has a

mass of 1 kg (which is included in M below). Additional masses are added to the piston and the indicated pressure (P) recorded. Note that P is in units of $10^3\,\text{N/m}^2$ (kN/m²). The masses are then removed and P recorded again. The readings are shown below:

M (kg)	P ($10^3\,\text{N/m}^2$)
1 (piston)	34
2	63
3	92
4	119
5	151
6	180
5	154
4	123
3	93
2	63
1	34

(a) Using equation (2.14) calculate the actual pressure (P_{ACT}) corresponding to M. (b) For each value of M obtain the gauge error ($P - P_{ACT}$). Draw a graph of M against ($P - P_{ACT}$). (c) What is the largest percentage gauge error, i.e. $100 \times (P - P_{ACT})/P_{ACT}$, and the average percentage gauge error? (d) Is there any evidence of hysteresis?

[(c) 14.4% and 5.3%; (d) yes, 154 > 151, 123 > 119, 93 > 92 $\times 10^3\,\text{N/m}^2$]

3

Stability of a floating body

There are many situations where civil engineers have to work from barges and pontoons floating on water, rather than from dry land. A typical example could be building a bridge across a wide river or estuary, where the girders forming the structure have to be floated out on barges and then lifted into position using floating cranes. Another example could be the construction of a marina or jetty in the sea. In these situations all of the construction activity may have to take place from floating barges and pontoons. Therefore it is essential that an engineer has an understanding of whether or not a pontoon will float or sink, and whether or not it is stable. The alternative to it being stable is being unstable, so that if a piece of construction plant moves across the deck the whole pontoon capsizes, tipping everything and everyone into the water. This would be extremely dangerous and very expensive and, of course, must be avoided at all costs. Hence the need to know something about how and why bodies float, and what makes them stable or unstable. The latter involves the position of the metacentre with respect to the centre of gravity of the floating body. Thus the sort of questions that are raised in this chapter are:

What makes a body float?

How can we determine whether or not a body is stable?

What controls the stability?

What is the metacentre and the metacentric height?

How can we increase the stability of a floating body?

3.1 Introduction

To start with, let us define what we mean by a barge and a pontoon. A barge is a vessel used for transporting freight, usually flat-bottomed and with or without its own power. A pontoon is a watertight float or vessel used where buoyancy is required in water. Henceforth we will not distinguish between barges and pontoons, the term pontoon will be used

Figure 3.1 Pontoons on the New Bedford River used for transporting plant during maintenance work. The pontoons can be fastened together to give a larger working platform, if required

to denote a flat-bottomed vessel which is rectangular in cross-section and in plan. Pontoons like this are often used in practice (Fig. 3.1), while their simple shape makes them ideal for the sort of calculations we want to conduct.

Now let us ask why something floats?

The answer is because of hydrostatic pressure, which is $P = \rho gh$ where h is the depth in the liquid. The force acting vertically upward on the body as a result of hydrostatic pressure is $F = P_{AV}A$ where P_{AV} is the average pressure and A is the horizontal area over which it acts (see Fig. 1.33). Combining these two equations gives $F = \rho ghA$ or $F = \rho gV$ where V (= hA) is the submerged volume of the body, or put another way, the volume of water it displaces. Since $F = \rho gV$, the buoyancy force is proportional to the mass (ρV) or the weight (ρgV) of water displaced by the body. For a floating body the buoyancy force, F, acting vertically upwards equals the weight of the body, W, acting vertically downwards, so it is often said that a floating body displaces its own mass or weight of water. That is why we refer to the **displacement** of a ship as being (say) 10000 tonnes. The larger the displacement, the bigger the ship.

A body can float only if its average density is less than that of the liquid in which it floats. A ship can float because the steel from which it is made is spread out to form a large hollow body that displaces a large volume of water, V. If the steel was melted down into a solid block with a density about 7.8 times that of water, the block would sink because its displacement would be far too small to generate a large enough buoyancy force to balance its weight. In other words, because V is now very small and $F = \rho gV$, F becomes small while

| Box 3.1 | Buoyancy |

1. A floating body displaces its own weight, W, of water (or alternatively its own mass, M, of water).

2. The buoyancy force, F, equals the weight of water displaced, thus $F = \rho g V$ where V is the volume of water displaced by the body. So for floating bodies, $F = W$.

3. The buoyancy force, F, acts vertically upwards through the centre of gravity of the displaced liquid, which is called the **centre of buoyancy** (B).

4. The weight of the body, W, acts vertically downwards through the **centre of gravity** of the body (G).

5. If $F = W$ then the body floats with a constant depth of immersion in the liquid.

6. If $F > W$ then the body rises, like a cork pushed under water.

7. If $F < W$ then the body sinks, like a stone.

8. For a rectangular pontoon, the depth of immersion (that is how deep it floats in the water) is $h = V/A$ where A is the plan area of the body.

W stays the same. This was discussed and explained in section 1.7. A summary of the important points you need to remember in this chapter is given in Box 3.1 and illustrated in Example 3.1.

EXAMPLE 3.1

An unloaded pontoon being used in a river estuary to transport construction equipment has a mass of 15 tonnes (1 tonne = 1000 kg). In plan the pontoon is 8 m long by 5 m wide. It is rectangular in section and has sides 1.5 m high. The water is saline with a density of 1025 kg/m³. (a) What is the depth of immersion of the unloaded pontoon? (b) What weight can be carried by the pontoon while still maintaining a freeboard of 0.5 m?

(a) A floating body displaces its own mass of water, in this case 15 tonnes or 15 000 kg. Remembering that mass density = mass/volume, then:

Volume of water displaced by the pontoon, V = mass/density of saline water

$$= 15\,000/1025$$

$$V = 14.63\,\text{m}^3$$

Depth of immersion, h = volume (V) of water displaced/plan area of pontoon (A)

$$= 14.63/(8 \times 5)$$

$$h = 0.37\,\text{m}$$

(b) If the freeboard is 0.5 m then the depth of immersion $h = (1.5 - 0.5) = 1.0\,\text{m}$.

Therefore volume of water now displaced, $V = h \times A = 1.0 \times (8 \times 5) = 40\,\text{m}^3$.

Thus the maximum amount of water that can be displaced by the loaded pontoon is 40 m³.

The mass of the displaced saline water = 40 × 1025 = 41 000 kg or 41 tonnes.

The pontoon's mass is 15 tonnes, so it can carry a mass of (41 − 15) = 26 tonnes.

Thus the weight that the pontoon can carry = 26 000 × 9.81 = 255.06 × 10³ N

3.2 Factors affecting the stability of a floating body

❝ What exactly do you mean by the 'stability' of a floating body? The pontoons in Fig. 3.1 do not look capable of overturning, so how could they become unstable? ❞

To answer this we will start by looking at the forces acting on a stable pontoon and then consider what would cause it to become unstable. The forces involved are basically those described in Box 3.1, so make sure you understand what these forces represent and where they act.

Now look at the forces acting on the simple pontoon in Fig. 3.2. We have the weight force, W, acting vertically down through the centre of gravity, G, of the pontoon. Since the pontoon is floating in water with a constant depth of immersion, it follows that there must be an equal and opposite force opposing the weight force. This is the buoyancy force, F, which acts vertically up through the centre of gravity of the displaced liquid. Since the pontoon is a simple rectangle, the shape of the displaced liquid is also a rectangle with its centre at the geometrical centre, which is called the centre of buoyancy, B. The buoyancy force, F, acts upwards through B. Note that W and F act colinearly (that is along the same line, a vertical one in this case) with G being situated some distance above B.

In Fig. 3.3 the pontoon is shown tilted. The tilt could be caused by a sudden gust of wind or a wave. As before, W acts vertically down through G (which has not moved) but F now acts through a point B*, not B. This is because F acts through the centre of gravity of the displaced liquid, which is now trapezoidal in shape with its centre of gravity at B*. As a result F and W are no longer colinear, but form a couple of forces that will return the pontoon to an even keel as in Fig. 3.2. This is called a **righting couple**. Because the pontoon is capable of righting itself when tilted it can be classified as **stable**.

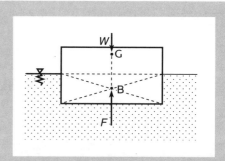

Figure 3.2 A pontoon floating on an even keel (no tilt) with W and F colinear

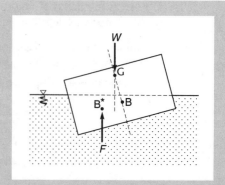

Figure 3.3 A pontoon floating with an imposed angle of tilt, showing the righting couple

There are three points worth noting here:

1. A couple is defined as two equal parallel forces acting in opposite directions but not in the same line. A couple produces a turning effect equal to one of the forces (either *W* or *F* in this case) multiplied by the perpendicular distance between the two lines of action of the forces. Thus a righting couple means a pair of forces that act in such a way as to return (or right) the pontoon to its original position with no tilt. Under these conditions the pontoon is stable.

2. The buoyancy force, *F*, does not change in magnitude when the pontoon is tilted because it depends upon the weight of water displaced (which equals the weight of the pontoon itself) and this has not changed.

3. Although the weight and total volume of water displaced do not change when the pontoon tilts, the volume displaced to either side of the centreline does. If you compare Figs 3.2 and 3.3, you will see that in the tilted position the volume of water displaced to the left of the centreline increases while the volume displaced to the right decreases. Since the buoyancy force is proportional to the volume of water displaced, it follows that the buoyancy force effectively increases on the left-hand side of the pontoon and decreases on the right-hand side. That is why the overall buoyancy force, *F*, acts through B* to the left of the centreline in Fig. 3.3, and why the pontoon rights itself. (REMEMBER!!!)

Now suppose that the pontoon has a large, relatively tall piece of equipment placed on it, drawn for simplicity as a rectangle in Fig. 3.4. The combined weight, *W*, of the pontoon and its load acts through the centre of gravity, G, which is relatively high. ***Note that as G becomes higher and the angle of tilt increases, W acts further and further to the left***. This means that at some point the movement of the buoyancy force, *F*, from B to B* is unlikely to be large enough to produce a righting couple. What we now have is the situation depicted in Fig. 3.4 where the line of action of *W* is outside (nearer the edge of the pontoon than) the line along which *F* acts. Thus *W* is trying to overturn the pontoon. The two forces *W* and *F* form an **overturning couple**. So once tilted, or tilted too far, the pontoon will overturn and capsize. Thus it is **unstable**.

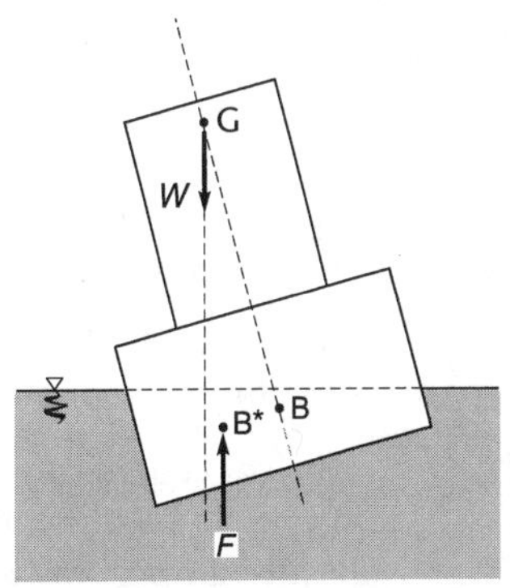

Figure 3.4 A pontoon with a raised G and an imposed angle of tilt, showing the overturning couple caused by *W* acting outside *F*

Now let us consider what is meant by the **metacentre**. A pontoon floating on an even keel has its centre of buoyancy at B and its centre of gravity at G. A line joining B to G would be as shown in Fig. 3.2, that is vertical and at 90° to the deck of the pontoon. Imagine the line BG extends upwards. Now consider the pontoon in its tilted position, as in Fig. 3.5. The centre of buoyancy has moved from B to B*. A line drawn vertically upwards through B* will intersect the line BG at the point labelled M in the diagram. This is called the **metacentre**. Provided G does not move, then for all relatively small angles of tilt: (REMEMBER!!!)

Box 3.2 ▶ **Remember**

From the above it is apparent that the height of the centre of gravity, G, of the pontoon is an important factor in determining whether it is stable and able to right itself when tilted, or unstable so that it capsizes when tilted. The higher G, the further to the left *W* will act when the pontoon is tilted, increasing the possibility of over-turning. Consequently:

1. If a pontoon or ship carries heavy ballast in its bottom holds then the centre of gravity will be relatively low, so it will be more stable and less likely to overturn.

2. If a pontoon or ship carries a heavy cargo on deck, then the centre of gravity will be relatively high so the risk of overturning is increased.

3. If a pontoon carries a crane which lifts something heavy off the deck and raises it to some significant height, this will increase the height of the centre of gravity of the pontoon and its cargo as a result of a redistribution of the weight, so increasing the risk of overturning. This effect will be magnified if the crane also moves its load towards the side, or over the side, of the pontoon since G will also move sideways so increasing the overturning couple.

1. The vertical line through B* will always pass through M. Consequently if the location of B* can be calculated, the position of M can be found graphically.

2. The distance of M above B is constant.

3. The distance GM is called the **metacentric height** of the pontoon.

The concept of the metacentre is an important one, and a difficult one. If you imagine a ship tilting from side to side, the mast swings through a sector of a circle, like the wind-screen wipers on a car. However, motion is relative. If the point M on the mast is considered to be stationary, then the ship would appear to swing like a pendulum beneath it. Thus the metacentre, M, is the point about which a ship or pontoon appears to rotate.

The metacentre and the metacentric height have an important application, which is in determining whether a vessel is stable or unstable. This is summarised in Box 3.3.

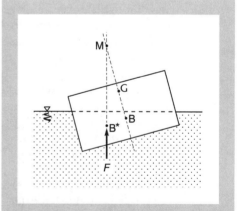

Figure 3.5 The extension of the line BG intersects the vertical line through B* at the metacentre M. The distance GM is called the metacentric height

Box 3.3 **▶Remember**

The way in which a pontoon or ship can be classified as stable or unstable is as follows:

1. If M is above G (so GM is +ve) then the pontoon is stable. If tilted, the pontoon will right itself and return to an even keel.

2. If M and G coincide so that GM = 0 then neutral stability exists so that if the pontoon is tilted it will continue to float with exactly that tilt; it will neither right itself nor overturn. Do not confuse this condition with a pontoon that is stable but has its weight unevenly distributed about its longitudinal centreline so that it floats with a slight angle of tilt but is capable of righting itself if tilted further.

3. If M is below G (so GM is −ve) then the pontoon is unstable. If tilted, the pontoon will overturn.

◀3.3▶ Calculation of the metacentric height, GM

❝ It ought to be apparent from Box 3.3 that to determine whether or not a floating body is stable or unstable we have to calculate the value of the metacentric height, that is the distance from G to M, or GM for short. The question is how? Another pertinent question is what controls the value of GM, and hence makes the pontoon stable or unstable? ❞

Well, fortunately, there are two ways of calculating GM. One is by carrying out a test on the vessel, which involves moving a known weight transversely across the deck and measuring the angle of tilt, from which data the value of GM can be calculated. However, the obvious disadvantage of this approach is that you have to build the ship and then find out if it is stable! This is not realistic, so in practice this test is used to determine (for example) whether or not a vessel has been overloaded, threatening its stability. The second way is to use a more theoretical approach to calculate the height of the metacentre M above the centre of buoyancy B, that is the distance BM. If we know BM we can obtain GM by further calculation. All of these calculations can be undertaken before building the vessel, so that is where we will start.

3.3.1 Theoretical determination of the height of the metacentre, M

When considering the stability of a floating body it is usual to assume that the angle of tilt, θ, is small. This is necessary to simplify the theory by making the assumption that θ radians $= \sin \theta = \tan \theta$. The validity of this assumption is demonstrated below.

If $\theta = 1°$ then $\theta = 0.01745$ radians while $\sin \theta = 0.01745$ and $\tan \theta = 0.01746$

If $\theta = 5°$ then $\theta = 0.08727$ radians while $\sin \theta = 0.08716$ and $\tan \theta = 0.08749$

If $\theta = 10°$ then $\theta = 0.17453$ radians while $\sin\theta = 0.17365$ and $\tan\theta = 0.17633$

If $\theta = 20°$ then $\theta = 0.34907$ radians while $\sin\theta = 0.34202$ and $\tan\theta = 0.36397$

If $\theta = 40°$ then $\theta = 0.69813$ radians while $\sin\theta = 0.64279$ and $\tan\theta = 0.83910$

If it is assumed that $\tan\theta = \theta$ radians then this results in errors of 0.1%, 0.3%, 1.0%, 4.3% and 20.2% at angles of 1, 5, 10, 20 and 40 degrees respectively. So when using the equation below, it should be kept in mind that the theory becomes less accurate at large angles of tilt, although in most situations it works well enough.

The derivation of the theoretical equation for the distance BM is presented as Proof 3.1 in Appendix 1. Study this, and you will see that it is obtained by considering the restoring moment (force × distance) that rights or restores a rectangular pontoon to an even keel when it is tilted. The equation is a simple one, namely that:

$$\mathbf{BM} = I_{ws}/V \tag{3.1}$$

where V is the volume of water displaced by the body, and I_{ws} is the second moment of area calculated about an axis through the centroid of the area of the body in the plane of the water surface. This axis must be at 90° to the direction in which the displacement or tilt occurs. In other words, for the rectangular pontoon in Fig. 3.6 the line X–X is the longitudinal axis about which I_{ws} is calculated. The tilt takes place at 90° to this so, for instance, when seen in plan as in Fig. 3.6, the left edge of the pontoon running parallel to X–X has a constant depth of immersion. This being so, then:

$$I_{ws} = lb^3/12 \tag{3.2}$$

where l is the length of the pontoon and b its breadth. Note that the dimension in which the tilt or displacement takes place, b, is raised to the cubic power, so altering b significantly affects the analysis. A common mistake with students is to attach the power to the wrong variable, so take care.

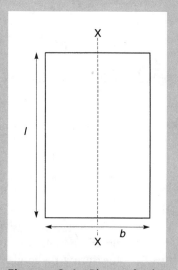

Figure 3.6 Plan of the pontoon showing its dimensions. The tilt takes place about the longitudinal axis X–X

It should now be apparent that BM depends only upon:

1. l and b, the dimensions of the pontoon which govern the value of I_{ws}, and
2. V, the volume of displaced water which depends only upon the weight of the pontoon (see Box 3.1).

It is also apparent that if the dimensions and weight of the pontoon do not change, then neither will BM. So for a particular vessel, BM is constant and has a fixed value. However, if the vessel is loaded or unloaded so that its weight changes, then this will alter BM (see Box 3.2).

We can now calculate BM, but how do we obtain GM?

The best way to approach this question is to draw a diagram and mark on it clearly the key reference points, as in Fig. 3.7. You should then be able to see that BM = BG + GM or:

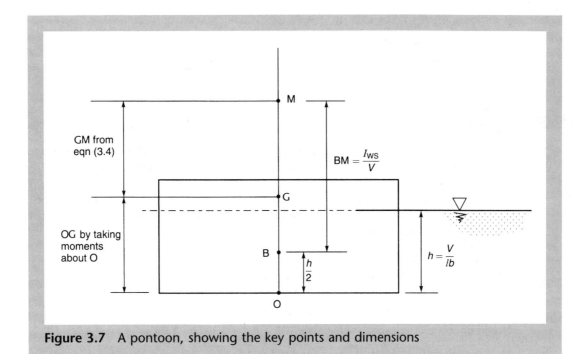

Figure 3.7 A pontoon, showing the key points and dimensions

$$\textbf{GM} = \textbf{BM} - \textbf{BG} \qquad\qquad (3.3)$$

So, if we can calculate BG, we can obtain GM and hence determine if the body is stable or unstable. Now B is the centre of buoyancy, and with the pontoon floating on an even keel B is located at a height equal to half the depth of immersion (that is $h/2$) above the point O on the bottom of the pontoon (Fig. 3.7). As shown in Box 3.1, $h = V/lb$. Thus we can calculate the distance OB, which is the height of B above the base.

Calculating the position of G is a little more difficult. If the weight of the various components that make up the vessel are known, then by taking moments about O the position of G can be obtained. Thus we obtain the distance OG. Since BG = OG − OB, substituting BG into equation (3.3) gives us GM. Do not forget that OG has to be found by taking moments, as in Example 3.3. `(REMEMBER!!!)`

Study Examples 3.2 and 3.3 and then try Self Test Question 3.1.

EXAMPLE 3.2

A loaded pontoon has a mass of 200 000 kg. Its length is 20 m and its width 9 m, and its overall centre of gravity is 5.6 m above the base of the pontoon. The pontoon floats in salt water with a density (ρ_s) of 1025 kg/m³. Is it stable?

Since it is floating, mass of pontoon, M = mass of sea water displaced = 200 000 kg.

Volume of sea water displaced, $V = M/\rho_s = 200\,000/1025 = 195.12\,\text{m}^3$

Depth of immersion, $h = V/lb = 195.12/(20 \times 9) = 1.08\,\text{m}$

Therefore height of centre of buoyancy B above base = $h/2 = 1.08/2 = 0.54\,\text{m}$

$BM = I_{ws}/V$ where $I_{ws} = lb^3/12$

BM = $(20 \times 9^3)/(12 \times 195.12) = 6.23$ m

Height of M above the base of the pontoon = $6.23 + 0.54 = 6.77$ m

M is above G (6.77 m > 5.6 m) so the pontoon is stable, and GM = $6.77 - 5.6 = 1.17$ m

EXAMPLE 3.3

A pontoon 15 m long, 7 m wide and 3 m high weighs 700×10^3 N unloaded and carries a load of 1600×10^3 N. The load is placed symmetrically on the pontoon so that its centre of gravity is on the longitudinal centre line at a height of 0.5 m above the deck (3.5 m above the base). The centre of gravity of the pontoon can be assumed to be on the longitudinal centreline at a height of 1.5 m above the base. The pontoon floats in saline water of density 1025 kg/m³. Calculate the metacentric height of the pontoon.

Total weight of pontoon and load = $(700 + 1600) \times 10^3 = 2300 \times 10^3$ N

Volume of sea water displaced, V = total weight/weight density
$$= 2300 \times 10^3/(1025 \times 9.81) = 228.74 \text{ m}^3$$

Depth of immersion, $h = V/lb = 228.74/(15 \times 7) = 2.18$ m

Height of centre of buoyancy above base, OB = $h/2 = 2.18/2 = 1.09$ m

Now determine the height of the centre of gravity above the base, OG, by taking moments about O:

$$2300 \times 10^3 \times OG = 1600 \times 10^3 \times 3.5 + 700 \times 10^3 \times 1.5$$
$$OG = (5600 \times 10^3 + 1050 \times 10^3)/2300 \times 10^3$$
$$OG = 2.89 \text{ m}$$
$$BG = OG - OB = 2.89 - 1.09 = 1.80 \text{ m}$$

Height of metacentre, M, above B is BM = I_{ws}/V where $I_{ws} = lb^3/12$ so:

$$BM = (15 \times 7^3)/(12 \times 228.74)$$
$$BM = 1.87 \text{ m}$$

Metacentric height GM = BM − BG = $1.87 - 1.80 = 0.07$ m

GM is +ve indicating stability but the value of 0.07 m is very small indicating that this is close to the condition of neutral stability, which would be unacceptable (see Self Test Question 3.1)

SELF TEST QUESTION 3.1

Repeat the calculations in Example 3.3, but this time:

(a) reduce the load carried by the pontoon to 1500×10^3 N, 1400×10^3 N and 1100×10^3 N and determine what effect this has on the value of GM;

(b) keep the load at 1600×10^3 N as in Example 3.3, but investigate the effect on GM of increasing the width of the pontoon to 7.5 m, 8.0 m and 9.0 m while assuming that its weight is unaffected at 700×10^3 N.

3.3.2 Experimental determination of the metacentric height, GM

It is common practice to carry out an experiment on a vessel to assess its stability (perhaps after it has been loaded) by calculating GM. This is a simple procedure utilising only a known movable weight positioned on the deck at approximately the middle of the longitudinal centreline and a pendulum hanging inside the vessel. The weight, often called a jockey weight (w_J), is moved from the centreline a known distance (δx) towards the side (Fig. 3.8). This moves the centre of gravity of the pontoon (parallel to the deck since the vertical weight distribution is unchanged) from G on the centreline to a new position G* and causes the vessel to tilt, the angle of tilt ($\delta\theta$) being measured by the pendulum.

The magnitude of GG* depends upon how far the jockey weight is moved and its size relative to the total weight of the pontoon. We can quantify this using the ratio of the weights and δx, thus:

$$GG^* = (w_J/W)\,\delta x \qquad (1)$$

where W is the total weight of the pontoon including the jockey weight.

If the movement of the jockey weight, δx, produces an angle of tilt, $\delta\theta$, then by geometry:

$$GG^* = GM \tan \delta\theta \qquad (2)$$

Since it is assumed that $\delta\theta$ is small we can use the approximation (described earlier at the start of section 3.3.1) that $\tan \delta\theta = \delta\theta$ radians. Therefore equating the right-hand sides of (1) and (2) above gives:

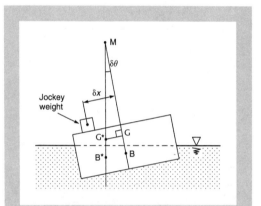

Figure 3.8 Movement of the jockey weight from the centreline moves G to G* and induces an angle of tilt, $\delta\theta$

$$GM\delta\theta = (w_J/W)\delta x$$

Replacing the small amounts $\delta\theta$ and δx by the limits $d\theta$ and dx and rearranging gives:

$$GM = \frac{w_J}{W}\frac{dx}{d\theta} \qquad (3.4)$$

It is important to remember that θ is in radians when using this equation, otherwise the calculations will be wrong. In practice, the jockey weight is often (REMEMBER!!!) moved in steps across the deck and the corresponding tilt measured so that a graph can be drawn of dx against $d\theta$. The slope of the graph gives the numerical value of $dx/d\theta$ which can be substituted in equation (3.4).

This simple test and equation can be used to assess the stability of a vessel, as shown in Example 3.4. Stability can also be studied in the laboratory using a model pontoon like that in Fig. 3.9. Here the centre of gravity of the pontoon can be increased by raising the height of the jockey weight. Some typical results are shown in Table 3.1. Column 1 shows the height of the jockey weight, and column 2 the corresponding measured height of the pontoon's centre of gravity from the base, OG (see Fig. 3.7). For each height, the jockey weight is moved at 15 mm intervals (dx) parallel to the deck and the tilt ($d\theta$) indicated by

Table 3.1 Results and calculated values for the model pontoon (after Markland (1994); reproduced by permission of TecQuipment Ltd)

Depth of immersion $h = 38.8$ mm so OB = 19.4 mm

BM = OG + GM – 19.4 mm

Height of jockey weight (mm)	Measured value of OG (mm)	GM (eqn (3.4)) (mm)	BM (mm)	Angle of tilt with $dx = 15$ mm (degrees)
105	58.7	45.7	85.0	2.7
165	67.1	38.3	86.0	3.2
225	75.4	30.8	86.8	3.9
285	83.7	22.9	87.2	5.2
345	92.0	16.0	88.6	7.5

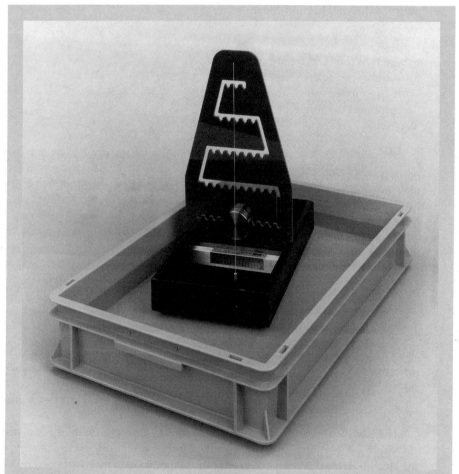

Figure 3.9 Investigation of stability using a model pontoon with a jockey weight that can be moved horizontally and vertically [*reproduced by permission of TecQuipment Ltd*]

the plumbline is recorded. The results can be plotted graphically and the gradient $dx/d\theta$ calculated (with θ in radians), then substituted into equation (3.4) to obtain the value of GM in column 3. The depth of immersion ($h = 38.8$ mm) is easily calculated from the weight and dimensions of the pontoon, thus OB = 19.4 mm and the values in column 4 are given by BM = OG + GM − OB. Note that the values of BM in the table are those obtained by experiment.

Some points to note from the table are as follows:

1. From equation (3.1), theoretically BM = I_{WS}/V and in this case has a value of 87.4 mm. This is constant since the weight and dimensions of the pontoon are constant. The average experimental value is 86.7 mm, the variation being due to experimental error.

2. Since M is a fixed point, as OG increases the value of GM decreases to compensate.

3. Reducing values of GM indicate reducing stabilty. The last column of the table shows that for a 15 mm horizontal displacement of the jockey weight the angle of tilt experienced increases as OG increases and GM decreases. With a high centre of gravity, a small movement of weight across the deck can result in an alarmingly large angle of tilt.

EXAMPLE 3.4

A vessel with a weight of 50×10^6 N tilts at an angle of 5° when a jockey weight of 300×10^3 N is moved 8 m across the deck from the centreline. What is the metacentric height of the vessel?

2π radians = 360° so 1 radian = 57.3°.

A tilt of 5° equals 5/57.3 = 0.087 radians

GM = $(w_J/W) \times dx/d\theta = (300 \times 10^3/50 \times 10^6) \times 8/0.087$

GM = 0.55 m

This indicates a reasonable stability: GM is usually not large (see section 3.4).

SELF TEST QUESTION 3.2

For the vessel in Example 3.4, suppose the same movement of the jockey weight had produced angles of tilt of 10° and 40°. What would you conclude about the stability of the vessel under these circumstances?

3.4 Period of roll

From the analysis and discussion above it would be sensible for you to conclude that the metacentric height, GM, should always be made as large as possible to ensure the stability of the vessel. Unfortunately, this is often neither possible nor desirable. A complicating factor is that the larger the value of GM, the more rapidly the vessel will roll from side to side when tilted by waves, or whatever. With something like a ferry or cruise ship, passenger comfort is an important consideration in addition to safety, and a vessel that rocks from side to side very rapidly would not be appreciated.

The relationship between GM and the periodic time of roll (that is the frequency at which the vessel rolls from side to side) is given by:

$$t = 2\pi \sqrt{\frac{I_M}{W \times GM}}$$

(3.5)

where t is the roll period (s), W is the weight of the vessel (N) and GM is the metacentric height (m). I_M is the moment of inertia ($kg\,m^2$) of the vessel calculated about an axis through the centre of mass. Note that I_M is the moment of inertia of the body with the units $kg\,m^2$; it reflects the way in which the mass of the body is distributed. On the other hand, I_{WS}, which was used earlier, is the second moment of area of the body in the plane of the water surface with the units m^4. Do not confuse I_{WS} and I_M.

The effect of GM on the period of roll can be easily demonstrated using equation (3.5), and Example 3.5 provides an illustration.

EXAMPLE 3.5

A vessel has a weight of $40 \times 10^6\,N$ and a moment of inertia of $55 \times 10^6\,kg\,m^2$. It is possible to construct the vessel in different ways so that its metacentric height is 2.0 m, 1.0 m, 0.5 m or 0.25 m. Assuming that the weight and moment of inertia remain unchanged, what is the roll period for each of the metacentric heights?

Using equation (3.5):

If GM = 2.0 m then $t = 2\pi \,(55 \times 10^6/40 \times 10^6 \times 2.0)^{1/2} = 5.2\,s$

If GM = 1.0 m then $t = 2\pi \,(55 \times 10^6/40 \times 10^6 \times 1.0)^{1/2} = 7.4\,s$

If GM = 0.5 m then $t = 2\pi \,(55 \times 10^6/40 \times 10^6 \times 0.5)^{1/2} = 10.4\,s$

If GM = 0.25 m then $t = 2\pi \,(55 \times 10^6/40 \times 10^6 \times 0.25)^{1/2} = 14.7\,s$

Although it may not be practical to adopt a very large metacentric height to ensure stability, there are other measures that can be employed. One is to use ballast in the bottom of the vessel to lower its centre of gravity (with pontoons, water is sometimes used). Another measure is to divide the vessel longitudinally into separate compartments. This is especially important with liquid cargoes (or water ballast) that have a free surface. Without the longitudinal dividing walls, if the vessel tilts then the liquid is free to flow from one side of the centreline to the other. This redistribution of mass increases the overturning moment by moving G* nearer to the side of the vessel. This can be a problem with vehicle ferries that have large, open lower holds, should water get inside the vessel.

The stability of a vessel can also be improved by adding triangular air tanks to its sides or by using a trapezoidally shaped hull (Fig. 3.10), for the following reason. In the plane of the water surface, as the vessel tilts to the left the trapezoidal hull increases the width of the vessel on the side which is down in the water, and decreases the width on the side that is rising. In other words $y_1 > y_2$ in Fig. 3.10. The larger the angle of tilt, the greater the respective increase and decrease in width. Since the buoyancy force is proportional to the volume

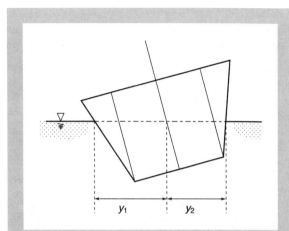

Figure 3.10 A vessel with sides that slope outwards has improved stability by virtue of the fact that its width and displacement increase on the side that is down in the water, so generating an increased buoyancy force that aids recovery. The width and displacement to the right of the centreline both decrease, which again helps the vessel to right itself

of water displaced, this means that the buoyancy force increases to the left of the longitudinal centreline and decreases to the right, helping the recovery of the vessel to an even keel. A similar thing happens with a vertically sided vessel, but the sloping sides make the effect more pronounced. Some ferry boats use this principle to achieve greater stability.

Summary

1. For floating bodies $W = F = \rho g V$ (or $M = \rho V$). Thus the starting point for many buoyancy problems is that a floating body displaces its own weight (or mass) of water. The buoyancy force acts at the centre of gravity of the displaced liquid (B), which is called the centre of buoyancy.

2. For a rectangular pontoon, the depth of immersion $h = V$ / plan area.

3. A tilted pontoon or ship is stable if the buoyancy force F acts nearer the side of the pontoon than W, which acts through the pontoon's own centre of gravity, G. Under these conditions a righting couple is formed, so if the edge of the pontoon is depressed it will rise up again (Fig. 3.3).

4. A tilted pontoon or ship is unstable if its weight W acts nearer the side of the pontoon than the buoyancy force F, forming an overturning couple. Under these conditions, if the side of the pontoon is depressed the pontoon will rotate and capsize (Fig. 3.4).

5. The metacentre, M, of the pontoon is located where the extension of the line BG and the

extension of the line of the buoyancy force for the tilted pontoon meet, as shown in Fig. 3.5.

6. For a pontoon or ship of fixed weight and dimensions, the distance between B and M is constant because it is given by:

$$BM = \frac{I_{ws}}{V} \qquad (3.1)$$

where I_{ws} is the second moment of area in the plane of the water surface (= $lb^3/12$ for a rectangular pontoon of length l and width b) and V is the volume of water displaced (which is constant if W and ρ are constant).

7. The distance between G and M is the metacentric height denoted by GM. If G is below M the pontoon is stable; if G is above M the pontoon is unstable; if G coincides with M neutral stability exists, and the pontoon floats with a permanent tilt if depressed. These three conditions equate to +ve, −ve and zero values of GM respectively.

8. GM can be calculated by moving a jockey weight (w_J) a distance dx across the deck of a pontoon or ship of total weight W (including w_J) and then measuring the angle of tilt (dθ radians) induced:

$$GM = \frac{w_J}{W} \times \frac{dx}{d\theta} \qquad (3.4)$$

Note that the height of G, and thus the value of GM and stability, can change depending upon whether heavy cargo is placed deep in the holds and light cargo on deck, or vice versa.

9. Although a large value of GM is desirable to safeguard against capsize, it can result in uncomfortably rapid rolling from side to side (see equation (3.5)), so compromise is sometimes necessary.

Revision questions

3.1 In terms of displacement and buoyancy force, explain why a block of polystyrene floats but a brick does not.

3.2 When a pontoon is loaded with a heavy cargo it settles deeper in the water. Obviously the buoyancy force must increase to offset the additional weight, because the pontoon continues to float. Explain why the buoyancy force increases by considering the hydrostatic pressure acting on the bottom of the pontoon.

3.3 An unloaded pontoon weighs 200 000 N and in plan is 12 m long by 7 m wide. It floats in sea water with a density of 1025 kg/m³. (a) What is the depth of immersion of the pontoon? (b) What is the distance between the centre of buoyancy (B) and the metacentre (M)?

[0.237 m; 17.245 m]

3.4 Explain what is meant by (a) the metacentre and (b) the metacentric height of a vessel.

3.5 (a) For a vessel of fixed dimensions and weight, is the height of the metacentre above the base (OM) a constant? (b) For a vessel of fixed dimensions and weight, is the metacentric height (GM) a constant? (c) If the weight of the vessel remains the same but the weight is redistributed vertically (say by loading it differently), how does this affect OM and GM? Are your answers to parts (a) and (b) still the same?

3.6 A pontoon has a length of 18 m and a width of 6 m when seen in plan. When loaded the pontoon will weigh 940 × 10³ N and have its centre of gravity at a height of 3.9 m above the base. The saline water in which the pontoon floats has a density of 1025 kg/m³. Determine whether or not the pontoon will be stable.

[Neutral stability, GM = 0]

3.7 A pontoon is 17.0 m by 6.5 m in plan and floats in water of density 1025 kg/m³. It has an unloaded weight of 900 × 10³ N and vertical sides that are 3.8 m high. The centre of gravity of the pontoon itself can be assumed to be at 1.9 m above the base. The pontoon carries two other weights, both positioned on the longitudinal centreline. The first weight of 230×10^3 N is positioned inside the pontoon and has its centre of gravity at a height of 1.2 m above the base. The second weight is placed on the deck. This 400×10^3 N weight has its centre of gravity 0.8 m above the deck, that is 4.6 m above the base. (a) What is the displacement of the pontoon (in m³) in its unloaded and loaded condition? (b) What is the depth of immersion of the pontoon, unloaded and loaded? (c) What is the value of BM when the pontoon is in its unloaded and loaded condition? (d) At what height above the base is the overall centre of gravity of the loaded pontoon? (e) What is the metacentric height of the pontoon in its unloaded and loaded state? (f) Is the pontoon stable in both its unloaded and loaded condition?

[(a) 89.5 m³, 152.2 m³; (b) 0.81 m, 1.38 m; (c) 4.35 m, 2.56 m; (d) 2.50 m; (e) 2.85 m, 0.75 m; (f) yes, yes]

3.8 A pontoon with a mass of 20 tonnes (1 tonne = 1000 kg) tilts to an angle of 10.5° when a mass of 0.5 tonne is moved 3.0 m towards the side from the centreline. What is the value of GM?

[0.409 m]

3.9 A pontoon loaded with construction equipment has a total weight of 1.2×10^6 N (including the jockey weight) and is 20 m long by 8 m wide in plan. It floats in saline water of density 1025 kg/m³. When a jockey mass of one tonne (1000 kg) is moved from the centreline 3.5 m transversely across the deck it causes a tilt of 5.5°. (a) What is the value of GM? (b) Including the jockey weight, what is the value of BM? (c) At what height above the base is the centre of gravity, G?

[(a) 0.30 m; (b) 7.15 m; (c) 7.23 m]

3.10 The metacentric height of a pontoon similar to that in Fig. 3.9 is to be determined experimentally from three sets of results. The height of the pontoon's vertically adjustable mass above its base is shown as y below, alongside the corresponding height of the pontoon's centre of gravity above its base OG (see Fig. 3.7). The jockey mass is moved in 15 mm increments from the longitudinal centreline, the value of x and the corresponding angle of tilt θ degrees being as below. The other values are as follows: total mass of pontoon $M = 2.600$ kg; mass of jockey weight $m_J = 0.200$ kg; width of pontoon $b = 0.202$ m, length of pontoon $l = 0.358$ m.

	x (mm)	Exp't 1 $\theta°$	Exp't 2 $\theta°$	Exp't 3 $\theta°$
Left	75	5.6	8.25	—
	60	4.4	6.75	—
	45	3.3	5.0	7.7
	30	2.2	3.25	5.3
	15	1.0	1.50	2.7
	0	0	0	0
	15	1.0	1.5	2.6
	30	2.1	3.5	5.4
	45	3.4	5.25	7.8
	60	4.5	7.0	—
Right	75	5.6	8.5	—
	y	0.129 m	0.234 m	0.300 m
	OG	0.050 m	0.069 m	0.083 m

For each of the three experiments, (a) draw a graph of θ against x using axes like those shown in Fig. Q3.10. (b) Use the graph to determine the value of dx/dθ and hence GM using equation (3.4). Remember to convert degrees to radians when you apply the equation. (c) Calculate the exprimental value of BM. (d) Calculate the theoretical value of BM using equation (3.1).

[(c) about 0.091 m, 0.090 m and 0.091 m; (d) 0.095 m]

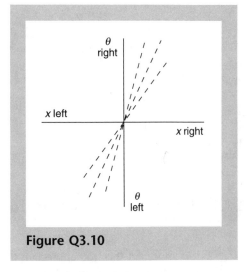

Figure Q3.10

3.11 A vessel has a weight of 200 000 N, a moment of inertia of 68 400 kg m² and a metacentric height of 1.5 m. (a) Calculate the period of roll of the vessel. (b) Is the period of roll acceptable?

[(a) 3.00 s; (b) rather short]

Fluids in motion

It is a gross oversimplification, but it is often said that there are only three equations in hydraulics: the continuity equation, the momentum equation and the energy equation. However, it is true that an awful lot of problems can be solved using only these equations, and they hold the key to many of the topics that follow later in the book. Therefore it is important to have a sound understanding of what they represent and how they are used. This chapter introduces the three equations and the basic principles required to understand and analyse what is happening in a system involving a moving liquid. The sort of questions that are answered include:

What are the significant differences between analysing static and moving liquids?

What is viscosity?

How can we visualise fluid flow?

What is the continuity equation and what is it used for?

What is the momentum equation and how is it applied?

What is the energy equation and how is it used?

Why do aeroplanes fly?

What is the ground effect associated with racing cars?

4.1 Introduction to the fundamentals

 66 It is important that you appreciate that we have a new ball game here. Hydrostatics is very simple from the mathematical point of view and the ancient Greeks were familiar with the basic principles. In hydrostatics the liquid is stationary so we do not have to worry about viscosity, turbulence or friction. Now that we are dealing with moving liquids these things have to be considered, but they are far too complex for a simple mathematical treatment. 99

66 That sounds serious. How are we going to get around the problem? **99**

The answer is that we are going to cheat! For simplicity, we will assume that viscosity, turbulence and friction do not exist and can be ignored. Similarly, we will ignore surface tension (which is rarely significant anyway) and we will assume that the liquid is incompressible. In other words we will assume that our **real liquid** behaves like an **inviscid** or **ideal liquid**, which has no viscosity, is not affected by turbulence, and is frictionless and incompressible. Now this means that many of the equations and answers that we get will not be accurate, *because we are ignoring things that do exist*. However, we can compensate for this by introducing experimentally derived coefficients into the equations, so that our simple equations for an ideal liquid give accurate answers for a real liquid. You could almost call these coefficients 'fiddle factors' since they convert the 'wrong' answers into accurate answers. The **coefficient of discharge** in section 5.1, is one such experimental fiddle factor.

One significant difference between a real and an ideal liquid is the fact that friction exists in the former but (by assumption) not in the latter. In reality, friction slows any liquid that flows over a boundary surface, resulting in relatively small velocities adjacent to the surface and higher velocities as the distance from the boundary increases. Thus the velocity across the diameter of a pipe, for example, is zero at the walls and highest in the centre. With an ideal liquid the velocity would be the same everywhere in the pipe because there is no friction. To allow for this, when working with real liquids we almost always use the mean velocity, V, which is calculated from Q/A where Q is the discharge and A the cross-sectional area of flow. Generally we will assume that the mean velocity occurs over the whole area of flow. If this leads to inaccuracies, it is possible to introduce coefficients to compensate. The **momentum coefficient** and **energy coefficient** are described in sections 4.6.5 and 4.8.1 respectively.

4.1.1 Understanding viscosity

66 I can understand how the approach you have described above would simplify things, and I know what friction is; it is just the resistance to motion experienced by a liquid flowing over a solid boundary. I can guess that energy will be needed to overcome friction and keep the liquid moving. I also know that turbulence is just a random motion, like eddies in the air when a large lorry drives past you at speed. However, I do not understand what viscosity is. Can you explain, please? **99**

You are right about friction. Remember that turbulence also requires energy to drive the eddies. Usually this comes from the main body of flow and represents a loss of energy compared to a fluid flowing smoothly with no turbulence. Viscosity is a measure of the internal friction of a fluid, or its resistance to flow and movement (more formally, viscosity is described as a measure of a liquid's resistance to shear stress). For example, cold treacle is very stiff and does not flow easily, while water is thin and runny. The difference is that water has a low viscosity and treacle a high viscosity.

One thing to note from Table 4.1 is that the thicker the liquid, the larger the numerical value of the dynamic viscosity. Take care to get this the right way around (think of a large number indicating a large resistance to move- REMEMBER!!!

Table 4.1 Viscosity and density of some fluids at 20°C

Fluid	Coefficient of dynamic viscosity, μ ($\times 10^{-3}$ kg/ms)	Density, ρ (kg/m^3)
Air	0.01815	1.2
Fresh water	1.005	998.2
Mercury	1.552	13546
Paraffin oil	1.900	800
Oil, s.a.e. 10	29	880–950
Oil, s.a.e. 30	96	880–950

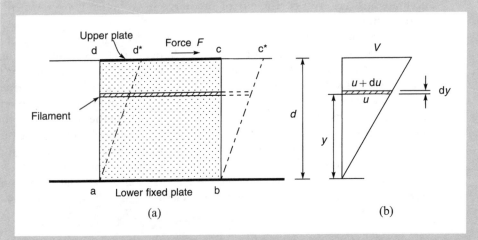

Figure 4.1 (a) Liquid between the two plates. (b) The velocity, V, caused by applying a force, F

ment or deformation of the liquid). We will define the coefficient of viscosity in more detail later on, and it will be clearer then why this is the case. You should also note that viscosity changes with temperature (which is why car engines use a multigrade oil, for instance), and so does density. At 100°C the dynamic viscosity of water falls to 0.284×10^{-3} kg/ms indicating that the liquid is getting thinner, and the density falls to 958.4 kg/m^3 showing that it is getting lighter.

Viscosity is the most important single property that affects the behaviour of a fluid. The more viscous the fluid, the thicker it is and the slower it deforms under stress. Newton investigated viscosity by sandwiching a liquid between two large horizontal plates (Fig. 4.1a). The lower plate is fixed but the upper plate (of area, A) is free to move in response to a force, F, pulling on it horizontally. The shear stress, τ, acting on the liquid in contact with the plate is:

$$\tau = F/A \tag{4.1}$$

Note that stress is similar to pressure, that is a force divided by an area, and so has the same units (N/m^2). However, shear stress acts parallel to the shear surface, whereas hydrostatic pressure acts at 90 degrees to a surface.

If the liquid sticks to the upper plate then the liquid in contact with it will have the same velocity as the plate itself, provided that there is no slip (how much slip there is depends in part upon the viscosity of the fluid). As the upper plate moves, the liquid that was contained within abcd deforms to abc*d*. Because the liquid is also sticking to the stationary bottom plate there is no movement or deformation of the liquid at this level. Between the two plates the velocity (and deformation) of each horizontal filament of liquid depends upon its distance above the fixed bottom plate, as shown by the velocity diagram in Fig. 4.1b. If V is the velocity of the upper plate and d the distance between the plates, Newton found that F is directly proportional to A and to V, but inversely proportional to d. In other words, the larger the area of the plate, the greater the force required to move it; a larger force is needed to move the plate quickly than to move it slowly, and the smaller the gap between the plates the larger the force required to move the top one. With respect to the latter, two flat glass plates with only a smear of water between them can stick together quite strongly. Expressing these relationships mathematically:

$$F \propto AV/d$$
$$\text{or} \quad F = \mu AV/d$$

where the constant, μ, is the coefficient of dynamic (or absolute) viscosity. Since F/A is the shear stress, τ, the above equation can be rearranged as:

$$\tau = \mu V/d$$

Now consider the velocity diagram in Fig. 4.1b. Take an infinitely thin filament of liquid at any height y above the bottom plate, and let the velocity at the bottom of the strip be u and the velocity at the top be $u + du$. If the thickness of the strip is dy, then the velocity gradient is du/dy (this is also called the **rate of shear strain** or the **rate of deformation**). It should be apparent from the diagram that du/dy is effectively the same as V/d, hence:

$$\boldsymbol{\tau = \mu du/dy} \tag{4.2}$$

Thus if a graph is drawn of τ against du/dy, the gradient of the line is the viscosity, μ (Fig. 4.2). A **Newtonian fluid** plots as a straight line (constant μ) through the origin. As mentioned above, μ is not really constant since its value changes with temperature, but at a particular temperature μ is constant. Air and water fall into this category. There are **non-Newtonian fluids** for which μ is not a constant at all but is a function of both temperature and rate of deformation, so the line is curved. We will consider only Newtonian fluids for which μ can be taken as constant.

4.1.2 The coefficients of dynamic and kinematic viscosity

Be careful not to confuse the coefficient of dynamic (absolute) viscosity, μ, with the coefficient of kinematic viscosity, v. Dynamic viscosity, μ, is the constant in equation (4.2). The kinematic viscosity, v, is the dynamic viscosity of a fluid divided by its density, that is $v = \mu/\rho$. Two examples are given below. Look carefully at the units if you are not sure which value of viscosity is being quoted.

Water: $\quad v = \mu/\rho = 1.005 \times 10^{-3}/998.2 = 1.007 \times 10^{-6} \, \text{m}^2/\text{s}$ at 20°C.

Oil, s.a.e. 30 $\quad v_O = \mu_O/\rho_O = 96 \times 10^{-3}/915 = 104.9 \times 10^{-6} \, \text{m}^2/\text{s}$ at 20°C.

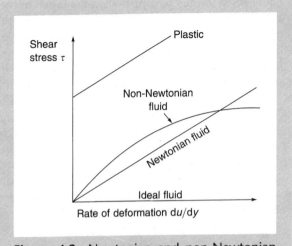

Figure 4.2 Newtonian and non-Newtonian fluids. Viscosity is a measure of resistance to shear stress

Now let us consider the units of viscosity in more detail, since they can be confusing. In some books you may find that the dynamic viscosity of water is quoted as $\mu = 1.14 \times 10^{-3}\,\mathrm{Ns/m^2}$ whereas in Table 4.1 it was given as $1.005 \times 10^{-3}\,\mathrm{kg/ms}$. These values are different numerically, and the units are different.

Which value is correct?

Well, the numerical value $1.005 \times 10^{-3}\,\mathrm{kg/ms}$ is correct if the temperature is 20°C. Remember the value increases with reducing temperature, so $1.14 \times 10^{-3}\,\mathrm{kg/ms}$ is correct at 15°C. What about the units? Well a Newton is the force required to give a mass of 1 kg an acceleration of $1\,\mathrm{m/s^2}$ (see section 1.1). Hence a Newton can be written as $\mathrm{kg\,m/s^2}$. Thus:

$$\frac{\mathrm{Ns}}{\mathrm{m^2}} = \frac{\mathrm{kg} \times \mathrm{m}}{\mathrm{s^2}} \times \frac{\mathrm{s}}{\mathrm{m^2}} = \mathrm{kg/ms}$$

So, $\mathrm{Ns/m^2}$ is the same as $\mathrm{kg/ms}$. Now let us consider the coefficient of kinematic viscosity. Remember that the kinematic viscosity of a liquid is its dynamic viscosity divided by its density, thus:

$$v = \frac{\mu}{\rho} = \frac{\mathrm{kg}}{\mathrm{m} \times \mathrm{s}} \times \frac{\mathrm{m^3}}{\mathrm{kg}} = \mathrm{m^2/s}$$

The units of kinematic viscosity are derived from the way in which it is defined: there is nothing mysterious about them.

One final point about viscosity. While it is one of the most important factors controlling the flow and behaviour of a fluid, it does not always appear in the equations that we will be deriving later. However, Examples 10.3 and 10.4 show that viscosity, in the form of the Reynolds number described in the next section, does indeed affect fluid flow.

(REMEMBER!!!)

Try this experiment for yourself

To get a better understanding of viscosity, try this simple little experiment for yourself. Fill a glass with water. Place the glass on a flat smooth surface like a kitchen worktop. Allow the glass to stand for about two minutes until the water has completely stopped moving.

Now place a matchstick on the surface of the water in the centre of the glass. If the liquid is static the match should not move. Make sure that the match is not touching the glass. Keeping the glass flat on the worktop, rotate the glass and watch what happens to the match. If you perform the experiment carefully, you should find that the match continues to point in the same direction, regardless of how you twist the glass (Fig. 4.3). The match resembles a compass needle: it always points in the same direction regardless of how you turn the glass.

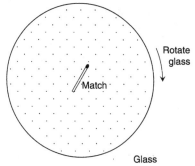

Figure 4.3

You can try the same experiment with a cup of coffee. Let a cup of black coffee rest for a few minutes, then add some milk or cream to one side. Watch the pattern made by the milk. Twist and rotate the cup. You should find that the milk does not rotate with the cup.

The explanation of what is happening is quite simple. Water, or coffee, is a relatively thin liquid with a low viscosity. By twisting the glass or cup, a high shear stress is being introduced along the side of the container. The water is too thin to transmit this motion to the main body of the liquid, so it stays in the same position, and the match continues to point in the same direction. Of course, if you put a highly viscous, thick liquid (like treacle) in the glass and put a match on top, then both the treacle and the match would turn with the glass. In this case the treacle is so viscous that there would be no slippage between the treacle and its container, so everything would rotate together.

4.2 Classifying various types of fluid flow

There are many different types of fluid flow, although they may not be immediately apparent to you. One of the first things you have to do when investigating a problem involving moving fluids is to define the type of flow that you are dealing with. Having done that, you will then have an idea of which equations can legitimately be applied to the problem. If you do not take care you could apply an equation which is not valid for the type of flow being experienced and obtain an answer which is wrong by a very serious margin.

4.2.1 Laminar and turbulent flow, and Reynolds number

Depending upon the problem, the first step of an investigation may be to decide if the flow is laminar or turbulent. **Laminar flow** is usually associated with slow moving, viscous fluids. It is relatively rare in nature, although an example would be the flow of water through an aquifer. Groundwater velocities may be as little as a few metres per year. **Turbulent flow** is much faster and chaotic, and is the REMEMBER!!! type usually encountered. A good example would be flow in a mountain stream.

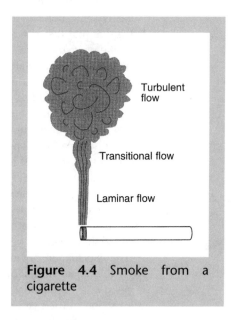

Figure 4.4 Smoke from a cigarette

To gain a better understanding of what laminar and turbulent flow look like, observe the smoke coming from the end of a cigarette. Near the tip the column of smoke usually forms a smooth cylinder and rises almost vertically. This illustrates laminar flow – a smooth, uniform flow where particles of smoke follow each other in regular succession. At some height above the tip, the column of smoke will start to waver and begin to break up. This is **transitional flow** between laminar and turbulent flow. Even higher from the tip, as a result of the disturbance of the smoke column by draughts and air currents, the smoke becomes a dispersed, billowing cloud of particles. The particles follow no regular path (as they do in laminar flow), and each particle may follow its own particular path. The motion of the particles is quite random. This is turbulent flow.

Whether the flow is laminar, transitional or turbulent is very important with respect to the flow of liquid through pipelines, since the characteristics of the three flow regimes are very different necessitating a different approach. For simple laminar flow conditions it is possible to analyse many problems mathematically. Unfortunately, most flows in nature are turbulent and the irregular pulsating nature of this type of flow is too complex for a mathematical treatment. For this reason it is necessary to rely on relationships based upon experiment, the equations for flow in a pipe providing a good example of this approach (section 6.5).

How can we determine which type of flow we are dealing with in a particular problem?

Osborne Reynolds found that the type of flow is determined by the size of the conduit (diameter, D), the density (ρ) of the liquid, its dynamic viscosity (μ), and the mean velocity (V). Remember that $V = Q/A$ (discharge/area of flow). These variables can be grouped together to form a dimensionless parameter, Re, called the **Reynolds number**:

$$\mathrm{Re} = \rho VD/\mu \tag{4.3}$$

This can also be written in terms of the kinematic viscosity since $v = \mu/\rho$ and hence Re = VD/v. For conduits other than full pipes the diameter is replaced by some other characteristic dimension (such as the depth of flow, D, in an open channel) so different Reynolds

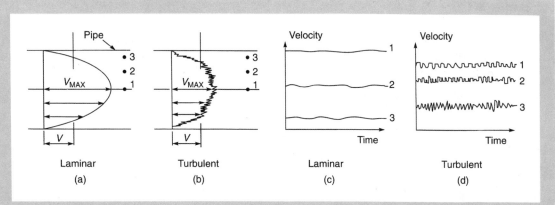

Figure 4.5 The difference between laminar and turbulent flow illustrated diagrammatically. (a) and (b) The variation of velocity across a pipe diameter. The length of the arrows represents the magnitude of the velocity. (c) and (d) Variation of velocity with time at three points across the diameter

numbers would be indicative of laminar, transitional and turbulent flow. As a very general guide for water:

	Pipes	Open channels
Laminar flow	Re < 2000	Re < 500
Transitional flow	Re = 2000 to 4000	Re = 500 to 2000
Turbulent flow	Re > 4000	Re > 2000

Example 4.1 illustrates the use of Reynolds number to determine whether a flow is laminar or turbulent.

It is quite important that you can visualise the difference between laminar and turbulent flow. Reynolds conducted a classic experiment in which he injected a thin stream of dye into a pipe carrying water (see section 6.5). In laminar flow the filament of dye remains intact: the particles move along slowly and uniformly, and at any point within a cross-section the particles follow one another and have the same path. In turbulent flow the particles move randomly, so the dye stream soon breaks up, and the dye spreads throughout the pipe. The difference between the two flows is illustrated by Fig. 4.5. Things to note are:

1. The velocity at the pipe wall is zero as a result of friction, with the largest velocity in the centre of the pipe. See section 6.5.3 for a description of the boundary layer.

2. In laminar flow the ratio of the maximum velocity V_{MAX} to the mean velocity V is roughly 2.0 (diagram a). In turbulent flow V_{MAX}/V has a value of about 1.7 (diagram b). The lower figure is the result of a less well ordered flow regime with a greater transverse velocity component, hence the spread of dye across the area of the pipe.

3. In laminar flow the velocity at any particular point in the pipe remains relatively constant from one moment to the next. In turbulent flow the velocity fluctuates with time (c and d).

EXAMPLE 4.1

A pipe of diameter 0.1 m carries water at the rate of 0.025 m^3/s. Taking the density of the water as 1000 kg/m^3 and its dynamic viscosity as 1.005×10^{-3} kg/ms, calculate the Reynolds number of the flow and determine whether it is laminar or turbulent.

$$Re = \rho VD/\mu \text{ where } A = \pi D^2/4 = \pi(0.1)^2/4 = 7.85 \times 10^{-3} m^2$$
$$V = Q/A = 0.025/7.85 \times 10^{-3} = 3.18 \, m/s$$
$$Re = 1000 \times 3.18 \times 0.1/1.005 \times 10^{-3} = 316\,400$$

The flow is clearly turbulent since $3.16 \times 10^5 \gg 4000$.

4.2.2 Steady and unsteady flow

Another significant way of classifying the flow is to determine whether it is **steady** or **unsteady**. The key concept here is whether or not the discharge is [REMEMBER!!!] changing **with respect to time**. In steady flow the discharge, Q, is constant with respect to time; for example, if there is a constant discharge through a pipe. However, in unsteady flow the discharge, Q, is not constant with respect to time and position. For example, a flood wave going down a river causes the discharge to change with time and location, so the flow is unsteady. With unsteady flow the cross-sectional area of flow and the velocity at any location also change, of course, whereas with steady flow they are constant.

If this definition of steady flow was strictly applied, then steady flow would be extremely rare. As explained above, turbulent flow is the most common and it is characterised by temporal fluctuations in velocity, so it could never truly be called steady. However, the definition is usually loosely interpreted so that if the mean velocity and discharge are not changing over a period of time the flow is said to be steady. Minor fluctuations are ignored.

The analysis of steady flow is relatively simple, and this is the type of problem mostly covered in this book. Unsteady flow is more complex, since it involves variables that change with time, and it is dealt with briefly in sections 8.12 and 11.9.

4.2.3 Uniform and non-uniform flow

The key concept here is whether or not the cross-sectional area of flow and mean velocity change **from one section to the next** along the length of the [REMEMBER!!!] conduit when the discharge is constant. For the flow to be **uniform** the area (depth and width) and the mean velocity must be the same at each successive cross-section. An example would be a pipe of constant diameter running full. It follows that **non-uniform flow** occurs where the cross-sectional area and mean velocity change from section to section, as would be the case with a pipeline of varying diameter.

Some hydraulic systems are described below. Classify each flow condition using the terms steady, unsteady, uniform and non-uniform.

(a) A long pipe in which the same pressure and velocity exist from one instant to another at all points along its length.

(b) An expanding pipe with a decreasing rate of flow.

(c) A wave travelling down a river channel.

(d) An expanding pipe with a constant rate of flow.

(e) A channel in which the depth of flow is the same at all sections along its length, and in which the discharge is the same at every section and does not vary with time.

4.3 Visualising fluid flow

66 I am having a bit of trouble visualising the flow through pipes, what with changing cross-sections that cause non-uniform velocities, and so on. Is there anything that that will help me to get a better picture of what is happening? **99**

Yes, there is a device that is used to visualise the flow. This involves drawing **streamlines** to depict the motion of the liquid as it passes through a conduit. Streamlines are referred to quite often in the text, usually in the context of explaining what is happening to the liquid. Figure 4.6 provides an illustration of some streamlines.

Strictly speaking, a streamline should only be drawn for laminar flow where all particles starting from the same point follow each other and have exactly the same path. Since most flows are turbulent, this does not happen: successive particles starting from the same point follow different paths because of the irregular, pulsating nature of turbulent flow. If the position of an individual particle was mapped at various moments in time and a line drawn through all the points where it had been, this would be the **pathline** of that particular particle. Each particle would have its own individual pathline, and each pathline would be different. An example would be the pathline of one particle of smoke in a billowing cloud of smoke.

As a further illustration of the difference between streamlines and pathlines, imagine that one hundred students are going to leave a lecture theatre to go to the canteen at the end of the corridor. Think of the students as individual particles, and the corridor as the conduit that they must travel along. To qualify as a streamline the students would literally have to follow in each others' footprints: they must all start at the same point, follow the exact same route at the same velocity, and arrive at exactly the same point. This would not happen. All of the students would arrive in the canteen, but at slightly different points at slightly different times, having walked on different sides of the corridor and travelled at slightly different speeds: they would not arrive conga fashion. Each student has an individual pathline.

At this point we have to be pragmatic. It is not possible to draw individual pathlines when investigating the flow of an almost infinitely large number of particles. What we are interested in is the average or general direction of motion. Consequently we do draw streamlines for turbulent flow, because they illustrate the general picture quite well and help us to understand what is happening. When drawing streamlines it may help you if you pretend that each particle is a car travelling at 100mph down a motorway. The car can only change direction smoothly and gently: there can be no 90° changes of direction. Also, the car must move

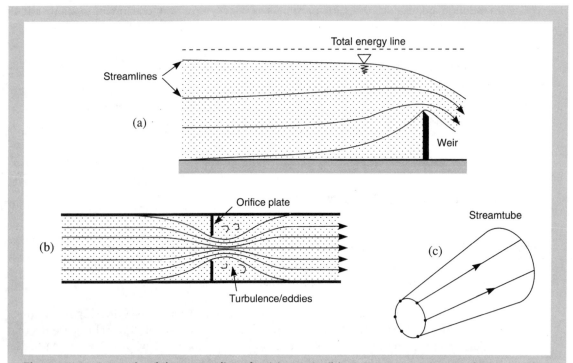

Figure 4.6 Example of the streamlines for (a) a weir, (b) a pipe constriction and (c) a streamtube. Streamlines are used to help visualise what is happening; if the streamlines are converging then the velocity is increasing. Streamtubes are used in the derivation of the equations in Appendix 1

parallel to solid boundaries like crash barriers: it should not hit the boundary at 90°. The same applies to streamlines. The more formal rules for drawing streamlines are as follows:

1. A streamline is a continuous line drawn through the fluid so that it represents the movement of any particle on the line. The direction of the velocity vector of a particle is always along the streamline.

2. Two particles on the same streamline follow the same course.

3. Streamlines cannot cross (if they did then a particle would theoretically have two velocity vectors).

4. Ideally streamlines have no width and no cross-sectional area.

5. One streamline always coincides with a free liquid surface and a solid boundary.

6. When starting to draw streamlines, make them equally spaced. Usually a relatively small number will suffice. Remember 5 above, and the analogy to cars on the motorway. Draw smooth lines that change direction gently. If the conduit changes in cross-sectional area then the spacing of the streamlines must change. The spacing may not become uniform again until some distance *after* the change in section.

7. Since the velocity vector of a particle at any point on a streamline is along it, there can be no flow across a streamline. In a sense the fluid is 'trapped' between two streamlines and can only escape by flowing along the conduit.

Figure 4.6(a and b) only depicts the motion two dimensionally; there could be movement into or out of the page as well. If streamlines are drawn through every point on the circumference of a very small area that forms part of the cross-section of a conduit, then a **streamtube** is formed (Fig. 4.6c). It follows from point 7 above that the fluid can only escape from the streamtube by flowing along it and out of the end. Therefore, a streamtube acts very much like a real pipe with a solid boundary wall. This is a convenient device which is used in Appendix 1 to derive the continuity, momentum and energy equations.

Streamlines can be more than aids to just visualising the flow. As will become clear later, it is possible to deduce the following quantitative relationships:

Parallel streamlines . . . . velocity constant . . . . pressure constant

Converging streamlines . . . . velocity increasing . . . . pressure decreasing

Diverging streamlines . . . . velocity decreasing . . . . pressure increasing

These relationships are illustrated by Fig. 4.7 which shows the two-dimensional flow around a cylinder. There is a uniform, laminar flow of water between two glass plates from left to right. Dye is injected into the flow at regular intervals to form evenly spaced streamlines. Because the flow is laminar the dye does not mix with the water. The central streamline hits the nose of the cylinder at right angles, causing the flow to momentarily stop. This is called a **stagnation point**. Since the flow is stationary at this point, the kinetic energy is converted into pressure energy (see section 4.7.2). This is indicated by the diverging streamlines, which are indicative of decreasing velocity and increasing pressure. Some of these effects can also be observed when water flows around bridge piers (section 9.3.1).

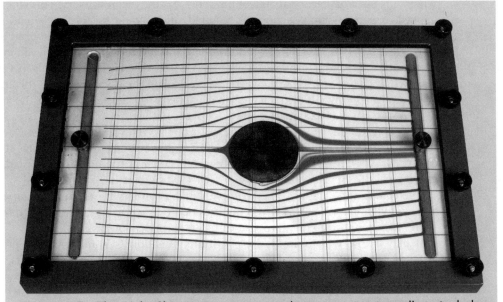

Figure 4.7 The Hele–Shaw apparatus uses dye to create streamlines to help visualise flow patterns. The flow is from left to right between two glass plates [*reproduced by permission of TecQuipment Ltd*]

Streamlines and streamtubes can be used quantitatively. If the streamlines are equally spaced in a uniform flow, then there must be an equal discharge between each pair of lines. If another set of lines, called **equipotential lines**, are drawn at 90° to the streamlines to form a series of squares, the result is a **flow net**. One example of a flow net is shown in Fig. 13.24b, where the equipotential lines represent the piezometric level or elevation of the groundwater surface in an aquifer. The spacing of the equipotential lines indicates how quickly head or pressure is lost and the relative permeability of the aquifer, while the quantity of flow between the streamlines enables the potential yield of the aquifer to be assessed. Flow nets can also be used to study seepage under dams and into excavations.

4.4 ▶ The continuity equation

The continuity equation is derived from first principles as Proof 4.1 in Appendix 1. In essence it is extremely simple, being based on the concept of the **conservation of mass**. This can be illustrated by considering the steady flow of water through a pipe. If there is no leakage or additional flow into the pipe, common sense dictates that the mass or quantity of water entering the pipe at cross-section 1 must equal that leaving the pipe at cross-section 2. If this was not the case, then matter would be either appearing magically out of thin air or being destroyed. If the volumetric flow rate in the pipe is Q m³/s and the density of the liquid is constant at ρ kg/m³ then the mass flow rate is ρQ kg/s. Thus at the two cross-sections the continuity equation based on the conservation of mass can be written as $\rho Q_1 = \rho Q_2$. It is normally more convenient to write this purely in terms of the volumetric discharge, i.e. $Q_1 = Q_2$ or:

$$Q_1 = A_1 V_1 = A_2 V_2 = Q_2 \tag{4.4}$$

where A is the cross-sectional area of flow and V is the mean velocity. This equation can be applied to as many cross-sections as required, and is valid for all flow situations (such as open channels), not just pipes. It has not been proved above that $Q = AV$, but a consideration of the units (m³/s = m² × m/s) shows that this relationship is both valid and logical.

The continuity equation is used whenever we need to calculate the mean velocity from a known discharge and area ($V = Q/A$), or to calculate the discharge from the known velocity and area ($Q = AV$). It also tells us what happens to the velocity when the area changes ($V_2 = V_1[A_1/A_2]$). For example, if the cross-sectional area of a pipe that is running full is halved, then the mean velocity of flow must double in order to discharge the same quantity of water through the reduced section. Of course, both sections must have the same discharge ($Q_1 = Q_2$). It can also be used to ensure that all of the flow is accounted for when a single pipe splits into two separate pipes ($Q_1 = Q_2 + Q_3$) as in Example 4.3.

The continuity equation is very simple, but it is one of the three most important equations in hydraulics (the other two being the energy equation and the momentum equation). With only these three equations you can solve many problems.

It may amuse you to apply the continuity equation to some everyday problems. For instance, if a dual carriageway is reduced from two lanes to one for roadworks, what speed should the vehicles travel at in order that the same number per hour can pass along the single lane? Well, the continuity equation would indicate that for the flow (Q) to be constant, if the area of flow (A) is halved then the velocity (V) should be doubled. That is the vehicles should travel at twice their normal speed through the roadworks (some drivers already appear to be working on this principle!). Of course, such a suggestion is dangerous and not practical for safety reasons.

Box 4.2 ▶ **The continuity equation**

Do not forget the continuity equation. Many students do! Perhaps because it is so simple? With most problems, it should be the first equation that you apply. It can be used to reduce the number of unknown variables (without it you may not be able to solve the problem at all). You may then apply the momentum and/or the energy equation. But remember, the continuity equation is applied first! (REMEMBER!!!)

More seriously, make sure that you understand Examples 4.2 and 4.3, and then do Self Test Question 4.2.

SELF TEST QUESTION 4.2

Water flows through the branching pipeline shown below. Given the following information, find the diameter of the pipe, D_2, required at section 2 to maintain continuity of flow.

| $D_1 = 0.70\,\text{m}$ | $D_2 = ?$ | $D_3 = 0.25\,\text{m}$ |
| $V_1 = 1.60\,\text{m/s}$ | $V_2 = 1.10\,\text{m/s}$ | $V_3 = 2.30\,\text{m/s}$ |

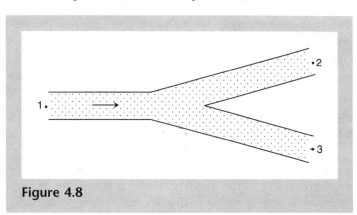

Figure 4.8

EXAMPLE 4.2

A pipe of diameter 0.2 m increases gradually to 0.3 m. If it carries 0.08 m³/s of water, what are the velocities at the two sections?

$$Q = A_1 V_1 = A_2 V_2$$
$$A_1 = \pi(0.2)^2/4 = 0.0314\,\text{m}^2$$
$$A_2 = \pi(0.3)^2/4 = 0.0707\,\text{m}^2$$
$$0.08 = 0.0314 V_1 \text{ so } V_1 = 2.548\,\text{m/s}$$
$$0.08 = 0.0707 V_2 \text{ so } V_2 = 1.132\,\text{m/s}$$

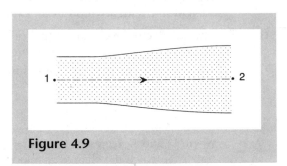

Figure 4.9

EXAMPLE 4.3

Water flows through a branching pipeline as shown in the diagram. If the diameter, D_2, is 250 mm, $V_2 = 1.77$ m/s and $V_3 = 1.43$ m/s:

(a) what diameter, D_3, is required for $Q_3 = 2Q_2$?

(b) what is the total discharge at section 1?

(a) $Q_3 = 2Q_2$ so $A_3V_3 = 2A_2V_2$
$(\pi D_3^2/4) \times 1.43 = 2 \times (\pi \times 0.25^2/4) \times 1.77$
$D_3 = 0.393$ m

(b) $Q_2 = A_2V_2$
$= (\pi \times 0.25^2/4) \times 1.77 = 0.087$ m³/s
$Q_3 = 2Q_2 = 2 \times 0.087 = 0.174$ m³/s
Total discharge $= Q_1 = Q_2 + Q_3$
$= 0.087 + 0.174$
$= 0.261$ m³/s

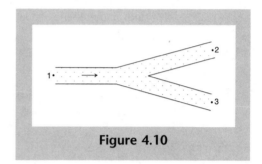

Figure 4.10

4.5 Understanding the momentum equation

❝ It is sensible to consider the momentum equation next, because you need it before you can derive the energy equation from first principles. However, I would guess that your understanding of what momentum actually represents is a bit shaky, so we will start by going back to basics. ❞

4.5.1 Understanding momentum and Newton's Laws of Motion

The simplest definition of momentum is: **momentum = mass × velocity**. Therefore a body has momentum by virtue of the fact that it is moving. If the velocity is zero, then the momentum is zero. Remember that we are dealing with mass, not weight. Even in the weightlessness of space a moving body has momentum.

At this point it is probably a good idea to revise Newton's Laws of Motion. They are:

Law 1 A body will remain in the same condition of rest or of motion with uniform velocity in a straight line until acted on by an external force.

Law 2 The rate of change of momentum of a body is equal to the force acting upon it and takes place in the line of action of the force. For a body of mass, M:

Force = rate of change of momentum (4.5)

$$F = (MV_2 - MV_1)/t$$
$$= M(V_2 - V_1)/t$$
$$= Ma \qquad (1.3)$$

where V_1 and V_2 represent the initial and final velocity of the body respectively, t is the time over which the change of velocity occurs, and a is the acceleration of the body.

Law 3 To every action there is an equal and opposite reaction.

To begin with, it is much easier to apply and to understand Newton's Laws when thinking of solid objects, like snooker balls, rather than liquids. The first law simply states that a stationary snooker ball will remain stationary unless it is acted upon by a force, such as being hit with a cue or by another ball. It goes on to say that once a ball is moving it will continue to move in a straight line until acted upon by an external force. In the case of a snooker ball the external force may be frictional resistance or may be imparted when the ball hits a cushion (see Law 3). Another example would be that a cricket ball will continue in a straight line unless deflected by a bat. The bat applies a force to the ball, but the ball applies an equal and opposite force to the bat (Law 3). If you do not believe this, would you let a cricket ball travelling at 90 mph hit you on the head?

Law 2 provides a means of calculating the force required to accelerate a body, or of calculating the force on something as a result of an impact. You met the equation $F = Ma$ in section 1.1.1. Accelereration, a, is the rate of change of velocity, so when we say that a car goes from 0 to 60 mph in 9 seconds we are quoting the value of $(V_2 - V_1)/t$ in equation (4.5). Thus if a 0.9 tonne (900 kg) car is to be accelerated from 0 to 26.8 m/s (i.e. 60 mph) in 9 seconds then the applied force would have to be $F = 900(26.8 - 0)/9 = 2680$ N. Obviously a car with twice the mass would require twice the force in order to have the same acceleration.

Because velocity is a vector quantity, so is momentum. For example, consider two snooker balls of identical mass, M, rolling towards each other in exactly opposite directions. Suppose the first ball has a velocity of 0.3 m/s from left to right, and the second ball a velocity of 0.8 m/s from right to left. Then:

$$\text{total momentum} = M_1V_1 - M_2V_2$$
$$= M \times 0.3 - M \times 0.8$$
$$= -0.5M \text{ kg m/s}$$

Note that the momentum of the second ball has a negative sign because it is travelling in the opposite direction to the first; it would have been positive if travelling in the same direction. This arises again in Boxes 11.1 and 11.2 when we consider impulse turbines and the impact of a jet of water on a curved vane.

Another way of looking at Newton's laws of motion is through the concept of the **conservation of momentum**. This concept states that, in a specified direction, momentum cannot be created or destroyed unless an external force is applied. As an illustration, consider the first snooker ball rolling across the table. If we ignore any resistance forces so that the ball has a constant velocity $V = 0.3$ m/s, then at successive points along its path its momentum would be $0.3M$ kg m/s. In other words, the momentum is constant unless some external force is applied to reduce or increase it. The concept is also applicable to problems that involve collisions. For example, it tells us that if the two snooker balls above collide and bounce off each other in a straight line, the total momentum after the impact would still be $-0.5M$ kg m/s. This concept is applied to hydraulic systems in section 4.6.4.

An interesting illustration of Newton's third law is as follows. Suppose that two people are skating on an ice rink. One is an adult, the other is a child who weighs exactly half as much as the adult, and so has half the mass. They face each other on the ice, then the child pushes hard against the chest of the adult. What do you think happens? The answer is that initially the lighter child will move backwards with twice the velocity $(2V)$ of the adult who will be moving in the opposite direction with velocity $-V$. Although the child tried to push the adult backwards, it is the child who is moving backwards with the greatest velocity. This illustrates that to every force, there is an equal and opposite force acting; or put another way, to every action there is an equal and opposite reaction.

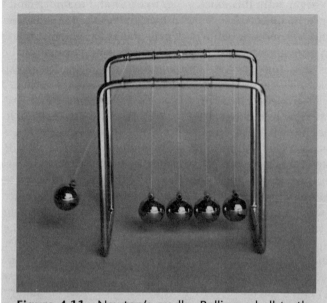

Figure 4.11 Newton's cradle. Pulling a ball to the side and letting go will transfer momentum through the others and knock the last ball out to the side, which then repeats the process in the opposite direction

Newton's cradle provides a very good illustration of Newton's laws and the conservation of momentum (Fig. 4.11). It consists of five metal balls suspended by wires from a cradle. When at rest, the balls are in a line touching each other. If the first ball is pulled to the side and then released, it hits the second so that momentum is transferred through the line of balls. The result is that the last ball in the line swings out by practically the same amount that the first was pulled to the side. The last ball then falls back and knocks the first out again, and so on. If two balls are initially pulled to one side and then released, the impact knocks the last two out of line.

 ❝ I have got that, but how can you apply the concept of momentum to liquids? The idea of two water jets hitting each other does not appear to be as sensible as two snooker balls colliding, or as simple to analyse. **❞**

Normally we are interested in problems where a body of moving liquid interacts with a solid surface, such as a pipe or a wall, and we want to calculate the force exerted on the surface. The best way to visualise the laws of solid mechanics being applied to a hydraulic system is to start from the beginning with equation (1.3), that is $F = Ma$. As just described, equation (1.3) is the same as equation (4.5):

$$F = M(V_2 - V_1)/t \tag{4.5}$$

However, there is nothing to stop us rearranging this equation so that:

$$F = \left(\frac{M}{t}\right)(V_2 - V_1)$$

Now this can be read as force equals mass per second multiplied by change in velocity. This is very convenient for hydraulic problems which involve a stream of liquid passing through a particular cross-section at a rate of Q m³/s, because the mass per second passing through the section is ρQ. You can prove this by analysing the units:

$$\frac{M}{t} = \rho Q = \frac{\text{kg}}{\text{m}^3} \times \frac{\text{m}^3}{\text{s}} = \text{kg/s}$$

Therefore we now have a form of the momentum equation suitable for hydraulic systems. Since $Q = A_1 V_1$ this can be written as either equation (4.6) or (4.7):

$$\boldsymbol{F = \rho Q(V_2 - V_1)} \tag{4.6}$$

$$\boldsymbol{F = \rho A_1 V_1(V_2 - V_1)} \tag{4.7}$$

❝ Equation (4.6/4.7) is the momentum equation written for a hydraulic system (a full derivation of the equation from first principles is given as Proof 4.2 in Appendix 1). Now there is one absolutely key concept inherent in the equation, which is that velocity is a vector quantity that has both *magnitude* and *direction*. Therefore the momentum equation must always be applied in a specified direction, such as along the x or y axis. All vector quantities that are not acting in the specified direction must be resolved until they are, and these values used in equation (4.6/4.7). This will be explained later, but first read Box 4.3 carefully. ❞ (REMEMBER!!!)

Box 4.3 ▶ **Velocity is a vector quantity**

Velocity is a vector, so magnitude and direction must be considered when applying the momentum equation. What this means in practice is that a force can arise in one of two ways, or as a combination of the two. These alternatives are listed below.

1. There is a change in velocity. An example could be a straight pipe along the x axis that reduces in diameter so that $V_{2x} > V_{1x}$. A force would be exerted on the pipe taper section since there is a change of momentum and $\Sigma F_x = \rho Q(V_{2x} - V_{1x})$.

2. There is a change in direction. Although the velocity in a pipeline of constant diameter does not change, when the pipe bends through an angle θ degrees the velocity component in (say) the x direction changes from V_1 initially to $V_1 \cos\theta$. *So a force is still exerted when there is no change in velocity but there is a change of direction.* A simple analogy would be a cricket ball glancing off your head at 90 mph. Even if the velocity of the ball was undiminished after impact, it would still exert a force on your head (and you would feel it!) because of the change in direction of the ball.

3. There is a change in velocity and direction.

Before we start to apply the momentum equation to various hydraulic problems, generally to calculate the force exerted by a moving body of water on some solid object, it is necessary to have an understanding of the concept of a control volume.

4.5.2 The concept of a control volume

❝ The control volume is a relatively simple device that helps us to apply the momentum equation to hydraulic systems. It is important that you understand it and get used to applying the momentum equation in a consistent manner that is compatible with the assumptions that have been made when deriving the equations and preparing the input data. ❞

A **control volume** is an imaginary, enclosed region within a body of flowing fluid, the shape of which can be selected to suit the problem under investigation. Suppose, for example, that we want to analyse the flow of water around the pipe bend shown in Fig. 4.12a, with the objective of calculating the resultant force, F_R, exerted by the water on the bend. The control volume for the pipe bend (Fig. 4.12b) has the same geometry as the real pipe. Fluid enters at one end of the control volume (it is conventional for the fluid to enter along the x axis) and leaves at the other end. Only the external forces acting on the control volume are shown, because inside it all of the dynamic forces cancel each other. However, when appropriate, the external forces should include the gravitational force in the form of the weight, W, of the fluid in the control volume. It is not appropriate in Fig. 4.12 because weight acts vertically and has no component in the horizontal plane of the bend, but W is included with the vertical bend in Fig. 4.13. When applying the momentum equation, the concept of the control volume can be summarised as:

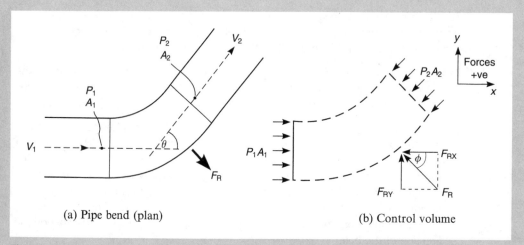

(a) Pipe bend (plan)

(b) Control volume

Figure 4.12 (a) A horizontal pipe bend which turns through an angle θ. (b) The imaginary control volume

| Box 4.4 | **Applying the control volume concept** |

The general steps necessary to set up a problem prior to analysis are:

1. Draw the hydraulic system, then draw the imaginary control volume which represents the part of the system to be analysed. If it helps, think of the control volume as cling-film enclosing part of the real system.

2. Use arrows to show the direction of travel of the liquid entering and leaving the control volume. Label them V_1 and V_2 to represent the velocities.

3. Label the axes, x and y for a two-dimensional problem in the horizontal plane, x and z for a problem in the vertical plane. The axes are positive in the initial direction of the fluid as it enters the control volume. For example, in Fig. 4.12b, x is positive from left to right and y in the upward vertical direction.

4. Draw the *external forces* acting *on* the control volume. This includes the external pressure forces (PA) acting on the ends of the pipe, and the resultant force, F_R.

5. Vector quantities such as velocity, pressure and the unknown resultant force must be resolved in the direction of the axes before the values are put into the momentum equation. If you do not know the direction in which F_{RX}, F_{RY} or F_{RZ} act initially, just guess. Having applied the momentum equation, your guess is correct if the answer obtained for F_{RX}, F_{RY} or F_{RZ} is positive and wrong if it is negative.

6. All *forces* acting in the same direction as the positive axes are positive, those acting in the opposite direction are negative. Use these signs when evaluating ΣF_X, ΣF_Y or ΣF_Z.

| The algebraic *sum* of the external forces acting *on* the fluid in the control volume in a *given direction*, ΣF | $=$ | The rate of change of momentum *in the given direction* as a result of the fluid passing through the control volume, $\rho Q(V_2 - V_1)$ | (4.8) |

Thus we ignore the equal and opposite internal forces acting on the control volume. If we write equation (4.8) mathematically for the x and y direction appropriate to Fig. 4.12b then:

$$\Sigma F_X = \rho Q(V_{2X} - V_{1X}) \tag{4.9}$$

$$\Sigma F_Y = \rho Q(V_{2Y} - V_{1Y}) \tag{4.10}$$

A more detailed explanation of how to apply the momentum equation to specific problems is given below.

4.6 Applying the momentum equation

The application of the momentum equation to the evaluation of the forces acting on pipe bends and nozzles on the end of pipelines is described below. The same principles can be applied to many other problems. The impact of a jet of water on a plate or vane is considered in section 11.2 since this forms the basis of impulse turbines.

4.6.1 Pipe bends

A common problem concerns the force exerted by the water flowing around a bend in a pipe. Because of the momentum of the moving liquid as it approches the bend (remember water weighs a tonne per cubic metre) its tendency is to continue moving in a straight line. To make it flow around the bend the pipe must exert a force on the water. This force is equal and opposite to the force that the water exerts on the pipe (Newton's first and third laws). With a large diameter pipeline the force can be considerable, and the pipeline may deform and leak if not supported. Therefore a concrete thrust block is often placed on the outside of the bend to hold the pipeline in place. If the force, F_R, is known then a thrust block of suitable size and weight can be designed from a consideration of the friction between its base and the soil and/or the passive earth pressure against its sides. Alternatively, the pipeline may be tied back to piles driven on the inside of the bend.

Let us start by considering a pipeline of uniform diameter which changes direction in the *horizontal* plane, that is on the centreline there is no change in elevation (Fig. 4.12). The control volume and the variables involved are shown in the diagram, and the procedure for obtaining the input data was described above. Assuming no loss of momentum and that there is no turbulence at the bend, for this problem equations (4.9) and (4.10) can be written specifically as:

x direction: $\quad P_1A_1 - P_2A_2\cos\theta - F_{RX} = \rho Q(V_2\cos\theta - V_1)$ $\qquad$ (4.11)

y direction: $\quad F_{RY} - P_2A_2\sin\theta = \rho Q(V_2\sin\theta)$ $\qquad$ (4.12)

Taking equation (4.11) first, the force due to pressure P_1 acting over the cross-sectional area of the pipe is P_1A_1 and this is positive because it acts from left to right along the x axis. P_2A_2 must be resolved into the x direction and the resulting component ($P_2A_2\cos\theta$) is negative because it acts from right to left. F_{RX} acts along the x axis in the negative direction. The mass of liquid flowing through the control volume per second is ρQ. The velocity of the water leaving the pipe, V_2, must be resolved into the x direction and the resulting component (V_{2X}) is $V_2\cos\theta$. The sign convention does not apply to velocities, only to forces, so $V_2\cos\theta$ remains positive. The velocity entering the control volume along the x axis (V_{1X}) is V_1, the negative sign in front of it in equation (4.11) being because we are calculating the change of momentum.

The same arguments are used to obtain equation (4.12). F_{RY} is acting vertically upwards in the positive direction. P_1A_1 has no component in the y direction ($\cos 90° = 0$) and does not appear in the equation. When P_2A_2 is resolved into the vertical direction, its component force ($P_2A_2\sin\theta$) acts downwards and so is negative. $V_2\sin\theta$ is the component of the exit velocity in the y direction (that is V_{2Y}). V_1 does not appear in the equation because it has no component when resolved through 90° into the y direction.

The resultant force, F_R, and its angle, ϕ, to the horizontal can be obtained from the equations below, enabling a suitable thrust block to be designed.

$$F_R = \left(F_{RX}{}^2 + F_{RY}{}^2\right)^{1/2}$$ $\qquad$ (1.15)

$$\phi = \tan^{-1}(F_{RY}/F_{RX})$$ $\qquad$ (1.16)

Note that because the pipe has a constant diameter the continuity equation gives $V_1 = V_2$, and if there is no loss of energy the energy equation gives $P_1 = P_2$. However, all of the terms must be included in equations (4.11) and (4.12) because they do not act colinearly and have to be resolved in the direction of the axes. If the pipeline changes diameter then

$V_1 \neq V_2$, and $P_1 \neq P_2$. Note also that in the above example gravitational forces can be ignored, because the pipeline bends in a horizontal plane. Thus the weight of the liquid in the control volume, W, does not appear in the equations because it acts vertically downwards and has no component in a horizontal plane. If, however, the centreline of the pipe changes elevation and the bend is in a vertical plane (Fig. 4.13) then the equations applicable to this system are:

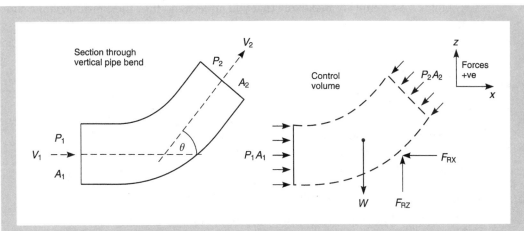

Figure 4.13 A vertical pipe bend which changes elevation, necessitating the introduction of the weight of water in the control volume, W, into the momentum equation written for the z direction

▶ **Applying the momentum equation**

It may help you to apply the momentum equation correctly if you remember that:

1. All of the terms on the left-hand side of the above equations are forces. Equation (1.2) shows that force = pressure × area, hence the terms P_1A_1, P_2A_2. P must be in N/m^2.

2. Remember the sign convention for *forces* when calculating ΣF.

3. Do not forget to resolve the vector quantities into the required direction, and that *does* include velocity.

4. The right-hand side of the equation, which is the rate of change of momentum, is the equivalent of a force. With ρQ in kg/s and $(V_1 - V_2)$ in m/s the units of this side of the equation are $kg\,m/s^2$, which is the Newton, the unit of force.

5. Take care with units: use only N (force), kg (mass), m (length) and s (time). Do **NOT** substitute values in kN (kiloNewtons) into the left side of the equation, since the right side is always in N, as explained in point 4.

x direction: $P_1A_1 - P_2A_2 \cos\theta - F_{RX} = \rho Q(V_2 \cos\theta - V_1)$ (4.11)

z direction: $F_{RZ} - W - P_2A_2 \sin\theta = \rho Q(V_2 \sin\theta)$ (4.13)

The equation for the x direction is unchanged, but there is a new equation for the z direction. The magnitude of the resultant force can be calculated, but the momentum equation gives no information regarding the location of the resultant, which must be found from an analysis of forces and moments.

Before moving on to the next section, study Example 4.4 and the general notes of guidance in Box 4.5 carefully to make sure that you understand how the control volume concept is applied.

EXAMPLE 4.4

A pipeline with a constant diameter of 0.3 m turns through an angle of 60°. The centreline of the pipe does not change elevation. The discharge through the pipeline is 0.1 m³/s of water, and the pressure at the bend is 30 m of water. Calculate the magnitude and direction of the resultant force on the pipe.

Step 1 Apply the continuity equation:
$Q = A_1V_1 = A_2V_2$ where $A_1 = A_2 = (\pi 0.3^2)/4 = 0.071\,\text{m}^2$
$0.1 = 0.071 \times V_1$
$V_1 = 1.41\,\text{m/s}$ $(= V_2)$

Step 2 Apply the momentum equation in the x and y directions, remembering the sign convention.
$P_1 = P_2 = \rho gh = 1000 \times 9.81 \times 30 = 294.30 \times 10^3\,\text{N/m}^2$

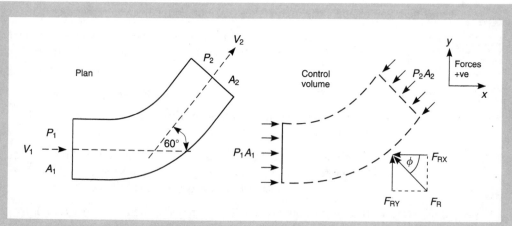

Figure 4.14 A 60° pipe bend in the horizontal plane with control volume and forces

For the x direction:

$$P_1A_1 - P_2A_2\cos\theta - F_{RX} = \rho Q(V_2\cos\theta - V_1)$$
$$294.30\times10^3\times0.071 - 294.30\times10^3\times0.071\cos60° - F_{RX} = 1000\times0.1(1.41\cos60° - 1.41)$$
$$20.90\times10^3 - 10.45\times10^3 - F_{RX} = 10^3\times0.1\times(-0.71)$$
$$F_{RX} = 10.52\times10^3\,\text{N}$$

(the +ve answer indicates that F_{RX} is acting in the direction assumed)

For the y direction:

$$F_{RY} - P_2A_2\sin\theta = \rho Q(V_2\sin\theta)$$
$$F_{RY} - 294.30\times10^3\times0.071\sin60° = 1000\times0.1(1.41\sin60°)$$
$$F_{RY} - 18.10\times10^3 = 10^3\times0.1\times1.22$$
$$F_{RY} = 18.22\times10^3\,\text{N}$$

(the +ve answer indicates that F_{RY} is acting in the direction assumed)

Step 3 Calculate the magnitude and direction of the resultant force, F_R.

$$F_R = \left(F_{RX}^2 + F_{RY}^2\right)^{1/2} = 10^3(10.52^2 + 18.22^2)^{1/2} = 21.0\times10^3\,\text{N}$$
$$\phi = \tan^{-1}(F_{RY}/F_{RX}) = \tan^{-1}(18.22/10.52) = 60°\text{ to the horizontal.}$$

This is the magnitude and direction of the external force exerted by the pipe on the water, as shown in Fig. 4.14. The water exerts an equal force on the pipe in the opposite direction.

4.6.2 Nozzles

The momentum equation can also be used to investigate the force exerted by a stream of moving liquid on a nozzle, such as that shown in Fig. 4.15. The liquid tries to force the nozzle off the end of the pipe, so it must be held in place by, for example, a number of bolts around the flange of the pipe. If the force, F_R, is calculated and it is known what tensile force one bolt can withstand, then a suitable connection with an appropriate number of bolts can be designed.

In this situation there is a change in velocity as a result of the reducing area of the nozzle, but no change in direction. The momentum equation only needs to be applied in the direc-

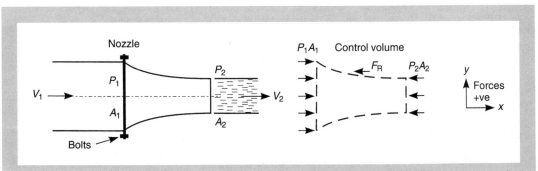

Figure 4.15 Pipe nozzle and control volume showing the external forces acting on the system

---

Box 4.6 ▶ **Try this experiment**

Get a balloon, blow it up, then let go of it. It will probably whizz around the room following quite an erratic path. The balloon is rocket powered. The increase in velocity (acceleration) of the air escaping from the balloon results in a force being applied to the balloon, which is why it flies. It follows an erratic path because the nozzle of the balloon is flexible and the flow of air from it is not perfectly symmetrical or steady, so there is a change of momentum (and hence a force) in more than one direction. A similar effect can be observed with a hose pipe which terminates in a nozzle. Unless held in position it will 'snake' around. It may take several firemen to hold the nozzle of a high pressure hose steady. Some garden sprinklers use the nozzle/rocket principle to make them revolve.

---

tion of motion (along the x axis) so the subscripts are omitted. Using the same sign convention as before, forces are positive if acting from left to right, so:

$$P_1A_1 - P_2A_2 - F_R = \rho Q(V_2 - V_1) \tag{4.14}$$

This can be simplified by assuming that because the nozzle is discharging to the atmosphere the pressure of the jet, P_2, will equal atmospheric pressure. Since we are using gauge pressure with atmospheric pressure as the datum (= 0) then $P_2 = 0$. The equation then becomes:

$$F_R = P_1A_1 - \rho Q(V_2 - V_1) \tag{4.15}$$

This enables the force, F_R, to be calculated quite easily, as shown in Example 4.5.

The force exerted by a nozzle is quite an interesting application of the momentum equation, since this is basically the principle of the rocket. A rocket is a good example of Newton's third law, that action and reaction are equal and opposite. If a fluid (air, gas, water) is accelerated out of a container at high velocity, then a force will be exerted in the opposite direction on the container e.g. F_R acts from right to left in Fig. 4.15.

Now read through Box 4.6, make sure you understand Example 4.5, and then try Self Test Question 4.3.

EXAMPLE 4.5

A pipeline reduces in diameter using a standard, symmetrical taper section as shown below. Given the following information, calculate the force exerted by the water on the taper section: $Q = 0.42\,m^3/s$, $D_1 = 0.60\,m$, $D_2 = 0.30\,m$, $P_1 = 25.30\,m$ water, $P_2 = 22.85\,m$ water, $\rho = 1000\,kg/m^3$.

Step 1 Apply the continuity equation:

$Q = A_1V_1 = A_2V_2$

$A_1 = (\pi 0.60^2)/4 = 0.283\,m^2$ and $A_2 = (\pi 0.30^2)/4 = 0.071\,m^2$

$0.42 = 0.283V_1$ so $V_1 = 1.48\,m/s$

$0.42 = 0.071V_2$ so $V_2 = 5.92\,m/s$

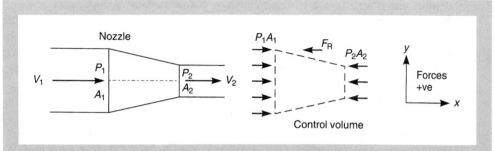

Figure 4.16 Standard reducer section in a pipe with control volume showing the external forces

Step 2 Apply the momentum equation in the direction of motion

$P_1 = \rho g h_1 = 1000 \times 9.81 \times 25.30 = 248.19 \times 10^3 \text{ N/m}^2$

$P_2 = \rho g h_2 = 1000 \times 9.81 \times 22.85 = 224.16 \times 10^3 \text{ N/m}^2$

$P_1 A_1 - P_2 A_2 - F_R = \rho Q(V_2 - V_1)$

$248.19 \times 10^3 \times 0.283 - 224.16 \times 10^3 \times 0.071 - F_R = 1000 \times 0.42(5.92 - 1.48)$

$70.24 \times 10^3 - 15.92 \times 10^3 - F_R = 1.86 \times 10^3$

$F_R = 52.46 \times 10^3 \text{ N}$ (from right to left)

The internal force exerted by the water on the taper is $52.46 \times 10^3 \text{N}$ from left to right.

SELF TEST QUESTION 4.3

For the situation shown in the diagram, calculate the force exerted by the water on the nozzle given that:

Discharge $Q = 0.65 \text{ m}^3/\text{s}$
Diameter $D_1 = 0.30 \text{ m}$ and $D_2 = 0.20 \text{ m}$
Pressure $P_1 = 129.4 \times 10^3 \text{ N/m}^2$
Density $\rho = 1000 \text{ kg/m}^3$

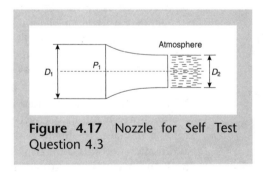

Figure 4.17 Nozzle for Self Test Question 4.3

4.6.3 Rocket and jet engines

At some time in their life many people will have blown up a balloon and then let it go. The result is usually that the balloon zooms about following an erratic path, as described in Box 4.6. This is Newton's second and third laws in action. These laws explain how and why rockets work in the vacuum of space (see also the next section regarding conservation of momentum). A common mistake used to be to assume that rockets would only work when they had something (air) to press against.

Unlike our simple balloon, rockets such as those used by NASA's Space Shuttle manufacture gas by burning solid or liquid fuels. However, once the gas has been produced the principle is the same. As the hot exhaust gas blasts backwards out of the rockets, the equal and opposite reaction pushes the Shuttle forwards. In principle rockets are deceptively simple, and can consist of little more than tanks to store the fuels, pumps to inject them into a combustion chamber, a method of ignition and an exhaust nozzle. There are no moving parts other than the pumps. However, one of the difficulties of rocket flight is maintaining a stable course. Some of the early rockets followed a path almost as erratic as that of our balloon.

A jet engine is similar to a rocket, and works on the same reaction principle. With the turbojet, air is sucked into the engine and compressed by a fan (the compressor). Kerosene fuel is then sprayed into the compressed air and ignited. This produces a large volume of hot, rapidly expanding gas, which is discharged at high speed backwards through the exhaust nozzle at the rear of the engine. As the gas is discharged it turns the blades of a turbine, which is connected by a drive shaft to the compressor. This shaft, with its compressor and turbine blades, forms the only moving part of the engine, which makes it ingeneous and relatively simple compared to a piston engine that has many moving parts. The turbofan engine is similar to the turbojet, but has another fan attached to the drive shaft at the front of the engine. This forces air around the central engine core as well as through it, so it has a larger discharge of air and exhaust gas than a turbojet, but at a slower speed. For slower, passenger aircraft turbofans are quieter, cooler and more efficient; turbojets suit faster aircraft such as fighters. Jet engines are most efficient at high altitudes where the air is thinner and provides less resistance to the discharging gas.

An interesting historical note is that by angling backwards the exhaust nozzles of a conventional piston engined propeller aircraft, the exhaust jets could give a fast airplane an additional thrust of up to 10% (Hunsaker and Rightmire, 1947). If you get the chance, look at the exhausts on the famous Spitfire. The thrust produced by a rocket or jet engine can be calculated using the momentum equation, in much the same way that the force on a nozzle was calculated in section 4.6.2. The thrust or force is proportional to the mass of the gas discharged multiplied by the acceleration imparted to it (i.e. $F = Ma$). It should be no surprise that the principle works equally well with water, and there are many boats, jetfoils and jetskis which are powered by high-velocity water jets.

For a beginner's guide to propulsion, try the following interactive and animated NASA/Glenn Research Center internet site: **http://www.grc.nasa.gov/WWW/K-12/airplane/shortp.html**. This covers aerodynamics, propulsion systems (rocket, propeller, gas turbine and ramjet), aircraft motion, thrust (rocket, propeller, turbojet, turbofan, turboprop and ramjet) and fundametals (i.e. Newton's laws). For more information on rocket motors try **http://www.im.lcs.mit.edu/rocket/intro.html**.

4.6.4 Conservation of momentum

Conservation of momentum provides an alternative way of looking at Newton's laws of motion. As described earlier, the law of the conservation of momentum states that a body in motion cannot gain or lose momentum unless some external force is applied. Previously the example of a snooker ball of mass M rolling at a constant 0.3 m/s was used to show that, without the application of an external force, at all points along its path it would have a momentum of 0.3 M kg m/s. When considering the flow of a continuous fluid it is easiest to compare the momentum flow rate $\rho Q V$ kg m/s per second at different cross-sections. Thus with a steady flow of fluid in a straight pipe of constant diameter, if no external force is applied, at two sections of the pipe the momentum flow rate would be the same so $\rho_1 Q_1 V_1 = \rho_2 Q_2 V_2$. Now consider what happens with the nozzle in Fig. 4.17. Here $\rho_1 = \rho_2 = 1000\,\text{kg/m}^3$, $Q_1 = Q_2 = 0.65\,\text{m}^3/\text{s}$ and (from Appendix 2) $V_1 = 9.196\,\text{m/s}$ and $V_2 = 20.691\,\text{m/s}$. This gives $\rho_1 Q_1 V_1 = 1000 \times 0.65 \times 9.196 = 5977.4\,\text{kg m/s}^2$ and $\rho_2 Q_2 V_2 = 1000 \times 0.65 \times 20.691 = 13\,449.2\,\text{kg m/s}^2$. Very clearly the momentum flow rate at the two cross-sections is not the same; to make it the same there has to be an external force acting at section 2 to balance the increase in momentum. This is, of course, the external reaction force F_R which acts from right to left, as in Fig. 4.15. It is this force that would thrust a rocket forward as a reaction to the jet discharging in the opposite direction. Thus the concept of the conservation of momentum shows why a rocket works, even in the vacuum of space.

The nozzle above concerned a fluid flowing in a straight line, but the same basic concept can be applied to Example 4.4 where water is flowing around a bend of constant diameter. The rate of momentum per second entering the system is $\rho_1 Q_1 V_1$ and the momentum per second leaving is $\rho_2 Q_2 V_2$. Since ρ and Q are constant and $V_1 = V_2$ it follows that there is no overall loss of momentum. This would be true even if there was a loss of energy and P_2 as a consequence was less than P_1. This fact means that the momentum equation can be applied to hydraulic systems where there is a loss of energy, because momentum and energy are not the same thing, as described earlier. However, note that in Example 4.4 the quantity of momentum entering and leaving in the x and y directions is not equal. For instance, the water entering the bend along the x axis has no momentum in the y direction, but there is momentum in the y direction after the bend. It is this inbalance in a particular direction that results in the component forces F_{RX} and F_{RY} according to the general equations:

$$\Sigma F_X = \rho Q (V_{2X} - V_{1X}) \tag{4.9}$$

$$\Sigma F_Y = \rho Q (V_{2Y} - V_{1Y}) \tag{4.10}$$

Thus it is not easy to apply the law of conservation of momentum to fluid systems, except through the application of equations such as (4.9).

4.6.5 The momentum coefficient

So far it has been assumed that the mean velocity $V = Q/A$, that is the discharge divided by the cross-sectional area of flow. However, in reality when liquid flows through a conduit the velocity at the boundary surface is zero and the velocity increases with distance from the boundary, so the velocity is not constant over the whole area of flow (Fig. 4.5). In many instances this fact is not particularly significant, but in some situations it may be. Furthermore, it is one way in which a real fluid differs significantly from an ideal one (see section 4.1).

To allow for the non-uniform distribution of velocity within the conduit a **momentum coefficient**, β, can be introduced into equation (4.6), and all other equations based on it, thus:

$$\Sigma F = \beta \rho Q (V_2 - V_1) \tag{4.16}$$

β always has a value greater than 1.00, but frequently not much greater, so it is often omitted. Some typical values are shown in Table 4.2 (section 4.8.1). β is defined as:

$$\beta = \Sigma (v^2 \mathrm{d}A) / V^2 A \tag{4.17}$$

where v is the local velocity over a small part of the area of flow, dA. The significance of the non-uniform distribution of velocity with respect to the momentum equation is illustrated by Example 4.6.

EXAMPLE 4.6

A jet of water emerges from a nozzle. (a) If there is uniform velocity, V, across the whole of the jet of area A, what is the momentum flow rate? (b) If the velocity over the central half of the jet (area $A/2$) is $1.5V$ and the velocity over the outer half is $0.5V$, what is the momentum flow rate now?

(a) Momentum = mass × velocity

Momentum flow rate = mass flow rate × velocity

$$= \rho QV \text{ and since } Q = AV$$

$$= \rho AV^2 \tag{4.18}$$

(b) From equation (4.18), for the revised condition the momentum flow rate is:

$$= \rho \times 0.5A(1.5V)^2 + \rho \times 0.5A(0.5V)^2$$

$$= 1.125\rho AV^2 + 0.125\rho AV^2$$

$$= 1.25\rho AV^2$$

There is a 25% increase in the rate of flow of momentum, despite the fact that the quantity of flow is the same in both cases, that is:

$$Q = 0.5A \times 1.5V + 0.5A \times 0.5V = 0.75AV + 0.25AV = AV \text{ as in part (a).}$$

Thus the velocity distribution affects the momentum flow rate.

4.7 The energy (or Bernoulli) equation

66 The energy equation, also known as the Bernoulli equation, is the third major tool that we can use to analyse a hydrodynamic system. The other two are the continuity equation and the momentum equation. Sometimes two, or perhaps all three, may be needed to solve a particular problem. **99**

66 I do not really understand what energy is. Can you explain please? Then can you tell me what the energy equation is and how it is used? **99**

4.7.1 Understanding energy and the energy equation

Energy is defined as the capacity for doing work. This is pretty much as in the common English usage: if you are tired and do not feel like working you might say that you haven't the energy or that you do not feel energetic. **Work** (done) is defined as a force multiplied by the distance moved in the direction of the force, and consequently has the units Nm.

Power is the rate of doing work, that is, the product of a force and the distance moved per second in the direction of the force (Nm/s). A good example is the power exerted by a jet of water on the turbine runner in section 11.2.3.

There are three ways that something can possess energy. Perhaps the easiest to understand is that a body can have energy as a result of being raised to some height, z. Thus if a car is driven to the top of a hill, it can freewheel down again and do work by virtue of its elevation. This is called the **potential energy** of the body. The following equation may be familiar to many students from school science classes:

$$\text{potential energy} = Mgz \tag{4.19}$$

where M is the mass of the body and g is the acceleration due to gravity. The product Mg is the weight of the body, W.

Another form of energy is **kinetic energy**, that is the energy possessed by a moving body. Quite obviously a moving vehicle has energy and the ability to do work. A car travelling at high speed can freewheel some distance up a hill, for example. Again the familiar form of the equation may be:

$$\text{kinetic energy} = \frac{1}{2}MV^2 \tag{4.20}$$

where V is the velocity of the body.

The third form of energy will be less familiar since it has no direct equivalent in solid mechanics. It is the energy of a fluid when flowing under pressure, so it is sometimes referred to as **pressure energy** for short. For example, water escaping from a high pressure water-main can be very damaging and scour out a large hole. If the water was not under pressure it would not do so. The origin of the pressure can be explained thus. If an incompressible liquid flows at a steady rate through a pipe of constant diameter running completely full, the continuity equation tells us that the mean velocity is constant along the pipeline. Therefore, if the liquid loses elevation suddenly the surplus potential energy cannot be converted into kinetic energy (because unlike a solid object, the velocity of the liquid cannot increase) so it has to be converted into pressure. If a liquid has a pressure P and acts over an area A then it is capable of exerting a force of PA. In travelling through a distance L the **flow work** done is PAL. Thus:

$$\text{pressure energy} = PAL \tag{4.21}$$

Liquids are capable of having potential, kinetic and pressure energy. Indeed, this is the basis of pumped storage hydroelectric schemes where water is pumped up to a high reservoir, which stores the potential energy of the water. When electricity is required, water is released from the reservoir into steep tunnels or pipes so that it gains kinetic and pressure energy, which can then be used to turn a turbine (see impulse and reaction turbines in section 11.1.3).

If we add together the three types of energy, we have the **total energy** possessed by the liquid. However, before we do this it is necessary to rewrite the equations to get them into a more convenient form to use with liquids. The first step is to replace the mass, M, in equations (4.19) and (4.20) with the corresponding weight, W. It is possible to substitute W straight into equation (4.19) for Mg, while in equation (4.20) M can be replaced by W/g. If we think of equation (4.21) as representing the pressure energy of a stream of moving liquid, then AL represents the volume of the liquid. If this volume has a weight, W, and the weight density of the liquid is ρg then $AL = W/\rho g$. Substituting this for AL in equation (4.21), then adding the three equations together to get the total energy of a body of liquid of weight W and density ρ gives:

$$Wz + \frac{1}{2}\frac{WV^2}{g} + \frac{PW}{\rho g}$$

This equation is written in terms of an unspecified weight, W (say 5 N). It would be much better to write the equation so that it represented the energy of a unit weight (1 N) of the liquid. All we have to do to obtain the equation in this form is to divide all the terms by W, thus:

$$z + \frac{V^2}{2g} + \frac{P}{rg} = \textbf{total energy per unit weight of fluid} \tag{4.22}$$

This is the energy or Bernoulli equation. Whenever possible it has been derived above using equations that you may already be familiar with. Perhaps the terms in the equation are those that you expected? You may have noticed that equation (4.22) is the hydrostatic equation (equation (1.22)) with an additional term to represent the kinetic energy of a moving liquid. Note that equation (4.22) says 'fluid' because **the Bernoulli equation also applies to gases** like air, and we will consider some examples involving air later. Now you should turn to Appendix 1 and look through the formal derivation of the Bernoulli equation (Proof 4.3), which is an application of the momentum equation.

❝ How can you actually apply the Bernoulli equation to a problem? **❞**

Excellent question. We have to be a little pragmatic in the way we apply the equation. The equation is frequently used to investigate how the energy varies between two points in a fluid. Strictly speaking, we should apply the Bernoulli equation only between two points on the same streamline. In reality, we cannot see streamlines. A compromise is to apply the equation between two points on (say) the centreline of a pipe (Fig. 4.18). This minimises the effect of different streamlines within the cross-section having different elevations and pressures. So when we talk about applying the Bernoulli equation between two cross-sections or two points, we should really be applying it to a streamline.

❝ Does the fact that we are not applying the Bernoulli equation to a streamline matter? Does it affect the accuracy of an analysis? **❞**

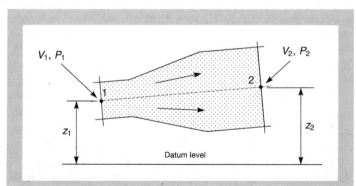

Figure 4.18 Application of the Bernoulli equation to a pipe

There are so many other assumptions and difficulties involved in applying the Bernoulli equation that this one is not especially significant. For instance, we are again assuming that V is the mean velocity of flow ($= Q/A$) and that the flow is uniform (later we will introduce a velocity distribution coefficient, α, to allow for non-uniform velocity). Of course, we still have the problems of friction, turbulence and energy losses to contend with. All of these add a degree of uncertainty, so part of the skill of conducting a successful analysis is to understand the equations, their limitations, and to know what assumptions can be justified. Experience and practice helps here. Now suppose that we want to analyse the problem in Fig. 4.18. We can apply Bernoulli to two points on the centreline, 1 and 2. If we assume that the total energy at points 1 and 2 is the same (that is there is no loss of energy, as with an ideal fluid) then, although energy may change from one form to another, the following is true:

$$z_1 + \frac{V_1^2}{2g} + \frac{P_1}{\rho g} = z_2 + \frac{V_2^2}{2g} + \frac{P_2}{\rho g} \tag{4.23}$$

With a real fluid, if we want to find out if there is an energy loss, quantify it, or if we know what the energy loss is and want to investigate its effect on the other variables in the equation, then equation (4.23) becomes:

$$z_1 + \frac{V_1^2}{2g} + \frac{P_1}{\rho g} = z_2 + \frac{V_2^2}{2g} + \frac{P_2}{\rho g} + \textbf{energy head losses} \tag{4.24}$$

There are equations that enable us to quantify energy losses in pipe expansions, contractions and bends, losses due to friction, and so on (see Chapter 6). These losses are added to the right-hand side of equation (4.24) so that all of the energy in the system is accounted for.

❝ Students often ask if P in the above equation should be in N/m² or m head. The answer is simple: P itself must be in N/m², but all three terms of the equation have the overall unit of metres. Thus they are often referred to as heads: elevation head, velocity head and pressure head. **❞**

(REMEMBER!!!)

$$z = \text{elevation} = \text{m}$$

$$\frac{V^2}{2g} = \left(\frac{\text{m}}{\text{s}}\right)^2 \times \frac{\text{s}^2}{\text{m}} = \text{m}$$

$$\frac{P}{\rho g} = \frac{\text{N}}{\text{m}^2} \times \frac{\text{m}^3}{\text{kg}} \times \frac{\text{s}^2}{\text{m}} = \frac{\text{kg} \times \text{m}}{\text{m}^2 \times \text{s}^2} \times \frac{\text{m}^3}{\text{kg}} \times \frac{\text{s}^2}{\text{m}} = \text{m}$$

Thus all three terms are measured in metres and can be called heads. The sum of the three terms is often called the 'total head' as an alternative to the 'total energy'.

4.7.2 Understanding the relationship between velocity and pressure

It is worthwhile spending a few moments thinking about the relationship between velocity and pressure. Suppose that the pipe in Fig. 4.18 was horizontal so that there was no change in elevation of its centreline. The z_1 and z_2 terms in equation (4.23) would then be equal, so if there is no loss of energy:

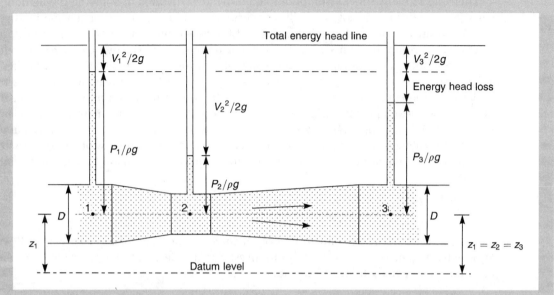

Figure 4.19 Venturi meter with its centreline horizontal. The liquid pressure is shown by the level in the piezometer tubes

$$V_1^2/2g + P_1/\rho g = V_2^2/2g + P_2/\rho g \tag{4.25}$$

This shows that if in Fig. 4.18 the **velocity decreases** between points 1 and 2 ($V_2 < V_1$), then the **pressure increases** ($P_2 > P_1$) so that the total energy remains the same on both sides of the equation. Although the equation shows mathematically the inverse relationship between velocity and pressure, this can be difficult to understand intuitively. For instance, Fig. 4.19 shows a Venturi meter, which consists of a tapered entrance section, a parallel sided throat, and then a gradually diverging exit section. The meter is commonly used to measure the flow of water in a pipe. It also demonstrates the Bernoulli equation nicely. If the centreline of the meter is horizontal the elevation, z, can be ignored. The pressure through the meter can be observed from the level of the water in the piezometer tubes.

The diagram shows that the pressure is initially high as water enters the meter. From a consideration of the continuity equation it is apparent that as the diameter of the meter reduces, the flow velocity, and hence the velocity head $V^2/2g$, must increase. Therefore, the pressure decreases, as shown by the water level in the middle piezometer. The velocity will be highest in the throat so this is where the pressure is lowest. As water leaves the throat and the velocity decreases, the pressure starts to rise again. To satisfy the continuity equation, the entrance and exit velocities must be equal since the pipe diameter (D) is the same, but the final pressure does not quite equal that at the entrance because of a small energy loss in the meter. **Note that the energy loss has to take the form of a reduction in pressure.**

Most students instinctively feel that the pressure ought to increase as the throat is approached due to the fact that the water is being forced or jammed through a narrow opening. In fact the opposite occurs. Usually fluids can flow into constrictions relatively easily without loss of energy, it is at the exit where the cross-section expands that most of

Box 4.7 ## Try this experiment for yourself

This is a simple experiment that only requires two sheets of A4 paper. Hold the sheets with the short sides vertical so that they are on either side of your mouth. Keep the sheets parallel to each other and about 50 mm apart. The next step is to blow between the two sheets of paper.

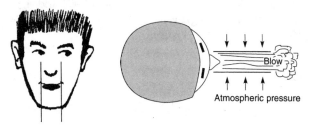

Atmospheric pressure

Figure 4.20 Blow between two sheets of paper, held vertically, a few centimetres apart

What do you think will happen?

Most people think that if you blow between the sheets of paper the air will force them apart. You should find that the opposite happens. Because the air blown between the sheets is travelling at a relatively high velocity, the air pressure is relatively low. In fact, the air pressure between the sheets is less than the pressure of the still air on the outside of the paper. Thus atmospheric pressure forces the sheets together. So you should find the two pieces of paper move together with no gap at all between them.

This experiment illustrates the so called 'ground effect' which helps Formula 1 racing cars to corner at well over 100 mph. Air is forced at high velocity through the small gap between the car's undertray and the road. This means that the air pressure in the gap is much lower than atmospheric pressure and so the car 'sticks' to the track, rather like your two pieces of paper sticking together. Of course, a Formula 1 car has a few other tricks as well. It has big sticky tyres, and wings on the front and rear to create even more down force. The latter is explained in Box 4.8.

Rear 'wing'

Front 'wing'

Figure 4.21 Air flow under a car causing the 'ground effect' to improve road-holding

the energy loss occurs. To minimise the loss, the meter expands gradually (more slowly than it contracts). The reason the loss occurs is because at the exit of the meter there is an **adverse pressure gradient** – that is the pressure is increasing in the direction of flow. Thus the pressure is trying to push water back into the meter (which is why the water slows in the first place). This causes turbulence, eddying, and a non-uniform distribution of velocity, all of which result in a loss of energy (see section 5.2.1).

Now, to help you get a better understanding of the relationship between velocity and pressure, try the experiments described in Boxes 4.7 and 4.8. These are designed to illustrate that *where there is a high velocity there is a low pressure*. Similarly, *where there is a low velocity there is a high pressure*.

Box 4.8 ▶ ## Now try this experiment as well

Everyone is familiar with the idea of an air gun where air can be used to propel a pellet quite a distance. You can achieve a similar effect if you roll up a piece of paper into a tube, put a ping-pong (table tennis) ball inside it, then blow hard down the tube. The ball can be blown a considerable distance if you get the size of the tube just right. This uses the same idea as a pea-shooter. Simple. Everyone knows what is going to happen.

Want to win £5 with a party trick?

OK. Get an ordinary funnel, like the ones used to pour wine into a bottle for example. Keep the narrow end pointing vertically down, put a ping-pong ball in the wide end, then challenge someone to blow through the narrow end hard enough for the ping-pong ball to hit the ceiling. If they can do it you give them £5, if they cannot even get the ball to leave the funnel they give you £5.

You should win £5 every time.

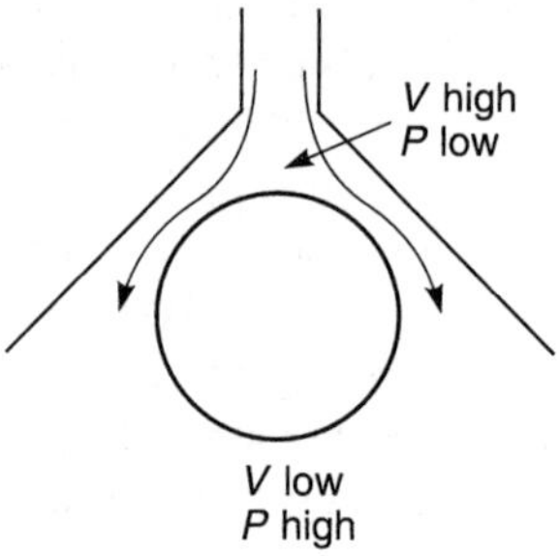

It is impossible to blow the ball out of the funnel. You can prove this by turning the funnel upside down, so that the wide end is now pointing at the ground. Of

V high
P low

V low
P high

Figure 4.22 Blowing through a funnel at a ping-pong ball

course the ball falls out! So hold it in place until *after* you have started blowing hard down the small end of the funnel. As long as you blow hard enough, the ping-pong ball will remain in the funnel. The reason is that as a result of blowing down the funnel the air velocity near the narrow end is relatively high, so the air pressure is low. At the wider end of the funnel the air is moving much more slowly, so the air pressure is higher. This difference in pressure is enough to keep a light ping-pong ball held in place. The same basic effect provides the lift to keep an aeroplane in the air. The aerofoil is shaped so that the flow-path over the wing is longer than underneath, so the velocity over the wing has to be higher than that underneath to maintain continuity of flow. As it starts to move, an aerofoil creates a starting vortex downstream of the wing. To balance this, another vortex, the bound vortex, forms around the aerofoil, as shown in the diagram. Thus when air with a relative velocity, V, flows past the wing there is a relatively high velocity on the top so the pressure is correspondingly relatively low compared to the underside of the wing. This generates the lift, which increases as V increases (see equation (4.29)), which is why aircraft take off into the wind. It is also argued that the shape and angle of attack of the wing are such that the air flowing off it is angled slightly downwards, so by Newton's third law there is a thrust (lift) in the upward direction. On a Formula 1 car the aerofoil has to be used upside down, of course, to generate downforce not lift.

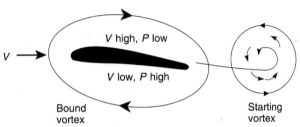

Figure 4.23 Flow of air over an aeroplane wing generating lift

4.8 Applying the energy equation

There are many problems to which the energy equation may be applied, so it is impossible to list them all. Some applications are illustrated by the examples in this chapter. Chapter 5 applies the equation to flow measurement using a Venturi meter, weir and orifice. Chapter 6 applies the equation to flow through pipes: an important application with respect to water supply.

66 When it comes to applying the energy/Bernoulli equation there are a few tricks that you should learn and remember. These will help you to analyse a hydraulic system. Remember, the initial objective should be to minimise the number of variables with an unknown value. If you have more unknown variables than equations then you cannot solve the problem. Remember that the continuity equation may give you the value of one or more of the variables, and sometimes the momentum equation may be of use also. Now as far as the energy equation itself is concerned, read the points in Box 4.9. Make sure that you understand all of them, and remember them. They can make the difference between being able to solve a problem, and not being able to solve it. Work through Example 4.7 to see how they are used. 99

(REMEMBER!!!)

Box 4.9 **Remember – when applying the energy equation**

1. Apply the Bernoulli equation in such a way as to minimise the number of unknown variables. If energy losses are ignored there are six variables (z_1, V_1, P_1, z_2, V_2, P_2). After the use of other equations, such as the continuity equation, there must be only one unknown to be able to solve the problem. *You* can select which two points to use in the analysis, so use two points where there is only one unknown between them. The notes below will help with the selection process.

2. Many problems involve a tank or a reservoir with a *free water surface*. Normally we work with gauge pressure which uses atmospheric pressure as a datum, so if a point is selected on the water surface the *pressure is atmospheric* and $P = 0$.

3. If a pipe or nozzle *discharges to the atmosphere* and the jet has a constant diameter then it can be assumed that the water pressure in the jet is the same as the surrounding atmosphere. So if gauge pressure is used, $P = 0$.

4. Following on from point 2 above, with *large tanks or reservoirs* the velocity on the water surface can be assumed to be zero, so $V = 0$. For example, if water is being drained out of a reservoir via a pipe, if the reservoir is described as 'large' this means that the velocity at the surface is not affected by the water entering the pipe and can be taken as zero.

5. If desired, the *datum* from which elevation is measured can be taken through the lower of the two points being used in the analysis, so either z_1 or $z_2 = 0$. This may not be important if the elevations are known, it simply results in one less term in the equation.

6. Use the five points listed above to mark on a drawing of the hydraulic system the known values and the unknown values. This should then enable you to select two points to which you can apply the Bernoulli equation. Example 4.7 provides a good illustration of the procedure.

EXAMPLE 4.7

Water is drained from a large reservoir using a syphon, as shown in the diagram. The end of the syphon pipe is 3.2 m below the water level in the reservoir. At the highest part of the syphon the centreline of the pipe is 2.3 m above the water surface. The pipe has a diameter of 200 mm and it discharges to the atmosphere. Assuming that the water level in the reservoir remains constant and that there are no energy losses, calculate the discharge through the syphon and the pressure head at the crest (the highest part) of the syphon.

Step 1 Mark the values of the known variables on a drawing of the system, as above. Using the notes in Box 4.9, on the surface of the reservoir, $V_1 = 0$. On the reservoir surface and at the end of the pipe the pressure is atmospheric so $P_1 = 0$, $P_2 = 0$.

Step 2 Decide which two points to use in the analysis. In this case both the velocity and pressure at the crest of the syphon are unknown, so this point can *not* be used in the analysis. All of the variables on the reservoir surface are known, while at the end of the pipe only the velocity is unknown. Therefore, first apply the energy equation to points 1 (reservoir surface) and 2 (end of pipe), then apply the equation between 1 and 3 (crest).

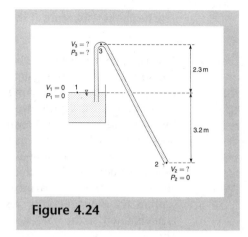

Step 3   Apply Bernoulli between points 1 and 2, ignoring losses of energy, to get V_2. Taking the datum level through point 2, $z_1 = 3.2$ m and $z_2 = 0$. The other values are as in Step 1.

$$z_1 + V_1^2/2g + P_1/\rho g = z_2 + V_2^2/2g + P_2/\rho g$$
$$3.2 + 0 + 0 = 0 + V_2^2/2g + 0$$
$$V_2 = (2g \times 3.2)^{1/2} = 7.924 \text{ m/s}$$

Figure 4.24

Step 4   Apply the continuity equation:

$$Q = A_2 V_2 = A_3 V_3 \text{ where } A_2 = A_3 = (\pi 0.2^2)/4 = 0.0314 \text{ m}^2$$

Therefore $V_3 = 7.924$ m/s (the same as V_2 since $A_2 = A_3$)

$$Q = 0.0314 \times 7.924 = 0.249 \text{ m}^3/\text{s}$$

Step 5   Apply the energy equation between points 1 and 3 to obtain $P_3/\rho g$, the pressure head in m.

$$z_1 + V_1^2/2g + P_1/\rho g = z_3 + V_3^2/2g + P_3/\rho g$$

Taking the datum level through point 1, $z_1 = 0$ and $z_3 = 2.3$ m

$$0 + 0 + 0 = 2.3 + (7.924)^2/2g + P_3/\rho g$$
$$P_3/\rho g = -2.300 - 3.200 = -5.500 \text{ m of water}$$

The pressure head at the crest is below atmospheric pressure by the equivalent of 5.500 m of water (see Fig. 1.8).

❝ Have you ever wondered how water can flow uphill when you syphon something? Or why the water keeps flowing through the syphon? Example 4.7 not only provides a good illustration of how to apply the energy equation, it also shows how and why a syphon works. ❞

The first thing to note is that if in step 3 the end of the pipe, and hence point 2 and the datum level, was at the same height as point 1 on the reservoir surface then we would have $z_1 = z_2 = 0$. As a result, $V_2^2/2g = 0$ so $V_2 = 0$. In other words, the syphon will not work if the end of the pipe is at the same level as the water surface in the reservoir, or above it. It only works when the pipe is below the water surface. Secondly, note that the pressure at the crest of the syphon is −5.5 m of water relative to the atmosphere. Atmospheric pressure is the equivalent of about 10.3 m of water or 101 × 10^3 N/m². So the pressure at the crest is about half way between a vacuum and atmospheric pressure.

The absolute pressure at the crest $= 1000 \times 9.81 \times (-5.5) + 101 \times 10^3$
$$= 47.0 \times 10^3 \, \text{N/m}^2$$

This is, of course, also less than atmospheric ($47.0 \times 10^3 < 101 \times 10^3 \, \text{N/m}^2$). If you are not too sure about negative gauge pressures and sub-atmospheric pressures, then go back to Fig. 1.8 and the accompanying text.

The reason the syphon works is as follows. Once the syphon is primed (i.e. the pipe is full of liquid), the water between points 2 and 3 will flow out as a result of gravity. As it does so, this will leave an empty pipe near the crest, that is a partial or complete vacuum. Since atmospheric pressure is acting on the surface of the reservoir, water is forced up the syphon to point 3; this is analogous to a water barometer which requires a water column about 10.3 m high to balance atmospheric pressure. With continuous flow through the syphon, the elevation of its crest above the reservoir surface means that a negative pressure is maintained, so the water keeps on flowing until something happens to break the syphon. This may happen if the crest is too high and the pressure falls below -7.5 m of water (e.g. if in step 5 of Example 4.7 the value of $z_3 = 2.3$ m is replaced by $z_3 > 4.3$ m) causing vapour or air to become trapped at the crest. A large vapour/air lock would stop the flow. Syphons are also discussed in sections 6.2 and 9.1.3.

❝ Before working through Examples 4.8 to 4.10, remember that whenever there is a change in the diameter of pipe there is a change in velocity, and so there must also be a change in pressure. In some circumstances the change in diameter may mean that the pressure does not quite vary in the way that you might expect. Study Examples 4.8 and 4.9 carefully. In the first, pressure reduces with increasing elevation, as would be expected from hydrostatics. In the second, the pressure rises with increasing elevation because of the change in cross-section. Be careful not to be caught out by something like this – do not always assume that something will happen without first proving that it does. Example 4.10 introduces an energy loss into the equation. ❞

REMEMBER!!!

EXAMPLE 4.8

Water flows through a pipeline of constant diameter that is inclined upwards. On the centreline of the pipe, point 1 is 0.3 m below point 2. The pressure at point 1 is $9.3 \times 10^3 \, \text{N/m}^2$. What is the pressure at point 2 if there is no loss of energy?

The diameter of the pipe is constant so $V_1^2/2g = V_2^2/2g$. Therefore the velocity heads cancel and they can be omitted from the Bernoulli equation which becomes: $z_1 + P_1/\rho g = z_2 + P_2/\rho g$. Taking the datum level through point 1, $z_1 = 0$ and $z_2 = 0.3$ m. Therefore:

$0 + (9.3 \times 10^3/1000 \times 9.81) = 0.3 + P_2/\rho g$
$P_2/\rho g = 0.948 - 0.300 = 0.648$ m of water
$P_2 = 0.648 \times 1000 \times 9.81 = 6.36 \times 10^3 \, \text{N/m}^2$

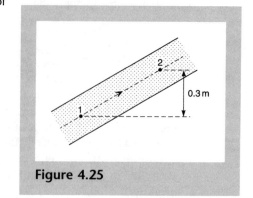

Figure 4.25

EXAMPLE 4.9

Water flows through an expanding pipeline that is inclined upwards. On the centreline of the pipe, point 1 is 0.3 m below point 2. The velocities are $V_1 = 3.1$ m/s and $V_2 = 1.7$ m/s. The pressure at point 1 is 9.3×10^3 N/m². What is the pressure at point 2 if there is no loss of energy?

$$z_1 + V_1^2/2g + P_1/\rho g = z_2 + V_2^2/2g + P_2/\rho g$$

Taking the datum level through point 1, $z_1 = 0$ and $z_2 = 0.3$ m. Therefore:

$$0 + (3.1^2/19.62) + (9.3 \times 10^3/1000 \times 9.81)$$
$$= 0.3 + P_2/\rho g + (1.7^2/19.62)$$
$$0.490 + 0.948 = 0.3 + P_2/\rho g + 0.147$$
$$P_2/\rho g = 0.991 \text{ m of water}$$
$$P_2 = 0.991 \times 1000 \times 9.81 = 9.72 \times 10^3 \text{ N/m}^2$$

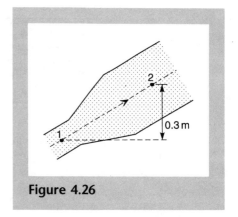

Figure 4.26

EXAMPLE 4.10

Water flows through a pipeline which reduces in cross-section. The centreline of the pipe is horizontal. If $V_1 = 1.54$ m/s, $V_2 = 2.65$ m/s, $P_1 = 20.00 \times 10^3$ N/m² and $P_2 = 16.89 \times 10^3$ N/m², what is the energy head loss between sections 1 and 2? Give the answer in m of water.

$$z_1 + V_1^2/2g + P_1/\rho g = z_2 + V_2^2/2g + P_2/\rho g + \text{head loss}$$

The centreline of the pipe is horizontal so $z_1 = z_2$ and these terms cancel. Therefore:

$$(1.54^2/19.62) + (20.00 \times 10^3/1000 \times 9.81)$$
$$= (2.65^2/19.62) + (16.89 \times 10^3/1000 \times 9.81)$$
$$+ \text{head loss}$$
$$0.121 + 2.039 = 0.358 + 1.722 + \text{head loss}$$
$$2.160 = 2.080 + \text{head loss}$$
energy head loss = 0.080 m of water

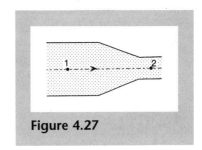

Figure 4.27

4.8.1 The energy coefficient

As mentioned earlier in connection with the momentum coefficient (section 4.6.5), there are situations where the distribution of velocity across the area of flow is distinctly non-uniform and the assumption that the mean velocity $V = Q/A$ is no longer accurate. To allow for the variation of velocity across the section the **energy coefficient**, α, can be introduced into the energy equation (this is sometimes called the **velocity distribution coefficient**). The coefficient is placed in front of the velocity head thus: $\alpha V^2/2g$. The coefficient always has a value of 1.00 or greater. In many situations, however, the value is near enough to unity for it to be taken as 1.00, and hence the energy equation is often written without α in front of the velocity head.

Table 4.2 Typical values of the energy and momentum coefficients

Condition	Typical α value	Typical β. value
Laminar flow in pipes (rare)	up to 2.00	—
Normal turbulent flow in pipes	1.01–1.10	1.02
Regular open channels, spillways	1.10–1.20	1.03–1.07
Natural streams	1.15–1.50	1.05–1.17
Flooded river valley	1.50–2.00	1.17–1.33

The only way to calculate α is to measure the velocity, v, within smaller subsections of the flow, dA, and then equate the total kinetic energy of the subsections to that of the total flow. This gives:

$$\alpha = \Sigma(v^3 \mathrm{d}A)/V^3 A \tag{4.26}$$

The calculation of α need not concern us greatly; in most of our calculations we will assume it has a value of 1.00 unless stated otherwise. However, as engineers you should be aware that there are situations where α can have a value of 2.0 or more. These, obviously, are situations where the flow is very uneven and the distribution of velocity varies greatly. Examples could include rapid changes of sections in pipes, or rivers in flood which flow over the floodplain so that there is a very complicated channel shape.

Table 4.2 illustrates that the range of α is larger and its variation is more significant than that of the momentum coefficient, β. Obviously, if the velocity head should be doubled ($\alpha = 2$) under some circumstance and this is not done, then the accuracy of the analysis will be affected. This is most likely to be important when rivers are in flood, possibly in the vicinity of channel obstructions, when velocities could perhaps reach as much as 3 m/s with a corresponding velocity head of 0.46 m. Under such conditions, doubling the velocity head can significantly influence the outcome of an analysis (see Example 4.11 and Self Test Question 4.4).

EXAMPLE 4.11

Water flows under a vertical lift gate as shown. The depth of water upstream is 3.4 m while downstream of the gate the depth is 1.2 m. The velocity distribution upstream is quite uniform so $\alpha_1 = 1.10$ but there is a significant variation downstream of the gate so $\alpha_2 = 1.55$. The channel is 4.0 m wide with a horizontal bed. Assuming no loss of energy, what is the discharge?

From the continuity equation $A_1 V_1 = A_2 V_2$

$(3.4 \times 4.0)V_1 = (1.2 \times 4.0)V_2$ so

$V_1 = 0.353 V_2$

In this case $z + P/\rho g$ is the height of the water surface above the bed (like the

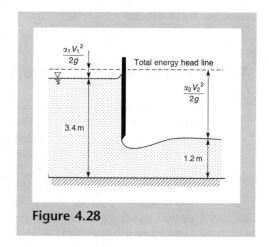

Figure 4.28

piezometric level in Fig. 4.19). Therefore for two points upstream and downstream the energy equation can be written as:

$$3.4 + \alpha_1 V_1^2/2g = 1.2 + \alpha_2 V_2^2/2g$$

Substituting from above:

$$3.4 + 1.10(0.353V_2)^2/2g = 1.2 + 1.55V_2^2/2g$$
$$3.4 + 0.0070V_2^2 = 1.2 + 0.0790V_2^2$$
$$V_2 = 5.53 \, \text{m/s}$$
$$Q = A_2V_2 = 1.2 \times 4.0 \times 5.53 = 26.54 \, \text{m}^3/\text{s}$$

SELF TEST QUESTION 4.4

Repeat the calculations in Example 4.11, but this time take $\alpha_1 = \alpha_2 = 1.00$. What is the percentage difference in the calculated value of Q?

SELF TEST QUESTION 4.5

A pipeline bends in the horizontal plane through an angle of 45° (Fig. 4.29). The diameter of the pipe changes from 0.6 m before the bend to 0.4 m after it. Water enters the bend at the rate of 0.5 m³/s with a pressure of 150×10^3 N/m². Assuming that there is no loss of energy, calculate the force exerted on the pipe bend by the water. (Hint: apply the continuity equation, energy equation and momentum equation, in that order. Many problems require all three equations to obtain a solution, as this one does).

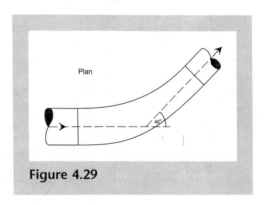

Plan

Figure 4.29

4.9 Drag and lift

When an ideal fluid with no viscosity flows past a body, it does not exert a force on it. However, when a viscous fluid flows past a body that is either partly or totally immersed in it, the fluid exerts a force on the body in the direction of flow. This is known as the **drag force**. A good example is the force exerted on the piers of a bridge by the water flowing past. The same argument applies if the fluid is stationary and the body is moving: a force is needed to propel an aircraft through the atmosphere. The force needed depends upon many factors, including the coefficient of drag, C_{DR}, and the cross-sectional area of the body presented to the flow, A:

$$\textbf{Drag force} = \frac{1}{2}C_{DR}\rho AV^2 \tag{4.27}$$

where V is the velocity of the fluid relative to the body and ρ is the density of the fluid. The value of C_{DR} is determined by shape, surface roughness and Reynolds number. For a flat plate, a graph of the variation of C_{DR} with Reynolds number and roughness has the same form as that for the variation of the pipe friction factor, λ, with the same two variables (Fig. 6.10). However, as a rough indication of the range, if a square plate is held normal to a stream of fluid then C_{DR} may be about 1.2, whereas it may be as little as 0.05 if the plate is held parallel to the flow (see Fig. 4.30). The coefficient is generally calculated from mea-

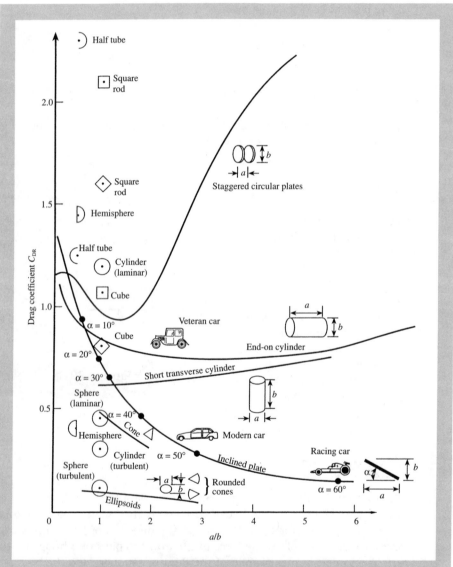

Figure 4.30 Approximate drag coefficient C_{DR} of various body shapes at Re $\approx 10^5$. Values are based on frontal area, except for the inclined plate where A is its full surface area. Flow is from left to right with respect to the body shape. Where not shown, a is the body dimension in the direction of flow and b is perpendicular to it [*after Douglas et al. (1995); reproduced by permission of Longman*]

surements of the drag force made in a wind tunnel, via equation (4.27). Car manufacturers now pay great attention to streamlining to reduce the coefficient of drag, because fuel efficiency depends upon the drag force which depends upon C_{DR}. The drag force has two components:

Total drag = pressure (or form) drag + surface (or skin friction) drag (4.28)

A flat plate, such as a large road sign hit perpendicularly by the wind, has a large form drag due to the difference in pressure on the two sides of the plate, while the friction drag is negligible. Such a body has a large wake (a region of eddying, disturbed fluid downstream) which is an indication of a large drag and severe disruption of the flow. Form drag is generally the most significant. The total drag usually falls as a body becomes more streamlined, that is as its cross-sectional area reduces and it becomes more rounded and more elongated in the direction of flow. An aerofoil, like an aeroplane wing, is one of the best examples of a streamlined body. It has a relatively large friction drag and a negligible form drag. With other less streamlined bodies the friction drag may be about the same magnitude as the form drag, perhaps larger.

Lift is similar to drag, but takes place perpendicularly to the flow. Hence the lift force on an aeroplane wing. If C_L is the coefficient of lift, then:

$$\text{Lift force} = \frac{1}{2}C_L \rho A V^2$$ (4.29)

The lift force and drag force can be combined to obtain the resultant force on the body using equations (1.15) and (1.16).

If parts of equations (4.27) and (4.28) look familiar, go back to equations (4.7) and (4.18). Can you see a similarity between the two phenomena?

4.10 Free and forced vortices

A vortex is 'free' if it occurs naturally without the input of external energy. For example, a vortex may occur because the flow has previously been caused to rotate by some kind of disturbance or because of some internal action. The free cylindrical vortex is the most familiar type, because this is what results when water flows down the plug hole of a sink. Free vortices may also occur as whirlpools in a river, in the casing of a centrifugal pump just outside the impeller or in a turbine casing as the water approaches the guide vanes. Other examples include flow around bends in open channels or ducts.

A cross-section through the water surface of a free vortex shows the characteristic hyperbolic profile in Fig. 4.31a. In plan, the streamlines are concentric circles (diagram b). The energy is constant along the streamlines, and there is no difference in total energy between one streamline and another or at points in a horizontal plane. The tangential velocity (V) is inversely proportional to the radius (r) from the vertical axis of the vortex, thus:

$$V = \frac{C}{r}$$ (4.30)

The constant (C) can be evaluated by measuring V at a known value of r. Note that a free vortex cannot extend to the axis of rotation, because theoretically as r approaches zero V approaches infinity. In reality this does not arise because the friction losses are proportional to V^2 and become increasingly important near the core.

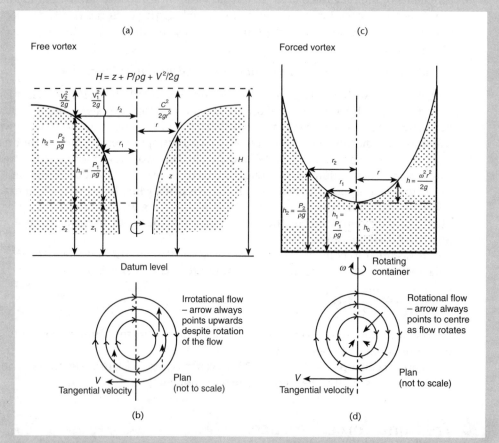

Figure 4.31 Free vortex (a) in cross-section and (b) in plan (not to scale) illustrating irrotational flow; forced vortex (c) in cross-section and (d) in plan (not to scale) illustrating rotational flow

The difference in elevation between two points on the free water surface at distances r_1 and r_2 from the centre can be found by applying the energy equation. Assuming hydrostatic pressure (i.e. $h = P/\rho g$) and that $z_1 = z_2$ then for the horizontal plane shown dashed on the left side of Fig. 4.31a:

$$P_1/\rho g + V_1^2/2g = P_2/\rho g + V_2^2/2g$$

From equation (4.30), $V_1 r_1 = V_2 r_2$ so $V_2 = V_1(r_1/r_2)$. Thus:

$$\frac{P_2}{\rho g} - \frac{P_1}{\rho g} = \frac{V_1^2}{2g} - \frac{V_1^2}{2g}\left(\frac{r_1}{r_2}\right)^2$$

or

$$\boldsymbol{h_2 - h_1 = \frac{V_1^2}{2g}\left[1 - \left(\frac{r_1}{r_2}\right)^2\right]} \tag{4.31}$$

Alternatively, as shown on the right side of Fig. 4.31a, the elevation of the water surface at any radius r is $z = H - V^2/2g$ or $z = H - C^2/2gr^2$ since equation (4.30) gives $V^2 = C^2/r^2$.

The free vortex is an example of **irrotational flow**. Although the flow follows a circular path around the central axis, small individual elements of water do not rotate and always point in the same direction. Thus if an arrow is attached to an individual element and the arrow initially points towards the top of the page as in Fig. 4.31b, even though the water circulates around the core the arrow will always point to the top. This can be demonstrated by using small floats with arrows attached.

Free vortices are often a nuisance, because they disrupt the flow. They may occur naturally at the overflow weir to a shaft spillway, where vertical anti-vortex piers are used to suppress their formation (Fig. 9.8); downstream of bridge piers and abutments where the wake vortex causes scour (Figs 9.17 and 9.18); at the entrance to a culvert; and in a sump supplying water to a pump (Fig. 11.22).

Forced vortices occur where a liquid is forced to rotate without any relative motion between elements. This motion can be created by a rotating paddle wheel, or the impeller of a centrifugal pump which uses a forced vortex to increase both the head and velocity of a liquid. Alternatively, a forced vortex results when a partially full container is rotated about a vertical axis as in Fig. 4.31c. After a short time there is no shear, the liquid behaves like a solid (e.g. a rotating CD or record on a turntable) and the water surface exhibits the characteristic parabolic profile. All elements of the liquid have the same angular velocity ω rad/s and thus at any radius r the tangential velocity V is:

$$V = \omega r \tag{4.32}$$

Note that the velocity increases with the radius, which is the opposite of the free vortex. You may already know that in circular motion the force on a body of mass M acting towards the centre of a circle is MV^2/r. Similarly a consideration of the forces acting on an element of liquid gives the general result:

$$\frac{dP}{dr} = \rho \frac{V^2}{r} \tag{4.33}$$

Replacing V^2 with $w^2 r^2$ using equation (4.32), integrating for points at distances r_1 and r_2 from the centre and assuming hydrostatic pressure, then for any horizontal plane with rotation about a vertical axis the difference in elevation of the free surface is:

$$\int_{P_1}^{P_2} dP = \rho \omega^2 \int_{r_1}^{r_2} r\, dr$$

$$P_1 - P_2 = \rho \omega^2 \left[\frac{r_2^2}{2} - \frac{r_1^2}{2} \right]$$

$$\frac{P_1}{\rho g} - \frac{P_2}{\rho g} = \frac{\omega^2}{2g}[r_2^2 - r_1^2]$$

$$h_2 - h_1 = \frac{\omega^2}{2g}[r_2^2 - r_1^2] \tag{4.34}$$

This is illustrated on the left side of Fig. 4.31c. Alternatively, if $r_1 = 0$ and h_0 is taken as the datum level, then at any radius r the elevation of the water surface is $h = w^2 r^2/2g$, as shown on the right of the diagram. Equation (4.34) can also be applied to a closed vessel: the pressure distribution would be the same, and the equation shows the height to which liquid would rise in a piezometer.

The forced vortex is an example of **rotational flow**. As the flow follows a circular path around the central axis, the small individual elements of water also rotate (Fig. 4.31d). Thus if arrows are attached to individual elements and they all initially point towards the centre, they will always point to the centre even though the water circulates around the core (just like drawing arrows on a CD or record).

Summary

1. To facilitate a simple analysis, ideal liquids are often assumed so that complexities due to viscosity, friction and turbulence are eliminated. Experimental coefficients can be introduced to compensate.

2. Viscosity is a measure of a liquid's internal resistance to movement and deformation. The coefficient of dynamic (or absolute) viscosity is the constant in equation (4.2).

    Dynamic viscosity = μ kg/ms (or Ns/m^2)
    Kinematic viscosity, $v = \mu/\rho$ m^2/s

3. In laminar flow all of the particles on a particular streamline follow the same path and have the same velocity. In turbulent flow all of the particles follow different, random paths and have different velocities. Reynolds number is Re $= \rho VD/\mu$ (equation (4.3)). For water, the flow is turbulent when Re > 4000 (pipes) or Re > 2000 (open channels).

4. The continuity equation is based on the conservation of mass between two cross-sections (A_1, A_2). In terms of the volumetric flow rate (Q m^3/s) and mean velocity of flow (V_1, V_2) it can be written as:

    $$Q = A_1 V_1 = A_2 V_2 \qquad (4.4)$$

5. Momentum = mass × velocity = MV. Newton's Laws state that:

    (a) unless acted upon by an external force a body remains at rest or continues to move in a straight line with a constant velocity;

    (b) in a particular direction, the force acting on a body equals the rate of change of momentum. Hence $F = Ma$ where a = acceleration = $(V_2 - V_1)/t$. For a hydraulic system, with $M/t = \rho Q$ this becomes:

    $$F = \rho Q(V_2 - V_1) \qquad (4.6)$$

    (c) to every action there is an equal and opposite reaction.

6. For hydraulic systems, remembering that force and velocity are vector quantities, the momentum equation is applied by considering the external forces acting on a control volume. Using a sign convention for forces, the algebraic sum of the external forces acting on the fluid in a control volume in a given direction equals the rate of change of momentum in that direction as a result of the fluid passing through the control volume. Thus:

    $$\Sigma F_X = \rho Q(V_{2X} - V_{1X}) \text{ and}$$
    $$\Sigma F_Y = \rho Q(V_{2Y} - V_{1Y}) \qquad (4.9/4.10)$$

7. Energy is the capacity for doing work. Work is the product of a force and the distance moved in the direction of the force (hence the units of energy and work are Nm). The total energy per unit weight of a fluid is the sum of the three components arising from elevation, velocity and pressure:

    $$z + V^2/2g + P/\rho g = \begin{array}{l}\text{total energy per}\\ \text{unit weight of fluid}\end{array}$$
    $$(4.22)$$

8. All components of equation (4.22) have the units of m and can be drawn as heads as in Fig. 4.19. The energy or Bernoulli equation above can be applied to two points on a streamline to investigate how the energy changes. The concept of the conservation of energy requires that:

    $$z_1 + V_1^2/2g + P_1/\rho g$$
    $$= z_2 + V_2^2/2g + P_2/\rho g + \text{energy head loss}$$
    $$(4.24)$$

    When applying the equation, *where appropriate* it can be simplified by drawing the elevation datum level through one of the points ($z = 0$), assuming that the velocity on the surface of a large reservoir is negligible ($V = 0$), and using atmospheric pressure as a datum so $P = 0$ on the surface of a reservoir or where a jet discharges to the atmosphere and has attained the pressure of its surroundings.

9. A body located in a real liquid experiences both a drag force (in the direction of flow) and a lift force (perpendicular to the direction of flow), thus:

$$\text{Drag force} = \tfrac{1}{2}C_{DR}\rho AV^2 \text{ and}$$
$$\text{Lift force} = \tfrac{1}{2}C_L\rho AV^2 \qquad (4.27/4.29)$$

The total drag force is the sum of the pressure (or form) drag and friction drag. Form drag is due to the difference in pressure on the two sides of the body. Friction drag is the result of the interaction between the body's surface roughness and the moving fluid.

Revision questions

4.1 Describe the differences between (a) an ideal fluid and a real fluid; (b) dynamic viscosity and kinematic viscosity; (c) laminar and turbulent flow; (d) steady and unsteady flow; and (e) uniform and non-uniform flow. (f) Describe what is meant by viscosity. (g) Define Reynolds number, Re. (h) Water with a kinematic viscosity of $1.007 \times 10^{-6}\,\text{m}^2/\text{s}$ flows along a 0.015 m diameter pipe at a velocity of 0.23 m/s. What is the value of Re, and is the flow laminar or turbulent?

[(h) 3430 – transitional]

4.2 Describe in words the meaning of the continuity equation. Explain what continuity of flow in a pipeline entails, and how this interacts with fluid pressure and velocity.

4.3 Two separate pipelines (1 and 2) join together to form a larger pipeline (3). It is known that $D_1 = 0.2\,\text{m}$, $D_2 = 1.0\,\text{m}$, $Q_2 = 0.23\,\text{m}^3/\text{s}$ and $Q_3 = 0.35\,\text{m}^3/\text{s}$. (a) what is the value of Q_1, V_1, and V_2? (b) If V_3 must not exceed 3.00 m/s, what is the minimum diameter, D_3, that can be used?

[(a) $0.120\,\text{m}^3/\text{s}$, 3.820 m/s, 0.293 m/s; (b) 0.385 m]

4.4 Describe in words what is meant by (a) momentum; (b) momentum flow rate; (c) the momentum equation (Newton's second law); (d) a control volume; (e) conservation of momentum.

4.5 A pipeline of constant 0.6 m diameter with its centreline in the horizontal plane turns through an angle of 75°. The pipeline carries water at the rate of 0.85 m³/s. A tapping at the bend indicates that the pressure is 41.3 m of water. Calculate the force exerted on the pipe bend by the water, and the direction in which it acts.

[$142.58 \times 10^3\,\text{N}$ at 52.5° to horizontal]

4.6 A straight pipeline with a horizontal centreline increases in diameter uniformly and symmetrically from $D_1 = 1.3\,\text{m}$ to $D_2 = 2.0\,\text{m}$. The flow rate through the pipeline is 4.114 m³/s and the corresponding pressures are $P_1 = 149.573 \times 10^3\,\text{N/m}^2$ and $P_2 = 153.523 \times 10^3\,\text{N/m}^2$. Calculate the force exerted on the pipe expansion by the water, and state clearly in which direction it acts.

[$276.518 \times 10^3\,\text{N}$, internally from right to left]

4.7 Describe in words what is meant by (a) energy; (b) work; (c) power; (d) the energy equation; and (e) conservation of energy. For (a) to (c) quote the units of measurement.

4.8 Water flows through a straight pipeline that reduces in diameter from section 1 to 2. The centreline of the pipe is horizontal. If $V_1 = 1.54\,\text{m/s}$, $P_1 = 20.00 \times 10^3\,\text{N/m}^2$, and $V_2 = 2.65\,\text{m/s}$, what is P_2 in N/m² and as the equivalent head of water? Assume no loss of energy.

[$17.68 \times 10^3\,\text{N/m}^2$; 1.802 m]

4.9 A horizontal pipeline terminates in a nozzle that discharges to the atmosphere. The pipeline has a diameter of 0.8 m and operates with a velocity of flow of 2.5 m/s. (a) What diameter nozzle is required to obtain a jet with a velocity of 7.0 m/s?

(b) What is the pressure of the water in the pipeline? (c) What is the force exerted by the water on the nozzle?

[(a) 0.478 m; (b) 21.37 × 10³ N/m²; (c) 5083 N left to right]

4.10 A pipeline that carries 0.5 m³/s of water bends in the horizontal plane through 45°, as in Fig. 4.29, but this time the pipeline is expanding. The water enters along the x axis from left to right where the diameter is 0.4 m, and leaves through a 0.6 m diameter pipe where the pressure is 150 × 10³ N/m². Assuming there is no loss of energy, calculate the force exerted by the water on the bend and state clearly in which direction this internal force acts. Is the force the same as in Self Test Question 4.5?

[32.39 × 10³ N; the internal force acts at 70.9° to the horizontal from top right to bottom left]

4.11 If the flow through a syphon can be maintained up to about −7.5 head of water, how high above the water surface would the crest of the syphon in Example 4.7 have to be before the syphon ceased to function?

[4.3 m]

4.12 (a) Describe what is meant by the terms drag force, lift force, form drag, coefficient of drag and coefficient of lift. (b) A body has a cross-sectional area of 2.3 m² and a coefficient of drag of 1.05. What is the drag force on the body when it is subjected to a stream of air of density 1.2 kg/m³ travelling at 8.9 m/s?

[(b) 114.8 N]

4.13 A kite weighs 12 N and has a surface area of 0.8 m². The tension in the kite string is 35 N when it flies in a wind of 8.6 m/s. The kite string is inclined at an angle of 45° to the horizontal, and the density of the air is 1.18 kg/m³. Assuming that the kite is a flat plate (so A = 0.8 m² in equations (4.27) and (4.29)), calculate the value of the coefficient of drag and the coefficient of lift (*do not forget to take the weight into consideration*).

[(a) 0.71; (b) 1.05]

5

Flow measurement

This chapter describes how the flow or discharge of a stream of liquid in a pipe or open channel can be measured using a variety of devices such as a Venturi meter, Pitot tube, orifice, sharp crested weir and velocity meter. The characteristics of each device, its advantages and disadvantages, and situations where it may be used are described. The theoretical discharge equations are derived, but because they ignore effects like viscosity, friction and turbulence, they have to be used with experimentally determined coefficients of discharge in order to obtain an accurate estimate of the flow rate. The definition and evaluation of these coefficients is fully described. The information presented enables questions like those below to be answered:

How can the discharge in a pipeline be measured?

How can the flow in an open channel be calculated?

Which method of measurement is the most appropriate in a given situation?

How do the measuring devices work?

How can the theoretical equations for discharge be derived?

How can the instruments be calibrated and the coefficients of discharge obtained?

5.1 Introduction

The principle behind almost all of the measuring devices described below is the energy equation (Bernoulli equation) described in the last chapter. The exception is the velocity or current meter, which is simply a rotating propeller that is driven by the flow of water (see section 5.7). The faster the flow, the faster it rotates. This is a simple device that, although it needs to be calibrated, does not have a theoretical basis. All of the other measuring devices are basically applications of the energy equation, and as such share the limitations of this equation.

Real liquids are very difficult to describe mathematically when they are flowing because friction, viscosity and turbulence should be considered. However, the complexity of these three parameters usually precludes the possibility of them being satisfactorily incorporated into a simple discharge equation. Consequently the only practical approach is to derive the discharge equation without considering them (that is, assume an ideal liquid) and then determine experimentally the value of the 'fiddle factor' that makes the inaccurate answers from the simplified theory agree with the actual answers obtained by experiment. The formal name for this factor is the **coefficient of discharge**, C_D, where:

C_D = **actual discharge/theoretical discharge**

or $C_D = Q_A/Q_T$ (5.1)

Much time and effort has been devoted to obtaining accurate C_D values for the various flow measuring devices. In many cases the appropriate value can be found in the British Standard, provided that the device has a standard configuration. If a non-standard device is used, it must be calibrated individually. Thus the accuracy of the discharge measurement depends upon the accuracy of the coefficient of discharge, as well as upon how well the theoretical equation describes the flow. The accuracy of the values substituted into the equation (such as cross-sectional area of flow or the head) are also very important.

The Venturi meter provides a good illustration not only of the principles of flow measurement and the ideas discussed above, but also of the energy equation itself. Many of the things that can be observed from the flow of water through a Venturi meter can be applied elsewhere. Consequently, we will start with the Venturi meter.

5.2 The Venturi meter

5.2.1 Understanding what happens in a Venturi meter

This device is ideal for measuring the discharge through a pipeline, since its streamlined shape presents a relatively small obstruction to the flow, resulting in a small loss of energy through the meter. This can be important in some situations. Described simply, the meter is a constriction in the pipeline (Fig. 5.1). It consists of a converging section, followed by a short parallel portion called the **throat**, then a section which diverges gradually at an angle of about 5 to 7°. Note that the angle of convergence (often 21°) is larger than the angle of divergence. The throat length is equal to its diameter. To complete the meter, there are tappings at the entrance to the converging section and at the throat which, when connected to a piezometer or manometer, enable the head difference, H, to be measured. As will be shown below, the discharge, Q, through the meter is proportional to $H^{1/2}$. So if H is measured, Q can be calculated. This is basically how the meter works: the constriction causes a change in pressure, which when measured enables the discharge to be calculated.

The Venturi meter in Fig. 5.1 is shown with a third piezometer tube at the end of the divergent section. This is not needed to measure the discharge, but is included to illustrate more clearly what happens when water flows through the meter. Let us start at the upstream end of the meter (section 1), and work our way through to the throat (section 2) and then to the exit (section 3). The subscripts relate to these sections. The continuity equation as applied to the meter can be written as:

$Q = A_1V_1 = A_2V_2 = A_3V_3$ (5.2)

where A is the cross-sectional area of the meter and V the mean flow velocity. Since $A_2 < A_1$ then $V_2 > V_1$. In other words, the velocity increases in the converging portion of

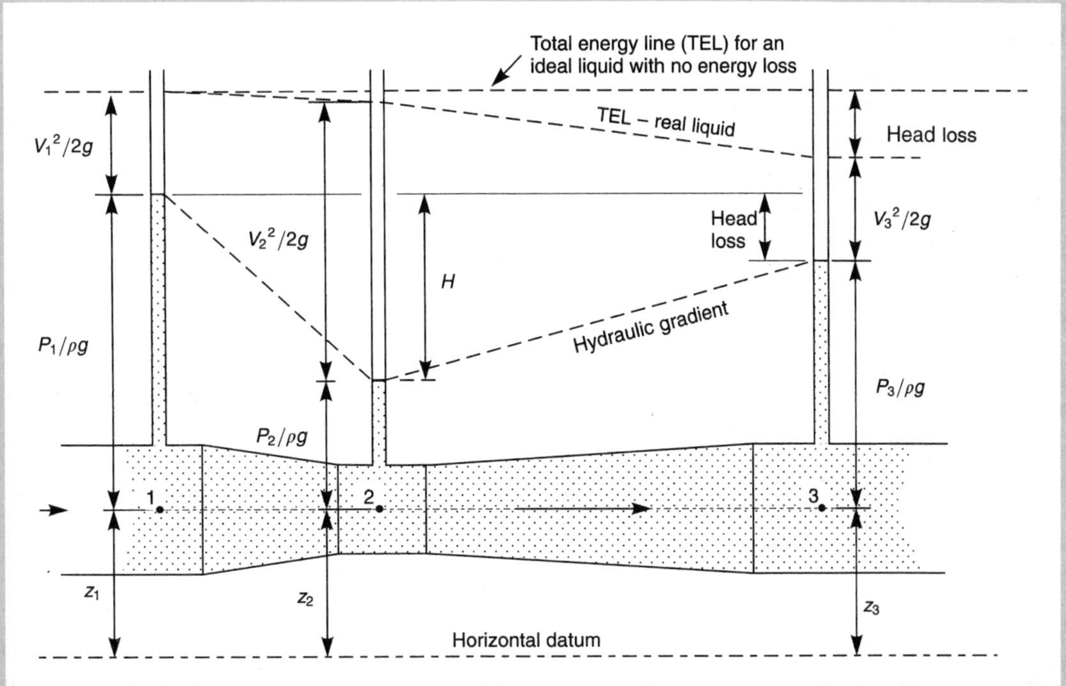

Figure 5.1 A Venturi meter, shown with a third piezometer at the exit instead of the usual two at the entrance and the throat. The centreline of the meter is horizontal

the meter, reaching a maximum in the throat. Now if the centreline of the meter is horizontal so that the elevation term z can be omitted from the energy equation, then:

$$V_1^2/2g + P_1/\rho g = V_2^2/2g + P_2/\rho g \qquad (5.3)$$

where P is the pressure (N/m²). Since $V_2 > V_1$ it follows from equation (5.3) that $P_2 < P_1$ or, in words, the pressure reduces in the converging part of the meter.

❝To students it always seems that the pressure ought to increase in the narrow part of the meter, not decrease. We discussed this in section 4.7.2, so go back and read that section again. Try the little experiments as well. You have to get used to the idea that a reducing section means an increase in velocity and a reduction in pressure.❞

There are some other important principles illustrated by the Venturi meter. If an arbitrary horizontal datum is chosen below the meter, then at any point on a streamline:

$$z + V^2/2g + P/\rho g = \textbf{total energy head}$$
$$\text{and} \quad z + P/\rho g = \textbf{the piezometric head or level}$$

If the total energy line (TEL) is drawn for the three piezometers in Fig. 5.1, it will be horizontal if there is no loss of energy (ideal liquid) because $V_1 = V_3$ and $P_1 = P_3$. However, if energy head losses are taken into consideration (real liquid) it will slope in the direction of

---

Box 5.1 ▶ **Remember**

This is an important point to be remembered for future reference. When liquid flows through a Venturi meter or pipeline, the continuity equation has to be obeyed, so any loss of energy appears as a reduction in pressure. For example, if water flows through a long pipeline of constant diameter at a constant rate, then the mean velocity must be the same at all points along the pipeline to maintain continuity of flow. Thus any loss of energy appears as a reduction in pressure head.

---

flow. With a real liquid the total energy line always falls in the direction of flow, because there must always be a loss of energy due to friction and turbulence. The energy line never goes up in the direction of flow, the only exception being if a pump is placed in the pipeline and energy is added. This leads to an interesting and important point. Continuity of flow has to be maintained through the meter: there is no other possibility, otherwise liquid would either be magically disappearing or being created. Thus V_1 must equal V_3 (with $A_1 = A_3$) and the velocity heads at sections 1 and 3 are the same. Consequently the energy head loss must appear as a fall in pressure head (piezometric level) between the two sections so $P_1 \neq P_3$.

If the piezometric levels at sections 1, 2 and 3 are joined together by a straight line as in Fig. 5.1, then this is referred to as the **hydraulic gradient**, or **hydraulic grade line**. This represents the height to which water will rise in a piezometer (or a stand-pipe in a pipeline) if a tapping is made. If the hydraulic grade line falls below the centreline of the meter (or the streamline that the energy equation applies to) then a negative pressure or suction occurs, resulting in air being drawn into the piezometer and pipeline. The meter should be selected or designed to avoid this happening.

Normally water flows in the direction of the hydraulic gradient, that is from high to low pressure. However, one exception to this general rule is the flow of liquid under pressure along a pipeline. Under such circumstances the direction of flow over a short length of the pipeline may be in the opposite direction to the hydraulic gradient. Consider the flow through the horizontal Venturi meter in Fig. 5.1. Between sections 1 and 2 the velocity increases and the pressure falls. Because the flow is from an area of high pressure (section 1) to low pressure (section 2), that is in the direction of the hydraulic gradient, there is little loss of energy despite the gradual reduction in diameter. Remember that the hydraulic gradient is the slope of the piezometric pressure line. Now look at what happens between sections 2 and 3. The flow is in the opposite direction to the hydraulic gradient, since $P_3 > P_2$. The flow is from left to right, but the high pressure at section 3 is trying to push the liquid back towards section 2, towards the lower pressure. This is often called an **adverse pressure gradient**. The effect of the adverse pressure gradient is to slow the flow, so the velocity $V_3 < V_2$. In fact, it is the adverse pressure gradient that is responsible for the reduction in velocity dictated by the continuity equation.

The adverse pressure gradient has one other major consequence. Try to imagine the flow from left to right through the meter, with a relatively high velocity through the throat section. This high velocity 'jet' then emerges into the expanding portion of the meter where the adverse pressure gradient is trying to push the flow back in the opposite direction. The result is some eddying and turbulence, which is synonymous with a loss of energy. It is generally true that increases in cross-sectional area always result in a significant energy loss. The more sudden the increase in section, the larger the energy loss. For this reason the

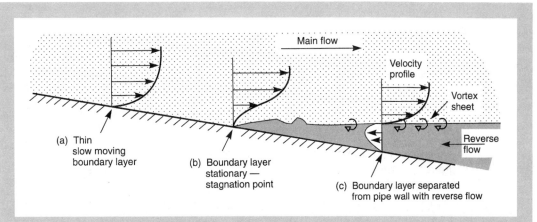

Figure 5.2 (a) At the pipe wall the liquid in the boundary layer has a very small forward momentum so as the pipe expands the adverse pressure gradient can stop (b) and then reverse the flow with the boundary layer peeling away from the wall (c). A vortex sheet forms between the main body of forward moving liquid and the reverse flow at the wall

Venturi meter expands at a small angle of between 5 and 7°, to reduce the energy loss. The angle of the converging part of the meter can be larger, about 21°, because energy losses in contractions tend to be relatively small as a result of the hydraulic gradient being in the same direction as the flow.

Looking at the flow in the expansion in more detail, what tends to happen is as follows. Near the centreline of the meter the liquid has a relatively high velocity as it emerges from the throat. At the pipe walls the liquid adheres to the boundary forming a **boundary layer** (Fig. 5.2a), that is a layer of liquid with a negligible or relatively small velocity (see section 6.5.3). This liquid has so little forward momentum that the adverse pressure gradient can stop it or even reverse the flow (Fig. 5.2b). The boundary layer may then peel away from the pipe wall, being replaced by liquid flowing in the reverse direction. This is termed **boundary layer separation** (Fig. 5.2c). Thus it is possible to draw an imaginary line that has the flow on either side going in opposite directions. When thought of in two or three dimensions this line is called a **vortex sheet**. The streams of liquid travelling in opposite directions interact, forming a series of discrete vortices (Fig. 5.2c). The separation of the boundary layer and the vortices results in a much more turbulent flow than would otherwise be the case and, of course, turbulence and vortices (or eddies) result in a loss of energy. So basically this is why an expanding flow is generally unstable and why it is usually associated with a larger energy loss than either parallel or converging flow. The more sudden the expansion, the larger the energy loss. Although the Venturi meter is designed to minimise energy losses by expanding very gradually downstream of the throat, there is still a small loss as a consequence of the effects just described being evident to some extent. As a result the distribution of velocity in the expanding section of a Venturi meter is distinctly non-uniform, so this could be a situation where the velocity distribution coefficient, α, should be used with the velocity head in the energy equation. Fortunately the problem of what α value to adopt is avoided by virtue of the fact that the discharge equation for the Venturi meter is obtained by applying the energy equation to sections 1 and 2, so the region where the energy loss occurs is avoided and the question does not arise.

Box 5.2 ▶ **For future use**

REMEMBER!!!

1. The total energy line always goes down in the direction of flow, never up (unless there is a pump in a pipeline).

2. Flow normally occurs in the direction of the hydraulic gradient, although the hydraulic gradient may rise over short distances giving an adverse gradient which can be overcome by the momentum of the fluid.

3. Expansions and diverging flow are usually associated with an energy loss resulting from turbulence and a non-uniform distribution of velocity. The more sudden the expansion, the greater the loss.

5.2.2 Derivation of the discharge equation for the Venturi meter

The equation can be derived by a straightforward application of the continuity and energy equations. With the centreline of the meter horizontal, equations (5.2) and (5.3) show that:

$$P_1/\rho g - P_2/\rho g = V_2^2/2g - V_1^2/2g \qquad (1)$$

$$A_1 V_1 = A_2 V_2 \qquad (2)$$

Giving either $V_2 = A_1 V_1/A_2$ or $V_1 = A_2 V_2/A_1$

Now the left-hand side of equation (1) is $(P_1 - P_2)/\rho g$ which is the difference in the piezometer readings, H, in Fig. 5.1. Equation (2) can be used to replace either V_2 or V_1 in equation (1), thus:

$$H = (A_1 V_1/A_2)^2/2g - V_1^2/2g \quad \text{or} \quad H = V_2^2/2g - (A_2 V_2/A_1)^2/2g$$

$$H = \frac{V_1^2}{2g}\left[(A_1/A_2)^2 - 1\right] \quad \text{or} \quad H = \frac{V_2^2}{2g}\left[1 - (A_2/A_1)^2\right]$$

$$V_1 = \sqrt{\frac{2gH}{\left[(A_1/A_2)^2 - 1\right]}} \quad \text{or} \quad V_2 = \sqrt{\frac{2gH}{\left[1 - (A_2/A_1)^2\right]}}$$

$$Q_T = A_1 V_1 \quad \text{or} \quad Q_T = A_2 V_2$$

$$Q_T = A_1\sqrt{\frac{2gH}{\left[(A_1/A_2)^2 - 1\right]}} \quad \text{or} \quad Q_T = A_2\sqrt{\frac{2gH}{\left[1 - (A_2/A_1)^2\right]}} \qquad (5.4)$$

To obtain the actual discharge the coefficient of discharge, C_D, is introduced into the equation:

$$Q_A = C_D A_1\sqrt{\frac{2gH}{\left[(A_1/A_2)^2 - 1\right]}} \quad \text{or} \quad Q_A = C_D A_2\sqrt{\frac{2gH}{\left[1 - (A_2/A_1)^2\right]}} \qquad (5.5)$$

Note that the two alternative equations are given simply to show that, depending upon whether the substitution is made for V_1 or V_2 near the beginning, two expressions can be derived. Students often think that only one can be correct. In fact they are both correct. It does not matter which you use, although a small error incurred when measuring the larger area A_1 would be less significant than the same error incurred with the smaller area A_2. If the meter is tilted instead of horizontal, almost the same discharge equation is used (. Proof 5.1, Appendix 1).

In the case of the Venturi meter, the coefficient of discharge, C_D, is needed to compensate for the assumption of an ideal liquid and for the fact that equation (5.3) assumes that there is no loss of energy through the meter when in fact there is. However, the energy loss between sections 1 and 2 is very small, which is why the Venturi meter has a C_D of about 0.97. This is much higher than the equivalent value (about 0.6) for weirs and orifices, for example. The value of C_D depends upon the design of the Venturi meter and the flow rate. It may be found that C_D increases with increasing discharge. This is often an unexpected result since energy losses also increase with increasing discharge, so it might have been expected that C_D would fall to compensate. This apparent discrepancy can be explained by the fact that the energy loss occurs mainly downstream of the throat, while the part of the meter upstream of the throat is used to measure the discharge. The increase in the C_D value indicates that in this part of the meter the actual and theoretical discharges are becoming closer, as indicated by equation (5.1). In other words, the effects of friction, viscosity and turbulence which are not allowed for in the theoretical equation, are becoming less significant in the converging part of the meter as the discharge increases.

An alternative to the Venturi meter for measuring the discharge in a pipe is the orifice meter described later in the chapter. Now study Example 5.1 and try Self Test Question 5.1.

SELF TEST QUESTION 5.1

The flow of water in a 150 mm diameter pipeline has to be measured using a horizontal Venturi meter. The normal operational velocity in the pipeline is about 2.3 m/s. What diameter must the throat of the Venturi be in order to obtain a differential head, H, of about 1.2 m of water?

EXAMPLE 5.1

Water flows along a horizontal pipeline of 100 mm diameter at an unknown rate. A Venturi meter installed in the pipeline indicates a piezometric head of 950 mm at the entrance and 200 mm at the throat. The throat diameter is 60 mm. If the $C_D = 0.97$, what is the discharge through the pipeline?

$$A_1 = \pi D_1^2/4 = \pi \times 0.10^2/4 = 0.00785 \text{m}^2$$
$$A_2 = \pi D_2^2/4 = \pi \times 0.06^2/4 = 0.00283 \text{m}^2$$
$$[(A_1/A_2)^2 - 1] = [(0.00785/0.00283)^2 - 1] = 6.694$$
$$H = (0.95 - 0.20) = 0.75 \text{m}$$

From equation (5.5):

$$Q_A = C_D A_1 \{2gH/[(A_1/A_2)^2 - 1]\}^{1/2}$$
$$= 0.97 \times 0.00785\{2 \times 9.81 \times 0.75/6.694\}^{1/2}$$
$$= 0.0113 \text{m}^3/\text{s or } 11.3 \text{l/s}$$

(Note that while it may be convenient to summarise an answer in litres per second, *never* use a value in l/s in your calculations – for the reasons described in the Introduction).

itot tube

tube provides a means of measuring the velocity at a point either within a pressurised pipeline or in an open channel. Often the Pitot tube can be moved across the conduit to obtain a picture of the variation in velocity over the whole cross-sectional area of flow, A. If the average velocity, V, is calculated then the discharge can be obtained from $Q = AV$.

Essentially the Pitot tube measures the velocity head, $V^2/2g$, of the flow at a point by turning the kinetic energy into the equivalent static head of liquid, H. The principle is illustrated by Fig. 5.3a. Point 1 is located in a region of undisturbed flow within the pipeline. The static pressure of the liquid at this point is shown by the height of the liquid in the piezometer above point 1. The Pitot tube is introduced at point 2. Some of the moving liquid collides with the nose of the Pitot tube and is brought to rest in front of the open end of the tube. This point of zero velocity is called a **stagnation point**. The impact causes a rise in pressure of the liquid within the Pitot tube, so that its piezometric height is a distance, H, above that in the static tube. By applying the energy equation, it will be shown below that the velocity at 2 is proportional to $H^{1/2}$. So by measuring H we can calculate V.

Applying the energy equation to the centreline of the pipe, assuming no loss of energy:

$$z_1 + V_1^2/2g + P_1/\rho g = z_2 + V_2^2/2g + P_2/\rho g$$

If the centreline of the pipe is horizontal then $z_1 = z_2$ and cancels. Assuming that the velocity, V_2, is zero at the stagnation point, then:

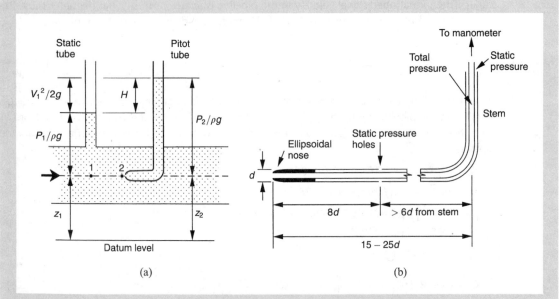

Figure 5.3 (a) Separate static tube and Pitot tube in a pipe. (b) Combined Pitot–static tube. The outer holes measure the static pressure, while the inner tube measures the combined pressure. When connected to a suitable manometer this enables the differential head, H, to be measured without the need for a separate static tube as in (a)

Box 5.3 ▶ **For future use**

This equation appears time and time again in hydraulics, so remember $V = (2gH)^{1/2}$ for future use (when appropriate!). There are circumstances where it is not appropriate, however, so it does not solve all problems.

$$V_1^2/2g = P_2/\rho g - P_1/\rho g$$

The difference in the piezometric levels $(P_2 - P_1)/\rho g$ is the differential head, H, thus:

$$\mathbf{V_1 = (2gH)^{1/2}}$$ (5.6)

For complete accuracy it is necessary to introduce a coefficient, C, into equation (5.6) to allow for any disruption of the flow caused by the tube:

$$\mathbf{V_1 = C(2gH)^{1/2}}$$ (5.7)

With a well designed tube the coefficient is almost, but not quite, unity. A typical value would be about 0.98 or 0.99. The sort of tube used in practice is shown in Fig. 5.3b. The nose of the tube is rounded so as to cause as little disruption to the flow as possible. The combined static and dynamic pressure is measured by the inner of the two tubes, the static pressure by the holes on the outer of the two tubes. This combined Pitot–static tube eliminates the need for the two tappings shown in Fig. 5.3a.

The Pitot–static tube is particularly suited to measuring air velocities. Indeed, the long tube sticking out of the nose of some aircraft is a Pitot tube. However, the drawback in normal earthbound use is that the Pitot–static tube has to be connected to a very sensitive manometer or other pressure measuring device, because equation (5.6) shows that a velocity of 0.5 m/s gives a value of H of only 13 mm. Under these conditions even a small error when measuring H can be significant, so it is most useful when the velocities involved are large.

EXAMPLE 5.2

A Pitot tube is used to measure the velocity in a pipeline. The stagnation pressure head is 2.666 m and the static pressure head is 1.815 m. If the coefficient of the meter is 0.98, what is the actual velocity?

$$H = (2.666 - 1.815) = 0.851\,\text{m}$$

Form equation (5.7), $V_1 = C(2gH)^{1/2} = 0.98(2 \times 9.81 \times 0.851)^{1/2}$

$$V_1 = 4.004\,\text{m/s}$$

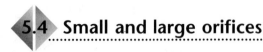

5.4 Small and large orifices

There are many different types of orifice, as will be seen later. However, one of the basic distinctions is between a large and a small orifice. These can be distinguished as follows:

Small orifice The diameter of the orifice is small compared to the head of water producing the flow, so the head at the top of the orifice is essentially the same as the head at the bottom. Consequently it can be assumed that the velocity of the jet emerging from the orifice is constant over its entire cross-section, that is from the top to bottom.

Large orifice The diameter of the orifice is large compared to the head of water, H, producing the flow. Thus H at the top of the orifice is significantly different from H at the bottom. Consequently over the cross-section of the emerging jet there is a significant variation in velocity, as dictated by $V = (2gH)^{1/2}$.

This distinction means that a different approach is needed to derive the discharge equations for small and large orifices. For a small orifice with V assumed constant over the cross-sectional area of the jet we can apply the energy equation. With a large orifice we must take into account the variation of V over the area of the jet by using an approach involving integration. A further distinction is that small orifices are generally round, but large orifices are frequently rectangular.

5.4.1 Free discharge through a small orifice

A small orifice is usually circular and may be located in the base or side of a tank. It can be used as a flow measuring device, or possibly as a flow control device. Sometimes short pipes and bridge waterway openings are treated as a small orifice (for simplicity, since the theoretical basis for doing so is often dubious).

Figure 5.4 shows a jet of water emerging from an orifice and discharging freely to the atmosphere. The jet is not affected by any downstream flow or water level. The discharge equation for the orifice can be obtained by applying the energy equation to a streamline connecting point 1 on the surface of the tank to point 2, which is located at the centre of the **vena contracta** some distance outside the plane of orifice. The reason for locating point 2 there is as follows. Inside the plane of the orifice the pressure of the water is the hydrostatic pressure $P = \rho gH$, where H is the depth above the centre of the orifice. It is much more convenient if point 2 is located where atmospheric pressure exists, so the pressure term can be eliminated from the energy equation. Such a point is located at the vena contracta. The vena contracta forms because the water just inside the plane of the orifice has a relatively large hydrostatic pressure, P. As the water flows through the orifice, the streamlines near the sides of the tank have a velocity component towards the centre so they continue to converge for some distance after they have passed though the orifice. At the same time the pressure decreases. The narrow part of the jet where the streamlines become parallel is the vena contracta, and this is located roughly $0.5D$ to $1.0D$ from an orifice of diameter D. The parallel streamlines indicated that the jet has stopped contracting and that the jet has attained the pressure of its surroundings, that is the atmosphere. After the vena contracta, the jet may start to expand or break up slightly. Note that the area of the jet at the vena contracta is less than the area of the orifice.

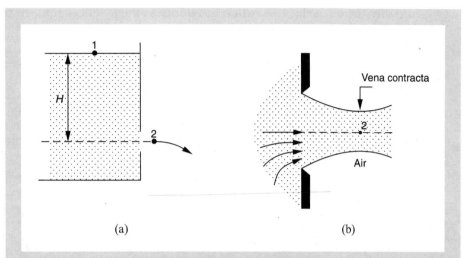

Figure 5.4 (a) An orifice in the side of a tank discharging freely to the atmosphere, and (b) the vena contracta caused by the contraction of the streamlines as they pass through the plane of the orifice. Note that point 2 is located at the centre of the vena contracta where it is assumed that atmospheric pressure exists, not in the plane of the orifice

Applying the energy equation between points 1 and 2 assuming no loss of energy gives:

$$z_1 + V_1^2/2g + P_1/\rho g = z_2 + V_2^2/2g + P_2/\rho g$$

If it is assumed that the volume of the tank is large and the discharge through the orifice is relatively small, at the surface the water will be unaffected by the flow so that $V_1 = 0$. With atmospheric pressure as the datum, $P_1 = 0$. Similarly, at the vena contracta $P_2 = 0$. If elevation is measured above point 2, which is assumed to be at the same elevation as the centre of the orifice, then $z_2 = 0$ and $z_1 = H$. Consequently the energy equation reduces to:

$$H = V_2^2/2g$$

or $\quad V_2 = (2gH)^{1/2}$ $\hspace{4cm}$ (5.8)

This is sometimes referred to as Torricelli's theorem, and this is the same as equation (5.6). The theoretical discharge, Q_T, can be obtained from the continuity equation, $Q_T = AV_2$. For convenience A is taken as the area of the orifice, thus:

$$Q_T = A(2gH)^{1/2} \hspace{4cm} (5.9)$$

This equation is not accurate, because the area of the jet at the vena contracta, a_J, is less than the area of the orifice, A. Thus it is necessary to introduce a **coefficient of contraction**, C_C, into equation (5.9) to allow for this. This coefficient is defined as:

$$C_C = a_J/A \hspace{4cm} (5.10)$$

The value of C_C can easily be determined by measuring the diameter of the jet. It is found to vary between about 0.60 and 0.97 depending upon the geometry of the orifice (see below). However, even if C_C is inserted into equation (5.9) an accurate value of Q still would not be obtained because it has been assumed that there is no energy loss, when in reality there is a slight reduction in velocity as the jet passes through the orifice. Thus the actual velocity of the jet at the vena contracta, v_J, is slightly less than the theoretical velocity $(2gH)^{1/2}$. Thus another coefficient, the **coefficient of velocity**, C_V, is introduced where:

$$C_V = v_J/(2gH)^{1/2} \tag{5.11}$$

The value of C_V can be determined either by measuring v_J with a Pitot tube and comparing it with the theoretical velocity, or by adopting the approach described in section 5.4.2. If both coefficients are introduced into equation (5.9), the actual discharge through the orifice, Q_A, is:

$$Q_A = C_C C_V A(2gH)^{1/2}$$

or $\qquad Q_A = C_D A(2gH)^{1/2} \tag{5.12}$

where $\quad C_D = C_C \times C_V \tag{5.13}$

Since C_V generally has a value close to unity, say about 0.95–0.99, it follows from equation (5.13) that *the coefficient of discharge of an orifice is primarily a coefficient of contraction*. Therefore it is not accurate to say that the coefficient of discharge is introduced into the equation to allow for energy losses, since these losses are very small. The value of C_D varies according to the configuration of the orifice. A rounded orifice causes a smaller contraction than a sharp edged one, and so has a higher C_D (Fig. 5.5). Sharp edged circular, square and rectangular orifices all have C_D values in the range 0.59–0.66, increasing as the diameter and head decrease (Brater *et al.*, 1996). The value of C_D depends upon many factors including edge configuration, orifice diameter and position, tank size and Reynolds number. It can be determined in a variety of ways, such as by measuring the actual discharge (with a weighing tank or by collecting a known volume of water in a given time) and then solving equation (5.12), or by using the approaches described in section 5.4.2 or Chapter 7.

Now work through Example 5.3 and Self Test Question 5.2.

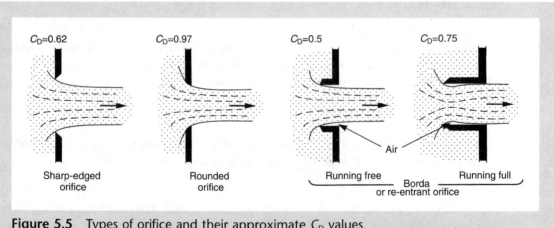

Figure 5.5 Types of orifice and their approximate C_D values

EXAMPLE 5.3

Water is contained in a large tank whose surface is open to the atmosphere. The water discharges freely to the atmosphere through an orifice 50 mm in diameter. The C_D of the orifice is 0.62. (a) What is the discharge if the head is maintained at a constant 2.50 m? (b) If the head is reduced by 50% to 1.25 m, what is the percentage decrease in the discharge?

(a) $Q = C_D A (2gH)^{1/2}$

  $A = \pi \times 0.05^2/4 = 0.00196 \, \text{m}^2$

  $Q = 0.62 \times 0.00196(2 \times 9.81 \times 2.50)^{1/2}$

   $= 0.0085 \, \text{m}^3/\text{s}$

(b) $Q = 0.62 \times 0.00196(2 \times 9.81 \times 1.25)^{1/2}$

   $= 0.0060 \, \text{m}^3/\text{s}$

  Therefore % reduction $= 100 \times (0.0085 - 0.0060)/0.0085 = 29.4\%$

  Note that the relationship between head and discharge is not linear since $Q \propto H^{1/2}$.

SELF TEST QUESTION 5.2

Suppose that the tank of water in Example 5.3a was covered and not open to the atmosphere, and that the air between the water surface and the lid of the tank had a pressure of $113.4 \times 10^3 \, \text{N/m}^2$ above atmospheric. Re-derive the equation for the discharge through the orifice taking into account this new circumstance, and calculate the new discharge through the orifice (assume that everything else remains unaltered).

5.4.2 The trajectory of a jet

One method of estimating the value of the coefficient of velocity, C_V, of an orifice is to measure the trajectory of the jet. After emerging from the orifice into the atmosphere the jet will follow a curved path as in Fig. 5.6. Using the vena contracta as the starting point, say that the centreline of the jet travels a distance x horizontally while falling through a vertical distance, y. Let t be the time required for the jet to travel from the vena contracta to the point x, y.

The relationship between horizontal velocity, distance and time is $v = x/t$ or

  $t = x/v$ (1)

The distance travelled vertically as a result of gravity (g) is:

$$y = \frac{1}{2} g t^2 \quad (2)$$

Substituting for t in (2) from (1) gives:

$$y = \frac{1}{2} g (x/v)^2$$

Hence $v = x(g/2y)^{1/2}$ (5.14)

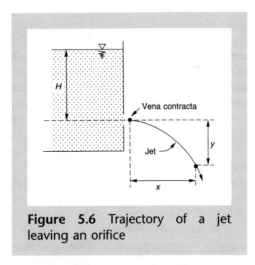

Figure 5.6 Trajectory of a jet leaving an orifice

If it is assumed that the velocity of the jet remains constant as it travels from the vena contracta to the point x, y then $v_J = v$ and the coefficient of velocity, C_V, can be obtained from:

$$C_V = v/(2gH)^{1/2}$$
$$= x(g/2y)^{1/2}/(2gH)^{1/2}$$

thus $\quad C_V = x/2(yH)^{1/2} \qquad (5.15)$

If desired, the diameter of the jet at the vena contracta, d_J, can be measured and the coefficient of contraction obtained from $C_C = (d_J/D)^2$ where D is the orifice diameter. The coefficient of discharge can then be obtained since $C_D = C_C \times C_V$. It is also possible to estimate the discharge from the orifice using equation (5.14) with $Q = a_J v$. This may be necessary if it is not possible to measure the head in the tank or the pressure above the liquid surface (see Self Test Question 5.2 and Example 5.5).

A couple of points to note are as follows:

1. A large head above the orifice will result in a high velocity jet that will travel a large distance horizontally for a given fall in the vertical direction. A smaller head would result in a smaller horizontal distance for the same vertical fall, since the jet velocity is reduced.

2. If the head in the tank falls (that is the liquid is not replaced), the jet velocity and the discharge through the orifice will decrease (see Chapter 7, Flow under a varying head).

EXAMPLE 5.4

Water discharges into the atmosphere through an orifice with a diameter of 25 mm. The head above the centre of the orifice is 1.42 m. The jet travels 1.25 m horizontally while falling through a vertical distance of 0.30 m. The diameter of the jet at the vena contracta has been measured as 20 mm. Determine the value of the coefficients.

From equation (5.15): $C_V = x/2(yH)^{1/2}$ where $x = 1.25$m, $y = 0.30$m and $H = 1.42$m

$C_V = 1.25/2(0.30 \times 1.42)^{1/2} = 0.96$

$C_C = a_J/A$ or $(d_J/D)^2$

$C_C = (0.020/0.025)^2 = 0.64$

$C_D = C_V \times C_C = 0.96 \times 0.64 = 0.61$

EXAMPLE 5.5

A very large tank containing a corrosive chemical liquid has been holed. The level of the liquid in the tank is unknown and cannot be measured. The liquid is discharging freely into the atmosphere. Observation of the jet indicates that the liquid travels about 1.95 m horizontally while falling through a distance of 0.25 m. Assuming that the hole in the tank is similar to a sharp edged orifice and that its diameter is about 20 mm diameter, estimate:

(a) the rate at which liquid was being lost when the observations were made;
(b) the head of liquid above the hole.

(a) From equation (5.14) the actual velocity of the jet $v = x(g/2y)^{1/2}$
  $x = 1.95$ m and $y = 0.25$ m, so $v = 1.95(9.81/2 \times 0.25)^{1/2}$
  $v = 8.64$ m/s
  Q = actual area of the jet × actual velocity
  $Q = C_c \times$ area of the hole × v (where C_c = about 0.60 for a sharp edged orifice)
  $Q = 0.60 \times (\pi \times 0.02^2/4) \times 8.64$
  $Q = 0.0016$ m³/s
  (This is only a rough estimate, but this is often better than no estimate at all.)

(b) From equation (5.11): actual velocity $= C_v(2gH)^{1/2}$
  Say C_v has a value of about 0.95, and from above the actual velocity, $v = 8.64$ m/s.
  $8.64 = 0.95(2 \times 9.81 \times H)^{1/2}$
  $H = 4.22$ m

5.4.3 Discharge through a submerged small orifice

This condition may arise when liquid flows through an orifice in the dividing wall between two tanks, for example (Fig. 5.7). However, an equation of the form derived below is also applied to many other situations, even if the theoretical justification for doing so is dubious. For instance, this pragmatic approach leads to the drowned orifice equation being applied

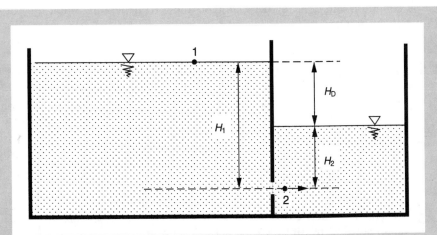

Figure 5.7 Flow through a submerged orifice in the dividing wall between two tanks. The pressure at the vena contracta is assumed to equal the hydrostatic head, H_2. The discharge through the orifice depends upon the differential head, H_D

to bridge waterways when the water level is above the top of the opening and the water-way is running full (see also sections 5.4.5 and 9.3.3).

The energy equation is again used to derive the discharge equation for a drowned orifice, but the assumptions are modified to suit the new conditions. The equation is applied to two points on a streamline: point 1 on the surface of the upstream reservoir, and point 2 at the centre of the vena contracta. Assuming no loss of energy and taking the datum level through point 2:

$$z_1 + V_1^2/2g + P_1/\rho g = z_2 + V_2^2/2g + P_2/\rho g$$

Considering point 1, the conditions are as before with $z_1 = H_1$, $V_1 = 0$ and $P_1 = 0$ (taking atmospheric pressure as the datum). At point 2, $z_2 = 0$, and V_2 is the value we are trying to determine. In this situation, however, we cannot assume that P_2 = atmospheric pressure since it is beneath the surface. A reasonable assumption is that the pressure at point 2 equals the hydrostatic pressure at the depth H_2, that is H_2 m of liquid. (REMEMBER!!!) Thus:

$$H_1 = V_2^2/2g + H_2$$
$$\text{Hence} \quad V_2 = (2g[H_1 - H_2])^{1/2}$$
$$\text{or} \quad V_2 = (2gH_D)^{1/2} \tag{5.16}$$

This is basically the same as equation (5.8), except that the differential head, H_D, is used instead of the head above the orifice. This is perfectly logical. If the water level in the two tanks is the same there will be no flow, so the head in equation (5.16) has to be the differential head because under these circumstances $H_D = 0$ whereas (REMEMBER!!!) $H_1 \neq 0$. The actual discharge is:

$$Q_A = C_D A (2gH_D)^{1/2} \tag{5.17}$$

where C_D is the coefficient of discharge and A is the area of the orifice, as before. For drowned sharp edged circular, square and rectangular small orifices C_D values are similar to the free condition and typically between 0.60 and 0.62 depending upon the circumstances (Brater et al., 1996).

5.4.4 The orifice meter

The orifice meter performs the same function as a Venturi meter, namely measuring the discharge in a pipeline, and is an alternative to the Venturi. The orifice meter is basically just a sharp edged orifice located in a pipe with tappings to measure the pressure on either side of it (Fig. 5.8). Its advantages over a Venturi are that it is cheaper, easily inserted at any flange in a pipeline, and compactness (although for accuracy it does require a certain length of straight pipe on either side of it). Its disadvantages are that it causes a considerable head loss due to the sudden downstream expansion of the flow, and its relatively thin edges can be damaged leading to reduced accuracy. Nevertheless, a properly installed orifice meter in accordance with British Standard 1042 ought to be accurate to ±2%.

The discharge equation for the orifice meter is exactly the same as that for the Venturi meter, and is derived in an identical manner. The only noteworthy points are that the orifice plate creates a vena contracta (the Venturi's convergent cone does not) and that the cross-sectional area of flow at the vena contracta is not equal to the area of the orifice. Although the area of the orifice, A, is used in discharge equation (5.5) for convenience, the contrac-

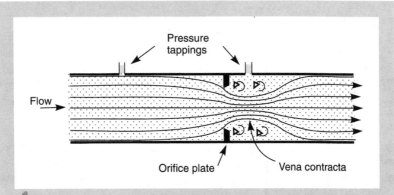

Figure 5.8 An orifice meter, consisting of an orifice plate installed at a flange in a pipeline, and two pressure tappings. Note that the pipe is full of water, with a sudden expansion of the flow downstream of the orifice from the width of the vena contracta to the full pipe width

tion of the flow and the energy loss in the subsequent expansion mean that a coefficient of discharge of around 0.65 is typical, compared to about 0.97 for the Venturi.

5.4.5 Free discharge through a large orifice

Whereas small orifices are usually round, large orifices can be almost any shape. In fact it is more convenient if they are not round, for reasons that will become apparent. Large orifices can be square or rectangular holes in a concrete wall or tank that are designed to allow water to overflow or escape in a controlled manner, or bridge waterways that are operating with the upstream face of the opening submerged, or partially open sluice gates.

As mentioned at the beginning of section 5.4, with a large orifice the head at the top of the opening is significantly different from that at the bottom, that is H_1 and H_2 respectively in Fig. 5.9. Thus the velocity at the top of the orifice is significantly different from that at the bottom since $V = (2gH)^{1/2}$. Because this is not a linear relationship, the average velocity of the jet cannot be obtained by simply averaging V_1 and V_2. Instead, the derivation of the discharge equation involves writing an expression to describe the discharge through a thin horizontal strip of the orifice at some particular depth, and then integrating this expression over the whole cross-sectional area of the orifice to obtain the total theoretical discharge.

Suppose that water discharges through the rectangular opening in Fig. 5.9. Consider a thin horizontal strip with a vertical height, δh, that extends the full breadth, b, of the orifice. The strip is at a depth, h, measured from the water surface. Then:

Area of the strip, $\delta A = b\delta h$
Velocity of flow through the strip $= (2gh)^{1/2}$
Discharge through the strip, $\delta Q = $ area × velocity
$$\delta Q = b\delta h(2gh)^{1/2}$$
Rearranging gives: $\qquad \delta Q = b(2gh)^{1/2}\delta h$

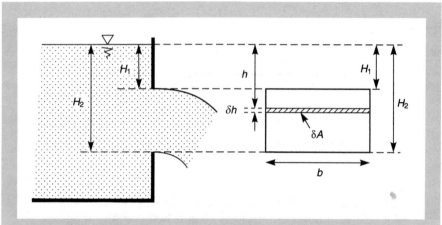

Figure 5.9 Discharge through a large orifice

To obtain the total theoretical discharge through the whole area of the orifice, integrate this expression to obtain the sum of all the horizontal strips as defined by the limits $h = H_1$ and $h = H_2$, that is the depth to the top and bottom of the opening respectively. Note that b and g are constants, so:

Total theoretical discharge, $Q_T = b(2g)^{1/2} \int_{H_1}^{H_2} h^{1/2} dh$

$$Q_T = \frac{2}{3} b(2g)^{1/2} \left[H_2^{3/2} - H_1^{3/2} \right] \tag{5.18}$$

To obtain the actual discharge, Q_A, a coefficient of discharge, C_D, has to be introduced:

$$Q_A = \frac{2}{3} C_D b(2g)^{1/2} \left[H_2^{3/2} - H_1^{3/2} \right] \tag{5.19}$$

One of the disadvantages of large orifices is that there is often no guidance as to the appropriate value of C_D, while its value may change with the degree of submergence of the orifice and its shape (see Example 5.6). Additionally it is impractical to run a laboratory test on a full size sluice gate or bridge, although model tests are possible (see Chapter 10).

Box 5.4 ▶ **Note for future use**

REMEMBER!!!

Look at the derivation of the discharge equation for a large orifice above. It consists of nothing more than saying $\delta Q = \delta AV$ where $V = (2gh)^{1/2}$ and then integrating between limits that define the top and bottom of the jet. Almost exactly the same method will be used to derive the equation for the discharge over a sharp crested weir. If you can understand what is being done and why, and if you can recognise the similarity between some of the measuring devices, then this will help you to remember how to derive the discharge equations.

EXAMPLE 5.6

Water from a large reservoir overflows through a series of rectangular openings that have a breadth of 4.0 m and a height of 2.0 m. The depth of water above the top of the openings is 0.9 m. (a) Calculate the theoretical discharge through each of the openings if they are treated as large orifices. (b) Calculate the theoretical discharge through each of the openings if they are considered to be small orifices, and obtain the percentage difference from the previous answer.

(a) Equation (5.18) is: $Q_T = \frac{2}{3} b(2g)^{1/2} \left[H_2^{3/2} - H_1^{3/2} \right]$

$b = 4.0$ m, $H_1 = 0.9$ m, and $H_2 = 2.9$ m

$Q_T = \frac{2}{3} \times 4.0 (2 \times 9.81)^{1/2} [2.9^{3/2} - 0.9^{3/2}]$

$Q_T = 11.812 [4.939 - 0.854]$

$Q_T = 48.252 \text{m}^3/\text{s}$

(b) If the opening is considered as a small orifice with the water surface 1.9 m above the centre of the opening, equation (5.9) is: $Q_T = A(2gH)^{1/2}$

$Q_T = 4.0 \times 2.0 (2 \times 9.81 \times 1.9)^{1/2}$

$Q_T = 48.845 \text{ m}^3/\text{s}$

Percentage difference $= 100 \times (48.845 - 48.252)/48.252 = +1.2\%$

The relatively small percentage difference perhaps explains why the small orifice equation is sometimes applied to situations where theory would suggest that the large orifice equation ought to be employed (although the error may be larger under different conditions). However, the real difficulty lies in knowing what value of C_D to use in order to obtain the actual discharge, which would be perhaps 0.6 (or even less?) of the values above. Obtaining accurate C_D values for a rectangular orifice (large or small) formed from concrete is difficult. As an example of the variation of the coefficient with respect to a bridge waterway flowing full with the opening submerged on the upstream side only, C_D can change from 0.2 to 0.3 just after submergence, to a maximum of about 0.5 when the depth of water upstream is 1.4 times the height of the opening. These coefficients were calculated from the small orifice equation, but using the total upstream head ($H + V^2/2g$) above the centre of the opening instead of H.

SELF TEST QUESTION 5.3

Water flows through a bridge waterway that can be assumed to have a perfectly rectangular opening 6 m wide and 3 m high. The discharge, as measured at a nearby gauging station, is 61.7 m³/s when the upstream water level is approximately 1.35 m above the top of the opening. On the downstream side of the bridge it can be assumed that the jet discharges freely into the river channel. Calculate the coefficient of discharge of the bridge under these conditions, first by assuming the opening is a large orifice, then by treating it as a small orifice.

5.5 ▶ Discharge over a sharp crested weir

Sharp crested weirs (or notches) are generally used to measure the discharge in small open channels where accuracy is required. Because such weirs can be accurate to ±2% or even ±1%, they are often used as measuring devices in hydraulics laboratories, but they also have practical applications. For instance, the seepage through a dam may be measured by channelling it over a sharp crested weir. Because the weirs have thin, sharp crests they are not suitable for measuring the discharge in large rivers where they would be prone to damage by the impact of floating debris. Concrete structures like the broad crested weir (see section 9.5) are used for this, and they operate on a totally different principle from those described here. The two types of weir should not be confused.

A sharp crested weir is usually formed from a sheet of brass or stainless steel. A notch is then cut out of the plate, the shape of which defines the geometry of the weir (Fig. 5.10). Common shapes for sharp crested weirs are rectangular, triangular and trapezoidal. The weir must have an accurately finished square upstream edge, a crest width of less than 2 mm with a bevel on the downstream side. With prolonged use the crest may become worn and rounded, and this can adversely affect the accuracy.

The weir plate must be installed with the upstream face vertical. Normally the length of the weir crest is less than the width of the channel ($b < B$) so, in plan, the flow has to contract to pass through it. Similarly, the crest is usually set above the bottom of the channel, so the streamlines have to rise upwards to pass over the weir (Fig. 5.11). The crest, or sill, of the weir has to be high enough for the water to fall freely into the downstream channel, so that the flow over the weir is not affected by the downstream water level. The water flowing over the crest, that is the **nappe** or **vein**, should spring clear of the downstream face of the weir plate. This is the 'free' condition in which both the upper and lower sur-

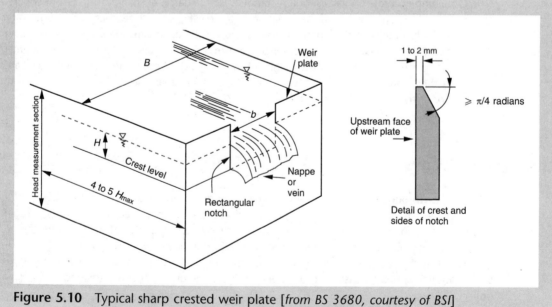

Figure 5.10 Typical sharp crested weir plate [*from BS 3680, courtesy of BSI*]

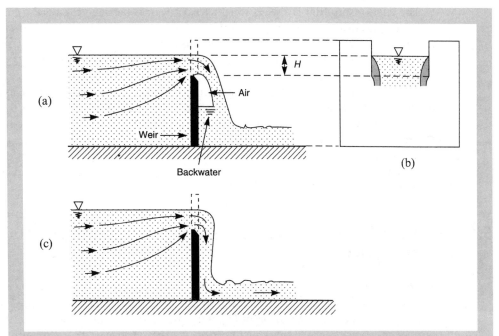

Figure 5.11 Free discharge condition in longitudinal section (a) and end view (b). (c) A clinging nappe in longitudinal section. There is no air under the nappe and the water clings to the weir plate

faces of the nappe are exposed to the atmosphere. This enables a convenient assumption to be made, namely that the pressure distribution throughout the nappe is close to atmospheric (remember, a similar assumption was made for the small orifice, that the pressure at the vena contracta was atmospheric).

If the sort of weir plate in Fig. 5.10 is used, then the free condition should exist naturally at all but the smallest of discharges. However, if the weir crest spans the full width of the channel so that $b = B$, this is called the **suppressed** condition. This maximises the discharge over the weir for a particular channel width, which may be desirable in some circumstances, but means that the air under the nappe is now trapped. Gradually the air becomes entrained in the nappe and is carried away. This leaves air at low pressure under the nappe, which enables the backwater to rise. When most or all of the air has been removed, the nappe collapses and adheres to the face of the weir plate, forming a **clinging nappe** as in Fig. 5.11c. This is undesirable because the weir will not function as an accurate measuring device like this. To prevent the formation of a clinging nappe with the suppressed condition, it is necessary to provide an air vent or pipe to admit air to the underside of the nappe.

If the discharge in a relatively wide channel is to be measured, then more than one weir plate may be used. In this case the individual weir plates would be attached to vertical posts. It may then be desirable to set the weir crests at different heights, forming a **compound weir**. This is often done to improve accuracy: when the discharge is small all the water passes over only the lowest crest, but as the flow increases the other weirs come into use.

The relationship that is always sought with a weir is between the head, H, over the weir crest and the discharge, Q. Note that H is always the **head above the weir crest** (not the total depth of water in the channel). Note also that because the weir has a smaller cross-sectional area of flow than the approach channel, the continuity equation $Q = AV$ dictates that the velocity over the weir crest must be higher than the velocity in the approach channel. This increase in velocity means that the velocity head increases so, assuming that the total energy line is horizontal, the water surface falls towards the weir as the flow accelerates. Consequently the head over the weir is usually measured some distance upstream where the velocity head is still relatively small. A distance of at least $4 \times H_{MAX}$ is desirable, where H_{MAX} is the maximum head that will occur over the weir. The head is usually measured in a stilling well or tube which is located to one side of the channel and connected to it by a pipe.

5.5.1 Derivation of the basic discharge equation for a rectangular weir

A rectangular weir can be thought of as a large rectangular orifice where the water surface has fallen below the top of the opening, so the weir equation can be obtained by adjusting that for the orifice. A comparison of Fig. 5.9 with Fig. 5.12a and equation (5.19) with equation (5.22) shows there is some justification for this. However, there are also some differences so the weir equation will be derived fully below.

In Fig. 5.12a the flow of water over a weir is shown in an idealised way with a horizontal water surface and nappe. The energy equation will be applied to points 1 and 2 on the streamline (or a cross-section through 1 and 2 as we use the mean velocity $V = Q/A$). Point 1 is some distance upstream of the weir. Point 2 is in the nappe as it passes over the weir crest, at a depth h below the water surface. There are a number of assumptions that are of importance with respect to the derivation:

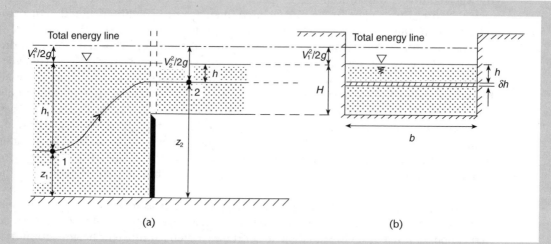

Figure 5.12 (a) Longitudinal section through a simplified weir and nappe. The energy equation is applied to points 1 and 2 on the streamline. (b) End view of the weir

(i) At section 1 upstream of the weir the velocity distribution is uniform. The approach velocity V_1 is relatively small compared to V_2, so below it is assumed $V_1 = 0$. If a hydrostatic pressure distribution is assumed then $h_1 = P_1/\rho g$. This essentially treats the upstream channel as a large reservoir.

(ii) The streamlines are horizontal as they pass over the weir crest.

(iii) The water in the nappe is surrounded by the atmosphere, so the nappe is assumed to be at atmospheric pressure. Thus with atmospheric pressure as the datum, $P_2 = 0$. Note that $P_2/\rho g = 0$ so there is no vertical line representing the pressure head at 2; there can only be the elevation head (z_2) and velocity head ($V_2^2/2g$).

(iv) The nappe is as wide as the weir crest, that is it also has a length b.

(v) There is no loss of energy.

Applying the energy equation to 1 and 2 using the channel bed as the datum level:

$$z_1 + P_1/\rho g + V_1^2/2g = z_2 + P_2/\rho g + V_2^2/2g$$
$$z_1 + h_1 + 0 = (z_1 + h_1 - h) + 0 + V_2^2/2g$$
$$V_2 = (2gh)^{1/2} \tag{5.20}$$

Thus the velocity in the nappe varies with depth, as in equation (5.8). If a thin horizontal strip of length b and thickness δh is taken across the nappe at a depth h, as in Fig. 5.12b then:

Area of the strip, $\delta A = b\delta h$
Velocity of flow through the strip $= (2gh)^{1/2}$
Discharge through the strip, δQ = area × velocity
$\qquad\qquad\qquad\qquad\qquad = b(2gh)^{1/2}\delta h$

To determine the total theoretical discharge, Q_T, the above expression must be integrated to obtain the sum of all the horizontal strips covering the entire depth of the nappe as defined by the limits $h = 0$ and $h = H$. Note that b and g are both constants.

$$Q_T = b(2g)^{1/2} \int_0^H h^{1/2} dh$$
$$Q_T = \frac{2}{3} b(2g)^{1/2} H^{3/2} \tag{5.21}$$

The $\frac{2}{3}$ arises from the integration. Equation (5.21) is the same as equation (5.18) with the upper limit H_1 omitted since this is now zero (the water surface). To obtain the actual discharge, a coefficient of discharge, C_D, is introduced so that:

$$Q_A = \frac{2}{3} C_D b(2g)^{1/2} H^{3/2} \tag{5.22}$$

A typical value for C_D is about 0.62. However, the value of the coefficient of discharge is found to vary slightly with discharge (see BS 3680: Part 4A). This is partly because the nappe contracts when seen in plan (Fig. 5.13b), resulting in the **effective length** of the weir changing with discharge. The greater the discharge, the larger the velocity, the greater the contraction of the flow at the sides of the weir, the smaller the effective length, L_E, of the weir crest. Consequently assumption (iv) above is not valid. For an accurate measurement of the discharge, L_E should be used in equation (5.22) instead of b. Francis discovered by experiment that for a rectangular weir with $b > 3H$ the **side contractions** average $0.1H$ for every side that is affected, where H is the head over the weir crest. Thus the effective length of the weir is:

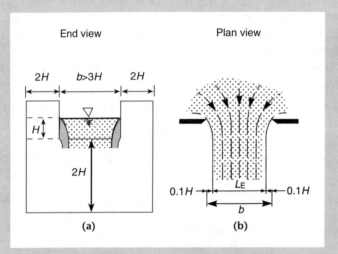

Figure 5.13 (a) End view showing the limiting or standard proportions of the weir used by Francis. (b) Plan view showing the end contractions (0.1H) and effective length of the weir, L_E

$$L_E = (b - 0.1nH) \tag{5.23}$$

where n is the number of side contractions. Thus $n = 2$ for a standard weir plate like that in Figs 5.10 and 5.13b, and $n = 0$ for a suppressed weir that has a crest length equal to the width of the channel. For a compound weir, consisting of a number of different weir plates, n may have a value larger than two (4, 6, etc.) if each weir results in two side contractions, so the effective length of the weir crest may be significantly less than the width of the channel. By using the effective length a more accurate measurement of the discharge should be obtained.

5.5.2 Discharge over a rectangular weir allowing for the velocity of approach

Assumption (i) above was that the water discharges over the weir from a large reservoir. This conveniently allowed us to assume that $V_1 = 0$ so that the term $V_1^2/2g$ could be omitted from equations (5.21) and (5.22). However, normally the water does not flow over the weir from a large reservoir, but from a relatively narrow channel within which water is flowing towards the weir. This means that the velocity of the water approaching the weir, V_1, may be significant, and it should be allowed for in the discharge equation (Fig. 5.12). The other assumptions listed in section 5.5.1 remain the same.

To allow for the velocity of approach we have to go back to the derivation of equation (5.20). If we now say that $V_1 \neq 0$, but everything else is as before:

$$z_1 + h_1 + V_1^2/2g = (z_1 + h_1 - h) + 0 + V_2^2/2g$$

$$V_2^2/2g = V_1^2/2g + h \tag{5.24}$$

$$V_2 = (2g[V_1^2/2g + h])^{1/2} \tag{5.25}$$

Thus the velocity through a horizontal strip of the nappe depends upon both the velocity head of the approaching flow $V_1^2/2g$ and h. Following a similar procedure as before, the discharge through the elemental strip of thickness, δh, is:

$$\delta Q = b(2g[V_1^2/2g + h])^{1/2}\,\delta h$$

To determine the total discharge this expression must be integrated between $h = 0$ and $h = H$ to obtain the sum of all the horizontal strips covering the entire depth of the nappe.

Total theoretical discharge, $Q_T = b(2g)^{1/2}\int_0^H [V_1^2/2g + h]^{1/2}\,dh$

$$Q_T = \frac{2}{3}b(2g)^{1/2}\left[(V_1^2/2g + H)^{3/2} - (V_1^2/2g)^{3/2}\right] \tag{5.26}$$

where H is the head of water over the weir crest. The actual discharge, Q_A, is:

$$Q_A = \frac{2}{3}C_D b(2g)^{1/2}\left[(V_1^2/2g + H)^{3/2} - (V_1^2/2g)^{3/2}\right] \tag{5.27}$$

where C_D is the coefficient of discharge. This is the equivalent of equation (5.22) but with the velocity of approach included. The problem is that without knowing V_1 we cannot use equation (5.27). Conversely, if V_1 is known there is no need for equation (5.27) because $Q_A = A_1 V_1$. The solution to this problem involves an iterative approach, as follows:

(1) Assume that the velocity of approach, $V_1 = 0$.

(2) Estimate the actual discharge using equation (5.22): $Q_A = \frac{2}{3}C_D b(2g)^{1/2}H^{3/2}$.

(3) Estimate the approximate value of the velocity of approach from $V_1 = Q_A/A_1$.

(4) Calculate the actual discharge, Q_A, from equation (5.27). This is the first iteration.

(5) Re-calculate V_1 as in step 3 using the new value of Q_A from step 4.

(6) Re-calculate the actual discharge, Q_A, as in step 4. This is the second iteration.

(7) Re-calculate V_1 as in step 3 using the new value of Q_A from step 6.

(8) Re-calculate the actual discharge, Q_A, as in step 6. This is the third iteration.

(9) Stop when there is no significant difference between successive values of Q_A.

It should be noted that the velocity of approach, V_1, is often small, in which case the velocity head is negligible. For instance, if $V_1 = 0.3\,\text{m/s}$, then $V_1^2/2g = 4.6\,\text{mm}$. This makes a small difference to the calculated discharge. Consequently it is not unusual for the velocity head to be neglected, or alternatively it may be incorporated into the coefficient of discharge if the weir is calibrated using equation (5.22). However, if the most accurate possible value of discharge is required (which may be the case in the laboratory, for example) or if the velocity of approach is large, then equation (5.27) gives a more accurate answer. This is illustrated by Example 5.7.

EXAMPLE 5.7

An experiment is being conducted in a laboratory channel. Accuracy is important. The channel is 0.40 m wide. The discharge down the channel is measured using a rectangular weir that has a crest 0.25 m long. The crest of the weir is set at a height of 0.10 m above the bottom of the channel. The head over the weir crest, measured at suitable distance upstream, is 0.19 m. The coefficient of discharge of the weir is 0.62. Ignoring side contractions:

(a) Calculate the discharge ignoring the velocity of approach.

(b) Calculate the discharge taking into account the velocity of approach.

(c) Calculate the percentage error in the discharge that results from ignoring the velocity of approach.

(a) Equation (5.22) is: $Q = \frac{2}{3}C_D b(2g)^{1/2}H^{3/2}$ where $C_D = 0.62$, $b = 0.25$m, $H = 0.19$m

$$Q = \frac{2}{3} \times 0.62 \times 0.25(2 \times 9.81)^{1/2}0.19^{3/2}$$

$$Q = 0.0379 \text{m}^3/\text{s}$$

(b) Using $Q = 0.0379 \text{m}^3/\text{s}$ as the first estimate of discharge, the velocity of approach is $V_1 = Q/A_1$ where A_1 is the cross-sectional area of the upstream channel (note that the depth in the channel $= 0.10 + 0.19 = 0.29$m).
Thus $V_1 = 0.0379/(0.40 \times 0.29) = 0.327$m/s.

From equation (5.27): $Q = \frac{2}{3}C_D b(2g)^{1/2}\left[(V_1^2/2g + H)^{3/2} - (V_1^2/2g)^{3/2}\right]$

$C_D = 0.62$, $b = 0.25$m, $H = 0.19$m and $V_1^2/2g = 0.327^2/(2 \times 9.81) = 0.0055$m.

$$Q_1 = \frac{2}{3} \times 0.62 \times 0.25(2 \times 9.81)^{1/2}\left[(0.0055 + 0.1900)^{3/2} - (0.0055)^{3/2}\right]$$

$$Q_1 = 0.4577[0.0864 - 0.0004]$$

$$Q_1 = 0.0394 \text{m}^3/\text{s} \text{ (this is a difference of 3.8\% on the first iteration)}$$

For the second iteration, re-calculate V_1 as $V_1 = 0.0394/(0.40 \times 0.29) = 0.340$m/s giving $V_1^2/2g = 0.0059$m.

$$Q_2 = 0.4577\left[(0.0059 + 0.1900)^{3/2} - (0.0059)^{3/2}\right]$$

$$Q_2 = 0.4577[0.0867 - 0.0005]$$

$$Q_2 = 0.0395 \text{m}^3/\text{s}$$

For the third iteration, $V_1 = 0.0395/(0.40 \times 0.29) = 0.341$m/s giving $V_1^2/2g = 0.0059$m. Hence $Q_3 = 0.0395 \text{m}^3/\text{s}$ as in the previous iteration.

(c) Percentage error from ignoring the velocity of approach
$= 100 \ (0.0395 - 0.0379)/0.0395 = 4.1\%$

The error is not large but in some circumstances it may be significant. Of course, the error increases with increasing velocity of approach.

5.5.3 Derivation of the discharge equation for a triangular weir

The triangular weir (or V notch) has several advantages over a rectangular weir. These stem from the fact that the width of the weir varies with the head (or discharge) over the weir. As the discharge decreases, so does the width of the weir, so preserving accuracy. Similarly, the reducing width means that for a particular discharge there is a larger head over a triangular weir than there would be over a rectangular weir (see Example 5.8). This increased sensitivity means that the velocity of approach can be ignored with a triangular weir, as can any contraction of the nappe. Thus the triangular weir is simpler to

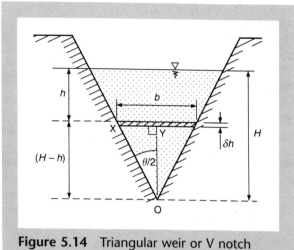

Figure 5.14 Triangular weir or V notch

use than a rectangular weir, and is ideal where small flows have to be measured accurately. Its main disadvantage is that with large discharges it requires a relatively big head over the weir crest, and thus may raise upstream water levels considerably.

The discharge equation for the triangular weir is derived using the same assumptions and procedure as for the rectangular weir. Taking a thin horizontal strip of breadth, b, and thickness, δh, across the nappe at a depth, h, from the water surface as in Fig. 5.14, then:

Area of the strip, $\delta A = b\delta h$
Velocity of flow through the strip $= (2gh)^{1/2}$
Discharge through strip, $\delta Q = $ area $\times$ velocity
$$= b(2gh)^{1/2}\delta h \qquad (1)$$

There is one additional step in the derivation for the triangular weir, and that is to write an equation for the variation of the width of the weir, b, with depth, h. This can be done by considering the triangle OXY. Thus:

$$\tan(\theta/2) = (b/2)/(H-h)$$
$$\text{or} \quad b = 2\tan(\theta/2)(H-h) \qquad (2)$$

Substituting for b in equation (1) above:

Discharge through strip, $\delta Q = 2\tan(\theta/2)(H-h)(2gh)^{1/2}\delta h$

To obtain the total theoretical discharge, integrate the above expression to obtain the sum of all of the horizontal strips covering the entire depth of the nappe, as defined by the limits $h = 0$ and $h = H$. Note that θ and g are constants.

Total theoretical discharge, $Q_T = 2\tan(\theta/2)(2g)^{1/2}\int_0^H (H-h)h^{1/2}\mathrm{d}h$

$$Q_T = 2\tan(\theta/2)(2g)^{1/2}\int_0^H (Hh^{1/2}-h^{3/2})\mathrm{d}h$$

$$Q_T = 2\tan(\theta/2)(2g)^{1/2}\left[\frac{2Hh^{3/2}}{3}-\frac{2h^{5/2}}{5}\right]_0^H$$

$$Q_T = 2\tan(\theta/2)(2g)^{1/2}\left[\frac{10H^{5/2} - 6H^{5/2}}{15}\right]$$

$$Q_T = 2\tan(\theta/2)(2g)^{1/2}\left[\frac{4H^{5/2}}{15}\right]$$

$$Q_T = \frac{8}{15}\tan(\theta/2)(2g)^{1/2}H^{5/2} \tag{5.28}$$

Note that the $\frac{4}{15}$ term in the square bracket arises from the integration, 15 being the lowest common denominator for 3 and 5. It often seems to be a strange fraction to appear in an equation, but this is the reason. To obtain the actual discharge C_D is introduced, as before:

$$Q_A = \frac{8}{15}C_D\tan(\theta/2)(2g)^{1/2}H^{5/2} \tag{5.29}$$

A typical value of C_D for a triangular weir is 0.58 if θ is between about 45° and 120°.

EXAMPLE 5.8

Water flows down a channel at a rate of 0.053 m³/s. What would be the head over a triangular and a rectangular weir at this discharge if the rectangular weir has a crest length of 0.3 m and the triangular weir has a total angle of 60°? Ignore the velocity of approach and the effect of side contractions when dealing with the rectangular weir and take the C_D for both weirs as 0.60.

Equation (5.22) for a rectangular weir is: $Q = \frac{2}{3}C_Db(2g)^{1/2}H^{3/2}$ where $Q = 0.053$ m³/s, $C_D = 0.60$, $b = 0.3$ m thus:

$$0.053 = \frac{2}{3} \times 0.60 \times 0.30(2 \times 9.81)^{1/2}H^{3/2}$$

$$H^{3/2} = 0.100$$

$$H = 0.215\,\text{m}$$

Equation (5.29) for a triangular weir is: $Q = \frac{8}{15}C_D\tan(\theta/2)(2g)^{1/2}H^{5/2}$ where $Q = 0.053$ m³/s, $C_D = 0.60$, $\theta/2 = 30°$ (note that the equation uses the half angle)

$$0.053 = \frac{8}{15} \times 0.60 \times \tan 30° (2 \times 9.81)^{1/2}H^{5/2}$$

$$H^{5/2} = 0.065$$

$$H = 0.335\,\text{m}$$

5.5.4 Trapezoidal and Sutro weirs

The trapezoidal or Cipolletti weir is simply a combination of a rectangular and a triangular weir, and its discharge equation can be derived accordingly (see Self Test Question 5.4). If the side slope of a Cipolletti weir is 1 horizontal to 4 vertical, then the increase in the length of the weir with increasing stage more or less balances the reduction in the effective crest length caused by the side contractions. Typically C_D is about 0.63.

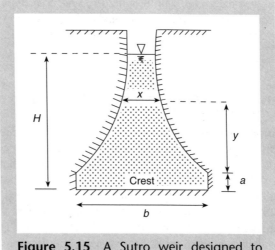

Figure 5.15 A Sutro weir designed to have a linear head discharge relationship, i.e., $Q \propto H^{1.0}$

The Sutro weir is designed for use in a rectangular channel as a flow meter, as a control for irrigation outlets, or with float regulated chemical dosing. Its distinguishing characteristic is that once the stage is above an arbitrary reference level the discharge (Q) is linearly proportional to the head over the weir (H), i.e. $Q \propto H^{1.0}$. The shape of the weir is as shown in Fig. 5.15. Above the horizontal crest there is a small rectangular section of height a. Above this reference level both sides curve inwards towards the centre (alternatively, one side can be vertical and one side curved). Sutro proposed a hyperbolic curvature of the form:

$$x = b\left[1 - \frac{2}{\pi}\tan^{-1}\sqrt{\frac{y}{a}}\right]$$

(5.30)

where b is the crest length, x is the width of the weir and y is the height above the reference level a. Note that $\tan^{-1}$ is in radians. The head–discharge relationship is:

$$Q_A = C_D b\sqrt{2ga}(H - a/3)$$

(5.31)

where C_D typically has a value of between 0.600 and 0.625 depending upon the value of a, b and whether the weir is symmetrical or unsymmetrical (French, 1986). To ensure $Q \propto H$ the nappe should be free so the tailwater level must be well below the weir crest; $H \geq 1.2a$ and never less than 0.03 m (to avoid the effect of viscosity and surface tension); the width $b \geq 0.15$ m, and the weir should be fully contracted.

SELF TEST QUESTION 5.4

(a) Using the same procedure as for a rectangular and triangular weir, derive from first principles the equation for the discharge over a trapezoidal weir of crest length L and sides that slope at an angle θ to the vertical. Ignore the velocity of approach and side contractions.

(b) A trapezoidal weir has a crest length of 0.20 m, sides that slope at 20° to the vertical, and a coefficient of discharge of 0.60. If the head of water over the crest of the weir is 0.17 m, calculate the actual discharge using the equation from part (a).

5.6 Calibration of flow measuring devices

If a flow measuring device such as an orifice plate or sharp crested weir is designed according to the British Standard (or the equivalent), the value of the coefficient of discharge, C_D, can often be obtained from the document. On the other hand, if the device is non-standard or is installed in a non-standard manner, the device should be calibrated. For most devices, calibration basically involves comparing the actual and theoretical discharge over the widest possible range of flows so that the average C_D can be calculated (equation (5.1)). The actual discharge can be determined by collecting either a given volume of water in a known time, or a given mass of water in a known time. If a mass of water is collected this can be turned into a volume, and hence the volume flow rate, by dividing by the mass density. The theoretical discharge is obtained by measuring the head of water and the dimensions of the device and substituting the values into the appropriate equation.

The reliability of the calculations is often improved if a graph of discharge against head is drawn. This enables the average coefficient to be calculated and also allows any obvious errors in the data to be spotted. However, a plot of actual discharge, Q_A, against the head, H, does not yield a straight line because Q_A is proportional not to H but to H raised to some power, as Table 5.1 shows.

Because Q_A is proportional to H^N (where N is the power in the last column of Table 5.1) this means that the discharge-head relationships are curves when plotted to a natural scale. It is useful to be aware of the shape of these curves when plotting experimental data. As a simple illustration, say that $H = 0, 1, 2, 3, 4, 5$ and 6. The corresponding values of Q are shown in Table 5.2 assuming that $Q = H^{1/2}$, $Q = H$, $Q = H^{3/2}$ and $Q = H^{5/2}$. These data are then shown plotted in Fig. 5.16 (note the different vertical scales).

Table 5.2 shows that the various relationships yield widely differing values of Q when $H = 6$. Thus the powers of H are not just a mathematical fact with no practical significance. Quite the reverse. The power determines the shape of the discharge-head curve, how sensitive a measuring device is, and how effectively it can cope with a wide range of discharges.

Table 5.1 Comparison of principal discharge equations

Device	Discharge equation	Q–H relationship
Venturi meter	$Q_A = C_D A_1 \sqrt{\dfrac{2gH}{\left[(A_1/A_2)^2 - 1\right]}}$	$Q_A \propto H^{1/2}$
Small orifice	$Q_A = C_D A \sqrt{2gH}$	$Q_A \propto H^{1/2}$
Large orifice	$Q_A = \frac{2}{3} C_D b (2g)^{1/2} [H_2^{3/2} - H_1^{3/2}]$	$Q_A \propto H^{3/2}$
Rectangular weir	$Q_A = \frac{2}{3} C_D b (2g)^{1/2} H^{3/2}$	$Q_A \propto H^{3/2}$
Triangular weir	$Q_A = \frac{8}{15} C_D \tan(\theta/2)(2g)^{1/2} H^{5/2}$	$Q_A \propto H^{5/2}$

Table 5.2 Illustration of variation of Q with H^N

H	$Q = H^{1/2}$	$Q = H$	$Q = H^{3/2}$	$Q = H^{5/2}$
0	0	0	0	0
1	1	1	1	1
2	1.414	2	2.828	5.657
3	1.732	3	5.196	15.588
4	2	4	8	32
5	2.236	5	11.180	55.902
6	2.449	6	14.697	88.182

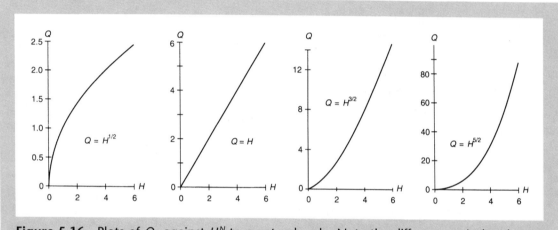

Figure 5.16 Plots of Q_A against H^N to a natural scale. Note the different vertical scales

The calibration process can be made easier if the graph is not drawn as Q_A against H (as in Fig. 5.16) but as Q_A against H^N. For example, with a Venturi meter a graph of Q_A against $H^{1/2}$ should plot as a straight line, whereas for a triangular weir a graph of Q_A against $H^{5/2}$ should result in a straight line. A straight line is, of course, easier to draw, extrapolate and analyse than a curve. The straight line plot must always pass through the origin, because if it did not this would indicate something impossible, like a discharge with no water present. As an illustration, take the data shown in Table 5.3 which were obtained from a small rectangular weir used in a hydraulics laboratory. The graph of Q_A against $H^{3/2}$ is shown in Fig. 5.17.

Figure 5.17 shows the experimental data with the best fit straight line drawn by eye. It is often found that either the data plots as a shallow curve, or that the best straight line through the plotted points does not pass through the origin. It is just possible to see in Fig. 5.17 that points A and B lie slightly below the line and point D just above. This could be due to error, or it could be that the constant or exponent used in the analysis is slightly inaccurate (this is explored in more detail later). Nevertheless, *the line must be drawn through the origin*. The discharge equation for a rectangular weir (Table 5.1) is:

$$Q_A = \frac{2}{3} C_D b (2g)^{1/2} H^{3/2}$$

$$(5.22)$$

Table 5.3 Actual discharge–head (Q_A–H) data for a small rectangular weir

Reading	Q_A ($\times10^{-3}\,m^3/s$)	H (m)	$H^{3/2}$
A	0.199	0.0260	0.0042
B	0.314	0.0346	0.0064
C	0.500	0.0463	0.0100
D	0.752	0.0604	0.0148

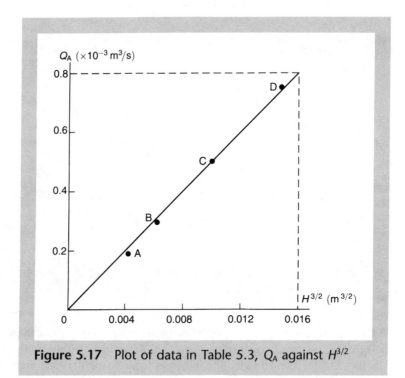

Figure 5.17 Plot of data in Table 5.3, Q_A against $H^{3/2}$

so $\quad C_D = \text{constant} \times Q_A/H^{3/2}$ (5.32)

where $\quad \text{constant} = \dfrac{3}{2b(2g)^{1/2}}$ (5.33)

In equation (5.32), the constant can be easily calculated while the value of $Q_A/H^{3/2}$ is the gradient of the line plotted in Fig. 5.17. Thus the average C_D can be calculated. For example, the line in Fig. 5.17 passes through the origin (0, 0) so:

gradient of line $= Q_A/H^{3/2} = 0.8 \times 10^{-3}/0.016 = 0.050$

with $b = 0.03$ m, the constant $= 3/[2 \times 0.030 \times (19.62)^{1/2}] = 11.288$

Thus $C_D = \text{constant} \times \text{gradient} = 11.288 \times 0.050 = 0.564$ (say 0.56)

Thus over the range of the experiment, equation (5.22) can be written as:

$$Q_A = (C_D/\text{constant}) \times H^{3/2}$$
$$= (0.564/11.288) \times H^{3/2}$$
$$Q_A = 0.050H^{3/2} \tag{5.34}$$

This analytical technique provides an easy way to calculate the average coefficient of discharge, from zero to the maximum recorded flow. Plotting the graph makes it easy to identify any points which are inconsistent with the other data, and which should be ignored. Evaluating the constant part of the equation avoids having to repeat the same calculation. However, one limitation of the technique is the assumption that the power (or exponent) of the discharge equation is as shown in Table 5.1. For instance, although normally $N = \frac{3}{2}$ for a rectangular weir, this is largely for convenience. There is no reason why the exponent, N, cannot be 1.45 or 1.53, say. However, if $N \neq 1.50$ then C_D may have a value which differs from the normal value (about 0.62). In other words, a number of alternative equations can be written that may fit the actual discharge–head data just as well, or even better, than the standard equation.

One method that can be used to investigate alternative forms of the discharge equation is to plot the data to a log scale, that is $\log Q_A$ against $\log H$ (Fig. 5.18). If this is done a straight line plot should be obtained, and:

$$\log Q_A = i + N\log H \tag{5.35}$$

where i is the intercept on the $\log Q_A$ axis (that is the value of $\log Q_A$ when $\log H = 0$) and N is the gradient of the line. If j is the antilog of i, then equation (5.35) can also be written as:

$$Q = jH^N \tag{5.36}$$

Thus j is the equivalent of 0.050 in equation (5.34), so C_D can be calculated since the other variables have known values. An illustration of this technique is provided by Example 5.9.

The concept that the discharge equation can have more than one form and still be accurate over the range of the data can be illustrated by comparing the discharges obtained from the 'standard' equation (that is equation (5.22), with C_D having the typical value of 0.62 and $b = 0.030\,\text{m}$), equation (5.34) (above) and equation (5.37) (in Example 5.9). The results are shown in Table 5.4.

$$Q_A = 0.055H^{1.5} \tag{5.22}$$

$$Q_A = 0.050H^{1.5} \tag{5.34}$$

$$Q_A = 0.061H^{1.568} \tag{5.37}$$

Table 5.4 shows quite clearly that the most inaccurate answers are obtained by adopting a typical value of the coefficient of discharge and then using the standard discharge equa-

Table 5.4 Comparison of calculated discharges (all $\times 10^{-3}\,\text{m}^3/\text{s}$)

	H (m)	Recorded discharge	Eqn (5.22)	Eqn (5.34)	Eqn (5.37)
A	0.0260	0.199	0.231	0.210	0.200
B	0.0346	0.314	0.354	0.322	0.312
C	0.0463	0.500	0.548	0.498	0.493
D	0.0604	0.752	0.816	0.742	0.748

tion (5.22). A more accurate answer is obtained by calibrating the weir, using either of the analytical techniques that led to equations (5.34) and (5.37). There is little to choose between these two equations which are, of course, only applicable in this precise form to the particular weir used to obtain the data in Table 5.4. However, note that calculations like those in Example 5.9 must be conducted very accurately, because even small errors in the gradient or the extrapolation can be significant.

The arguments applied above in connection with a rectangular weir work just as well for other weirs and measuring devices. However, a Pitot–static tube would probably have a specified coefficient when purchased and would not need to be calibrated. Similarly, velocity meters like those described below are already calibrated when purchased, although they need frequent recalibration. This can only be done by specialist hydraulics laboratories since it generally involves towing the meter at a fixed speed through a tank of water and noting the number of revolutions per second of the propeller. By altering the towing speed, a calibration curve can be built up relating revolutions per second to velocity. The very small pygmy meters are particularly troublesome, and need constant attention and recalibration if accurate results are required.

EXAMPLE 5.9

Using a log–log plot determine the discharge equation and C_D value from the data in Table 5.3.

Reading	$Q_A (\times 10^{-3}\,m^3/s)$	$H(m)$	$\log Q_A$	$\log H$
A	0.199	0.0260	−3.701	−1.585
B	0.314	0.0346	−3.503	−1.461
C	0.500	0.0463	−3.301	−1.334
D	0.752	0.0604	−3.124	−1.219

From Fig. 5.18, exponent N = gradient of line = $-(3.722 - 3.095)/ -(1.600 - 1.200)$

$$= -0.627/-0.400$$

$$N = 1.568$$

Between $\log H = -1.2$ and $\log H = 0$ change in $\log Q_A = 1.568 \times -1.200 = -1.882$. Therefore intercept on $\log Q_A$ axis (when $\log H = 0$) = $-(3.905 - 1.882) = -1.213$

$i = -1.213$

$j = \text{antilog}(-1.213) = 0.0612$

$Q_A = jH^N$ (5.36)

$Q_A = 0.0612H^{1.568}$ (5.37)

Also $j = \frac{2}{3}C_D b(2g)^{1/2}$ so $0.0612 = \frac{2}{3}C_D \times 0.03 \times 4.429$

$C_D = 0.69$

5.7 Velocity meters

The velocity meter is different from the other devices described above in that there is really no theory involved. The principle is simply that the speed of rotation of a particular propeller depends upon the velocity of the water. Velocity meters (also called current meters) can be obtained in different sizes and varying designs. Three horizontal axis propeller meters are shown in Fig. 5.19. The largest is used for gauging flood flows and is generally positioned in a river by means of a suspension cable. It may be used with a sinker weight, to stop it being carried downstream, and tail fins, to keep it pointed directly into the flow. The intermediate size is the sort often employed for gauging a river under normal conditions, and it is usually positioned via a hand-held wading rod. It is pointed directly into the flow manually. The small pygmy meter is used for measuring velocities in laboratory channels and hydraulic models.

Some meters (usually the medium sized ones) can be obtained with more than one propeller. This makes the meter more versatile. For instance, a small propeller can be used if the depth of water is small, or the pitch of the propeller can be selected to suit the velocity of the flow. This may be necessary to prevent the propeller rotating either too slowly or too quickly. Each rotation of the propeller is converted into an electrical pulse, with the total number of pulses during a particular time interval being displayed by an electrical counter. The velocity can be obtained from an equation or chart provided by the manufacturer.

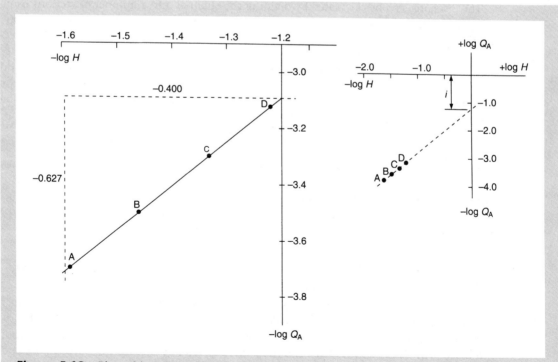

Figure 5.18 Plot of log Q_A against log H

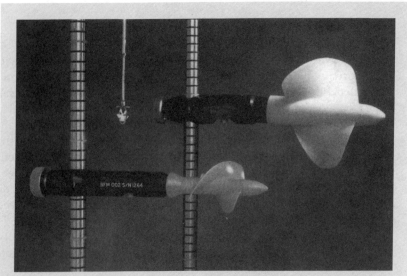

Figure 5.19 Three velocity meters of different size (about 10 mm, 25 mm and 125 mm diameter). The largest is used for river gauging during flood, the smallest for measuring velocities in the laboratory

The discharge can be calculated by dividing the channel into a number of sections using imaginary vertical lines, measuring the mean velocity in each section, and then multiplying the mean velocity by the area of the section. The total discharge is the sum of the sectional discharges. One method of obtaining the mean velocity is to measure the velocity on two adjacent vertical dividing lines that form the boundary of a particular section and then take the average. The depth of the water may be measured at the same time as the velocity, so the average of the depths multiplied by the width of the section (w) gives the area of the section. Thus the sectional discharge is obtained from the product of the mean velocity and the area:

$$q_n = \left(\frac{v_n + v_{n+1}}{2}\right)\left(\frac{d_n + d_{n+1}}{2}\right)w \qquad (5.38)$$

where subscript n is the number of the section, q is the sectional discharge, v the point velocity, d the depth on the dividing vertical and w the width of the section. There are, of course, a number of problems that make it difficult to get an accurate answer from this equation. These are:

1. The velocity can only be measured on a limited number of verticals across the width of the channel. The spacing used depends upon the channel width and the conditions.

2. On a particular dividing vertical, the velocity can only be measured at a limited number of depths. If the water is relatively shallow or the measurement has to be made in a hurry, it may be assumed that the average velocity occurs at 0.4 of the depth measured from the bed. This is not always true. Alternatively, the velocity may be measured at 0.2 and 0.8 of the depth and the average value used (Fig. 5.20).

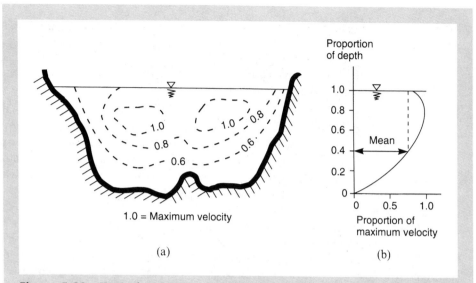

Figure 5.20 Typical variation of relative velocity in a river channel. (a) Cross-section showing isovels (contours of equal relative velocity), and (b) the velocity profile on a typical vertical section

3. In turbulent flow the magnitude and direction of the flow change quite rapidly with time.

4. It is difficult to measure the area of the sections accurately, particularly if the bed is covered in stones or boulders and the channel is far from rectangular.

Thus both the velocity and area are estimates that are likely to involve a significant error. Consequently it would not be surprising for a river gauging to contain an error of ±5% to ±15%, possibly up to 20% or 25% during a flood. The accuracy depends to some extent upon the site, how often it is used, the equipment employed, and the time available to complete the measurement. A more accurate answer may be obtained under controlled laboratory conditions. BS 3680 should be consulted for more details of gauging procedures and errors.

Summary

1. A coefficient of discharge (C_D) is used with most flow measuring devices. It is defined as:

$$C_D = Q_A/Q_T \qquad (5.1)$$

where Q_A is the actual measured discharge and Q_T is the discharge obtained using the theoretical equation. The C_D compensates for: the assumption of an ideal liquid (i.e. no viscosity, friction or turbulence and no energy loss); inadequacies in the theory; the height of a weir (p); the effect of surface tension; and the contraction of a jet. See Example 10.3 and Revision Question 10.3.

2. The equation for the discharge through a Venturi meter is derived by applying the energy equation to point 1 at the entrance and point 2 at the throat. This gives an equation for V_1. The continuity equation gives $Q_T = A_1 V_1$ and hence:

$$Q_A = C_D A_1 \sqrt{\frac{2gH}{[(A_1/A_2)^2 - 1]}} \qquad (5.5)$$

3. The energy equation leads to equation (5.8), Torricelli's theorem, namely $V = (2gh)^{1/2}$ where V is the velocity of flow at a depth h in a liquid (typically a jet or nappe). This and the continuity equation enables the discharge equation of many flow measuring devices to be derived: $Q_A = C_D A V$ where A is the cross-sectional area of the orifice/weir. This is illustrated by 4–7 below.

4. With a small orifice the jet velocity does not vary significantly from the top to bottom of the opening, so the head above the centre of the orifice (H) can be used to obtain the mean V. From 3 above:

$$Q_A = C_D A \sqrt{2gH} \qquad (5.12)$$

For a drowned orifice the differential head (H_D) is used instead of H, so $Q_A = 0$ when $H_D = 0$.

5. With a large rectangular orifice the head at the top (H_1) and bottom (H_2) of the orifice are significantly different, as are V_1 and V_2. Considering an elemental strip of breadth b and thickness δh, the discharge through

the strip is $\delta Q = b \delta h (2gh)^{1/2}$ and integrating between limits $h = H_1$ and $h = H_2$ gives:

$$Q_A = \frac{2}{3} C_D b (2g)^{1/2} [H_2^{3/2} - H_1^{3/2}] \qquad (5.19)$$

6. Ignoring the velocity of approach, the equation for the discharge over a sharp crested rectangular weir is derived as in 5 above, but in this case the water level is below the top of the orifice so $H_1 = 0$ and $H_2 = H$, the head over the weir, thus:

$$Q_A = \frac{2}{3} C_D b (2g)^{1/2} H^{3/2} \qquad (5.22)$$

This assumes the approach velocity in the upstream channel (V_1) is zero. If $V_1 \neq 0$ then the total head of the flow approaching the weir is increased by $V_1^2/2g$, so the discharge is increased and:

$$Q_A = \frac{2}{3} C_D b (2g)^{1/2} [(V_1^2/2g + H)^{3/2} - (V_1^2/2g)^{3/2}] \qquad (5.27)$$

7. The equation for a sharp crested triangular weir with a half-angle of $\theta/2$ is again based on the discharge through an elemental strip, $\delta Q = b \delta h (2gh)^{1/2}$. The width of the weir at any depth h from the water surface, i.e. at a height of $(H - h)$ above the weir crest, is $b = 2 \tan(\theta/2)(H - h)$. So after integrating:

$$Q_A = \frac{8}{15} C_D \tan(\theta/2)(2g)^{1/2} H^{5/2} \qquad (5.29)$$

Revision questions

5.1 Describe and explain what is meant by (a) coefficient of discharge; (b) hydraulic gradient; (c) piezometric level; (d) hydraulic grade line; (e) adverse pressure gradient; (f) boundary layer; (g) boundary layer separation; and (h) a vortex sheet.

5.2 (a) Where does most of the energy loss occur in a Venturi meter, and why is this the case? (b) A Venturi meter is being calibrated in the laboratory.

The meter has a diameter of 75 mm at the entrance and 50 mm at the throat. The differential head is 1.574 m. The flow rate is obtained by measuring the time required to collect a certain quantity of water. The average of a number of such measurements gives 0.614 m³ of water collected in 55.82 s. At this discharge, what is the value of the coefficient of discharge, C_D?

[(b) 0.90]

5.3 A combined Pitot–static tube has a coefficient of 0.98. The differential head reading is 0.874 m of water when it is positioned on the centreline of a pipe of constant diameter. (a) What is the velocity of flow at this point? (b) If the stagnation pressure head is 2.942 m of water, what is the pressure head of the flowing water?

[(a) 4.058 m/s; (b) 2.068 m of water]

5.4 (a) Explain the hydraulic difference between a small orifice and a large orifice. (b) With respect to a small orifice, define what is meant by the coefficient of velocity and the coefficient of contraction. Give typical values of each coefficient. (c) A small orifice with a diameter of 0.012 m discharges water under a head of 1.43 m. If C_D is 0.59, what is the actual discharge?

[(c) 0.00035 m³/s]

5.5 The tank in question 5.4 is covered with an airtight lid and the air space above the water is pressurised so that the flow increases to 0.00093 m³/s. If the other details remain the same, what is the pressure of the air?

[83.12 × 10³ N/m² above atmospheric]

5.6 A jet of water discharges from a small orifice. The trajectory of the jet is measured, and it is found to travel 2.7 m horizontally while dropping vertically through a distance of 0.9 m. (a) Calculate the velocity of the jet. (b) If the coefficient of velocity of the orifice is 0.98, calculate the head producing the flow.

[(a) 6.303 m/s; (b) 2.108 m]

5.7 (a) List the factors that control the discharge through a drowned orifice. (b) Water flows between two tanks through a drowned orifice that has a coefficient of discharge of 0.80. The head measured above the centre of the orifice is 2.45 m in the first tank and 1.13 m in the second tank. The orifice has a diameter of 15 mm. Calculate the flow rate between the tanks at the instant these measurements were taken.

[0.00072 m³/s]

5.8 (a) What procedure is used to derive the equation for the discharge through a large orifice? Is it the same as for a small orifice, and if not, why not? (b) Water discharges through a vertical sluice gate that can be considered to be a large orifice.

The sluice gate is 4.0 m wide and it is raised 1.3 m from the bed. The head of water above the top of the opening is 2.6 m, giving a total depth of 3.9 m above the bed. If the coefficient of discharge of the opening is 0.50, calculate the discharge.

[(b) 20.73 m³/s]

5.9 What is meant by (a) a clinging nappe; (b) a suppressed weir; (c) a compound weir; (d) the velocity of approach; and (e) end or side contractions? (f) A rectangular weir and a triangular weir are located in the same channel with their crests at the same level. Both weir plates have an opening 0.3 m wide at the top and 0.3 m deep, both weirs have a head of 0.25 m over their crest, and both have a C_D of 0.60. Ignoring the effect of side contractions and the approach velocity for the rectangular weir, calculate the proportion of the total combined discharge that passes over the triangular weir.

[25%]

5.10 Water approaches a rectangular weir through a channel 2.00 m wide and in which the depth of flow is 0.57 m. The weir has a crest length of 0.90 m located centrally in the channel. The crest is set at a height of 0.30 m above the bed. The C_D is 0.61 and the head over the crest is 0.27 m. (a) Ignoring the side contractions, calculate the discharge without considering the velocity of approach, and then with the approach velocity included. (b) Repeat the above calculations, but this time allowing for the side contractions as well.

[(a) 0.227 m³/s, 0.230 m³/s; (b) 0.214 m³/s; 0.216 m³/s]

5.11 A small sharp crested triangular weir with a half angle $(\theta/2)$ of 15° is calibrated in a laboratory. The actual discharge, Q_A, over the weir is measured by collecting a known mass (M) of water in a time, T. The corresponding head over the weir crest, H, is as shown below. (a) Plot a graph of Q_A against $H^{5/2}$ and use it to calculate the average value of the coefficient of discharge. (b) Plot a graph of $\log Q_A$ against $\log H$ and use it to evaluate (i) the exponent of the discharge equation; (ii) the constant of the discharge equation; and (iii) the corresponding coefficient of discharge.

[(a) 0.64; (b) (i) 2.41; (ii) 0.317 so $Q_A = 0.317H^{2.41}$; (iii) 0.50]

M (kg)	T (s)	H (m)
30	45.2	0.0769
15	36.7	0.0638
7.5	34.6	0.0488
15	93.8	0.0430
7.5	96.2	0.0320

Meter location	Horizontal distance (m)	Total depth of water (m)	Point velocity (m/s)
Right abutment	0	0.59	0.660
vertical 1	0.14	0.61	0.780
2	0.64	0.60	1.064
3	1.14	0.59	0.968
4	1.64	0.59	0.806
5	2.14	0.59	0.595
6	2.64	0.52	0.448
7	3.14	0.47	0.203
Left abutment	3.28	0.43	0

5.12 The table to the right shows the results from a river gauging conducted between the abutments of a bridge. Using the mean section method, calculate the discharge of the river (the width of the sections is obtained from the difference in the horizontal distances).

[1.34 m^3/s]

Flow through pipelines

Much of Britain's water supply is obtained from rainfall that has been collected in upland reservoirs and then piped to the consumer via a treatment works. Thus flow through pipelines is an important part of hydraulics. These pipelines flow full under pressure, and are analysed by applying the energy equation. This is a totally different approach from the analysis of gravity flow in open channels, such as rivers and partially full sewers where there is a free water surface at atmospheric pressure (see Chapter 8). This chapter begins by revising the concept of head and the application of the energy equation to situations involving a significant loss of energy. The energy equation is then used to calculate the flow rate when, for example, a reservoir discharges to the atmosphere through a long pipeline. The analysis of the flow between two reservoirs, flow in branching pipelines and flow in two or more parallel pipelines is then outlined.

The second part of the chapter develops in more detail the theory used in the first part. It starts by recalling the terms used to classify the flow in a pipe, describes the classic series of experiments conducted by Reynolds and explains the development of the equations for the evaluation of pipe friction. The influence of boundary layer, pipe roughness and Reynolds number on the form of the equations is outlined, and their empirical nature emphasised. Equations for the head loss due to pipe expansions and contractions are derived. The information presented enables questions like those below to be answered:

Why does water flow through a pipeline, and what controls it?

How can the loss of energy due to friction be evaluated?

What is the loss of energy at a sudden expansion or contraction of the pipeline?

If a reservoir discharges to the atmosphere via a pipeline, how is the flow rate obtained?

How can we calculate the flow in a pipeline connecting two reservoirs?

How can branching and parallel pipelines be dealt with?

What actually controls pipe friction? How does it vary?

Which equation for friction loss should be used under what circumstances?

6.1 Introduction

The energy equation (or Bernoulli equation) is applied to two points connected by a stream-line, often the centreline of the pipe. For two points, A and B, it can be written as:

$$z_A + \alpha_A V_A^2/2g + P_A/\rho g = z_B + \alpha_B V_B^2/2g + P_B/\rho g + \text{energy head losses} \qquad (4.24/6.1)$$

where z is the elevation of the point above a datum level (m), α is the velocity distribution coefficient (dimensionless, assumed = 1.0 for simplicity and therefore omitted below), V is the mean velocity of flow (m/s), P is the pressure (N/m^2), ρ is the mass density of the liquid (kg/m^3) and g is the acceleration due to gravity (9.81 m/s^2). The product ρg is the weight density, w, of the liquid (N/m^3). Thus all of the terms in equation (6.1) have the units of metres:

$$z = \text{elevation in metres above a datum} = m$$

$$\frac{V^2}{2g} = \left(\frac{m}{s}\right)^2\left(\frac{s^2}{m}\right) = m$$

$$\frac{P}{w} = \left(\frac{N}{m^2}\right)\left(\frac{m^3}{N}\right) = m$$

Because all of the terms can be represented by a length in metres, they are called **heads**: elevation (or potential), velocity and pressure head, respectively. The heads can be drawn as vertical lines whose length above the centreline of the pipe is proportional to the magnitude of the term, as shown in Fig. 6.1 (and Figs 4.19 and 5.1). At any point on the streamline:

$$z + P/\rho g = \text{ piezometric head} \qquad (6.2)$$

$$z + V^2/2g + P/\rho g = \text{ total head} \qquad (6.3)$$

The piezometric head or level represents the height to which water will rise in a piezome-ter (called a **stand-pipe** when dealing with pipelines). The line obtained by drawing the variation in piezometric level along the pipeline is called the **hydraulic grade line** (HGL). The hydraulic grade line represents the height to which water would rise in a stand-pipe at any point. It lies below the **total head line** (THL) by an amount equal to the velocity head, as apparent from equations (6.2) and (6.3) and Fig. 6.1.

The slope of the total head line and the hydraulic grade line is important. If there is no loss of energy in equation (6.1), then the total head line will be horizontal. If there is a gradual loss of energy (for instance due to pipe friction), then the total head line will fall in the direction of flow. The gradient of the total head line is proportional to the rate at which energy (head) is being lost; it is sometimes called the **friction gradient**, S_F. Steeply sloping lines indicate a large loss, and almost horizontal lines a small loss. If there is a sudden loss of energy (say, due to a sudden change in pipe diame-ter), then this will result in a vertical downwards step in the total head line.

REMEMBER!!!

The slope of the hydraulic grade line equals that of the total head line since it lies below the total head line by an amount equal to the velocity head, which is constant in a pipe of uniform diameter. If the velocity in the pipeline suddenly increases (say, due to a reduction in pipe diameter), there will be a vertical downward step in the hydraulic grade line. However, if there is a sudden reduction in velocity (due to an increase in pipe diameter for example) then there will be a sudden upward step in the hydraulic grade line. Therefore, the hydraulic grade line is different from the total energy line in that it can rise

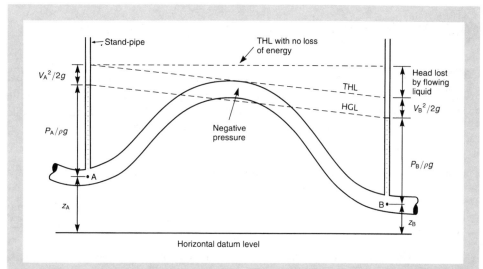

Figure 6.1 Elevation head, pressure head and velocity head at two points on a pipeline. The hydraulic grade line (HGL) lies below the total head line (THL) by an amount equal to the velocity head. A negative pressure occurs if the pipe centreline rises above the HGL

in the direction of flow under some circumstances, whereas the total head line must always fall in the direction of flow.

The gradient of the hydraulic grade line (for instance, 1 in 100) is called the **hydraulic gradient**, although the hydraulic grade line itself is often referred to as the hydraulic gradient. Generally the hydraulic gradient is in the direction of flow, the exception being when there is an increase in cross-sectional area, when it may rise. Remember that when a conduit expands there is usually a significant energy loss as a result of the **adverse hydraulic gradient**. Thus a sharp rise in the hydraulic grade line may be accompanied by a sharp fall in the total head line. Go back and read section 5.2.1.

Box 6.1 ▶ **For future use** REMEMBER!!!

1. The total head line always falls in the direction of flow (the exception being if there is a pump in the pipeline). The steeper the gradient of the total head line, the greater the rate of loss of energy.

2. The hydraulic grade line generally falls in the direction of flow but it can also go up, such as at an increase in section. As a memory aid, think of the line representing the top of the pipeline drawn in longitudinal section. The hydraulic grade line changes in the same way as this line. For example, if the pipeline diameter suddenly increases, then the line representing the top of the pipe REMEMBER!!! would go up, and so would the hydraulic grade line.

3. Adverse hydraulic gradients, as may occur at a sudden expansion, often result in large head losses.

6.2 Understanding reservoir — pipeline flow

❝ I have never really understood what causes water to flow through a pipeline. Suppose there is a reservoir up in the hills and the water has to be transferred to somewhere 50 km away. What causes the water to flow through the pipeline, and to keep on flowing through it, despite all the friction and other energy losses? ❞

❝ OK, good question. It is the difference in elevation, Z, that causes the flow: the larger Z, the larger the flow through a given pipeline. Of course, this only works if the end of the pipeline where the water is discharged is below the water level in the reservoir, as in Fig. 6.2. ❞

There are some other important points for you to remember. These are:

1. We are dealing with pipes flowing completely full under pressure here. Continuity of flow must be maintained at all points along the pipeline with $Q = AV$, even if the pipeline slopes uphill over part of its length.

2. Pipes flowing full under pressure are analysed using the Bernoulli equation, that is in a totally different way from open channels or pipes which run partially full with a free water surface that is at atmospheric pressure (see Chapter 8). Open channel flow is the result of gravity, not pressure, so the channel must always slope downhill.

3. This chapter is concerned with the flows and head losses that occur naturally in pipelines. In other words, there are no pumps in the pipeline to increase the discharge or to lift the flow vertically for whatever reason, nor any partially closed valves to restrict the flow.

As stated above, pressure pipelines can rise over part of their length, they do not have to slope continuously downhill (Fig. 6.2). The pipeline can even rise above the water level in the reservoir (like going over a hill between the reservoir and the discharge point) because the flow will be maintained as a result of syphonage, provided that the end of the pipeline is below the surface level of the reservoir (see Example 4.7). Once the pipeline is full with the syphon 'primed' and flow starts, the pressure at the top of the syphon is below atmospheric pressure, that is the pressure head is negative. This effectively sucks water up the pipeline. In fact, negative pressures occur at any place where the pipeline rises above the hydraulic grade line (Fig. 6.1). Syphonic action is possible up to around −7.5 m of water (atmospheric pressure equals about 10.3 m of water) but, after this, vaporisation of the liquid can be expected with an air lock forming at the crest, so air relief valves should be provided at high points along the pipeline to allow trapped air to escape.

Another interesting and practical point is that when water is flowing through a pipeline the head that causes the flow (that is Z in Fig. 6.2) is gradually lost through friction and other losses, as indicated in Fig. 6.1. So when water is flowing, the pressure head $P_B/\rho g$ is relatively small. However, if a valve is closed at the end of the pipeline so that there is no flow through it, then $V_B^2/2g = 0$ and there will be no head loss (there are no energy losses in a static liquid). Consequently, the pressure head $P_B/\rho g$ in Fig. 6.1 will increase, while at the downstream end of the pipeline in Fig. 6.2 the pressure will be the full static head, Z. This is another factor that governs the design of a pipeline, because the pipe

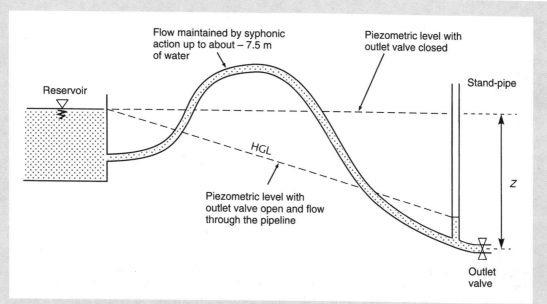

Figure 6.2 Flow from a reservoir discharging to the atmosphere. Note that when the water is flowing the pressure in the pipeline is much less than the full static head experienced when the outlet valve is shut

material and joints must be strong enough to withstand the stresses imposed by the large static head. This caused great problems in the past when pipes were made from materials such as wood, clay and cast iron. Even today it is generally best to keep the head to 70 m or less, otherwise problems may be encountered with the operation of valves and with leakage. To limit the static pressure that occurs when valves are shut, **break pressure tanks** can be introduced along the pipeline, effectively splitting it into sections (see Twort *et al.*, 1994).

 66 I now understand what causes the water to flow through the pipeline, but I do not understand what controls the flow. How does the size of the pipe, its length and so on come into it? Can you explain please? 99

A good way to approach this question is to compare what happens when water flows out of a reservoir through a small orifice to what happens when it flows out of a reservoir through a long pipeline. In a way, the difference between the two situations is the key to understanding pipeline problems.

Free discharge through a small orifice was considered in section 5.4.1. If we apply the Bernoulli equation to a streamline joining point A on the surface of the reservoir to point B at the centre of the vena contracta (Fig. 6.3a), then:

$$z_A + V_A^2/2g + P_A/\rho g = z_B + V_B^2/2g + P_B/\rho g \qquad (6.4)$$

If the reservoir is large, then $V_A = 0$. If atmospheric pressure is the datum, then $P_A = P_B = 0$. If the datum for elevation passes through point B in the vena contracta, then $z_B = 0$ and z_A

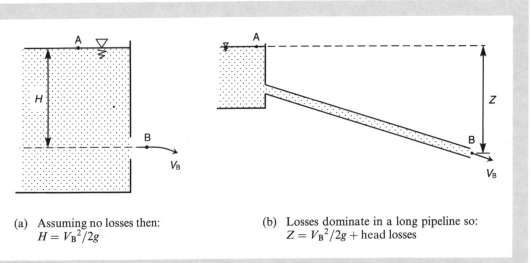

(a) Assuming no losses then:
$H = V_B^2/2g$

(b) Losses dominate in a long pipeline so:
$Z = V_B^2/2g + \text{head losses}$

Figure 6.3 (a) Discharge through a small orifice. The Bernoulli equation is applied to a streamline joining points A and B assuming no loss of energy. (b) Discharge through a pipeline. The Bernoulli equation is again applied to a streamline joining points A and B, but this time the energy losses are dominant and are included in the equation

is the head above the centreline of the orifice, H. If it is assumed that there is no loss of energy then equation (6.4) reduces to:

$$H = V_B^2/2g \tag{6.5}$$

Now let us consider the application of the Bernoulli equation to the situation in Fig. 6.3b, where a large reservoir discharges to the atmosphere via a long pipeline. Because the pipeline is long (tens or even hundreds of kilometres perhaps) it is reasonable to assume that the energy loss through friction and other factors is going to be very large. Consequently we must write the Bernoulli equation with the losses included:

$$z_A + V_A^2/2g + P_A/\rho g = z_B + V_B^2/2g + P_B/\rho g + \textbf{head losses} \tag{6.6}$$

The same logic as above applies, namely that $V_A = 0$, $P_A = P_B = 0$, $z_B = 0$, so:

$$z_A = V_B^2/2g + \textbf{head losses} \tag{6.7}$$

If z_A is written as Z, the difference in elevation between points A and B, then equation (6.7) can be written as:

$$Z = \textbf{head losses} + V_B^2/2g \tag{6.8}$$

The term $V_B^2/2g$ is the velocity head at the end of the pipeline. It must be included because it is the energy remaining after the flow has passed through the pipeline. Some of the principal losses are shown in Table 6.1, others are shown in Table 6.4. *All losses should be included.*

REMEMBER!!!

Table 6.1 Head losses in pipelines

Type of loss	Equation to calculate head loss	Example
Friction (Darcy equation)	$h_F = \dfrac{\lambda L V^2}{2gD}$	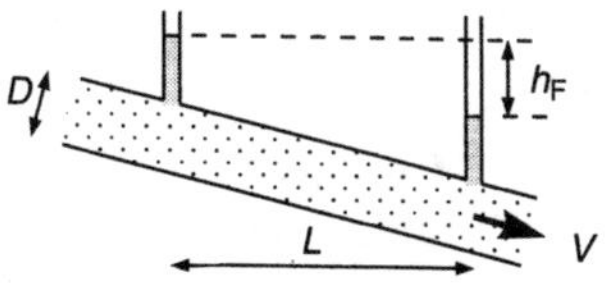
Sudden expansion (sharp increase in diameter or exit from pipeline)	$h_L = \dfrac{(V_1 - V_2)^2}{2g}$	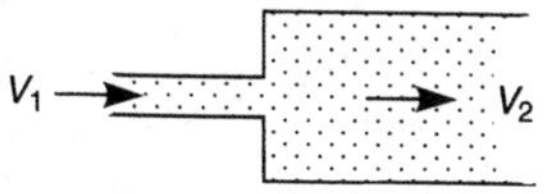
Sudden contraction (sharp entrance to pipeline or reduction in diameter)	$h_L = 0.5\dfrac{V_2^2}{2g}$	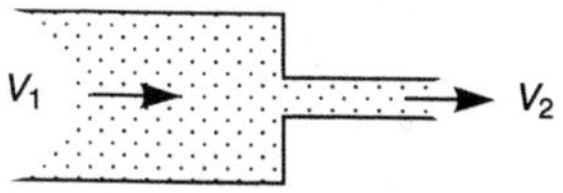

Notation: h_F = head loss due to friction

h_L = minor head loss

λ = Darcy friction factor indicating pipe roughness (dimensionless)
(Note: in the UK $\lambda = 4f$, in the USA $\lambda = f$)

L = pipe length (m)

V = mean velocity (m/s)

D = pipe diameter (m)

g = acceleration due to gravity (m/s^2)

subscript 1 = upstream velocity before change of section

subscript 2 = downstream velocity after change of section.

Box 6.2 ▶ **Remember**

It is important to realise that the velocity head $V_B^2/2g$ in equation (6.8) may often be quite small, so effectively Z = head losses. Thus for a given value of Z it is the losses in Table 6.1 and Table 6.4 that control the flow and determine the discharge through the pipeline.

It is possible to categorise the losses in equation (6.8) and Table 6.1 into **friction losses** and other **minor losses**. In long pipelines friction may be the most important loss, whereas in short pipes minor losses may be very significant. These losses can be defined as follows:

Friction losses   The head loss resulting from friction between the moving column of water and the walls of the pipe. Obviously, friction forces oppose the movement of water through the pipeline.

Minor losses   The head loss resulting from changes in pipe diameter and the cross-sectional area of flow, pipe bends and fitting (for example valves and junctions). See also section 6.6 and Table 6.4.

Table 6.1 contains some information worth highlighting. You should try to remember these basic relationships.

REMEMBER!!!

Pipe friction: $h_F \propto \lambda$ the rougher the pipe, the greater the head loss
$h_F \propto L$ the longer the pipe, the greater the head loss
$h_F \propto V^2$ the higher the velocity, the greater the head loss
$h_F \propto 1/D$ the larger the diameter, the **smaller** the head loss

Changes in $h_L \propto V^2$ the larger the velocity, or change in velocity, the greater the
diameter: head loss

An illustration of how the information above is used is given in Example 6.1. Work through the example carefully, and make sure that you understand where the head losses occur and why, and why some lines are drawn more steeply than others. If we wanted to analyse this problem numerically, we could do so by applying the Bernoulli equation to point A on the surface of the large upper reservoir and point B on the surface of the large lower reservoir, thus:

$$z_A + V_A^2/2g + P_A/\rho g = z_B + V_B^2/2g + P_B/\rho g + \text{head losses}$$

In this case $V_A = 0$, $V_B = 0$, $P_A = P_B = 0$. If the datum for elevation is the water surface of the lower reservoir then $z_A = Z$, the difference between the two reservoir levels (note that it really makes no difference where the datum is since $z_A - z_B$ always equals Z). Thus:

$Z = \text{head losses}$ (6.9)

This equation emphasises the point made in Box 6.2 that, for a given head (Z), it is the losses which actually control the flow through a pipeline. The losses experienced as the water flows between the reservoirs via pipes 1 and 2 (represented by subscripts 1 and 2) are listed in Example 6.1. These losses can be evaluated using the equations in Table 6.1, so:

$Z = \text{entrance loss} + \text{friction loss } 1 + \text{expansion loss} + \text{friction loss } 2 + \text{exit loss}$

$$Z = 0.5V_1^2/2g + \lambda_1 L_1 V_1^2/2gD_1 + (V_1 - V_2)^2/2g + \lambda_2 L_2 V_2^2/2gD_2 + V_2^2/2g \qquad (6.10)$$

Note that the last term is the velocity head in pipe 2. The velocity head is lost when the water flows out of the pipeline (within which the velocity is V_2) into the large reservoir where the velocity is assumed to be zero ($V_B = 0$ above). This can be considered as a sudden expansion with the downstream velocity in Table 6.1 being equal to zero, so that the head loss is the total velocity head in such circumstances.

Box 6.3 ▶ **And do not forget**

The flow through pipes 1 and 2 has to be the same so the continuity equation can be written as $Q = A_1V_1 = A_2V_2$. Thus a substitution can be made for either V_1 or V_2 in equation (6.10). For example, V_1 can be replaced with (A_2V_2/A_1) and the equation solved for V_2. So do not forget that the continuity equation holds the key to solving these problems.

REMEMBER!!!

❝ Example 6.2 shows how equation (6.10) can be solved numerically and the head losses evaluated. Example 6.3 illustrates the same sort of procedure applied to a pipeline discharging to the atmosphere. Example 6.4 involves a situation where the minor losses are starting to become much more significant. Note the difference in head (3.0 m and 0.8 m) needed to obtain the required discharge through a 0.9 m and 1.2 m diameter pipeline. This emphasises the relationships highlighted earlier in the chapter in connection with Table 6.1. Work through the examples and then try Self Test Question 6.1. ❞

SELF TEST QUESTION 6.1

Water flows from one large reservoir to another via a pipeline which is 0.9 m in diameter, 15 km long, and for which $\lambda = 0.04$. The difference in height between the water surface levels in the two reservoirs is 50 m.

(a) Ignoring the minor losses in the pipeline, calculate the flow rate between the two reservoirs.

(b) Assuming that the pipeline entrance and exit are sharp and that the minor losses are as in Table 6.1, calculate the discharge now. What is the difference between the answers to parts (a) and (b)?

Remember that brief guide solutions to the Self Test Questions can be found in Appendix 2, but do try the questions yourself before consulting the solution. This is how you find out how much you understand, and how to overcome any difficulties that you may have on your way to solving the problems.

(REMEMBER!!!)

EXAMPLE 6.1

Water flows between two large reservoirs. The entrance to the pipeline from the upper reservoir is sharp. At the mid-point between the reservoirs the diameter of the pipeline suddenly doubles. The exit from the second pipe to the lower reservoir is also sudden so that all of the velocity head is lost. Sketch the total head line and the hydraulic grade line.

Note 1: The question says the reservoirs are **large**, so the velocity on the surface can be assumed to be zero. The elevation of the water surface can be assumed to be constant.

Note 2: The question says that the entrance is **sharp**, the change in diameter is **sudden**, and the expansion into the lower reservoir is **sudden**. This is telling you that the head losses have to be taken into consideration and should not be ignored.

The reservoirs, pipeline, total head line (THL) and hydraulic grade line (HGL) are shown in Fig. 6.4.

The head losses are:

1. The entrance loss at the sharp entrance to the pipeline, shown as a vertical step in the THL.

2. The friction loss in the smaller pipe, shown by the gradual fall of the THL and the parallel HGL.

3. The loss at the sudden expansion of the pipeline, shown by a vertical step downwards in the THL. There is a corresponding vertical step upwards in the HGL since the doubling of the pipe diameter results in a smaller velocity head, so the HGL and THL are closer together.

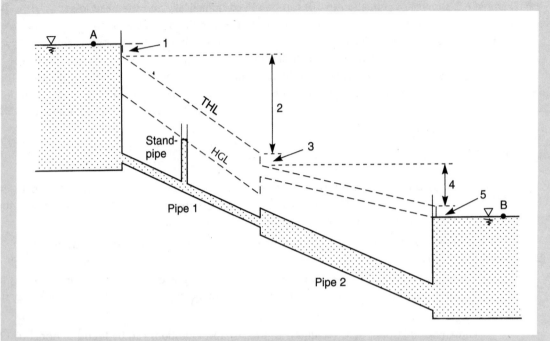

Figure 6.4 Flow between two reservoirs showing the THL and HGL and the head losses (numbered)

4. The friction loss in the larger diameter pipeline. The larger diameter results in a smaller head loss, so the gradient of the THL and HGL is less steep than for the smaller pipe.

5. The exit loss at the sudden expansion, shown as a vertical fall in the THL.

Note 3: The THL must start and finish at the elevation of the water surface in the reservoirs.

Note 4: The HGL is below the THL by an amount equal to the velocity head, and also finishes at the elevation of the water surface in the lower reservoir, since $V = 0$ in a large reservoir.

EXAMPLE 6.2

The difference in elevation (Z) between the water surface of the reservoirs in Example 6.1 is 53 m. Both pipe 1 and pipe 2 are 1.0 km in length. Pipe 1 has a diameter of 0.3 m and pipe 2 a diameter of 0.6 m, and the friction factors λ_1 and λ_2 are both 0.04. Evaluate the losses and calculate the discharge.

In this case the equation governing the flow between the two reservoirs is equation (6.10):

Z = entrance loss + friction loss 1 + expansion loss + friction loss 2 + exit loss

$$Z = 0.5V_1^2/2g + \lambda_1 L_1 V_1^2/2gD_1 + (V_1 - V_2)^2/2g + \lambda_2 L_2 V_2^2/2gD_2 + V_2^2/2g \qquad (6.10)$$

From the continuity equation $V_1 = (A_2/A_1)V_2$ which can be written as $V_1 = (D_2/D_1)^2 V_2$ thus $V_1 = (0.6/0.3)^2 V_2$ or $V_1 = 4V_2$. Substituting for V_1 in equation (6.10) gives:

$$Z = 0.5(4V_2)^2/2g + \lambda_1 L_1 (4V_2)^2/2gD_1 + (4V_2 - V_2)^2/2g + \lambda_2 L_2 V_2^2/2gD_2 + V_2^2/2g$$

$$53 = \left[0.5 \times 16V_2^2/19.62\right] + \left[0.04 \times 1000 \times 16V_2^2/19.62 \times 0.3\right] + \left[9V_2^2/19.62\right]$$
$$+ \left[0.04 \times 1000 \times V_2^2/19.62 \times 0.6\right] + \left[V_2^2/19.62\right]$$

$$53 = 0.408V_2^2 + 108.733V_2^2 + 0.459V_2^2 + 3.398V_2^2 + 0.051V_2^2 \qquad (1)$$

$$53 = 113.049V_2^2$$

$$V_2^2 = 0.4688$$

$$V_2 = 0.685 \, \text{m/s}$$

$$V_1 = 4V_2 = 2.740 \, \text{m/s}$$

$$Q = A_1 V_1 = (\pi \times 0.3^2/4) \times 2.740 = 0.194 \, \text{m}^3/\text{s}$$

CHECK:
From equation (1) above:

entrance loss = $0.408V_2^2 = 0.408 \times 0.4688 =$	0.191 m
friction loss in pipe 1 = $108.733V_2^2 = 108.733 \times 0.4688 =$	50.974 m
expansion loss pipe 1 to 2 = $0.459V_2^2 = 0.459 \times 0.4688 =$	0.215 m
friction loss in pipe 2 = $3.398V_2^2 = 3.398 \times 0.4688 =$	1.593 m
exit loss to lower reservoir = $0.051V_2^2 = 0.051 \times 0.4688 =$	0.024 m
TOTAL HEAD LOSS =	52.997 m OK

NOTE:
The problem is dominated by the friction losses. Even if a 0.6 m diameter pipeline was used along the entire 2 km, friction would still dominate with about 52.4 m of the head being lost through friction (see Revision Question 6.3). The minor losses (entrance, expansion and exit) are relatively small in this particular example because the pipeline is long, but would become more significant if the pipeline was shorter (see Example 6.4) and if the other losses in Table 6.4 were included. The individual losses tabulated above can be drawn to scale on a diagram like Fig. 6.4 if required.

EXAMPLE 6.3

A large reservoir discharges to the atmosphere via a pipeline. The pipeline is 0.9 m in diameter for the first 300 m, then reduces to 0.6 m diameter for the remaining 550 m. The end of the pipeline is 27 m below the water surface level in the reservoir. The entrance to the pipeline is sharp, and the reduction in pipeline diameter is sudden. Taking λ as 0.04 for the first pipe and 0.05 for the second, calculate the discharge from the pipeline and evaluate all of the losses.

This is the situation described by equation (6.8):

$$Z = \text{head losses} + V_B^2/2g \qquad (6.8)$$

In this case $V_B = V_2$. Starting at the upstream end of the pipeline and working to the downstream end, taking the losses from Table 6.1 gives:

$$Z = \text{entrance loss} + \text{friction loss pipe 1} + \text{loss at contraction} + \text{friction loss pipe 2} + V_2^2/2g$$

$$Z = 0.5V_1^2/2g + \lambda_1 L_1 V_1^2/2gD_1 + 0.5V_2^2/2g + \lambda_2 L_2 V_2^2/2gD_2 + V_2^2/2g \qquad (1)$$

From the continuity equation, $V_1 D_1^2 = V_2 D_2^2$ so $V_1 = (D_2/D_1)^2 V_2$
Hence $V_1 = (0.6/0.9)^2 V_2 = 0.444 V_2$
Substituting $0.444 V_2$ for V_1 in equation (1) gives:

$$Z = 0.5(0.444V_2)^2/2g + \lambda_1 L_1(0.444V_2)^2/2gD_1 + 0.5V_2^2/2g + \lambda_2 L_2 V_2^2/2gD_2 + V_2^2/2g$$

$$27 = \left[0.5 \times 0.444^2 V_2^2/19.62\right] + \left[0.04 \times 300 \times 0.444^2 V_2^2/19.62 \times 0.9\right] + \left[0.5V_2^2/19.62\right]$$
$$+ \left[0.05 \times 550 V_2^2/19.62 \times 0.6\right] + \left[V_2^2/19.62\right]$$

$$27 = 0.005V_2^2 + 0.134V_2^2 + 0.025V_2^2 + 2.336V_2^2 + 0.051V_2^2 \qquad (2)$$

$$27 = 2.551 V_2^2$$

$$V_2^2 = 10.584$$

$$V_2 = 3.253 \, \text{m/s}$$

$$Q = A_2 V_2 = (\pi \times 0.6^2/4) \times 3.253 = 0.920 \, \text{m}^3/\text{s}$$

$$V_1 = 0.444V_2 = 0.444 \times 3.253 = 1.444 \, \text{m/s}$$

CHECK:
From equation (2) above:

entrance loss = $0.005V_2^2 = 0.005 \times 10.584 =$	0.053 m
friction loss in pipe 1 = $0.134V_2^2 = 0.134 \times 10.584 =$	1.418 m
loss at contraction of pipeline = $0.025V_2^2 = 0.025 \times 10.584 =$	0.265 m
friction loss in pipe 2 = $2.336V_2^2 = 2.336 \times 10.584 =$	24.724 m
velocity head at exit = $0.051V_2^2 = 0.051 \times 10.584 =$	0.540 m
TOTAL HEAD LOSS =	27.000 m OK

EXAMPLE 6.4

Water has to be discharged at the rate of 1.8 m³/s from a large storm water detention tank into the sea via a submerged outfall. The pipes have a friction factor λ of 0.04 and the entrance to the 130 m long pipeline is sharp. Calculate the elevation of the water surface in the detention tank (above sea level) required to obtain the necessary discharge if the pipeline is constructed with a uniform diameter of either 0.9 m or 1.2 m.

This problem can be treated as one involving flow between two large reservoirs, thus equation (6.9) is applicable. The entrance to the pipeline is sharp and the velocity head will be lost at exit, thus:

Z = entrance loss + friction loss
   + exit loss

$$Z = 0.5V^2/2g + \lambda L V^2/2gD + V^2/2g \qquad (1)$$

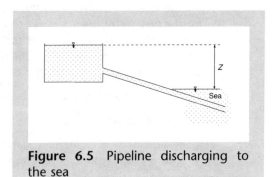

Figure 6.5 Pipeline discharging to the sea

0.9 m diameter pipeline

From the continuity equation $Q = AV$, so $V = 1.8/(\pi \times 0.9^2/4) = 2.829$ m/s

Substituting in equation (1):

$$Z = [0.5 \times 2.829^2/19.62] + [0.04 \times 130 \times 2.829^2/19.62 \times 0.9] + [2.829^2/19.62]$$

$$Z = 0.204 + 2.357 + 0.408$$

$$Z = 2.969\,\text{m}$$

Thus the water level in the detention tank must be about 3.0 m above maximum sea level to ensure a discharge of 1.8 m³/s through a 0.9 m diameter pipeline.

1.2 m diameter pipeline

From the continuity equation $Q = AV$, so $V = 1.8/(\pi \times 1.2^2/4) = 1.592$ m/s

Substituting in equation (1):

$$Z = [0.5 \times 1.592^2/19.62] + [0.04 \times 130 \times 1.592^2/19.62 \times 1.2] + [1.592^2/19.62]$$

$$Z = 0.065 + 0.560 + 0.129$$

$$Z = 0.754\,\text{m}$$

Thus the water level in the detention tank must be about 0.8 m above maximum sea level to ensure a discharge of 1.8 m³/s through a 1.2 m diameter pipeline.

Note: With a diameter of 1.2 m the minor losses comprise 26% of the total head loss. The losses in Table 6.4 should also be included in equation (1), when appropriate.

 ## 6.3 Branching pipelines

 Figure 6.6 shows an upper reservoir that is connected to two lower reservoirs via a pipeline that branches at point B. How can this problem be analysed? In fact, the problem is little different from those covered earlier in the chapter. The key to solving problems involving branching pipelines is outlined in Box 6.4.

Box 6.4 ▶ **Branching and parallel pipelines**

To solve the earlier pipeline problems we applied the Bernoulli equation to two points connected by a single streamline. We wrote the equation so that it included all of the head losses encountered by the flow as it followed the streamline between the two points.

To solve problems involving branching pipelines (or pipelines in parallel, as in the next part of the chapter) apply the Bernoulli equation in exactly the same way. However, remember that the streamline cannot split or follow two different paths. Thus for the problem in Fig. 6.6 we can write the Bernoulli equation for a streamline joining A to B to C, and then again for a streamline joining A to B to D. This gives us two equations that, when combined with the continuity equation, allow us to calculate the discharge in all three pipes.

REMEMBER!!!

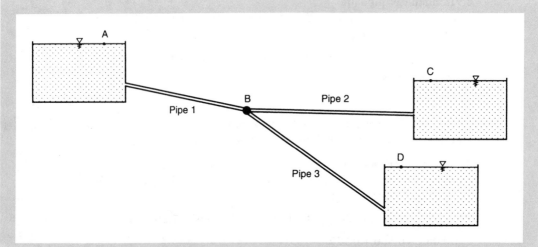

Figure 6.6 A branching pipeline connecting three reservoirs. The solution involves considering two streamlines, one joining A to B to C and the other A to B to D, so giving two equations. The continuity equation, $Q_1 = Q_2 + Q_3$ provides a third so the three unknowns, Q_1, Q_2 and Q_3 can be calculated

The procedure for solving problems involving branching pipelines is best illustrated by an example. This is Example 6.5. Study it carefully, then see if you can do Self Test Question 6.2.

EXAMPLE 6.5

Three large reservoirs are joined by a branching pipeline exactly as in Fig. 6.6. The elevation of the water surface in the reservoirs is A = 680 m OD, C = 640 m OD and D = 590 m OD. These levels do not change. Details of the three pipelines are:

pipeline	length	diameter	λ
1	0.5 km	1.2 m	0.04
2	0.3 km	0.9 m	0.06
3	0.4 km	0.6 m	0.05

Assume that all minor losses are negligible, so only friction losses need be considered. Determine the discharge through each pipeline if the flow is not controlled by valves.

First, apply the Bernoulli equation to a streamline joining A to C assuming only a friction loss:

$$Z_{AC} = \lambda_1 L_1 V_1^2 / 2gD_1 + \lambda_2 L_2 V_2^2 / 2gD_2$$

where $Z_{AC} = 680 \text{ m OD} - 640 \text{ m OD} = 40 \text{ m}$.

$$40 = \left[0.04 \times 500 \times V_1^2 / 19.62 \times 1.2\right] + \left[0.06 \times 300 \times V_2^2 / 19.62 \times 0.9\right]$$

$$40 = 0.849 V_1^2 + 1.019 V_2^2$$

$$1.019 V_2^2 = 40 - 0.849 V_1^2$$

$$V_2 = \left(39.254 - 0.833 V_1^2\right)^{1/2} \qquad (1)$$

Now apply the Bernoulli equation to a streamline joining A to D, as above:

$$Z_{AD} = \lambda_1 L_1 V_1^2 / 2gD_1 + \lambda_3 L_3 V_3^2 / 2gD_3$$

where $Z_{AD} = 680\,m\,OD - 590\,m\,OD = 90\,m$.

$$90 = \left[0.04 \times 500 \times V_1^2 / 19.62 \times 1.2\right] + \left[0.05 \times 400 \times V_3^2 / 19.62 \times 0.6\right]$$

$$90 = 0.849V_1^2 + 1.699V_3^2$$

$$1.699V_3^2 = 90 - 0.849V_1^2$$

$$V_3 = \left(52.972 - 0.500V_1^2\right)^{1/2} \qquad (2)$$

The continuity equation for a branching pipeline can be written as $Q_1 = Q_2 + Q_3$ or:

$$A_1 V_1 = A_2 V_2 + A_3 V_3$$

$$\left(\pi D_1^2 / 4\right)V_1 = \left(\pi D_2^2 / 4\right)V_2 + \left(\pi D_3^2 / 4\right)V_3$$

$$D_1^2 V_1 = D_2^2 V_2 + D_3^2 V_3$$

$$1.2^2 V_1 = 0.9^2 V_2 + 0.6^2 V_3$$

$$1.440V_1 = 0.810V_2 + 0.360V_3$$

$$V_1 = 0.563V_2 + 0.250V_3 \qquad (3)$$

This completes the hydraulic analysis. The remaining part of the question concerns the solution of the equations. This can be done by trial and error (as below), graphical means or by using appropriate computer software.

Substituting the expressions for V_2 and V_3 in equations (1) and (2) into equation (3) gives:

$$V_1 = 0.563\left(39.254 - 0.833V_1^2\right)^{1/2} + 0.250\left(52.927 - 0.500V_1^2\right)^{1/2} \qquad (4)$$

This equation must be solved by trial and error. However, from equation (4), for the solution to be real (+ve) then $0.833V_1^2 < 39.254$ and $0.500V_1^2 < 52.927$. This gives $V_1 < 6.8$ and $10.3\,m/s$ respectively. Try $V_1 = 5.0\,m/s$ in equation (4) and see if the left-hand side (LHS) equals the right-hand side (RHS), as required for a valid solution.

$$RHS = 0.563(39.254 - 0.833 \times 5.0^2)^{1/2} + 0.250(52.927 - 0.500 \times 5.0^2)^{1/2}$$
$$= 2.417 + 1.590 = 4.007\,m/s$$

The RHS = 4.007 not 5.0 m/s so this is not the answer. Repeating the calculation with $V_1 = 4.5\,m/s$ and summarising the results in a table gives:

LHS: Try $V_1 = 5.0\,m/s$ RHS = 4.007 m/s   (LHS − RHS) = +0.993
            $V_1 = 4.5\,m/s$ = 4.300 m/s   = +0.200
            $V_1 = 4.3\,m/s$ = 4.402 m/s   = −0.102

Now RHS > LHS so V_1 lies between 4.3 and 4.5 m/s. V_1 can be found by interpolation: (LHS − RHS) changes by 0.200 + 0.102 = 0.302 when V_1 changes by 4.5 − 4.3 = 0.2 m/s.

Therefore 0.102 is equivalent to (0.102/0.302) × 0.2 = 0.068 m/s.

Thus $V_1 = 4.3 + 0.068 = 4.368\,m/s$.

Check the solution by substituting $V_1 = 4.368\,m/s$ in equation (4):

$$RHS = 0.563(39.254 - 0.833 \times 4.368^2)^{1/2} + 0.250(52.927 - 0.500 \times 4.368^2)^{1/2}$$
$$= 2.721 + 1.647 = 4.368\,m/s \quad OK$$

From equation (1):

$$V_2 = \left(39.254 - 0.833 V_1^2\right)^{1/2} = \left(39.254 - 0.833 \times 4.368^2\right)^{1/2} = 4.833\,\text{m/s}$$

From equation (2):

$$V_3 = \left(52.972 - 0.500 V_1^2\right)^{1/2} = \left(52.972 - 0.500 \times 4.368^2\right)^{1/2} = 6.590\,\text{m/s}$$

Now apply the continuity equation to the individual pipelines to calculate the discharge:

$$Q_1 = A_1 V_1 = \left(\pi \times 1.2^2/4\right) \times 4.368 = 4.940\,\text{m}^3/\text{s}$$
$$Q_2 = A_2 V_2 = \left(\pi \times 0.9^2/4\right) \times 4.833 = 3.075\,\text{m}^3/\text{s}$$
$$Q_3 = A_3 V_3 = \left(\pi \times 0.6^2/4\right) \times 6.590 = 1.863\,\text{m}^3/\text{s}$$

CHECK
Apply the continuity equation to the pipe branches.

$$Q_1 = Q_2 + Q_3$$

From above $Q_1 = 4.940\,\text{m}^3/\text{s}$

$$Q_2 + Q_3 = 3.075 + 1.863 = 4.938\,\text{m}^3/\text{s} \quad \text{OK}$$

SELF TEST QUESTION 6.2

Three reservoirs are connected as in Fig. 6.6. Details of the three pipelines are shown below. The elevation of the water surface in the reservoirs A, C and D is constant at respectively 250 m OD, 220 m OD and 190 m OD. The pipelines are long so it can be assumed that friction losses will dominate and that minor losses can be ignored. Assuming that the flow is not controlled by means of valves, calculate the discharge through each of the pipes.

pipeline	length (km)	diameter (m)	λ
1	17.5	1.2	0.05
2	5.3	0.9	0.04
3	6.4	0.9	0.04

6.4 Parallel pipelines

If two reservoirs are connected by two or more pipelines laid parallel to each other (Fig. 6.7) then Box 6.4 again provides the key to the solution. If the Bernoulli equation is applied to a streamline joining two points, A and B, on the water surface of the reservoirs, the stream-line can only pass through one of the pipes. Thus a streamline can be drawn connecting A to B via pipeline 1, while another streamline joins A to B via pipeline 2. So an equation is obtained from each application. Of course, each equation is formulated to take into account the losses in the pipeline, so if the pipelines have a different diameter or roughness this results in different equations.

When solving this type of problem a couple of points to note are:

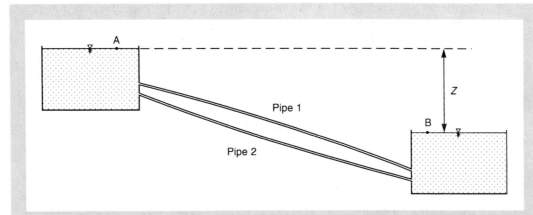

Figure 6.7 Flow between two reservoirs via 'parallel' pipelines. The Bernoulli equation can be applied to points A and B which are connected by streamlines passing either through pipe 1 or pipe 2

1. When applying the Bernoulli equation to the two pipelines, the head causing the flow, Z, is the same in both equations since A and B have the same elevation however the water flows between them.

2. The flow in each pipeline is unaffected by the flow in the other.

These two factors make so called 'parallel' pipeline problems relatively easy to solve (of course, the pipelines do not really have to be physically parallel). Example 6.6 illustrates how to calculate the discharge through the two pipes for a given head. Since the Bernoulli equation can be applied twice (once to each pipe) and there are only two unknowns (Q_1 and Q_2) it is very easy to obtain a solution. Example 6.7 shows how to calculate the head needed to obtain a specified total discharge between the two reservoirs. In this case there are three unknowns: the discharges in the two pipelines (Q_1 and Q_2) and the head, Z, so three equations are needed to obtain a solution. These can be obtained by applying the Bernoulli equation to the two pipelines and then the continuity equation, which states that the total discharge $\Sigma Q = Q_1 + Q_2$.

Study these two examples carefully and then try Self Test Question 6.3.

A variation on the above problems is where a pipeline joining two reservoirs splits so that there are 'parallel' pipelines over part of the distance (Fig. 6.8 in Example 6.8). Because the head causing the flow is the same for both of the lower parallel pipelines, this type of question is easier to solve than the branching pipeline problem in Example 6.5. Work through Example 6.8 and confirm this for yourself.

EXAMPLE 6.6

Two reservoirs are connected by two pipelines as shown in Fig. 6.7. The difference in the elevation of the water surface between the upper and lower reservoir, Z, is constant at 150 m. Pipeline 1 has a diameter of 1.2 m, while pipeline 2 has a diameter of 0.9 m. The pipelines are each 43 km long with the roughness factor $\lambda = 0.04$. Since the pipelines are long, assume that friction losses will dominate and that minor losses can be ignored. Calculate the discharge through each of the pipelines.

Applying the Bernoulli equation to a streamline joining A to B via pipeline 1 gives:

$Z = \text{head losses} = \lambda_1 L_1 V_1^2 / 2gD_1$

$150 = 0.04 \times 43000 \times V_1^2 / 19.62 \times 1.2$

$V_1 = 1.433 \, \text{m/s}$

Applying the Bernoulli equation to a streamline joining A to B via pipeline 2 gives:

$Z = \text{head losses} = \lambda_2 L_2 V_2^2 / 2gD_2$

$150 = 0.04 \times 43000 \times V_2^2 / 19.62 \times 0.9$ 97.406

$V_2 = 1.241 \, \text{m/s}$

Applying the continuity equation to obtain the discharges:

$Q_1 = A_1 V_1 = (\pi \times 1.2^2 / 4) \times 1.433 = 1.621 \, \text{m}^3/\text{s}$

$Q_2 = A_2 V_2 = (\pi \times 0.9^2 / 4) \times 1.241 = 0.789 \, \text{m}^3/\text{s}$

Total flow between the reservoirs, $\Sigma Q = Q_1 + Q_2 = 1.621 + 0.789 = 2.410 \, \text{m}^3/\text{s}$

EXAMPLE 6.7

Two reservoirs are connected by two parallel pipelines with diameters of 0.8 m and 0.6 m. The pipelines are both 2.6 km long with a λ value of 0.05. The combined discharge through the pipelines must be at least 1.7 m³/s. (a) When the total discharge is 1.7 m³/s, what is the flow rate in the individual pipelines? (b) If the larger pipeline has to be taken out of service for maintenance, what proportion of the total discharge will be lost? (c) What is the minimum difference in the elevation of the surface water level in the two reservoirs required to maintain a flow of 1.7 m³/s through the pipelines?

(a) Ignoring minor losses and applying the Bernoulli equation to each of the pipelines:

$$Z = \lambda_1 L_1 V_1^2 / 2gD_1 \qquad (1)$$

$$Z = \lambda_2 L_2 V_2^2 / 2gD_2 \qquad (2)$$

From above it is apparent that: $\lambda_1 L_1 V_1^2 / 2gD_1 = \lambda_2 L_2 V_2^2 / 2gD_2$
Since $\lambda_1 = \lambda_2$ and $L_1 = L_2$, cancelling gives: $V_1^2/D_1 = V_2^2/D_2$ so $V_1^2 = (D_1/D_2)V_2^2$
Putting $D_1 = 0.8$ m and $D_2 = 0.6$ m gives: $V_1^2 = (0.8/0.6)V_2^2$

$$V_1^2 = (1.333)V_2^2$$

$$\text{thus:} \quad V_1 = 1.155V_2 \qquad (3)$$

Now applying the continuity equation, total discharge $\Sigma Q = Q_1 + Q_2$ (4)

$1.7 = A_1 V_1 + A_2 V_2$

$1.7 = (\pi \times 0.8^2 / 4)V_1 + (\pi \times 0.6^2 / 4)V_2$

$1.7 = 0.503V_1 + 0.283V_2$

Substituting for V_1 above from equation (3) gives:

$$1.7 = 0.503(1.155V_2) + 0.283V_2$$

$$1.7 = 0.581V_2 + 0.283V_2$$

$$V_2 = 1.968\,\text{m/s}$$

Putting $V_2 = 1.968\,\text{m/s}$ in equation (3) gives: $V_1 = 1.155V_2$

$$V_1 = 1.155 \times 1.968$$

$$V_1 = 2.273\,\text{m/s}$$

Therefore $Q_1 = A_1V_1 = (\pi \times 0.8^2/4) \times 2.273 = 1.143\,\text{m}^3/\text{s}$

$$Q_2 = A_2V_2 = (\pi \times 0.6^2/4) \times 1.968 = 0.556\,\text{m}^3/\text{s}$$

(b) Proportion of flow in larger pipe = $(1.143/1.7) \times 100 = 67\%$

(c) From equation (1) the differential head is: $Z = \lambda_1 L_1 V_1^2 / 2gD_1$

$$Z = 0.05 \times 2600 \times 2.273^2/19.62 \times 0.8$$

$$Z = 42.8\,\text{m}$$

The minimum head required to maintain a combined discharge of $1.7\,\text{m}^3/\text{s}$ is $42.8\,\text{m}$.

EXAMPLE 6.8

Two reservoirs are connected by a pipeline that splits into two branches as shown in Fig. 6.8. The difference in the elevation of the water surface between the upper and lower reservoir, Z, is constant at $150\,\text{m}$. The roughness factor of all the pipes is $\lambda = 0.04$. The branch in the pipeline occurs $23\,\text{km}$ from the upper reservoir, the lower pipelines being each $20\,\text{km}$ long. The diame-

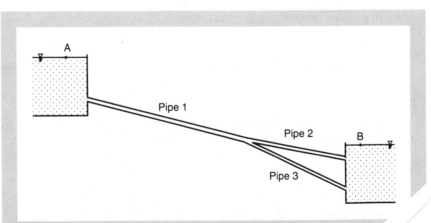

Figure 6.8 Two reservoirs connected by a single pipeline th~ into two parallel pipelines. By applying the Bernoulli equ~ streamline joining A to B firstly via pipes 1 and 2 and via pipes 1 and 3 we get two equations. The contin~ $= Q_2 + Q_3$, provides the third, so we can solve f~

ters of pipelines 1, 2 and 3 are respectively 1.5 m, 0.9 m and 1.0 m. Ignoring minor losses, calculate the discharge through the three pipelines.

The pipeline is long so friction losses will dominate, allowing minor losses to be ignored. Applying the Bernoulli equation to the streamline joining A to B via pipes 1 and 2 gives:

$$Z = \lambda_1 L_1 V_1^2 / 2gD_1 + \lambda_2 L_2 V_2^2 / 2gD_2$$

$$150 = \left[0.04 \times 23\,000 \times V_1^2 / 19.62 \times 1.5\right] + \left[0.04 \times 20\,000 \times V_2^2 / 19.62 \times 0.9\right]$$

$$150 = 31.261 V_1^2 + 45.305 V_2^2 \qquad (1)$$

Now applying the Bernoulli equation to the streamline joining A to B via pipes 1 and 3:

$$Z = \lambda_1 L_1 V_1^2 / 2gD_1 + \lambda_3 L_3 V_3^2 / 2gD_3$$

The first two terms of this equation are the same as in equation (1) above.

$$150 = 31.261 V_1^2 + \left[0.04 \times 20\,000 \times V_3^2 / 19.62 \times 1.0\right]$$

$$150 = 31.261 V_1^2 + 40.775 V_3^2 \qquad (2)$$

The continuity equation provides the third equation needed to solve for three variables:

$$Q_1 = Q_2 + Q_3 \text{ or } A_1 V_1 = A_2 V_2 + A_3 V_3$$

$$D_1^2 V_1 = D_2^2 V_2 + D_3^2 V_3 \qquad \text{(the } \pi/4\text{'s have been cancelled)}$$

$$1.5^2 V_1 = 0.9^2 V_2 + 1.0^2 V_3$$

$$2.250 V_1 = 0.810 V_2 + 1.000 V_3 \qquad (3)$$

This effectively ends the hydraulic analysis, the remaining calculations being concerned with solving the equations. This is relatively simple because the first two terms of equations (1) and (2) are identical. Subtracting equation (2) from equation (1) gives:

$$0 = 45.305 V_2^2 - 40.775 V_3^2$$

$$V_3^2 = 1.111 V_2^2$$

$$V_3 = 1.054 V_2 \qquad (4)$$

Substituting for V_3 in equation (3) gives: $2.250 V_1 = 0.810 V_2 + 1.000(1.054 V_2)$
$$2.250 V_1 = 1.864 V_2$$
$$V_1 = 0.828 V_2 \qquad (5)$$

Substituting for V_1 in equation (1) gives: $150 = 31.261(0.828 V_2)^2 + 45.305 V_2^2$
$$150 = 21.432 V_2^2 + 45.305 V_2^2$$
$$150 = 66.737 V_2^2$$
$$V_2 = 1.499 \text{ m/s}$$

From equation (4): $V_3 = 1.054 V_2 = 1.054 \times 1.499 = 1.580 \text{ m/s}$
From equation (5): $V_1 = 0.828 V_2 = 0.828 \times 1.499 = 1.241 \text{ m/s}$

Therefore: $\quad Q_1 = A_1 V_1 = (\pi \times 1.5^2/4) \times 1.241 = 2.193 \text{ m}^3/\text{s}$
$$Q_2 = A_2 V_2 = (\pi \times 0.9^2/4) \times 1.499 = 0.954 \text{ m}^3/\text{s}$$
$$Q_3 = A_3 V_3 = (\pi \times 1.0^2/4) \times 1.580 = 1.241 \text{ m}^3/\text{s}$$

CHECK: $\quad Q_1 = Q_2 + Q_3$
$$Q_1 = 0.954 + 1.241 = 2.195 \text{ m}^3/\text{s} \quad \text{OK}$$

SELF TEST QUESTION 6.3

Two reservoirs 5.4 km apart are to be connected by three pipelines whose diameters have to be determined. Operational requirements mean that the discharge in the pipelines must be $Q_1 = 0.4\,m^3/s$, $Q_2 = 0.6\,m^3/s$, and $Q_3 = 1.1\,m^3/s$. The difference in the water levels in the reservoirs is 23 m and for all pipes $\lambda = 0.06$. Ignoring minor losses, calculate the required diameters.

HINT: Use the last expression for head loss (below) in your solution. This is obtained as follows: $V = Q/A$, so $V^2 = Q^2/(\pi D^2/4)^2$. Substituting for V^2 in the Darcy equation:

$$h_F = \frac{\lambda L V^2}{2gD}\text{ becomes }h_F = \frac{\lambda L Q^2}{2gD(\pi D^2/4)^2}\text{ which gives }h_F = \frac{\lambda L Q^2}{12.1D^5}$$

SELF TEST QUESTION 6.4

Two large reservoirs with a difference in water level of 27 m are connected by a pipeline that splits into two branches (as in Fig. 6.8) after a distance of 10 km. Ignoring minor losses, calculate the discharge in each of the three pipelines if the details of the pipelines are:

Pipeline	Diameter (m)	Length (km)	λ
1	0.90	10	0.04
2	0.75	21	0.07
3	0.60	23	0.05

6.5 The development of the pipe friction equations

In the above examples we have been using the Darcy equation to evaluate the head loss due to friction. This provided a simple means of quantifying the loss, and therefore a suitable introduction to the problem. However, it has taken over 100 years to develop the theory and analytical techniques needed to analyse pipeflow effectively. The key to understanding this complex phenomenon is experimental investigation, as conducted by Reynolds and Nikuradse.

6.5.1 Laminar and turbulent flow

One of the most important experimental contributions was made by Reynolds. The concept of Reynolds number and its use to classify various types of flow was introduced in section 4.2. The Reynolds number (Re) was defined in equation (4.3) as:

$$Re = \rho VD/\mu \tag{4.3}$$

where ρ is the mass density of the liquid (kg/m³), V its mean velocity (m/s), μ its dynamic viscosity (kg/ms), and D the diameter of the pipe (m). The Reynold number can also be expressed as $Re = VD/v$ since kinematic viscosity $v = \mu/\rho$. By calculating the dimensionless Reynolds number of the flow (as in Example 4.1) its nature can be determined. For water in pipes:

Laminar flow   Re < 2000
Transitional flow   Re = 2000 to 4000
Turbulent flow   Re > 4000

In laminar flow the viscous effects dominate and all of the streamlines or pathlines are parallel to each other and the flow is very smooth, uniform and steady, whereas in turbulent flow the pathlines are random and the flow is uneven and, at a particular point in the pipeline, the velocity fluctuates from one instant to the next (Fig. 4.5). There are other major differences which will be described shortly. Between laminar and turbulent flow there is an ill-defined transition region. The boundaries of this region depend upon several factors, such as whether the velocity is increasing or decreasing, so the values above are only an approximate guide. Sometimes turbulent flow may not be experienced until Re > 10000. A value of 10000 may sound quite large, but in reality it is not. In domestic plumbing Re may be around 25000. Prove this for yourself by conducting the experiment described in Box 6.5. In commercial pipelines operating under much larger pressures with higher velocities Re may be of the order of 100000 or even 1000000. Therefore turbulent flow is the type most commonly encountered by engineers. Viscous, sluggish laminar flow is relatively rare in nature.

REMEMBER!!!

Box 6.5 ▶ **Try this experiment**

Work out the Reynolds number of the flow in the pipe supplying your kitchen tap.

You probably know where the stopcock is for the cold water supply (usually under the sink), so you can measure the pipe diameter. Allow a couple of mm for the wall thickness so you get the internal diameter. Work out the internal cross-sectional area of flow (A). Next, measure the time it takes to collect a given volume of water in a bucket. You can calibrate the bucket by filling a one litre fizzy drink bottle six times and emptying it into the bucket. Mark the water level. Then put the empty bucket under the tap and measure how many seconds (T) it takes to fill up to the mark. The flow rate from the tap (Q) is $0.006/T$ m^3/s. The mean velocity of flow $V = Q/A$. You should normally find V is between 1 and 2 m/s depending upon the water pressure.

Now calculate the Reynolds number of the flow in the pipe, as in Example 4.1. You should find that Re is somewhere around 20000 to 30000, well into the turbulent flow range.

❝ How did Reynolds discover that flow could be classified as laminar below 2000 and turbulent above 4000? It is not exactly obvious just from watching water flow through a pipe. ❞

Around 1884 Reynolds conducted a classic series of experiments that involved allowing liquid to escape from a large tank through a pipe with a bell-mouth entry (Fig. 6.9). A thin stream of dye was injected into the bell-mouth and observed as it travelled along the pipe. The velocity of the flow in the pipe was varied using a valve at the downstream end. Reynolds performed many experi-

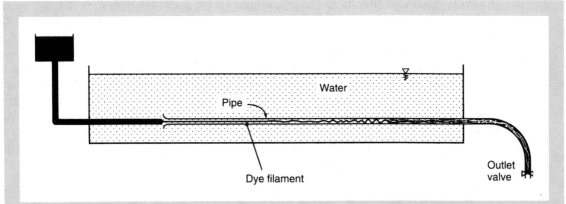

Figure 6.9 Reynolds experiment. The liquid escapes from the large tank via the pipe (glass tube), into which a stream of dye is injected. By varying the flow through the pipe, the critical velocity at which the dye stream starts to break up can be determined

ments to determine the critical velocity at which the stream of dye started to break up. He repeated the procedure using pipes of different diameter and liquids of different density and viscosity. He discovered that if he combined the variables into the dimensionless grouping that bears his name, the stream of dye never broke up if Re was less than 2000 and that it would always break up if Re was greater than 10000. In the transitional range Reynolds discovered that other factors appeared to be involved, like the roughness of the internal pipe wall. Nevertheless, Reynolds had provided a reliable means of determining whether a flow is laminar or turbulent. This is important because:

in laminar flow $\quad h_F \propto V$

in turbulent flow $\quad h_F \propto V^2$

REMEMBER!!!

Since turbulent flow is associated with relatively high velocities, squaring the velocity gives a rapidly increasing head loss which must be allowed for in design calculations. For instance, if $V = 1, 2, 3, 4, 5$ and 6 then the head loss in turbulent flow is proportional to 1, 4, 9, 16, 25 and 36 respectively. There are other, less obvious differences between laminar and turbulent flow that will be explained below.

Around 1841 Poiseuille developed an equation for the head loss due to friction in **laminar flow**:

$$h_F = 32vLV/gD^2 \qquad (6.11)$$

where v is the kinematic viscosity (m^2/s), V is the mean velocity, L the length of pipe, g the acceleration due to gravity and D the diameter of the pipe. Note that the head loss is proportional to V, as stated earlier. *One of the important characteristics of laminar flow is that the roughness of the pipe has no effect on it, so there is no term to denote pipe roughness in the Poiseuille equation.* On the other hand, pipe roughness is important in turbulent flow, so the Darcy equation for the head loss includes the pipe friction factor, λ. Note also that the head loss is now proportional to V^2 not V:

$$h_F = \lambda LV^2/2gD \qquad (6.12)$$

This is the Darcy equation for the head loss due to friction in **turbulent flow** that was used in the first part of the chapter. This equation was presented around 1850 by Darcy, Weisbach and others. It can be derived in several ways: by considering the shear stresses at the pipe boundary, by dimensional analysis (using the principles in Chapter 10), or by the simplified method shown in Proof 6.1 in Appendix 1. The advantage of the simplified derivation is that it has some things in common with the derivation of the Chezy equation, which helps to relate ideas across chapter headings and show that we are applying the same concepts to different problems. Study Proof 6.1 and ensure that you can follow all of the steps.

For **laminar flow** conditions the Poiseuille and Darcy equation for h_F can be equated, thus:

$$\frac{32 \nu L V}{g D^2} = \frac{\lambda L V^2}{2 g D}$$

Cancelling and rearranging gives $\lambda = 64\nu/DV$, and since $Re = DV/\nu$ then:

$$\lambda = 64/\text{Re} \tag{6.13}$$

This gives some indication that λ is not a simple coefficient depending only on the pipe material. Its value also depends upon the Reynolds number, which includes the properties of the liquid: viscosity and density, and also the flow velocity. Hence λ is an overall coefficient that represents the combined effect of many variables. Its value, for a particular type of pipe and conditions, can be determined by measuring the discharge (and hence velocity) and head loss, h_F, over a certain length, L, of pipe and then solving equation (6.12) to determine λ. However, it was soon realised that the head loss in turbulent flow is not exactly proportional to V^2 as indicated in the Darcy equation, but to some slightly lesser power. This and the complex nature of λ evident in turbulent flow meant that further investigation was required.

6.5.2 Rough and smooth pipes

As the nineteenth century closed and the twentieth began, the understanding of laminar flow was quite well advanced, but what governed the value of λ in turbulent flow, and why, was still unknown. Around 1913 Blasius examined the experimental data and identified two different types of pipe friction in turbulent flow:

smooth turbulent pipe friction – viscosity effects dominate with $\lambda \propto \text{Re}$;

rough turbulent pipe friction – viscosity and pipe roughness effects are important.

In fact we now know that a third intermediate category, **transitional turbulent pipe friction**, is needed (as explained later). Blasius presented the following equation to enable the friction factor to be calculated for **smooth turbulent pipeflow**:

$$\lambda = 0.316/\text{Re}^{0.25} \tag{6.14}$$

The question of rough pipe friction remained unresolved until about 1930 when Nikuradse provided an insight into the behaviour of turbulent flow in rough pipes. Nikuradse manufactured pipes with an artificial but uniform roughness by gluing similarly sized sand grains onto the internal surface of the pipe. By varying the size of the sand he was able to control the height of the protrusions from the surface of the pipe (k) relative to the pipe diameter (*D*). By this means Nikuradse created pipes with *k/D* values from 1/1014 to 1/30. Then by monitoring the flow of water through the pipes he obtained the relationship between λ and Re. The result was the Nikuradse diagram which plots the relationship between pipe friction (λ), Reynolds number (Re) and the relative roughness of the pipe (*k/D*). The equivalent diagram for commercial pipes with an irregular roughness is shown in Fig. 6.10. This illustrates many of the significant features of pipeflow, namely:

1. At the left of the diagram, which corresponds to Re < 2000, **laminar flow** exists. Laminar flow is not affected by surface roughness, so there is only one line. This is as expected from equation (6.13), which was derived from the Poiseuille equation.
2. Moving towards the right, the laminar flow region of the diagram gives way to an

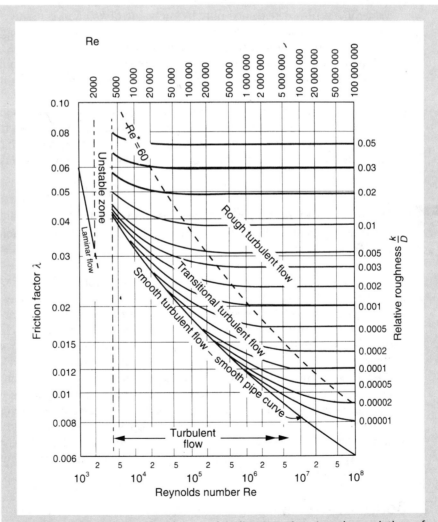

Figure 6.10 A generalised (Moody) diagram showing the variation of λ with Re and *k/D*

unstable zone in which Re is between 2000 and 4000. This is the transition zone between laminar and turbulent flow.

3. When Re > 4000 there is a diagonal line that forms the lower envelope to the curves above. This is the **smooth pipe curve** that represents turbulent flow in a hydraulically smooth pipe, or put another way, **smooth turbulent flow**.

4. Above the smooth pipe curve is a series of lines representing different pipe roughnesses (k/D). Following these curves from left to right, it is apparent that as Re increases the line representing the roughest pipe ($k/D = 0.05$) breaks away from the smooth pipe curve first, followed in sequence by the other curves representing relatively smoother pipes. The curved nature of these lines initially indicates that λ is decreasing as Re increases and that conditions are not constant. This represents the transition from smooth turbulent flow to fully developed turbulent flow in hydraulically rough pipes, that is **transitional turbulent flow**.

5. Where the curves become horizontal on the right of the diagram it is apparent that the value of λ is now determined only by the value of k/D and is unaffected by Re, which means that the viscosity of the liquid (which is included in Re) no longer influences the flow. This is the region of turbulent flow in hydraulically rough pipes, or **rough turbulent flow**. Note that the value of Re at which the curves become horizontal differs according to the relative roughness of the pipe. This is the region in which it is assumed that h_F is proportional to V^2.

6. The complex nature of the diagram means that it is very difficult to obtain any equation for pipe friction that covers the whole range of flows, and impossible to obtain a simple one. Thus simple equations of the form $h_F = \text{constant} \times V^N$ generally have a very limited range and are only valid for one of the three types of flow, such as transitional turbulent flow.

 ❝ How can the flow in a particular pipe that is manufactured from a particular material be described as smooth under some conditions and rough under others? I don't understand this. **❞**

I suppose in a couple of words the answer is 'relative roughness'. You see, if you or I fell down onto some tarmac and skinned our knees and elbows we might say that the damage was inflicted because the tarmac is rough (relative to our skin). Yet a car driving over the same flat tarmac surface would find the surface smooth. It is a question of scale, to some extent, and sensitivity as to whether a surface is considered rough or smooth. Velocity also comes into it, because if a car moved very slowly, so that it was barely moving, over a cobbled street there would be little vibration, because of the trivial velocity and the cushioning effect of the shock absorbers and springs. However, at higher speeds the vibration and disturbance to the ride caused by the cobbles would increase. Additionally, if the cobbles were made larger so that the road surface was bumpier then the ride would get even rougher. So it is back to the concept of the size of the protrusions from the surface, or relative roughness. To explain properly smooth, transitional and rough turbulent flow we need to have a basic understanding of boundary layer.

6.5.3 Boundary layer

The concept of boundary layer was developed at the beginning of the twentieth century, about the same time that the problem of turbulent flow in pipes was being investigated, and was based on a great deal of experimental and theoretical research. This concept is very important, because it describes what happens when a fluid passes over a solid boundary, which is what pipe flow is all about. If a real liquid passes over a solid surface then there is a loss of energy due to friction between the liquid and the surface. Less obvious is the fact that at the boundary itself the velocity of the liquid is zero, while the velocity is reduced for some distance from the surface. Thus the frictional retarding force affects not only the thin film of liquid in contact with the surface, but also the flow some distance away from it as the frictional effect is transmitted through the liquid by viscous shear. This layer of affected liquid adjacent to the boundary surface is called the **boundary layer**.

To explain boundary layer in more detail, suppose that we have a large body of liquid that is far away from any boundary surface so that it is moving with a uniform, undisturbed velocity, U. Now suppose we introduce a very thin plate into the liquid, pointing directly into the flow. Friction between the plate and the liquid will cause the thin film of liquid in contact with the surface to stop. However, as the perpendicular distance from the surface of the plate increases, the velocity increases until it reaches the undisturbed velocity U. If a surface is drawn joining all the points where the liquid velocity is $0.99U$, this line represents the thickness of the boundary layer, δ. This is the solid line in Fig. 6.11, which shows in two dimensions the growth of the boundary layer on one side of the plate. The diagram also shows the undisturbed velocity vector diagram to the left as the flow approaches the plate, and the velocity vector diagram at two sections through the boundary layer. *It is important that you clearly understand that the liquid is flowing forwards through the boundary layer (as shown by the velocity vectors); it is only at the boundary itself where the velocity is zero.* REMEMBER!!!

As the liquid travels over the plate in Fig. 6.11, more of the flow becomes affected by the retarding forces caused by the interaction between the liquid and the surface, so the boundary layer increases in thickness. However, the rate at which the thickness increases is not

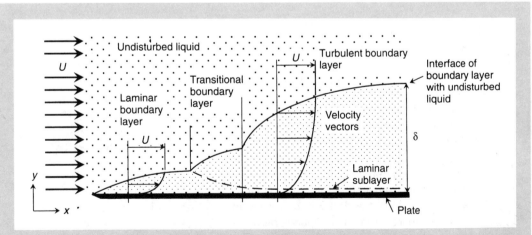

Figure 6.11 Boundary layer on one side of a flat plate introduced into a liquid having an undisturbed velocity, U. The boundary layer extends outward from the surface of the plate to the point where the velocity is $0.99U$, as indicated by the solid line. The flow in the boundary layer is at first laminar, then transitional and turbulent, corresponding to the regions marked on the diagram

uniform, and Fig. 6.11 shows three distinct regions of the boundary layer: the laminar boundary layer, the transitional boundary layer and the turbulent boundary layer. In these three regions the flow is respectively laminar, transitional and turbulent. The different nature of the flow within these regions explains the different thicknesses of the boundary layer. ***The key concept to grasp here is that it is the relatively high-velocity undisturbed flow that is moving the liquid in the boundary layer along with it and maintaining the forward motion.***

In the laminar region of the boundary layer the drag force that maintains the forward flow of the liquid is provided by the normal viscous shearing action between the layers of the liquid. In this region it helps to think of the liquid in the boundary layer above the plate as consisting of a number of thin, separate, horizontal layers, like a pack of playing cards. For forward motion to be maintained, the force exerted by the high-velocity undisturbed flow on the top layer must be transmitted downwards from layer to layer by viscous drag. However, if you try the experiment described in Box 6.6, you will find that this method of transmitting forward motion soon breaks down as the depth increases. If this was the only means by which the forward motion could be imparted, then adjacent to the plate surface a progressively increasing thickness of stationary liquid would occur. This does not happen, because there is another mechanism that takes over and keeps the boundary layer moving forwards.

As the viscous shearing action described above and in Box 6.6 begins to breaks down, a fairly thick layer of slow moving or stationary liquid is created close to the plate. Passing over the top of this stationary layer is a relatively fast stream of liquid with a velocity close to U. These are just the conditions under which eddies will occur (just as a lorry passing at speed through still air will cause a disturbance behind it). What happens is that the eddying action causes some particles of liquid from the fast moving layer to be directed randomly down into the slow moving layer. Thus momentum is transferred from the fast moving layer

---

Box 6.6 ◢ **Try this**

Take a pack of fairly clean playing cards, press gently down on the top card and slide it horizontally over the others in the deck. You should find that the top card drags some of the others below with it. This is effectively the sort of viscous shearing action that drags the flow along in the laminar boundary layer. However, you should find that only the cards in the top half of the deck can be moved in this way; there is too much slippage between the cards to transmit the drag force to the bottom of the pack so they remain stationary, suggesting that this method of imparting forward motion is only effective over relatively small depths.

Figure 6.12 Investigating shearing action with a pack of playing cards

to the slow moving layer, effectively pushing it forward and maintaining its motion. Conversely, some particles from the slow moving layer will find their way back into the fast moving stream. Since this process is random, the motion in the boundary layer is turbulent with different particles following different paths and having different velocities, so at any point the velocity varies from moment to moment. This is the type of flow experienced in the turbulent boundary layer. Note, however, that a very thin laminar sublayer is retained even in the turbulent part of the boundary layer (Fig. 6.11). At the actual surface of the plate the liquid velocity may still be zero.

We now have an explanation of Fig. 6.11. As the liquid moves over the flat plate a relatively thin laminar boundary layer forms first in which the forward motion is maintained by the viscous shearing forces. As the flow passes over more and more of the plate the frictional effects increase and the boundary layer thickens until the viscous shear mechanism breaks down, giving transitional flow in the boundary layer. After that, the transfer of momentum from the fast flowing upper layers of liquid to the slow moving lower layers keeps the boundary layer moving forwards with a turbulent eddying motion. There is, though, a laminar sublayer which has a vital significance with respect to pipeflow.

Pick up a piece of A4 paper and imagine that on one side of it there is a boundary layer like that on the surface of the plate depicted in Fig. 6.11. Roll it up so that the leading edge of the surface forms a circle, and the paper forms a cylinder with the boundary layer on the inside. What you now have is a pipe, with a three-dimensional boundary layer on the inside. In longitudinal section this looks like Fig. 6.13. Note that it is some distance into the pipe before the boundary layer becomes fully established, but once this has occurred all of the flow through the pipe takes place inside the boundary layer. In other words, there is no part of the flow which is unaffected by friction between the liquid and the pipe walls. That is why boundary layer is very important with respect (REMEMBER!!!) to pipeflow.

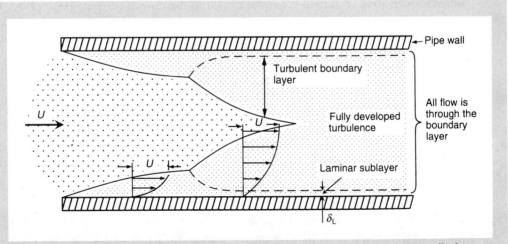

Figure 6.13 The boundary layer on the flat plate in Fig. 6.11 has been rolled up to form a pipe. The boundary layer now extends over the whole cross-section so all of the flow occurs in the boundary layer

Of course, the flow along the centreline of the pipe in Fig. 6.13 is the least affected by wall friction and has the highest velocity, while the flow in the laminar sublayer is stationary, or almost so. This brings us back to the question 'how can the same pipe be smooth sometimes and rough at others?' The answer lies with the laminar sublayer, the thickness of which (δ_L) can be calculated from the following equation:

$$\delta_L = 32.8 D/\text{Re } \lambda^{1/2} \qquad\qquad (6.15)$$

where D is the pipe diameter (m), λ the dimensionless friction factor and Re the dimensionless Reynolds number of the flow. This shows that the thickness of the laminar sublayer is inversely proportional to the Reynolds number. In other words, with a small value of Re there is a relatively thick laminar sublayer, whereas with a large Re the sublayer is relatively thin. If the liquid density and viscosity are constant and the same pipe diameter is considered, then Re $= \rho VD/\mu$ can be reduced to Re $\propto V$. Thus for a particular liquid flowing in a particular pipe, equation (6.15) tells us that if the velocity is low the boundary layer is relatively thick, while at high velocities it is relatively thin. For example:

if $D = 0.6$ m, Re $= 10\,000$ and $\lambda = 0.04$ then $\delta_L = 32.8 \times 0.6/10000 \times 0.04^{1/2} = 0.010$ m

if $D = 0.6$ m, Re $= 100\,000$ and $\lambda = 0.04$ then $\delta_L = 32.8 \times 0.6/100000 \times 0.04^{1/2} = 0.001$ m

if $D = 0.6$ m, Re $= 1\,000\,000$ and $\lambda = 0.04$ then $\delta_L = 32.8 \times 0.6/1000000 \times 0.04^{1/2}$
$\qquad = 0.0001$ m

Thus the thickness of the laminar sublayer may typically vary from a few mm to a tenth of a mm. We now come back to the idea of relative roughness. If the protrusions on the inside of a pipe have a height, k, which is less than δ_L, the thickness of the laminar sublayer, then the pipe effectively behaves as if it was smooth (Fig. 6.14a). The protrusions are entirely within the stationary or very slow moving laminar sublayer which effectively smooths out the pipe surface and prevents eddies from forming (remember that in laminar flow the roughness of the surface is irrelevant). This gives rise to smooth turbulent flow in the pipe, with the pipe being referred to as **hydraulically smooth**.

If k is slightly larger than δ_L so that the protrusions just extend through the laminar sublayer, then some disturbance may be caused to the flow in the pipe (Fig. 6.14b). This is transitional turbulent flow. However, if the protrusions completely penetrate the boundary layer ($k \gg \delta_L$) then eddying will result in the pipeline giving rise to rough turbulent flow. Under these conditions the pipe is considered to be **hydraulically rough**.

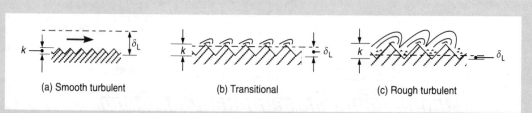

(a) Smooth turbulent   (b) Transitional   (c) Rough turbulent

Figure 6.14 The effect of the height of the protrusions (k) on the inside of a pipe relative to the thickness of the laminar sublayer (δ_L). (a) The protrusions lie within the sublayer resulting in smooth turbulent flow. (b) The protrusions just penetrate the sublayer giving transitional turbulent flow. (c) The protrusions are much higher than the sublayer, causing turbulence and resulting in rough turbulent flow [*after Webber (1971)*]

❝ So it is the thickness of the laminar sublayer relative to the physical roughness of the inside of the pipe that determines whether a pipe is classed as hydraulically smooth or hydraulically rough. Note that it does not depend entirely on the physical roughness since the thickness of the laminar sublayer varies inversely with the Reynolds number (or velocity) of the flow. Now go back to Fig. 6.10 and relate the three types of turbulent flow, called smooth turbulent, transitional turbulent and rough turbulent, to the three conditions shown in Fig. 6.14. Make sure that you understand the relationship. By the way, 'smooth turbulent flow' sounds like a contradiction of terms, but it means turbulent flow that occurs in a pipe that behaves hydraulically as though it was smooth. ❞

(REMEMBER!!!)

6.5.4 The Colebrook–White equation

After Nikuradse published his results the search began for an equation that would describe the various lines on his diagram. Prandtl analysed the velocity distribution in the boundary layer and formulated this expression for frictional resistance in hydraulically smooth pipes:

$$\frac{1}{\sqrt{\lambda}} = 2 \log \frac{\text{Re} \sqrt{\lambda}}{2.51} \tag{6.16}$$

This is often referred to as the **smooth pipe equation**. Next an equation was developed for the horizontal lines to the right of Fig. 6.10 to describe the behaviour of hydraulically rough pipes:

$$\frac{1}{\sqrt{\lambda}} = 2 \log \frac{3.7 D}{k} \tag{6.17}$$

This is the **rough pipe equation**. It contains no velocity or discharge term, nor the Reynolds number since the horizontal nature of Nikuradse's lines in the rough turbulent zone indicates that λ is independent of Re. It is simply an equation relating the Darcy friction factor, λ, to k. Unfortunately, when new most commercial pipes do not operate in the rough turbulent zone but in the more complex transitional turbulent flow region where viscosity and the Reynolds number still have an influence on λ. Colebrook and White investigated the flow in this region by blowing air over isolated sand grains inside a pipe. By combining equations (6.16) and (6.17) they obtained an expression that covered the whole range of turbulent flows in Fig. 6.10 with reasonable accuracy, including the important transition zone. The Colebrook–White equation is:

$$\frac{1}{\sqrt{\lambda}} = -2 \log \left(\frac{k}{3.7 D} + \frac{2.51}{\text{Re} \sqrt{\lambda}} \right) \tag{6.18}$$

The first term in the bracket in this expression is from the rough pipe equation (6.17) and dominates equation (6.18) at high Re values, while the second term appears in the smooth pipe equation (6.16) and becomes significant when Re is small. Consequently the equation is valid over the whole range of flows. In combination with the Darcy equation this enables the k values of various commercial surfaces to be calculated, and some typical values are shown in Table 6.2. Equation (6.18) can also be used to calculate λ for a given k, D and Re. To obtain an equation for the mean flow velocity in the pipe, V, the expression for λ in equation (6.18) must be substituted into the Darcy equation (6.12). The resulting expression is:

$$V = -2\sqrt{2gDS_F} \, \log\left(\frac{k}{3.7D} + \frac{2.5v}{D\sqrt{2gDS_F}}\right)$$

(6.19)

where S_F is the friction slope (that is the gradient of the total head line). S_F is equal to h_F/L, where h_F is the head lost due to friction in a pipe of length, L. When the pipe has a constant diameter (and hence the velocity is constant) S_F equals the hydraulic gradient. Thus S_F in equation (6.19) may be referred to as either the friction slope or the hydraulic gradient.

Equation (6.19) is easy to solve if D and S_F are already known and we want to calculate V. Conversely, if we know the discharge the pipeline has to carry and want to calculate the required diameter then the equation is difficult to solve since D appears in it three times while V depends upon the unknown D. This means the equation must be solved by trial and error (one way to avoid this may be to take the standard pipe sizes and calculate the velocity and discharge that would occur in each). Equation (6.18) also has to be solved by trial and error since λ appears on both sides of the expression. However, Moody gave the following equation which can be solved directly for λ:

$$\lambda = 0.0055\left[1 + \left(20\,000\,\frac{k}{D} + \frac{10^6}{Re}\right)^{1/3}\right]$$

(6.20)

For Re values between 4×10^3 and 1×10^7 and for k/D up to 0.01 this expression is claimed to be accurate to within 5%. This is usually good enough since the roughness of all pipes can be expected to change with age anyway (see below). If greater accuracy is needed then the value obtained from equation (6.20) can be used as the starting figure in a trial and error solution of equation (6.18).

The solution of some of the above equations by trial and error may sound laborious, but in fact it is a simple matter to write a short program for a desk-top personal computer that will solve them in a couple of seconds. Often it is not necessary as Examples 6.9 and 6.10 show, and even if it is, there are other ways around the problem. One is to obtain the solution to equation (6.19) from the charts or tables described in section 6.5.5 below. Another is to use a more pragmatic and less complicated equation that is reasonably accurate for a particular flow condition (section 6.5.6). Yet another alternative is to obtain the value of λ from a Moody diagram, which is the generalised version of the Nikuradse diagram that is applicable to commercial pipes (Fig. 6.10). For instance, if we take Re = 3.22×10^6 and $k/D = 0.000125$, as in Example 6.10, then Fig. 6.10 shows that λ has a value of about 0.013 and that the flow is in the transition zone. This is practically the same as the result obtained using equation (6.20), and exactly the same answer as that arising from the solution of equation (6.18). Therefore it could be concluded that the generalised diagram is accurate enough for most purposes.

EXAMPLE 6.9

Water flows through a 1.2 m diameter pipeline constructed from precast concrete pipes with 'O' ring joints. The hydraulic gradient (S_F) is 1 in 250 and v is $1.005 \times 10^{-6}\,m^2/s$. Calculate the mean velocity of flow using the Colebrook–White equation.

From Table 6.2, for precast concrete pipes with 'O' ring joints $k = 0.15\,mm = 0.00015\,m$. $S_F = 1/250 = 0.004$. Using equation (6.19):

Table 6.2 Typical roughness values (k) of surfaces (clean, new and in normal condition unless stated otherwise)*

Type of pipe	k (mm)
Perspex, glass, aluminium, brass, copper, lead, alkathene	0.003
Spun bitumen or concrete lined metal pipes	0.03
Steel: uncoated	0.03
Steel: coated	0.06
Uncoated cast iron	0.3
Old tuberculated water mains: up to 20 years old – slight attack	0.6
– moderate attack	1.5
– severe attack	15
80–100 years old – slight attack	3
– moderate attack	6
– severe attack	60
Precast concrete pipes with 'O' ring joints – normal condition	0.15
Spun precast concrete pipes with 'O' ring joints – normal condition	0.15
Glazed or unglazed clay with spiggot and socket joints and 'O' ring seals:	
< 0.15 m diameter	0.03
> 0.15 m diameter	0.06
UPVC with: chemically cemented joints	0.03
'O' ring seals at 6–9 m intervals	0.06
Sewer rising mains – mean operating velocity 1.0 m/s	0.3
Sewer rising mains – mean operating velocity 2.0 m/s	0.06
Brickwork: well pointed	3
in need of pointing	15
Straight uniform artificial earth channel	60
Straight natural channel free of obstructions	300

* After Hydraulics Research, 1990.

$$V = -2\sqrt{2gDS_F}\,\log\left(\frac{k}{3.7D} + \frac{2.5v}{D\sqrt{2gDS_F}}\right)$$

$$V = -2\sqrt{19.62 \times 1.2 \times 0.004}\,\log\left(\frac{0.00015}{3.7 \times 1.2} + \frac{2.5 \times 1.005 \times 10^{-6}}{1.2\sqrt{19.62 \times 1.2 \times 0.004}}\right)$$

$$V = -0.614\log(0.0000338 + 0.00000682)$$

$$V = -0.614 \times -4.391$$

$$V = 2.696 \text{m/s}$$

EXAMPLE 6.10

A 1.2 m diameter pipeline must discharge 3.049 m³/s when flowing full. If the viscosity and pipes are as in Example 6.9, calculate the friction factor, λ, and the required hydraulic gradient.

$$V = 3.049 \times 4/(\pi \times 1.2^2) = 2.696\,\text{m/s}$$
$$Re = VD/v = 2.696 \times 1.2/1.005 \times 10^{-6} = 3.219 \times 10^6$$
$$k/D = 0.00015/1.2 = 0.000125$$

The friction factor can either be found by solving equation (6.18) by trial and error or by solving equation (6.20) directly. Equation (6.20) is:

$$\lambda = 0.0055\left[1 + \left(20\,000\frac{k}{D} + \frac{10^6}{Re}\right)^{1/3}\right]$$

$$\lambda = 0.0055\left[1 + \left(20\,000 \times 0.000125 + \frac{10^6}{3.219 \times 10^6}\right)^{1/3}\right]$$

$$\lambda = 0.0055\left[1 + (2.5 + 0.311)^{1/3}\right]$$

$$\lambda = 0.0133$$

The Darcy equation gives $h_F = \lambda L V^2/2gD$ so $S_F = \lambda V^2/2gD$ (since $S_F = h_F/L$).

$$S_F = 0.0133 \times 2.696^2/19.62 \times 1.2$$
$$S_F = 0.0041 \text{ or 1 in 244}$$

Solving equation (6.18) by trial and error gives $\lambda = 0.0130$, about 2% difference. Note that the hydraulic gradient obtained above should have been 1 in 250 since this example is effectively Example 6.9 in reverse.

6.5.5 Hydraulics Research charts and tables

The Hydraulics Research 'Charts for the hydraulic design of channels and pipes' are widely used (Hydraulics Research, 1990). They provide a very simple means of calculating the head loss in a pipeline. The charts are very useful for design purposes since they show the relationship between pipe diameter (D), discharge (Q), velocity (V) and the head loss due to friction (h_F), which is the hydraulic gradient. They are based on equation (6.19), and provide an easy way of obtaining a solution. The charts (also available as tables) cover the range of k values corresponding to most types of commercial pipe in good, normal or poor condition (see Table 6.2). They can be applied to pipes running either full or part-full; different charts can be used for non-circular sections such as trapezoidal open channels. A typical chart for a pipe is shown in Fig. 6.15 and its use is explained in Example 6.11.

EXAMPLE 6.11

As in Examples 6.9 and 6.10, water flows through a pipeline of 1.2 m diameter that has a hydraulic gradient of 1 in 250 and a roughness k of 0.15 mm. Find the velocity of flow and the discharge using the Hydraulic Research chart (Fig. 6.15), assuming the pipeline is flowing full.

Step 1: Find the chart that relates to $k = 0.15$ mm.

Step 2: Express the gradient in terms of m per 100 m. Thus 1/250 is 0.4 m per 100 m.

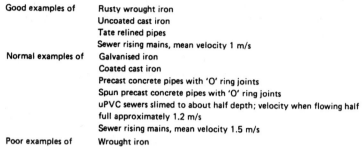

Good examples of	Rusty wrought iron
	Uncoated cast iron
	Tate relined pipes
	Sewer rising mains, mean velocity 1 m/s
Normal examples of	Galvanised iron
	Coated cast iron
	Precast concrete pipes with 'O' ring joints
	Spun precast concrete pipes with 'O' ring joints
	uPVC sewers slimed to about half depth; velocity when flowing half full approximately 1.2 m/s
	Sewer rising mains, mean velocity 1.5 m/s
Poor examples of	Wrought iron
	Coated steel
	Clayware (glazed or unglazed) with sleeve joints
	Sewer rising mains, mean velocity 2 m/s

Discharge *Q* (l/s) *for pipes flowing full*

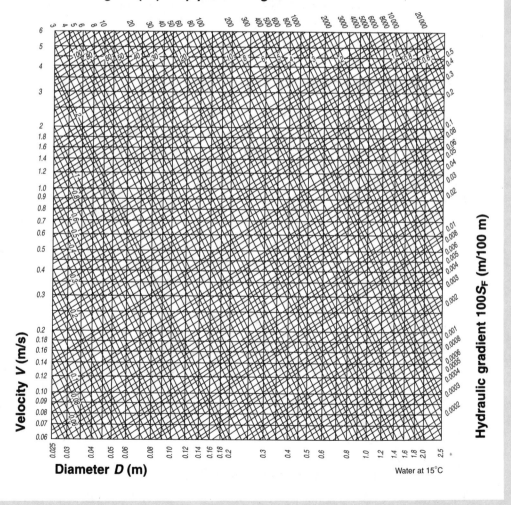

Figure 6.15 Hydraulics Research chart for *k* = 0.15 mm [*courtesy HR, Wallingford*]

Step 3: Find the 0.4 m/100 m hydraulic gradient line in the right-hand margin of the chart. Note that these lines slope from top right towards bottom left at the same angle as the numbers are printed in the margin. (For future reference, hydraulic gradients > 0.5 m/100 m have the number printed within the chart about 10 mm from the top.)

Step 4: Find the line representing $D = 1.2$ m in the bottom margin of the chart. The numbers are printed vertically to correspond with the vertical lines representing diameter.

Step 5: Find the point of intersection between the 0.4 m/100 m line and $D = 1.2$ m line.

Step 6: From this intersection point move horizontally across the chart and read the velocity printed in the left-hand margin (note, horizontal lines, horizontal numbers in the margin). Thus $V = 2.65$ m/s, about the same as in Example 6.9.

Step 7: From the intersection point move up the line sloping from bottom right towards top left, and read the discharge from the sloping numbers: $Q = 3100$ litres/s (3.100 m³/s). Again, this is more or less as in the previous examples.

6.5.6 Simplified or empirical formula

The solution of the Colebrook–White equation may require a time-consuming trial-and-error approach. With complicated pipeline systems this is not practicable. Therefore, engineers often use a simpler equation, even if it is less acurate. The important first step is to decide what type of flow will be encountered in the pipeline: smooth turbulent, transitional turbulent or rough turbulent. A convenient way of doing this is to employ a modified version of the Reynolds number which is called the **Reynolds roughness number**, Re* where:

$$\mathbf{Re} * = \mathbf{Re}\left(\frac{k}{D}\right)\sqrt{\frac{\lambda}{8}} \tag{6.21}$$

The value of the Reynolds roughness number indicates the type of turbulent flow as follows:

smooth turbulent   Re* < 4

transitional turbulent   Re* = 4 to 60

rough turbulent   Re* > 60

Having calculated Re* the equation appropriate to the type of flow must be used. As mentioned earlier, many commercial pipes when new operate in the transitional turbulent flow regime. However, taking the three types of flow in order, the expressions most popular with engineers are the Blasius equation, the Hazen–Williams equation and the Manning equation.

The Blasius equation for smooth turbulent flow

For Re up to 100000 and water at 15°C with $v = 1.14 \times 10^{-6}$ m²/s the combination of the Darcy and Blasius equations (6.12 and 6.14) gives an equation for mean velocity.

Darcy: $\quad S_F = \lambda V^2/2gD$ $\qquad$ Blasius: $\quad \lambda = 0.316/\mathrm{Re}^{0.25}$

Hence: $\quad S_F = 0.316V^2/2gD\,\mathrm{Re}^{0.25}$ $\quad$ where $\mathrm{Re}^{0.25} = D^{0.25}V^{0.25}/v^{0.25}$

$\qquad\quad S_F = 0.316V^2(1.14 \times 10^{-6})^{0.25}/19.62DD^{0.25}V^{0.25}$

$\qquad\quad S_F = 0.000526V^{1.75}/D^{1.25}$

$\qquad\quad V^{1.75} = D^{1.25}S_F/0.000526$

or in terms of simple fractions $\quad V^{7/4} = D^{5/4}S_F/0.000526$

so $\qquad V = D^{5/4 \times 4/7}S_F^{4/7}/(0.000526)^{4/7}$

$\qquad\quad \boldsymbol{V = 75D^{5/7}S_F^{4/7}}$ $\hfill (6.22)$

The Hazen–Williams equation for transitional turbulent flow

This is a formula widely used in the water industry for pipes in the transition zone:

$$\boldsymbol{V = 0.355C_{HW}D^{0.63}S_F^{0.54}} \hfill (6.23)$$

where C_{HW} is the Hazen–Williams coefficient. The equation is reasonably accurate for pipes with a diameter (D) over 0.15 m, velocities below 3 m/s and C_{HW} over 100. The value of C_{HW} varies with velocity, pipe diameter and material (see Table 6.3 and Twort *et al.*, 1994).

The Manning equation for rough turbulent flow in pipes

The Manning equation is most commonly used for the analysis of flow in open channels (see Chapter 8) but it can also be applied to pipelines. For a pipe flowing full the hydraulic radius (R = cross-sectional area/wetted perimeter) is $D/4$. Thus equation (8.8) becomes:

$$\boldsymbol{V = (0.397/n)D^{2/3}S_F^{1/2}} \hfill (6.24)$$

where n is the Manning roughness coefficient as shown in Table 8.1.

Table 6.3 Typical values of the Hazen–Williams coefficient (C_{HW}) for non-aggressive and non-sliming water at a velocity of 1 m/s

Type of pipe	Pipe diameter (m)		
	0.15	0.60	1.20
Coated cast iron – 60 years old – moderate corrosion ($k = 5.0$ mm)	80	92	96
Coated cast iron – 30 years old – moderate corrosion	90	102	107
Coated cast iron – 30 years old – slight corrosion	106	118	120
Coated spun iron – smooth and new ($k = 0.05$ mm)	142	148	148
Concrete – imperfect interior finish – tunnel linings ($k = 1.25$ mm)	102	110	113
Concrete – smooth interior with good joints ($k = 0.50$ mm)	117	125	128
Concrete – perfect interior with smooth joints ($k = 0.25$ mm)	126	132	134
Spun bitumen or cement lined pipes or PVC	148	152	153

EXAMPLE 6.12

For the conditions described in Examples 6.9 and 6.10, calculate the Reynolds roughness number and determine which of the simplified empirical equations can be applied.

Reynolds roughness number $Re^* = Re(k/D)\,(\lambda/8)^{1/2}$
From Example 6.10, $Re = 3.219 \times 10^6$, $k/D = 0.000125$, $\lambda = 0.013$

$Re^* = 3.219 \times 10^6 \times 0.000125 \times (0.013/8)^{1/2}$

$Re^* = 16$

Between $Re^* = 4$ and $Re^* = 60$ the flow is in the transition turbulent zone, so the Hazen–Williams equation would be appropriate.

EXAMPLE 6.13

For the pipeline and conditions described in Example 6.9, calculate the mean velocity using the Hazen–Williams equation.

$V = 0.355 C_{HW} D^{0.63} S_F^{0.54}$

From Table 6.3 a 1.2 m diameter concrete pipe with a smooth interior and good joints ($k = 0.15$) would have a C_{HW} of **about** 140 (note that this is probably not the exact equivalent). As before $S_F = 1$ in $250 = 0.004$.

$V = 0.355 \times 140 \times 1.2^{0.63} \times 0.004^{0.54}$

$V = 2.827\,m/s$

This is 5% higher than the answer obtained using the more complicated Colebrook–White equation (Example 6.9). Such an 'error' is usually acceptable (see section 6.5.7).

6.5.7 Deterioration of pipelines and changing roughness

Irrespective of the accuracy of the above equations, an analysis can become inaccurate in the long-term if the interior roughness of the pipe changes by a large amount due to corrosion, encrustation (also called tuberculation) or sliming. Partly for this reason, Tables 6.2 and 6.3 were included showing the variation of k and the Hazen–Williams coefficient with age. For example, a 0.6 m diameter coated cast iron pipe which is 30 years old with slight corrosion has $C_{HW} = 118$, while $C_{HW} = 92$ for a 60 year old pipe of the same diameter with moderate corrosion. This is a 22% difference. Obviously the change from a new, uncorroded pipe to a moderately or severely corroded 60 year old pipe would be even larger, perhaps 45 or 50%. Since V is directly proportional to C_{HW} both the velocity and the discharge in the pipeline may diminish significantly over a period of time. This should be remembered when designing a new system.

The severity of the deterioration process is illustrated by Fig. 6.16b. Here deposits on the inside of the pipeline have drastically reduced its diameter, in addition to altering its roughness. Deterioration and encrustation of this sort mean that the internal surfaces of water pipelines may have to be scraped and relined from time to time.

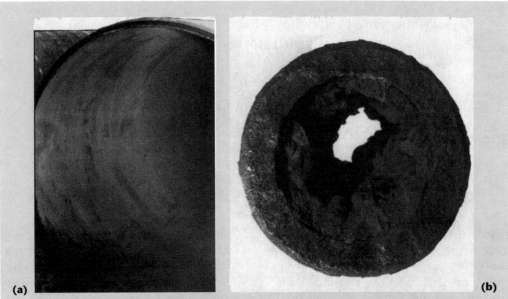

Figure 6.16 (a) A new 800 mm diameter ductile iron pipe with cement mortar lining. (b) An old, 50 mm diameter pipe showing severe tuberculation. Note the reduced bore as well as increased roughness

 ## 6.6 Head losses at changes of section

In the first part of the chapter we used the expressions in Table 6.1 to evaluate the head losses at changes of section. Now we will have a closer look at where there equations come from. In doing so we will use some of the material covered in earlier chapters.

6.6.1 Head loss at a sudden expansion

Expansions and diverging flow are usually associated with an energy loss: the more sudden the expansion, the greater the loss. Therefore, a very sudden expansion like that in Fig. 6.17 would certainly result in a significant loss of energy. With short pipelines in particular where minor losses can be significant, this loss must be evaluated and included when we write the Bernoulli equation for the system, as in Example 6.2.

Consider water flowing through the expansion in Fig. 6.17. The flow in the smaller pipe will have a relatively high velocity. On emerging into the larger cross-section the stream will slow (in accordance with the continuity equation $A_1V_1 = A_2V_2$) and gradually expand to fill the larger pipe, as shown by the streamlines in the diagram. In accordance with the Bernoulli equation, as the velocity decreases the pressure will increase so as to keep the total energy roughly constant. Thus a higher pressure exists in the larger pipe than in the smaller, so the **adverse pressure gradient** is trying to push the water back into the smaller pipe. This causes the reduction in velocity required by the continuity equation, and results in turbulence and eddying of the water occupying the corners of the expansion. Between the

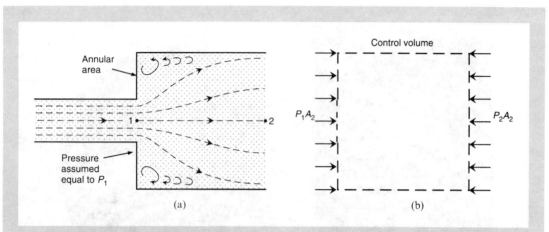

Figure 6.17 (a) A sudden expansion in a pipe flowing full of water. As the diameter increases the velocity reduces but the pressure increases, giving an adverse pressure gradient resulting in turbulence and eddying in the corners of the expansion. A vortex sheet may form separating the eddying water in the corners from the relatively high-velocity stream flowing along the pipeline. (b) The control volume for the expansion, showing the *external forces* acting on it

---

Box 6.7 ▶ **Recall**

When we apply the momentum equation to a control volume, only the *external forces* acting *on* the control volume are considered (the equal and opposite internal forces are ignored). We also adopt a sign convention, with the external forces generally being positive in the original direction of flow. If we obey these rules then for a particular direction (like the x direction) we can write the momentum equation for the control volume as: $\Sigma F_X = \rho Q(V_{2X} - V_{1X})$. Refer back to section 4.5.2 if you do not understand this.

---

live stream of water which is travelling through the larger pipe and the body of eddying water in the corners of the pipe, a separation boundary will form coinciding approximately with the two outer streamlines drawn in Fig. 6.17a. On one side of the separation boundary the water is moving forwards with a relatively high velocity, and on the other the water is relatively stationary. These are just the conditions under which a vortex sheet will form (Fig. 5.2c). All of this turbulence and eddying extracts energy from the flow and causes the energy loss.

So where does the expression for the head loss at the expansion, $h_L = (V_1 - V_2)^2/2g$, come from?

The answer is from a combination of the continuity, momentum and energy equations (where else?). Before we look at this in detail, read Box 6.7.

When applying the momentum equation to the control volume in Fig. 6.17b, we only need to consider the direction along the centreline of the horizontal pipe so the subscript that indicates the direction will be omitted. One additional piece of information is that experiments have shown that the pressure exerted by the water in the larger pipe on the annular area marked in Fig. 6.17a is P_1, the same as the pressure in the smaller pipe. Thus the pressure P_1 can be considered to act internally over the whole of the larger area, A_2. Therefore, the term for the equal and opposite external force acting on the control volume is $P_1 A_2$. This looks like a typographical error at first, but it isn't. Note that there is no other force acting on the annular area; this assumption is a means of eliminating from the equation the resultant force (the equivalent of F_R in equation (4.14)) by evaluating it experimentally. So effectively $P_1 A_2 = P_1 A_1 + F_R$. Thus $\Sigma F = \rho Q(V_2 - V_1)$ can be written as:

$$P_1 A_2 - P_2 A_2 = \rho Q(V_2 - V_1) \tag{1}$$

Note that the weight of the water in the control volume is ignored since it has no component along the horizontal centreline of the pipe. From the continuity equation, $Q = A_2 V_2$ so if we replace Q in equation (1) and divide both sides of the equation by ρg, then:

$$A_2 \frac{(P_1 - P_2)}{\rho g} = \frac{\rho A_2 V_2}{\rho g}(V_2 - V_1)$$

cancelling gives: $\quad \dfrac{(P_1 - P_2)}{\rho g} = \dfrac{V_2}{g}(V_2 - V_1) \tag{2}$

Now applying the Bernoulli equation to a horizontal streamline along the centreline of the pipe, ignoring friction losses between the water and the pipe walls but including the head loss at the expansion, h_L, then:

$$P_1/\rho g + V_1^2/2g = P_2/\rho g + V_2^2/2g + h_L$$

Thus $\quad h_L = \dfrac{(P_1 - P_2)}{\rho g} + \left[\dfrac{V_1^2}{2g} - \dfrac{V_2^2}{2g}\right] \tag{3}$

The expression for $(P_1 - P_2)/\rho g$ in equation (2) can be substituted for the equivalent expression in equation (3) giving:

$$h_L = \frac{V_2}{g}(V_2 - V_1) + \left[\frac{V_1^2}{2g} - \frac{V_2^2}{2g}\right]$$

$$h_L = \frac{2V_2^2}{2g} - \frac{2V_1 V_2}{2g} + \frac{V_1^2}{2g} - \frac{V_2^2}{2g}$$

$$h_L = \frac{V_1^2}{2g} - \frac{2V_1 V_2}{2g} + \frac{V_2^2}{2g}$$

$$\boldsymbol{h_L = \frac{(V_1 - V_2)^2}{2g}} \tag{6.25}$$

If you have to derive this equation it may help to remember the general expansion $(a - b)^2 = (a - b)(a - b) = a^2 - 2ab + b^2$.

Equation (6.25) gives the head or energy loss at a sudden or abrupt expansion. You should be aware that we can reduce the magnitude of the loss by making the expansion more gradual by using a tapered section of pipe (sometimes called a diffuser section) as in the Venturi meter. If the angle of the diffuser walls to the horizontal is about 20°, then the head loss may be reduced to about half that for a sudden expansion, perhaps even less. Remember that a pipe discharging into a large reservoir can be treated as a sudden expansion. If the water in the reservoir is static then $V_2 = 0$ so $h_L = V_1^2/2g$. In other words, the entire velocity head of the water in the pipe is lost, which is logical if the water becomes static.

We can also arrange equation (6.25) into a more general form. The continuity equation gives us $V_2 = A_1V_1/A_2$ so replacing V_2 in equation (6.25) gives:

$$h_L = \frac{\left(V_1 - \dfrac{A_1}{A_2}V_1\right)^2}{2g}$$

$$h_L = \left(1 - \frac{A_1}{A_2}\right)^2 \frac{V_1^2}{2g}$$

$$\text{or} \quad h_L = \left(1 - \frac{D_1^2}{D_2^2}\right)^2 \frac{V_1^2}{2g} \tag{6.26}$$

This has the merit of expressing the head loss as $KV_1^2/2g$, that is as a multiple of the velocity head in the smaller pipe, where K is the multiple (such as 0.2 or 0.5). This is the general form in which all head losses are usually expressed. By assuming various diameters, we can now get an idea of the relative magnitude of the head loss. For example, if $D_1 = 0.6\,\text{m}$ and $D_2 = 1.2\,\text{m}$ then $K = 0.56$ but if $D_1 = 0.6\,\text{m}$ and $D_2 = 0.9\,\text{m}$ then $K = 0.31$. Examples of some typical K values for different losses are given in Table 6.4. The table also gives some l/D values. This is because in practice a convenient way to allow for head losses at valves, expansions, contractions, and so on, is to add for each loss an additional length of pipe, l, to the true physical length of the pipeline. The length, l, is defined as 'the length of straight pipe of diameter D needed to give the equivalent loss of head'.

6.6.2 Head loss due to a sudden contraction

With a contraction the pressure gradient is in the direction of flow so the water enters the contraction quite smoothly with little loss of energy. However, as the streamlines in Fig. 6.18 show, the live stream of water continues to contract for some distance into the smaller pipe forming a vena contracta, then gradually expands to fill the pipe. It is this expansion that causes the energy loss, so this is what must be analysed to obtain an expression for the head loss. We start by recognising that in this situation V_1 in equation (6.25) is the velocity of the live stream (or jet) at the vena contracta, V_J, where the cross-sectional area of flow is a_J. If we apply the continuity equation then $V_J a_J = A_2 V_2$ so:

$$V_J = A_2 V_2/a_J$$

Now equation (5.10) defined the coefficient of contraction as $C_C = a_J/A$ so:

$$V_J = V_2/C_C$$

Substituting this expression into equation (6.25) gives:

$$h_L = \frac{(V_2/C_C - V_2)^2}{2g}$$

If we say that a typical value for C_C is 0.6 then $h_L = (V_2/0.6 - V_2)^2/2g = (1.67V_2 - V_2)^2/2g$ giving $h_L = 0.45V_2^2/2g$. However, this is usually rounded up to:

Table 6.4 Head losses caused by changes in geometry and fittings to pipelines*

Loss		K	approx. l/D
Sharp edged entrance		0.50	22
Slightly rounded entrance		0.25	11
Bell-mouth entrance		0.05	2
Footvalve and strainer (entrance to a pipeline to a pump)		2.50	113
Bend – with $r/D = \frac{1}{2}$,	22.5° bend	0.20	9
	45° bend	0.40	18
	90° bend	1.00	45
Bend – with $r/D = 1$,	22.5° bend	0.10	5
	45° bend	0.20	9
	90° bend	0.40	18
Bend – with $r/D = 8$ to 50,	22.5° bend	0.05	2
	45° bend	0.10	5
	90° bend	0.20	9
Tapers – contraction with flow from large to small diameter			negligible
– expansion with inlet : outlet diameter 4 : 5		0.03	1
3 : 4		0.04	2
1 : 3		0.12	6
Gate valve – fully open		0.12	6
– quarter closed		1.00	45
– half closed		6.00	270
– three-quarters closed		24.00	1080
Sudden enlargement or sudden exit loss		1.00	45
Bell-mouth outlet		0.20	9

* Losses are expressed in term of K where $h_L = KV^2/2g$ and V is the velocity in the smaller pipe or as *approximate* l/D where l is the length of straight pipe required to give the equivalent head loss in a pipe of diameter D. The radius of a bend is denoted by r.

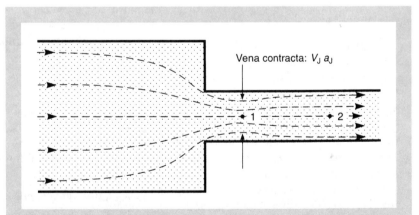

Figure 6.18 The flow enters the contraction with minimal head loss but continues to contract forming a vena contracta. The loss occurs from the vena contracta onwards as the live stream expands to fill the pipe

$$h_L = \frac{0.5V_2^2}{2g}$$

(6.27)

Note that as with equation (6.26), the head loss is expressed as a multiple, K, of the velocity head in the smaller pipe. The value of K would be smaller if a tapered section of pipe (known as a reducer or confusor) was used to reduce the pipe diameter gradually. Remember also that the entrance to a pipeline from a reservoir can be considered as a sudden contraction. In such cases the head loss can be reduced by providing a rounded entrance to the pipeline, often called a '**bell-mouth**' entry. Such an entry may have a much larger C_C (see Fig. 5.5) so this can reduce K to 0.05 compared to the 0.5 shown in equation (6.27). Table 6.4 gives some other examples of K values.

Summary

1. Head losses control the flow though pipelines. These losses may be due to:

   (a) Friction, $h_F = \lambda LV^2/2gD$ (Darcy equation) where λ = pipe roughness, L = length, V = mean flow velocity and D = pipe diameter. In long pipelines, friction losses dominate.

   (b) Minor losses due to a sudden expansion when $h_L = (V_1 - V_2)^2/2g$ or sudden contraction when $h_L = 0.5V_2^2/2g$.

2. With losses defined as in 1 above and discharge through one pipeline only, the energy equation gives:

   Z = pipeline head losses $+ V^2/2g$

   (a) With a pipeline discharging to the atmosphere, Z is the difference in elevation between the water surface in the reservoir and the end of the pipeline; $V^2/2g$ is the residual velocity head as water is discharged.

   (b) With a pipeline connecting two reservoirs, Z is the difference in elevation between the water surfaces of the reservoirs; $V^2/2g$ is the velocity head lost as the water exits the pipe and becomes static in the second reservoir (equivalent to a sudden expansion with $V_2 = 0$). Don't include $V^2/2g$ twice.

3. With branching pipelines, apply the energy equation to a streamline joining each reservoir in turn, so that an equation is obtained for each branch. For example, with reference to Fig. 6.6:

   Z_{AC} = losses in pipe 1 + losses in pipe 2

   Z_{AD} = losses in pipe 1 + losses in pipe 3

   and $Q_1 = Q_2 + Q_3$

   These three equations enable V_1, V_2 and V_3 to be calculated, and thus Q_1, Q_2 and Q_3.

4. With two parallel pipes between two reservoirs as in Fig. 6.7, again apply the energy equation to a steamline passing through each pipeline in turn. Assuming only friction losses:

   Z = losses in pipeline 1

   Z = losses in pipeline 2

   These two equations can be solved directly for V_1 and V_2, and hence Q_1 and Q_2.

5. The flow of water in pipes can be classified as follows:

Laminar flow (Re < 2000)	$h_F \propto V$
e.g. Poiseuille eqn (6.11)	$h_F = 32vLV/gD^2$
Turbulent flow (Re > 4000)	$h_F \propto V^2$
e.g. Darcy eqn (6.12)	$h_F = \lambda LV^2/2gD$

Turbulent flow can be further classified depending upon whether the pipe roughness (protrusions of height k) lie within the laminar sublayer (smooth), just penetrate (transitional) or are much higher (rough) as in Fig. 6.14.

Smooth turbulent flow – λ depends on Re and k/D

Transitional turbulent flow – λ depends on Re and k/D

Rough turbulent flow – λ depends only on k/D

The Colebrook–White equation covers the whole range of pipeflow and is the basis of Fig. 6.15.

$$\frac{1}{\sqrt{\lambda}} = -2\log\left(\frac{k}{3.7D} + \frac{2.51}{\mathrm{Re}\,\sqrt{\lambda}}\right) \qquad (6.18)$$

6. Simplified or empirical equations for pipe-flow are:

Smooth turbulent flow – Blasius eqn (6.22):

$$V = 75D^{5/7}S_F^{4/7}$$

Transitional turbulent flow – Hazen–Williams eqn (6.23):

$$V = 0.355C_{HW}D^{0.63}S_F^{0.54}$$

Rough turbulent flow – Manning eqn (6.24):

$$V = (0.397/n)D^{2/3}S_F^{1/2}$$

Revision questions

6.1 Define what is meant by (a) energy (refer back to Chapter 4); (b) total energy; (c) head; (d) total energy; (e) total head line; (f) piezometric level; (g) hydraulic gradient; (h) hydraulic grade line; and (i) minor loss.

6.2 Explain where and why a negative pressure may occur in a pipeline. Use a couple of sketches to illustrate your answer. Does a negative pressure cause operational problems?

6.3 (a) The friction and minor losses are tabulated in Example 6.2. Draw the losses to scale, as in Fig. 6.4 in Example 6.1. (b) Repeat Example 6.2 using a 0.6 m diameter pipeline over the entire 2.0 km length between the reservoirs and recalculate all of the losses. What proportion of the total head loss is the result of friction in the 0.6 m pipeline?
[0.197 m; 52.410 m; 0.393; about 99%]

6.4 A reservoir must discharge to the atmosphere via a short pipeline. The entrance to the pipeline is sharp, and the diameter is 0.3 m for the first 10 m. The pipeline then expands suddenly to 0.45 m diameter for the last 10 m. For both pipes $\lambda = 0.06$. If 0.5 m³/s of water must be discharged from the pipeline, determine the height that the water level in the reservoir must be above the centre of the outlet pipe (a) ignoring minor losses; (b) taking both friction and minor losses into account; and (c) if the smaller pipeline has three 90° bends with $r/D = \frac{1}{2}$ in addition to the losses in (b).
[6.28 m; 8.34 m; 15.99 m with $K = 1.0$]

6.5 Water flows from one large reservoir to another via a pipeline that is 0.9 m in diameter, 10 km long with $\lambda = 0.04$. The difference in elevation of the water level in the reservoirs is 50 m. Taking into consideration all of the losses (and assuming that the entrance to the pipeline is sharp) calculate the discharge through the pipeline.
[0.943 m³/s]

6.6 Water flows between two reservoirs via two separate pipes (subscripts 1 and 2). The difference in elevation of the water level in the reservoirs is 50 m. Details of the pipelines are:

$L_1 = 10$ km, $\lambda_1 = 0.04$, $D_1 = 0.3$ m,

$L_2 = 10$ km, $\lambda_2 = 0.04$, $D_2 = 0.6$ m.

Ignoring the minor losses, calculate the discharge in each pipe and hence the total discharge between the reservoirs.
[0.061 m³/s; 0.343 m³/s; 0.404 m³/s]

6.7 Two reservoirs have a difference in water surface level of 6.0 m. The pipeline connecting them has a sharp entrance and is initially straight (pipe 1) but then splits at a junction into two branches (pipes 2 and 3 as in Fig. 6.8). Details of the pipes are as follows:

pipe	length	diameter	λ	
1	20 m	0.6 m	0.05	straight – no bends
2	35 m	0.4 m	0.05	4 bends with $K = 0.20$ each
3	15 m	0.4 m	0.05	2 bends with $K = 0.10$ each

The head loss at the junction is $0.4 V_1^2/2g$. Using Tables 6.1 and 6.4, calculate the discharge in each of the three pipelines.

$$[1.09\,\text{m}^3/\text{s}; \ 0.45\,\text{m}^3/\text{s}; \ 0.64\,\text{m}^3/\text{s}]$$

6.8 Three large reservoirs are joined by a branching pipeline as in Fig. 6.6. The elevation of their water surfaces is constant at A = 480 m OD, C = 390 m OD and D = 310 m OD. Details of the three pipelines are:

Pipe 1 – $L_1 = 15$ km, $\lambda_1 = 0.02$, $D_1 = 0.8$ m;
Pipe 2 – $L_2 = 11$ km, $\lambda_2 = 0.03$, $D_2 = 0.6$ m;
Pipe 3 – $L_3 = 6$ km, $\lambda_3 = 0.04$, $D_3 = 0.5$ m.

Assuming that the minor losses are negligible, determine the discharge in each pipeline.

$$[0.79\,\text{m}^3/\text{s}; \ 0.35\,\text{m}^3/\text{s}; \ 0.44\,\text{m}^3/\text{s}]$$

6.9 Explain briefly what is meant by (a) a boundary layer; (b) the laminar sublayer within the boundary layer; (c) a turbulent boundary layer; (d) smooth turbulent flow; (e) rough turbulent flow; (f) a hydraulically smooth surface; (g) a hydraulically rough surface; (h) tuberculation (of a pipe).

6.10 Water (viscosity $v = 1.005 \times 10^{-6}\,\text{m}^2/\text{s}$) flows through a 0.4 m diameter pipeline with a mean velocity of 3.49 m/s. The surface roughness of the pipe, $k = 0.15$ mm. (a) Calculate λ using the Moody approximation (equation (6.20)). (b) Calculate λ using the Colebrook–White equation (6.18). (c) Using the Darcy equation (6.12) with the value of λ from part (b), calculate the head loss caused by friction per 100 m length of pipe. (d) Using Fig. 6.15 find the hydraulic gradient (friction loss) per 100 m.

$$[\text{(a) } 0.0166; \text{ (b) } 0.0161; \text{ (c) } 2.499\,\text{m}; \text{ (d) } 2.5\,\text{m}]$$

6.11 Water flows through a 0.050 m diameter pipe for which $k = 0.003$ mm. The mean velocity of the flow is 0.14 m/s and the viscosity of the water is $v = 1.005 \times 10^{-6}\,\text{m}^2/\text{s}$. (a) Calculate λ using the Moody equation. (b) Calculate the Reynolds roughness coefficient. (c) Calculate the friction gradient, S_F, and hence the head loss per 100 m using the Blasius, Hazen–Williams or Manning equation according to the result of part (b).

$$[0.034; \ 0.027; \ 0.071\,\text{m}/100\,\text{m}]$$

Flow under a varying head – time required to empty a reservoir

A problem sometimes faced by engineers is calculating the time required to empty a reservoir. This may be a purpose-built flood storage reservoir located upstream of a town or city to store excess riverflow during a storm, and then discharge it safely back to the river channel after the flood has subsided. It is important that the reservoir empties as quickly as possible, because if another flood occurs while the reservoir is still full the flow cannot be stored and flooding will occur downstream. So, when designing the dam it is important to know how long it takes for the reservoir to empty; if it is too long the spillway or outlet must be made larger. Alternatively, the problem may involve a water supply reservoir that has been damaged by either natural causes or terrorist action, as happened in the former Yugoslavia in 1993. Whether or not the dam will collapse may depend upon how quickly the hydrostatic force on it can be reduced, that is the time to empty the reservoir. On a more modest scale, the problem could involve the time taken to empty an oil storage tank or a water distribution reservoir. In the laboratory, by measuring the time required to empty a tank it is possible to calculate the coefficient of discharge of, say, an orifice. Therefore the sort of questions answered in this chapter include:

What is meant by 'flow under varying head' and why is it significant?

What variables are involved?

How can we calculate the time required to empty a tank of uniform surface area?

How can the time required to empty a tank be used to calculate C_D?

What do we do if the surface area of the tank varies with depth?

How can we estimate the time required to empty an irregularly shaped reservoir?

What happens when water flows from one tank to another?

7.1 Introduction

In Chapter 6 it was assumed that when water was discharged through a pipeline the head in the large reservoir remained at its original level. This chapter is concerned with the time required to empty a reservoir assuming that the liquid stored in it is not being replaced. In other words, we assume that there is no flow into the reservoir so the water level in the reservoir will fall as its content is gradually discharged through the outlet. This is what is meant by *flow under varying head*.

As an introduction to the problem of flow under varying head, go back and look at the discharge–head relationships given in Table 5.1. These show that:

For a small orifice: $Q_A \propto H^{1/2}$

For a rectangular sharp crested weir or broad crested weir: $Q_A \propto H^{3/2}$

For a triangular sharp crested weir: $Q_A \propto H^{5/2}$

What this actually means is illustrated by Table 5.2 and Fig. 5.16, which show the non-linear nature of the relationships. In other words, you cannot assume that if you halve the head the discharge will be half. For instance, from Table 5.2 you can see that for $Q \propto H^{5/2}$ when $H = 6\,m$ then $Q = 88.2\,m^3/s$, but when $H = 3\,m$ then $Q = 15.6\,m^3/s$.

REMEMBER!!!

What has this got to do with emptying a reservoir?

Using the data in Table 5.2 for $Q \propto H^{5/2}$, suppose that the reservoir is full with $H = 6\,m$ and we want to calculate the time to empty it. The important point is that you **cannot** say that when $H = 6\,m$ then $Q = 88.2\,m^3/s$ and when it is empty ($H = 0$) then $Q = 0$, so the average discharge will occur at $H = 3\,m$ with $Q = (88.2 + 0)/2 = 44.1\,m^3/s$. This is wrong. Table 5.2 shows that at $H = 3\,m$ the value of Q is $15.6\,m^3/s$. The larger the power of H, the greater the curvature of the $Q_A - H$ line, and the larger the error incurred by assuming an 'average' discharge (this is explored later in Examples 7.5 and 7.6). Thus any calculation of the time required to empty the reservoir based on this average discharge will also be wrong. In other words, we must take into account the curvature of the $Q - H$ line when we calculate the time required to empty the reservoir.

How do we do this?

The way in which we do this is similar to the way in which we allowed for the fact that when water flows over a sharp crested weir the discharge through a horizontal strip of the nappe varies according to the depth, h, since $V = (2gh)^{1/2}$. In section 5.5.1 we considered the incremental discharge through an element of liquid, then integrated (between limits denoting the position of the water surface) to obtain the total discharge. Similarly, we can solve our reservoir problem by using integration between limits denoting the initial and final water levels; all we need is an equation to integrate. The equation we use is the one appropriate to the outflow device, so if it is an orifice we use the orifice equation, if it is a sharp crested rectangular weir, broad crested weir or spillway then we use the equation for that device.

7.2 Time to empty a reservoir of uniform cross-section

This section deals with reservoirs or tanks that when seen in plan have the same cross-sectional area, regardless of the depth of water. What happens when the area changes with the water level is considered later in the chapter.

The starting point for most reservoir storage problems is:

Change in storage = Inflow − Outflow (7.1)

This reappears later as equation (12.1). Here we are assuming that inflow = 0 so equation (7.1) can be rewritten as:

−Change in reservoir volume = Volume discharged through the outlet

The −ve sign indicates that the storage volume decreases as water is discharged. This equation provides the starting point for the solution of all of the problems that follow. Referring to Fig. 7.1 and expressing it in symbols gives:

$$-A_{WS}\delta h = Q_A \delta t$$ (7.2)

where A_{WS} is the plan area of the water surface in the reservoir (m²), δh is the fall in head (m), Q_A is the actual flow rate though the outlet device (m³/s) and δt is the time(s) corresponding to the fall in head δh. The negative sign indicates that δh decreases as δt increases. Note that equation (7.2) is perfectly logical if you think of the units:

$$m^2 \times m = m^3/s \times s$$
$$m^3 = m^3$$

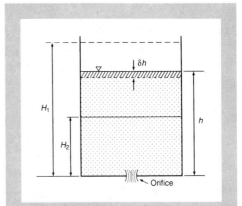

Figure 7.1 The water level falls by an amount δh in time δt as the water discharges through the orifice

This provides a good way of remembering the equation.

Now suppose we want to calculate the time for the water level in the tank in Fig. 7.1 to fall from an initial height H_1 to H_2. At some arbitrary time, t, the head above the outlet is denoted by h. Note that h is the head above the outlet (not the depth below the water surface) since the discharge equations use the head above the orifice or weir crest. Let δt be the time taken for the water level to fall (from h) by a very small amount δh. If the outlet device is an orifice with $Q_A = C_D A(2gh)^{1/2}$ then equation (7.2) becomes:

$$-A_{WS}\delta h = C_D A(2gh)^{1/2}\delta t$$

or $$\delta t = -\frac{A_{WS}}{C_D A(2g)^{1/2}h^{1/2}}\delta h$$ (7.3)

This represents a general relationship for the time (δt) to lower the water level in the reservoir by a very small amount (δh) through an orifice that has a head, h, of water above it. To obtain the time, T, to lower the water level from an initial height H_1 at time T_1, to a final height H_2 at time T_2, we integrate equation (7.3) between these limits, thus:

$$T = \int_{T_1}^{T_2} dt = -\frac{A_{WS}}{C_D A(2g)^{1/2}}\int_{H_1}^{H_2} h^{-1/2}dh$$

> **Box 7.1** **Summary of the general procedure**
>
> When the area of the water surface, A_{ws}, is constant regardless of the water level then almost all simple reservoir emptying problems can be tackled as follows:
>
> 1. Remember that: – change in volume of reservoir = volume discharged through outlet.
>
> 2. This can be written mathematically as: $-A_{ws}\delta h = Q_A \delta t$.
>
> 3. Substitute the appropriate discharge equation for Q_A in 2 above. If it is an orifice, use the orifice equation; if it is a weir, use the relevant weir equation.
>
> 4. Rearrange the expression to get an equation for δt.
>
> 5. Integrate the equation in step 4 to get an expression for the time, T, for the water level in the reservoir to fall from level H_1 to level H_2.

$$T = -\frac{A_{ws}}{C_D A(2g)^{1/2}}[2h^{1/2}]_{H_1}^{H_2}$$

$$T = \frac{2A_{ws}}{C_D A\sqrt{2g}}\left[H_1^{1/2} - H_2^{1/2}\right] \tag{7.4}$$

A few important points to remember with respect to equation (7.4) are:

1. This equation applies only to a reservoir emptying through an orifice.

2. The number 2 in the numerator of equation (7.4) arises from the integration. When trying to memorise the formula, the most common mistake made by students is to omit the 2.

3. To get rid of the negative sign the limits of the integration have been reversed so $[H_1^{1/2} - H_2^{1/2}]$ appears in equation (7.4) not $[H_2^{1/2} - H_1^{1/2}]$.

Example 7.1 illustrates numerically how the equation is applied to calculate the time to empty or partially empty a reservoir that has a uniform cross-section in plan. Example 7.2 shows how the general procedure can be applied to a problem involving a weir, in this case to calculate the coefficient of discharge. Many problems can be solved using this general procedure – try Self Test Question 7.1 for yourself.

SELF TEST QUESTION 7.1

If the tank in Example 7.2 has a triangular weir with a total angle of 60° (that is the half angle $\theta/2 = 30°$) in its side instead of the rectangular weir:

(a) Using the procedure in Box 7.1, derive the expression for the time to lower the water level from H_1 to H_2.

(b) If the triangular weir has a coefficient of discharge of 0.60, how long would it take for the water level to fall from 0.3 m to 0.1 m now?

EXAMPLE 7.1

Water from a vertically sided rectangular tank 3 m by 2 m in plan discharges through a 50 mm diameter orifice located in the base. The C_D of the orifice is 0.61. If the depth of water in the tank is initially 1.5 m, how long will it take for the tank to empty?

Since this question concerns water draining through an orifice, equation (7.4) **does** apply.

$$T = \frac{2A_{ws}}{C_D A\sqrt{2g}}\left[H_1^{1/2} - H_2^{1/2}\right]$$

$A_{ws} = 3 \times 2 = 6\,\text{m}^2$, $A = \pi(0.05)^2/4 = 0.00196\,\text{m}^2$, $H_1 = 1.5\,\text{m}$ and $H_2 = 0$.

$$T = \frac{2\times6}{0.61\times0.00196\times\sqrt{19.62}}[1.5^{1/2} - 0]$$

$$T = 2775\,\text{s}$$

or $\quad T = 46.25$ minutes

EXAMPLE 7.2

A sharp crested rectangular weir with a crest length of 0.2 m is situated in the side of a tank that has vertical walls and which measures 6 m by 3 m in plan. If the head over the weir reduces from 0.3 m to 0.1 m in 136 seconds, what is the coefficient of discharge of the weir?

In this case equation (7.4) does **not** apply because the outlet device is a weir, not an orifice. Therefore adopt the procedure described in Box 7.1.

– Change in reservoir volume = volume discharged through outlet

$$-A_{ws}\delta h = Q_A \delta t$$

hence $\quad \delta t = -\dfrac{A_{ws}}{Q_A}\delta h \qquad$ (1)

For a rectangular weir, at any head, h, above the crest $Q_A = \frac{2}{3}C_D b(2g)^{1/2}h^{3/2}$ so substitution in (1) gives:

$$\delta t = -\frac{A_{ws}}{\frac{2}{3}C_D b(2g)^{1/2}h^{3/2}}\delta h$$

Integrating this equation between limits representing the initial head over the crest, H_1, at time T_1, and the final head, H_2, at time T_2:

$$T = -\frac{A_{ws}}{\frac{2}{3}C_D b(2g)^{1/2}}\int_{H_1}^{H_2} h^{-3/2}dh$$

$$T = -\frac{A_{ws}}{\frac{2}{3}C_D b(2g)^{1/2}}[-2h^{-1/2}]_{H_1}^{H_2}$$

$$T = +\frac{2A_{ws}}{\frac{2}{3}C_D b(2g)^{1/2}}\left[\frac{1}{H_2^{1/2}} - \frac{1}{H_1^{1/2}}\right] \qquad (7.5)$$

We are given $T = 136\,\text{s}$, $A_{ws} = 6 \times 3 = 18\,\text{m}^2$, $b = 0.2\,\text{m}$, $H_1 = 0.3\,\text{m}$ and $H_2 = 0.1\,\text{m}$ so:

$$136 = \frac{2\times18}{\frac{2}{3}\times C_D \times0.2\times(19.62)^{1/2}}\left[\frac{1}{0.1^{1/2}} - \frac{1}{0.3^{1/2}}\right]$$

$$136 = \frac{36}{0.591\,C_D}[3.162 - 1.826]$$

$$C_D = 81.381/136$$

$$C_D = 0.60$$

7.3 Time to empty a reservoir of varying cross-section

❝ How do you deal with a reservoir that does not have a constant area when seen in plan? The flood storage reservoirs and water supply reservoirs that you were talking about at the beginning of the chapter would be built in a natural valley and would not have vertical sides. What do we do now? ❞

There are two ways to get around this. If the reservoir or tank has sides that slope at a constant angle (like a cone or a pyramid, for instance) we can express the variation of the water surface area with the water level mathematically, and then incorporate this variation in the equation for the time to empty the tank. This is described in section 7.3.1 below. On the other hand, if the sides of the valley or container cannot be described mathematically, we have to adopt a less rigorous approach and estimate the time. This procedure is described in section 7.3.2.

7.3.1 Reservoirs that have regular side slopes

The procedure for solving these problem remains as in Box 7.1, but with the addition of an expression to describe the variation of the area of the water surface, A_{WS}, with the depth of water in the reservoir, h. This could be written mathematically as $A_{WS} = f(h)$ where $f(h)$ means 'a function of h'. Because A_{WS} is now expressed as a function of h, we will have to combine this with the h term(s) from the discharge equation before integrating with respect to h. This is a similar procedure to that used in section 5.5.3 to allow for the variation of the width of a triangular weir with the depth of water. If this does not make much sense, then work through Example 7.3 and compare it with Example 7.2.

EXAMPLE 7.3

A tank that increases in size with height has a square cross-section in plan, with sides 4 m long at the plane of the water surface and 1 m long at the base of the tank which is 4.5 m below the water surface. Thus the tank forms an inverted pyramid with the apex cut off (Fig. 7.2). In the base of the tank is an orifice of 100 mm diameter with a C_D of 0.60. How long will it take to empty the tank if the water level is initially 4.5 m above the orifice?

As with the triangular weir, we need a relationship between the head of water, h, and the breadth, b of the tank (and then the area, A_{WS}).

Now $b/2 = 0.5$ m when $h = 0$.
And $b/2 = 2.0$ m when $h = 4.5$ m.

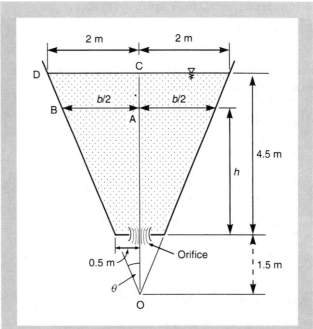

Figure 7.2 The tank forms part of an inverted pyramid with the apex cut off

Thus 1.5 m change in $b/2$ equals 4.5 m change in h.

So 0.5 m change in $b/2$ equals $4.5 \times (0.5/1.5) = 1.5$ m change in h. Therefore, the apex of pyramid is 1.5 m below the base of the tank.

Let b be the breadth of the tank corresponding to the head, h, as shown in Fig. 7.2. Considering the triangles OAB and OCD then:

$\tan \theta = (b/2)/(h + 1.5)$ from OAB, and
$\tan \theta = 2.0/(4.5 + 1.5)$ from OCD, thus
$(b/2)/(h + 1.5) = 2.0/(4.5 + 1.5)$
$(b/2) = (2.0/6)(h + 1.5)$
$b = 2 \times 0.333(h + 1.5)$
$b = (0.666h + 1.000)$

Since the tank is square, $A_{WS} = b^2 = (0.666h + 1.000)^2$
$A_{WS} = (0.666h + 1.0)(0.666h + 1.000)$
$A_{WS} = (0.444h^2 + 0.666h + 0.666h + 1.000)$
$A_{WS} = (0.444h^2 + 1.333h + 1.000)$ \hfill (1)

Now adopting the general procedure of Box 7.1: $-A_{WS}\delta h = Q_A \delta t$

Since the outlet device is an orifice, at any head, h: $Q_A = C_D A(2gh)^{1/2}$

Therefore: $-A_{WS}\delta h = C_D A(2gh)^{1/2} \delta t$

or: $\delta t = -\dfrac{A_{WS}}{C_D A(2g)^{1/2} h^{1/2}} \delta h$

Now substituting for A_{ws} from equation (1) above:

$$\delta t = -\frac{(0.444h^2 + 1.333h + 1.000)}{C_D A (2g)^{1/2} h^{1/2}} \delta h$$

Rearranging the equation and integrating with respect to h to obtain the time, T, required to reduce the water level from $h = 4.5\,\text{m}$ to $h = 0\,\text{m}$:

$$T = -\frac{1}{C_D A (2g)^{1/2}} \int_{h=4.5}^{h=0} (0.444h^2 + 1.333h + 1.000) h^{-1/2} dh$$

$$T = -\frac{1}{C_D A (2g)^{1/2}} \int_{h=4.5}^{h=0} (0.444h^{3/2} + 1.333h^{1/2} + 1.000h^{-1/2}) dh$$

$$T = +\frac{1}{C_D A (2g)^{1/2}} \left[\frac{2 \times 0.444 h^{5/2}}{5} + \frac{2 \times 1.333 h^{3/2}}{3} + 2 \times 1.000 h^{1/2} \right]_0^{4.5}$$

$$T = +\frac{1}{C_D A (2g)^{1/2}} [0.178 h^{5/2} + 0.889 h^{3/2} + 2.000 h^{1/2}]_0^{4.5}$$

Now $C_D = 0.60$, $A = \pi(0.1)^2/4 = 0.00785\,\text{m}^2$ so:

$$T = \frac{1}{0.60 \times 0.00785 \times (19.62)^{1/2}} [(0.178 \times 4.5^{2.5} + 0.889 \times 4.5^{1.5} + 2.000 \times 4.5^{0.5}) - (0)]$$

$T = 47.932 [(7.646 + 8.486 + 4.243) - (0)]$

$T = 47.932 [20.375 - 0]$

$T = 977\,\text{s}$

$T = 16.28\,\text{minutes}$

7.3.2 Reservoirs that have variable side slopes

Most dams are built in natural locations where the side slopes of the valley are likely to be irregular and to change with elevation. In such cases it is not usually possible to express mathematically the variation of water surface area (A_{ws}) with water depth, h. Consequently the technique of writing an equation for an elemental strip and then integrating is no longer appropriate. Instead we have to fall back on a less accurate method of estimating the time to empty the reservoir. This is summarised in Box 7.2.

Box 7.2 ▶ **Approximate procedure for estimating T**

If the surface area of the reservoir varies irregularly with the water depth then:

1. Split the reservoir into horizontal slices.

2. Estimate the volume of water, Vol_i, in each slice, where i is the number of the slice.

3. Estimate the average discharge, q_i, through the outlet corresponding to each slice; the best accuracy appears to be achieved by calculating the discharge at the two heads representing the top and bottom of the slice and then taking the average.

4. Calculate the time for the water in the slice to be discharged, $t_i = Vol_i/q_i$.

5. Obtain the total time to discharge all of the water from $T = \Sigma t_i$.

Some points to note when using this approximate technique are as follows:

a. For a natural valley, the contours of a map form a convenient means of splitting the reservoir into horizontal slices. If the area bounded by one contour is a_1 and the area enclosed by the next contour is a_2, then the volume of the slice can be estimated as $Vol_1 = c(a_1 + a_2)/2$ where c is the contour interval. The areas can be measured with a planimeter or suitable computer software. Example 7.4 illustrates the procedure.

b. Generally, the more slices the greater the accuracy.

c. The nearer the power of H in the discharge equation to 1.0, the greater the accuracy.

d. The method of splitting the reservoir into slices can be used as a means of quickly checking the answer obtained by integration. However, as mentioned above, the fewer the slices and the higher the power of H, the less likely it is that the two answers will agree closely. Consequently this check may be very accurate with an orifice in a tank of constant plan area, but inaccurate for a triangular weir. REMEMBER!!! Examples 7.5 and 7.6 illustrate this point.

It is important that you know when a simplified check using the approximate procedure will be accurate and when it will not. However, it is difficult to assess the accuracy of a simple check solution to complicated problems like that in Example 7.4; this has to be done using simpler problems. Therefore, study Examples 7.5 and 7.6 carefully, so you can appreciate the errors that might arise in Example 7.4. Then try Self Test Question 7.2.

SELF TEST QUESTION 7.2

Solve the question in Example 7.3 using the procedure in Box 7.2. Split the reservoir into three horizontal slices each 1.5 m thick. To calculate the volume of the slices you need to know that:

$$\text{Volume of a pyramid} = \frac{1}{3}\text{ area of the base} \times \text{vertical height}$$

What is the percentage difference between the two answers?

EXAMPLE 7.4

A large reservoir is situated in a natural valley. Water is discharged from the reservoir to the river channel below the dam through two parallel 0.8 m diameter pipelines, both of which have their centreline at the point of discharge at 110 m OD. The pipelines are very short so friction losses may be neglected, but the entrances to the pipelines are sharp so the minor loss should be included. It can be assumed that the entrances to the pipelines remain submerged at all times. The areas enclosed between the dam and the contours of the valley sides (Fig. 7.3) are:

150 m OD contour:	60 600 m²
140 m OD contour:	22 400 m²
130 m OD contour:	3 200 m²
120 m OD spot height:	zero – lowest point of valley at dam

The reservoir has to be emptied for maintenance work (the inflow has been diverted around the reservoir and can be ignored). Although great accuracy is not necessary, an estimate of the time taken to empty the reservoir is required. Produce a suitable estimate.

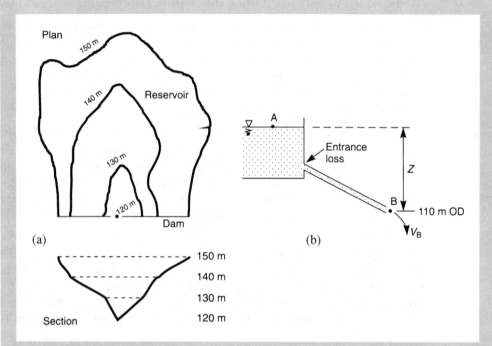

Plan

Reservoir

A

Entrance loss

Z

B

110 m OD

V_B

(a)

(b)

150 m
140 m
130 m
120 m

Dam

Section

Figure 7.3 (a) Contours and cross-section of the valley, and (b) details of the outlet pipes

Step 1. Split the reservoir into horizontal slices. The volumes between the contours in Fig. 7.3 will be used for this purpose.

Step 2. Estimate the volumes between the contours.

$$Vol_1 = 10(60600 + 22400)/2 = 415000 \, m^3$$
$$Vol_2 = 10(22400 + 3200)/2 = 128000 \, m^3$$
$$Vol_3 = 10(3200 + 0)/2 = 16000 \, m^3$$

Step 3. Estimate the average discharge, q_i, through the outlet corresponding to each slice. Since the pipes are parallel and have their outlets at the same level, they are identical and operate independently of each other (see Chapter 6). Applying the energy equation to points A and B on a streamline, assuming that atmospheric pressure = 0 and that the velocity on the surface of a large reservoir = 0:

$$Z = V_B^2/2g + \text{entrance loss}$$
$$Z = V_B^2/2g + 0.5V_B^2/2g \qquad \text{(for continuity, pipe velocity = exit velocity, } V_B)$$
$$Z = 1.5V_B^2/2g$$

$$V_B = (2gZ/1.5)^{1/2}$$
$$Q = A_T V_B = A_T(2gZ/1.5)^{1/2} \qquad \text{where } A_T \text{ is the total cross-sectional area of}$$
$$\text{both pipes} = 2 \times \pi (0.8)^2/4 = 1.005 \, m^2$$

$$Q = 1.005(19.62Z/1.5)^{1/2}$$
$$Q = 3.635Z^{1/2}$$

Water surface elevation	Z	Q
150m OD	40m	22.99 m³/s
140	30	19.91
130	20	16.26
120	10	11.49

Thus
$$q_1 = (22.99 + 19.91)/2 = 21.45 \, \text{m}^3/\text{s}$$
$$q_2 = (19.91 + 16.26)/2 = 18.09 \, \text{m}^3/\text{s}$$
$$q_3 = (16.26 + 11.49)/2 = 13.88 \, \text{m}^3/\text{s}$$

Step 4. Calculate the time for the water in each slice to be discharged from $t_i = Vol_i/q_i$.

$$t_1 = 415\,000/21.45 = 19\,347 \, \text{s} = 5.37 \, \text{hrs}$$
$$t_2 = 128\,000/18.09 = 7076 \, \text{s} = 1.97 \, \text{hrs}$$
$$t_3 = 16\,000/13.88 = 1153 \, \text{s} = 0.32 \, \text{hrs}$$

Step 5. Calculate the total time from $T = \Sigma t_i$

$$= 5.37 + 1.97 + 0.32$$
$$= 7.66 \, \text{hrs}$$

EXAMPLE 7.5

The water in a rectangular, vertically sided tank in a laboratory has to be drained over a triangular sharp crested weir that has a total angle of 60° (that is $\theta/2 = 30°$) and a coefficient of discharge of 0.60. The plan area of the tank is 20 m by 6 m, and the water surface must be reduced from an initial value of 0.8 m above the crest to 0.2 m. (a) Calculate the time required to do this using the procedure in Box 7.1. (b) Estimate the time required using the approximate method in Box 7.2, and find the percentage difference to the answer in part (a).

(a) $\delta t = -\dfrac{A_{ws}}{Q_A} \delta h$

where at any head, h, over the weir crest $Q_A = \frac{8}{15}C_D(2g)^{1/2}\tan(\theta/2)h^{5/2}$ (1)

$$T = -\frac{A_{ws}}{\frac{8}{15}C_D(2g)^{1/2}\tan(\theta/2)} \int_{H_1}^{H_2} h^{-5/2}dh$$

$$T = -\frac{A_{ws}}{\frac{8}{15}C_D(2g)^{1/2}\tan(\theta/2)} \left[-\frac{2h^{-3/2}}{3}\right]_{H_1}^{H_2}$$

$$T = -\frac{30A_{ws}}{24C_D(2g)^{1/2}\tan(\theta/2)} \left[\frac{1}{H_2^{3/2}} - \frac{1}{H_1^{3/2}}\right]$$

With $A_{ws} = 20 \times 6 = 120 \, \text{m}^2$, $C_D = 0.60$, $\theta/2 = 30°$, $H_1 = 0.8$ m and $H_2 = 0.2$ m:

$$T = \frac{30 \times 120}{24 \times 0.60 \times (19.62)^{1/2}\tan 30°} \left[\frac{1}{0.2^{1.5}} - \frac{1}{0.8^{1.5}}\right]$$

$$T = 97.758 \, [11.180 - 1.398]$$

$$T = 956 \, \text{s} \quad \text{or} \quad 15.9 \, \text{mins}$$

(b) Assume three slices each 0.2 m high. Since the tank is rectangular and vertically sided:
$$Vol_1 = Vol_2 = Vol_3 = 20 \times 6 \times 0.2 = 24\,m^3$$

With the values given the discharge equation (1) for the triangular weir becomes:

$$Q_A = \frac{8}{15} \times 0.60\,(19.62)^{1/2}\,\tan 30°h^{5/2}$$

$$Q_A = 0.818h^{5/2} \qquad\qquad (2)$$

Now using equation (2) to calculate the average discharge, q_i, for each slice:

Height above crest, h	Q_A
0.8 m	0.468 m³/s
0.6	0.228
0.4	0.083
0.2	0.015

Thus: $q_1 = (0.468 + 0.228)/2 = 0.348\,m^3/s$
$q_2 = (0.228 + 0.083)/2 = 0.156\,m^3/s$
$q_3 = (0.083 + 0.015)/2 = 0.049\,m^3/s$

Now calculating the time to drain the slices: $t_1 = 24/0.348 = 68.9\,s$
$t_2 = 24/0.156 = 153.8\,s$
$t_3 = 24/0.049 = 489.8\,s$

Time to reduce water level from 0.8 m to 0.2 m = $T = \Sigma t_i = 68.9 + 153.8 + 489.8$
$= 713\,s$

Percentage error involved in approximation = $100(956 - 713)/956 = -25.4\%$

Note that the even cruder approximation $T = \dfrac{20 \times 6 \times 0.6}{\frac{(0.468+0.15)}{2}} = 298\,s$ gives a -69% error.

Compare this with the answer to Example 7.6.

EXAMPLE 7.6

A large tank which is circular in plan has a diameter of 12 m, and a height of 15 m. The tank stores light oil for an industrial plant. During maintenance work someone drills a 10 mm diameter hole through the wall of the tank at ground level. Assuming that the hole behaves like a small orifice and has a coefficient of discharge of 0.80: (a) using equation (7.4), calculate the time it will take for the tank to empty; (b) estimate the time using three 5 m slices and the approximate method in Box 7.2; (c) calculate the percentage error incurred with the approximate method, and compare this to the error in Example 7.5.

(a) Equation (7.4) is: $T = \dfrac{2A_{WS}}{C_D A(2g)^{1/2}}[H_1^{1/2} - H_2^{1/2}]$

$A_{WS} = \pi\,(12)^2/4 = 113.097\,m^2$, $C_D = 080$, $A = \pi\,(0.010)^2/4 = 78.539 \times 10^{-6}\,m^2$, $H_1 = 15\,m$ and $H_2 = 0$

$$T = \frac{2 \times 113.097}{0.80 \times 78.539 \times 10^{-6} \times (19.62)^{1/2}}[15^{1/2} - 0]$$

$$T = 812\,749[3.873]$$

$$T = 3147800\,s \text{ or } 874.4\,hrs \text{ or } 36.43\,days$$

(b) Volume of each slice, $Vol = 113.097 \times 5 = 565.485 \, m^3$

At any head, h, the discharge through the orifice is:

$Q_A = C_D A (2gh)^{1/2}$

$Q_A = 0.80 \times 78.5 \times 10^{-6} (19.62)^{1/2} h^{1/2}$

$Q_A = 278.169 \times 10^{-6} h^{1/2}$

Head in tank, h	Q_A
15 m	$1.077 \times 10^{-3} \, m^3/s$
10	0.880×10^{-3}
5	0.622×10^{-3}
0	0

$q_1 = 10^{-3}(1.077 + 0.880)/2 = 0.979 \times 10^{-3} \, m^3/s$

$q_2 = 10^{-3}(0.880 + 0.622)/2 = 0.751 \times 10^{-3} \, m^3/s$

$q_3 = 10^{-3}(0.622 + 0)/2 = 0.311 \times 10^{-3} \, m^3/s$

Now calculating the time to drain the slices from $t_i = Vol_i / q_i$

$t_1 = 565.485/0.979 \times 10^{-3} = 577600 \, s$ or 160.45 hrs

$t_2 = 565.485/0.751 \times 10^{-3} = 753000 \, s$ or 209.16 hrs

$t_3 = 565.485/0.311 \times 10^{-3} = 1818300 \, s$ or 505.08 hrs

Therefore total time is $T = \Sigma t_i = 874.7$ hrs.

(c) Percentage error $= 100(874.4 - 874.7)/874.4 = -0.03\%$

In this case the curvature of the Q_A–H line is much shallower than the equivalent line for the triangular weir, since the powers of H are 0.5 and 2.5 respectively. The effect of this can be seen by comparing the rate at which Q_A reduces with head in Examples 7.5 and 7.6. This point was emphasised in the introduction to the chapter. So, if an approximate method is to be used, the engineer should appreciate when it will be reasonably accurate and when it will not. In this particular example a quick estimate can be obtained by dividing the total volume of the tank ($= 565.485 \times 3 = 1696.455 \, m^3$) by the average of the first and final values of Q_A, that is $(1.077 \times 10^{-3} + 0)/2 = 0.539 \times 10^{-3} \, m^3/s$. This gives a time to empty the tank of $1696.455/0.539 \times 10^{-3} = 874.3$ hrs, which is little different from the other methods. However, this approach did not work with the previous example, so take care!

7.4 ▶ Flow between two tanks

If water flows out of one tank into another then we have a situation where the head in the first tank is falling while the head in the second is rising. Consequently the head difference between the two tanks is constantly changing so there is flow under a varying head.

Suppose that water flows from tank 1 to tank 2 via a submerged orifice in the dividing wall (Fig. 7.4). As described in section 5.4.3 the flow rate depends upon the differential head between the water levels in the two tanks. If the two tanks have an equal plan area (A_{WS}), when the water level falls by an amount δx in the first tank it must rise by δx in the second tank. Thus the change in the differential head, $\delta h_D = 2\delta x$. If the two tanks do not have equal areas, then the ratio of the areas must be used to calculate the respective change in water levels, and hence δh_D. After this, the procedure is as described in Box 7.1. Example 7.7 provides an illustration.

EXAMPLE 7.7

Two tanks have a common wall in which an orifice with a C_D of 0.80 and a diameter of 0.10 m is located. Tank 1 is square with sides 4 m in length and contains water to a height of 4.5 m above the centre of the submerged orifice. Tank 2 is square with sides 2 m long and initially has a 0.5 m head of water above the centre of the orifice. Calculate the time it will take for the water levels in the two tanks to become equal.

If the water level in tank 1 falls by δx then the increase in water level in tank 2 (δy) is given by the ratio of the areas:

$$\delta y = \frac{(4 \times 4)}{(2 \times 2)} \delta x$$

$$\delta y = 4\delta x$$

Thus the change in the differential head producing flow is:

$$\delta h_D = \delta x + \delta y$$
$$\delta h_D = \delta x + 4\delta x$$
$$\delta h_D = 5\delta x$$

so $\quad \delta x = \delta h_D/5$ $\qquad$ (1)

Now applying the procedure in Box 7.1:

– Change in volume of tank 1 = amount discharged through orifice

$$-A_{WS}\delta x = Q_A\delta t$$

Substituting for δx from equation (1) above and putting $A_{WS} = 4 \times 4 = 16 \, \text{m}^2$:

$$-16 \times \delta h_D/5 = Q_A\delta t$$

For a submerged orifice, if the differential head at a particular instant is h_D, then the actual discharge, Q_A is: $Q_A = C_DA(2g)^{1/2}h_D^{1/2}$

Thus: $\quad -16 \times \delta h_D/5 = C_DA(2g)^{1/2}h_D^{1/2}\delta t$

or: $\qquad \delta t = -\dfrac{16}{5C_DA(2g)^{1/2}h_D^{1/2}} \delta h_D$

With $C_D = 0.80$, $A = \pi(0.10)^2/4 = 0.00785 \, \text{m}^2$ then:

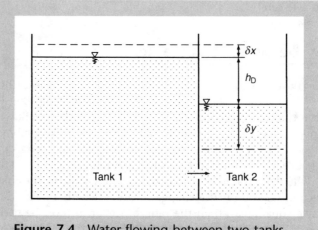

Figure 7.4 Water flowing between two tanks

$$\delta t = -115 h_D^{-1/2} \delta h_D$$

For the water levels in the two tanks to become the same the differential head, h_D, must change from its initial value of 4.0 m to zero. So integrating between these limits to obtain the time, T, required for this to happen:

$$T = -115.0 \int_{4.0}^{0} h_D^{-1/2} dh_D$$

$$T = 115.0 \left[2 h_D^{-1/2} \right]_0^{4.0}$$

$$T = 115.0 \left[(2 \times 4.0^{-1/2}) - 0 \right]$$

$$T = 460 \, s$$

SELF TEST QUESTION 7.3

Repeat Example 7.7 keeping the data the same with the exception of the sizes of the tanks and the heads. This time assume that the water flows from a small square tank with sides 2 m long and an initial head of 20 m into a square tank with sides 8 m in length and an initial head of 1.0 m. Both heads are measured above the centre of the orifice. How long will it take for the water levels in the two tanks to become equal?

Summary

1. The starting point for most reservoir storage problems is:

   Change in storage = Inflow − Outflow   (7.1)

   Thus if the inflow = 0, this becomes:

   −Change in reservoir volume = Volume discharged through the outlet

   or $-A_{WS} \delta h = Q_A \delta t$ (7.2)

   Q_A is replaced by the weir/orifice/pipe discharge equation appropriate to the outlet, for example:

   orifice $Q_A = C_D A (2gh)^{1/2}$ (5.12)

   rectangular weir $Q_A = 2/3 \, C_D b (2g)^{1/2} H^{3/2}$ (5.22)

2. The time T required to partially or completely empty the reservoir is found by integration:

   $$T = \int_{T_1}^{T_2} dt = \frac{A_{WS}}{Q_A} \int_{H_1}^{H_2} h^{-N} \, dh$$

   where T_1 and H_1 are the initial conditions and T_2 and H_2 the final condition.

3. Remember that $Q \propto H^N$, so the relationship between discharge and head becomes less linear as N becomes larger (e.g. triangular weir $N = 5/2$). This means approximate solutions (e.g. averaging the initial and final discharge) become less accurate. However, with irregularly shaped reservoirs an approximate solution may be necessary. This can be obtained by splitting the reservoir into an appropriate number of horizontal slices and following the procedure in Box 7.2.

4. With flow between two tanks, the ratio of the reservoir areas has to be used to calculate the change in the differential head (h_D) governing the discharge. After that, the procedure is as above.

Revision questions

7.1 (a) Explain what is meant by 'flow under varying head' and (b) explain why these problems are best solved using an approach that involves integration, rather than just averaging the initial and final discharge rates.

7.2 A tank of water is 5.6 m by 4.3 m in plan with vertical sides. Water from the tank discharges to the atmosphere through a 200 mm diameter orifice in the base and is not replaced. Over a period of 5 min 7 s the water level drops from 1.9 m to 0.7 m above the orifice. What is the value of the coefficient of discharge of the orifice.

[0.61]

7.3 Water from a vertically sided rectangular water service reservoir 30 m long by 20 m wide discharges to the atmosphere through a 1.2 m diameter (D) emergency draw-off pipe which terminates at a level 19.0 m below the base of the reservoir. The actual length of the pipeline is 110 m, and its friction factor $\lambda = 0.03$. The head losses at the pipe entrance, valves and bends are estimated to be the equivalent of an additional length of pipe equal to 100D (see Table 6.4). The depth of water in the reservoir is 3.0 m. (a) Derive an equation for the time to reduce the water level from H_1 to H_2. (b) Use the equation to determine the time required to empty the reservoir in an emergency. (c) Split the reservoir into three horizontal slices each 1.0 m thick and estimate the time required to empty the reservoir using the procedure in Box 7.2. (d) What is the percentage difference between the answer to parts (b) and (c)?

[(b) 206.0 s; (c) 206.3 s; about 0.1%]

7.4 A tank is rectangular in plan measuring 10 m by 4 m at the base. The cross-section of the tank is trapezoidal. The 10 m long sides slope outwards from the base at an angle of 45°, while the ends of the tank are vertical. A triangular weir with $\theta/2 = 50°$ is cut out of one of the vertical ends with the bottom of the V being at the same level as the base of the tank. Its C_D is 0.58. The water level in the tank is initially 1.3 m above the base of the tank, and has to be reduced to 0.1 m. Assuming that the inflow to the tank is stopped: (a) calculate the time needed to reduce the level in the tank by the required amount using integration, as in Example 7.3; (b) estimate the time using three horizontal slices 0.4 m thick and calculate the percentage error incurred compared to the answer in (a); (c) estimate the time required to reduce the water level using two horizontal slices 0.6 m thick. What is the percentage error now?

[(a) 561 s; (b) 164 s, about −71%; (c) 104 s, about −81%]

7.5 Water flows between two adjacent parts of a vertically sided service reservoir through a submerged 0.3 m diameter orifice ($C_D = 0.60$) in a common dividing wall. The first reservoir measures 45 m by 30 m in plan and initially contains water to a height of 5.0 m above the centre of the orifice. The second tank is 20 m by 30 m in plan and the initial water level can be assumed to be at the same height as the centre of the orifice. Assuming that there is no other flow into or out of the reservoirs: (a) how long will it take for the two water levels in the tanks to become equal? (b) what is the final depth of water in the tanks, measured above the centre of the orifice?

[9881 s; 3.462 m]

7.6 A reservoir has a surface area of 0.400 km² when full to spillway level. However, the surface area varies irregularly with reservoir level, shown as a head over the spillway below.

Head over spillway, H	Surface area
1.00 m	0.420 km²
0.75	0.407
0.50	0.404
0.25	0.402
0.00	0.400

The reservoir overflows over a 50 m long (= b) spillway whose discharge is given by $Q = 1.6bH^{3/2}$ where H is the head over the spillway. Assuming that there is no additional flow into the reservoir, using four horizontal slices estimate how long it would take for H to fall from 1.0 m to 0 m.

[8.17 hrs]

Flow in open channels

A significant part of Civil Engineering is that concerned with land drainage, much of which involves flow in open channels. These channels are very common: rivers, canals, pipes flowing partially full and irrigation ditches are all examples. Consequently it is very important that a channel can be designed to carry a particular discharge or, alternatively, that the discharge in a channel can be calculated from measurements of the bed slope, the width and the depth of flow.

The flow in open channels can be subcritical or supercritical, with critical depth representing the boundary between the two. The difference between these types of flow must be understood, otherwise an incorrect analysis will be conducted.

This chapter introduces the concept of using critical depth as a means of establishing a head–discharge relationship, and Chapter 9 shows how broad crested weirs and throated flumes use this to measure river discharge. However, before constructing anything that obstructs a river channel it must be determined how water levels will be affected, otherwise flooding and property damage could occur. The questions answered in this chapter include:

What exactly is an open channel?

What is meant by uniform flow and non-uniform flow?

How can the discharge or depth of flow in an open channel be calculated?

What channel proportions are the most efficient and maximise the discharge?

What is specific energy and critical depth, and why are they significant?

How can two depths of flow be possible in the same channel at the same discharge?

What is the difference between subcritical and supercritical flow, and does it matter?

How can we analyse gradually varying non-uniform flow?

How can we calculate the water surface profile and obtain the depth of flow?

8.1 Fundamentals

8.1.1 Definition of an open channel, uniform flow and non-uniform flow

An open channel is a conduit through which liquid flows with a free surface as a result of gravity. The pressure at the surface of the liquid is constant at all points along the length of the channel, and this pressure is usually atmospheric. A pipe which is partly full and which has a free surface is an open channel. It is important to distinguish this situation from a pipe flowing full under pressure, such as a pipeline discharging from a reservoir. In Fig. 6.2, as a result of the head, Z, flow can be maintained even if the pipeline slopes uphill for long distances. This is not the case in open channel flow.

What is meant by uniform flow?

Uniform flow was described in general terms in section 4.2.3. For uniform flow in an open channel everything must be constant, that is the discharge (Q), depth (D), breadth (B) and mean velocity (V) of the flow are the same at all cross-sections along the length of the channel. These variables are related by the continuity equation, $Q = AV = BDV$.

Since the depth and velocity do not vary along the length of the channel both the water surface and the total energy line are parallel to the bed (Fig. 8.2a), and hence their gradi-

Figure 8.1 An artificial, concrete lined open channel below Balderhead Reservoir. There are three reservoirs in the valley. The channel carries the flow from an upstream reservoir around the one below

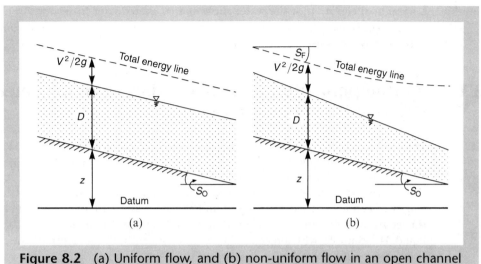

Figure 8.2 (a) Uniform flow, and (b) non-uniform flow in an open channel

ent is the same as the **bed slope**, S_O. Thus the loss of energy head can be determined by measuring either the fall of the bed or the water surface, since:

total energy head $= z + D + V^2/2g$ (8.1)

and D and $V^2/2g$ are constant at a particular discharge. Thus equation (8.1) shows that the total energy varies directly with the elevation of the bed, z, above the datum.

Truly uniform flow is quite rare in natural channels which tend to vary in width, depth and slope, but in the absence of any practical alternative the overall flow is often considered to be uniform. The conditions required for uniform flow are only really likely to be encountered in man-made channels and conduits, such as a concrete lined open channel.

So what is meant by non-uniform flow?

Non-uniform flow is basically the opposite of uniform flow. Although the discharge passing all the cross-sections along the length of a channel may be constant, the depth, breadth and mean velocity of flow may change *gradually* from section to section. Thus the water surface is not parallel to the bed and, since V and $V^2/2g$ are not constant, the total energy line is not parallel to either the bed or the water surface (Fig. 8.2b). This means that any loss of energy head in the channel must be calculated from the fall of the energy line. In a straight channel of constant section, the slope of the energy (REMEMBER!!!) line depends upon the rate of loss of energy through friction, that is the **friction gradient**, which is denoted by S_F.

Non-uniform flow is prevalent in natural streams and rivers. Most rivers vary in width, depth and bed slope. Over short lengths, it may even be possible for the river bed to have an uphill slope, that is a gradient that opposes flow. So, in practice, when solving problems involving non-uniform flow, the bed slope is often relatively meaningless since it can change rapidly both across and along the channel. A good habit to get into (with either uniform or non-uniform flow) is to plot the total energy line. It must always slope downwards in the direction of flow. If it slopes upwards in the direction of flow then this indicates that V, $V^2/2g$ and possibly the water depth have been incorrectly assessed, and the calculations should be amended accordingly.

Gradually varying non-uniform flow and the calculation of the surface profile (i.e. the longitudinal elevation of the water surface) are covered later in section 8.11. However, we will first consider uniform flow at constant depth in simple prismatic and compound channels where we can use the bed slope, S_O, as in Fig. 8.2a.

8.1.2 Wetted perimeter, hydraulic radius, hydraulic mean depth and Froude number

Two variables that are used constantly in open channel hydraulics are the wetted perimeter, P, and the hydraulic radius, R. You can make things easier for yourself by remembering from the outset what these two variables are.

The **wetted perimeter** of a cross-section perpendicular to the direction of flow is the length of contact between the liquid and the sides and base of the channel. It is literally the length of the wetted perimeter, that is the length of the perimeter in contact with the liquid. For example, with a rectangular channel of width, B, and depth of flow, D, the wetted perimeter $P = B + 2D$.

The **hydraulic radius**, R, is defined by:

$$R = A/P \tag{8.2}$$

where A is the cross-sectional area of flow (m^2) and P is the wetted perimeter (m). It is important to realise that R is not the same as the actual depth of flow, D. This can be illustrated by calculating the hydraulic radius of channels of various geometries, as in Examples 8.1 and 8.2. It is interesting to note that a half-full pipe and a full pipe have the same hydraulic radius. This is because a full pipe has exactly twice the cross-sectional area of flow as a half-full pipe, and exactly twice the wetted perimeter as well. Thus the ratio A/P has the same value in both cases. The same argument would also apply to a square or rectangular culvert flowing half-full and full. However, remember that a pipe flowing full under *pressure* is not an open channel.

The depth of uniform flow in a long channel of constant section is often called the **normal depth**, D_N. This is the depth that is assumed in the depth–discharge equations in section 8.2. The subscript 'N' is often omitted because this is the standard condition, thus $D = D_N$. Note that 'normal' does not mean 'usual'. Upstream of a channel obstruction the depth that usually occurs will be greater than D_N.

The **hydraulic mean depth**, D_M, represents an attempt to define the mean depth of flow in an irregular, non-rectangular channel where the depth varies across the width of the cross-section. This is defined as:

$$D_M = A/B_S \tag{8.3}$$

where B_S is the surface width of the water in the channel. With a rectangular channel where the width is constant regardless of the depth of flow and $B = B_S$ then the actual depth, D, and the hydraulic mean depth, D_M, are the same. In non-rectangular channels they have different values and should not be confused.

The hydraulic mean depth is often used in connection with the **Froude number**, F. This is a dimensionless parameter that tells us something about the type of flow in the channel. You will be familiar with the idea that there are subsonic and supersonic aeroplanes, the latter being the ones that can break the sound barrier which is represented by Mach 1. The Froude number is similar to the Mach number, but applies to water, not air. It is used to define the types of flow that can occur in a channel, or to determine which type exists, as follows:

F < 1.00 **subcritical flow** (a relatively deep, slow flow)

F = 1.00 **critical flow** (often a transitional flow)

F > 1.00 **supercritical flow** (a relatively shallow, fast flow)

Subcritical and supercritical flow are completely different in character, so the first step in many practical investigations should be to calculate the Froude number, F, where:

$$F = \frac{V}{\sqrt{gD_M}} \qquad\qquad (8.4)$$

V is the mean velocity of flow obtained from $V = Q/A$, and g is the acceleration due to gravity. The first part of this chapter concerns subcritical, uniform flow. The Froude number will be discussed in more detail in section 8.6. However, it should be appreciated that while the Froude number provides a good guide to the flow conditions in a simple rectangular channel, when applied to complex problems (like a river in flood with overbank flow) it is wise not to have too much faith in the result.

Examples 8.1 and 8.2 illustrate some of the points discussed above.

EXAMPLE 8.1

A trapezoidal channel has a bottom width of 5.0 m and its sides slope at an angle of 45°. If the depth of flow is 2.0 m, calculate the area of flow A, the wetted perimeter P, and the hydraulic radius R. If the discharge in the channel is 13.3 m³/s, calculate the Froude number, F.

$\tan 45° = X/2.0$ so $X = 2.0\tan 45° = 2.0$ m

Hence $B_s = 2.0 + 5.0 + 2.0 = 9.0$ m

$A = \frac{1}{2}(5.0 + 9.0)2.0 = 14.0$ m²

Let length of wetted side slopes $= Y$

$\cos 45° = 2.0/Y$

so $Y = 2.0/\cos 45° = 2.828$ m

$P = 2.828 + 5.000 + 2.828 = 10.656$ m

$R = A/P = 14.0/10.656 = 1.314$ m

$F = V/(gD_M)^{1/2}$

where $D_M = A/B_s = 14.0/9.0 = 1.556$ m

$V = Q/A = 13.3/14.0 = 0.950$ m/s

$F = 0.950/(9.81 \times 1.556)^{1/2} = 0.24$

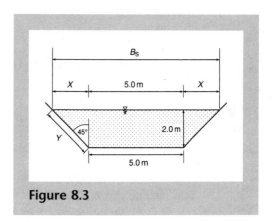

Figure 8.3

The flow is subcritical since 0.24 < 1.

EXAMPLE 8.2

Derive the theoretical expressions for the wetted perimeter and hydraulic radius of a rectangular channel, a pipe flowing half-full and a full pipe.

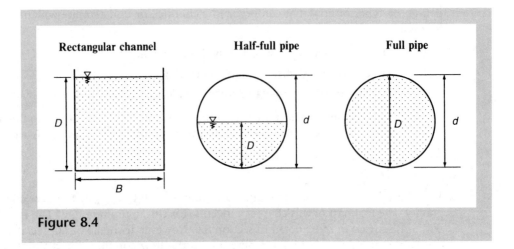

Figure 8.4

$A = BD$

$P = B + 2D$

$R = A/P$

$\quad = \dfrac{BD}{B + 2D}$

$A = \dfrac{1}{2} \times \pi d/4$

$P = \pi d/2$

$R = A/P$

$\quad = \dfrac{\pi d^2}{8} \times \dfrac{2}{\pi d}$

$\quad = d/4 \quad (= D/2)$

$A = \pi d^2/4$

$P = \pi d$

$R = A/P$

$\quad = \dfrac{\pi d^2}{4} \times \dfrac{1}{\pi d}$

$\quad = d/4$

8.1.3 Understanding why uniform flow occurs

66 I do not really understand why uniform flow occurs.
Why should the depth of flow in a channel be constant?
It seems an unlikely thing to happen. Can you explain please? **99**

You probably understand more about open channel flow than you realise. To demonstrate this fact, I will ask you a few questions. You answer them. Think of a river. **What causes the water to flow and to keep on flowing?**

Easy. It flows downhill.

But why does it flow downhill?

Because of gravity. The water seeks the lowest level possible by flowing down the slope.

Yes, but you have not completely answered the question. A book on a flat table top is affected by gravity, but it does not move. So try again, why does the water flow downhill?

Because the weight of the water has a component that acts down the slope of the channel. In the case of the book, the table top is perpendicular to the vertical direction in which gravity acts, so the component of weight in the plane of the table is zero because $\cos 90° = 0$.

Correct. Of course, if one side of the table is tipped up sufficiently so that the surface is no longer horizontal, then the book will slide off down the slope. However, if the table top is tipped only very slightly, friction keeps the book in place. Now consider this. If a car is parked on the top of a very steep hill and its handbrake fails, it will run away down the hill. At first it will move very slowly but, because the hill is very steep, it will accelerate and go faster and faster. **Why doesn't the water in a river accelerate and go faster and faster?**

Because there is friction between the moving water and the bed and banks of the river. This resistance force opposes the movement of the water. In the case of uniform flow, where the velocity is constant along the length of the channel, it follows that the force causing motion (the component of the weight of the water acting down the channel) must exactly balance the resistance force in the form of friction.

In a nutshell, that is why uniform flow occurs in an open channel, and this is the whole basis of the Chezy equation which is used to calculate the depth–discharge relationship.

8.2 Discharge equations for uniform flow

8.2.1 The Chezy equation

This equation was developed by a French engineer around the year 1768. The fundamental basis of the Chezy equation is as just described above, namely that:

force producing motion = friction force resisting motion

$$W \sin S_O = K A_P V^N \tag{8.5}$$

where for a length (L) of the channel, W is the weight of the water contained in it, S_O is the bed slope, K is a coefficient of roughness, A_P is the area of contact between the water and the perimeter of the channel, V is the mean velocity, and N is an exponent which usually has a value of 2 for turbulent flow. The full derivation of the Chezy equation is given in Appendix 1 (Proof 8.1), and this should be studied carefully. However, a few points are worth noting here.

Firstly, most channels have very small bed gradients so it can be assumed that $\sin S_O = \tan S_O = S_O$ where S_O is the bed slope expressed as a fraction, that is 1/800 for example. This can be easily confirmed by looking at Fig. 8.5. If the vertical distance is 1 m and the horizontal distance is 800 m, then from geometry the slope distance is 800.0006 m. This makes $\sin S_O$ = 1/800.0006 and $\tan S_O$ = 1/800. Thus there is no significant difference between $\sin S_O$ and S_O

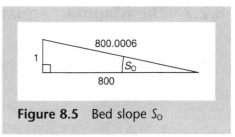

Figure 8.5 Bed slope S_O

under these conditions. With both the Chezy and Manning equations the bed slope S_O can be taken simply as $1/X$ where X is the horizontal distance (see Example 8.3).

The second thing to note is the similarity between the way that the friction forces in an open channel ($K A_P V^N$) and in a pipe are evaluated. The development of the Darcy pipe friction formula used a similar treatment of the friction forces, as can be seen by comparing Proofs 6.1 and 8.1 in Appendix 1.

If equation (8.5) is developed as in Proof 8.1, then the Chezy equation can be obtained as:

$$V = C\sqrt{RS_0} \qquad (8.6)$$

where V is the mean velocity and C is the Chezy coefficient. It is important to realise that the Chezy coefficient depends upon the Reynolds number and boundary roughness. Usually tables of typical values of C for various types of surface can be found in textbooks. It is also important to realise that C does have units, so the value used for C must be consistent with the system of units in use. The units of C can be found from equation (8.6) by substituting into it the known units of the other variables (S_0 is a dimensionless fraction) thus:

$$\frac{m}{s} = C \times m^{1/2} \quad \text{thus} \quad C = \frac{m}{s \times m^{1/2}} = m^{1/2}/s$$

Since C has the units $m^{1/2}/s$ it should never be thought of as a dimensionless coefficient.

EXAMPLE 8.3

A rectangular channel is 6.3 m wide and has a depth of flow of 1.7 m. The slope of the channel bed is 1/1000. Using the Chezy equation with $C = 49 \, m^{1/2}/s$, calculate the mean velocity and discharge in the channel.

$V = C(RS_0)^{1/2}$

where $R = A/P = (6.3 \times 1.7)/(6.3 + 2 \times 1.7) = 10.71/9.70 = 1.104 m$

$V = 49(1.104 \times 1/1000)^{1/2}$

$\quad = 1.628 m/s$

$Q = AV$

$\quad = (6.3 \times 1.7)1.628$

$\quad = 17.436 m^3/s$

8.2.2 The Manning equation

One of the most widely used discharge equations is that attributed to Manning in 1889. This was developed from empirical observations. Manning found that the Chezy coefficient, C, could be expressed as:

$$C = R^{1/6}/n \qquad (8.7)$$

where n is known as the Manning coefficient and represents the roughness of the surface. Typical values of Manning's n are given in Table 8.1. If the right-hand side of equation (8.7) is substituted into the Chezy formula, then the result is:

$$V = (1/n)R^{2/3}S_0^{1/2} \qquad (8.8)$$

$$\text{and} \quad Q = (A/n)R^{2/3}S_0^{1/2} \qquad (8.9)$$

where V is again the mean velocity, R the hydraulic radius, Q the discharge and A the cross-sectional area of flow. This is the **metric** form of the Manning equation. (For English units

Table 8.1 Typical values of Manning's n for different types of surface

	Conduit type, surface roughness and channel alignment	n (s/m$^{1/3}$)
Canals	Earth, straight	0.018–0.025
	Earth, meandering	0.025–0.040
	Rock, straight	0.025–0.045
Lined	Perspex	0.009
channels	Glass	0.009–0.010
	Cement mortar	0.011–0.015
	Concrete	0.012–0.017
	Dressed, jointed stone	0.013–0.020
Rivers	Earth, straight	0.020–0.025
	Earth, poor alignment	0.030–0.050
	Earth, with weeds and poor alignment	0.050–0.150
	Stones 75–150 mm diameter, straight, good condition	0.030–0.040
	Stones 75–150 mm diameter, poor alignment	0.040–0.080
	Stones >150 mm, boulders, steep slope, good condition	0.040–0.070
Floodplain	Short grass	0.025–0.035
	Long grass	0.030–0.050
	Medium to dense brush, in winter	0.045–0.110
Pipes	Cast iron	0.010–0.014
	Concrete	0.011–0.015

the 1 in equation (8.8) is replaced by 1.486. Take care, some old or American texts may give the English form of the equation.) Remember that n is not dimensionless, it has units which can be determined by substituting the known units (S_O is a dimensionless fraction) into equation (8.8):

$$\frac{m}{s} = \frac{1}{n} \times m^{2/3} \text{ thus } n = \frac{s \times m^{2/3}}{m} = s/m^{1/3}$$

Chow (1981) and French (1986) gave comprehensive details and information on the evaluation of Manning's n for open channels taking into account the bed material, degree of channel irregularity, variations in shape and size, relative effect of channel obstructions, vegetation growing in the channel and the degree of meandering. Some of these factors are already included in the values quoted in Table 8.1 and care must be taken not to compensate twice for the same thing. However, when necessary, n from Table 8.1 can be modified using Table 8.2 and the equation below, then n' used in equation (8.8) or (8.9).

$$n' = (n + \Delta n_1 + \Delta n_2 + \Delta n_3 + \Delta n_4) \times C \tag{8.10}$$

Chow and French also presented photographs of different types of river channel to illustrate what a particular n (or n') value looks like. Clearly, if an inappropriate roughness value is used in the Manning equation then an inaccurate estimate of the discharge will be obtained.

To avoid making serious errors, it is a good idea to remember that the range of Manning's n is from about 0.009 (perspex and glass) to 0.200 s/m$^{1/3}$ (flow through dense willow trees in leaf). Almost all values normally have one zero after the decimal point. Take care not to omit it. Only use values that do not have one zero after the decimal point after thinking about it, and when there is a reason for doing so. Remember, choosing the value of n often involves guesswork and can significantly affect the outcome of an analy- (REMEMBER!!!) sis, as Example 8.4 illustrates.

Table 8.2 Approximate factors for the modification of the basic Manning's n value for unlined natural channels

	Modification factor	Degree to which channel affected		
		Low	Moderate	Severe
Δn_1	Degree of irregularity of bed material Low = good condition as Table 8.1; severe = bank erosion, collapsed banks, or jagged, irregular surface in rock channels, etc.	0.000	0.010	0.020
Δn_2	Variation of cross-section Low = uniform channel; severe = significant changes in shape or area resulting in the main flow frequently shifting from side to side	0.000	0.005	0.014
Δn_3	Effect of obstructions such as debris and boulders Low = uniform, unobstructed channel; severe = cross-sectional area reduced, turbulence significantly increased, discharge reduced	0.000	0.018	0.050
Δn_4	Effect of vegetation on reducing the discharge Low = flow depth $D > 3\times$ vegetation height; severe when $D = 0.5\times$ vegetation height, or when trees and bushes in full foliage are in channel with hydraulic radius < 4 m	0.005	0.015	0.100
C	Channel meander ratio (Σmeander lengths/Σstraight lengths) Low = ratio 1.0–1.2; severe > 1.5	1.00	1.10	1.30

Example: straight river with bed of 75 mm diameter stones in good condition has $n = 0.030$ s/m$^{1/3}$ (Table 8.1). For a river with similar bed but where irregularity is low, variation of section is moderate, there are some minor obstructions, some vegetation that has a minor effect and moderate meanders, then equation (8.10) gives $n' = (0.030 + 0.000 + 0.005 + 0.010 + 0.005) \times 1.10 = 0.055$ s/m$^{1/3}$. Note that Table 8.1 gives the value for 75–150 mm stones with poor alignment as 0.040–0.080 s/m$^{1/3}$, so take care not to allow for the same channel defects twice.

EXAMPLE 8.4

A rectangular river channel 4.6 m wide carries water at a depth of 0.6 m. The slope of the channel is 1 in 400. The channel has a poor alignment and the bed is covered with stones about 75 mm to 150 mm in size. Using the range of n values in Table 8.1, investigate the range of discharge that results from the application of the Manning equation to the channel. What is the percentage difference in the flows, calculated as a proportion of the smaller value?

$$A = 4.6 \times 0.6 = 2.76 \text{m}^2$$
$$P = 0.6 + 4.6 + 0.6 = 5.8 \text{m}$$
$$R = A/P = 2.76/5.8 = 0.476 \text{m}$$
$$S_O = 1/400 = 0.0025$$

From Table 8.1, $n = 0.040$ to 0.080 s/m$^{1/3}$

with $n = 0.040$

$V = (1/n)R^{2/3}S_O^{1/2}$

$V = (1/0.040)0.476^{2/3}0.0025^{1/2}$

$\quad = 25.000 \times 0.610 \times 0.050$

$\quad = 0.76\,\text{m/s}$

$Q = AV = 2.76 \times 0.76 = 2.10\,\text{m}^3/\text{s}$

with $n = 0.080$

$V = (1/n)R^{2/3}S_O^{1/2}$

$V = (1/0.080)0.476^{2/3}0.0025^{1/2}$

$\quad = 12.500 \times 0.610 \times 0.050$

$\quad = 0.38\,\text{m/s}$

$Q = AV = 2.76 \times 0.38 = 1.05\,\text{m}^3/\text{s}$

$$\% \text{ difference} = 100(2.10 - 1.05)/1.05 = 100\%$$

Therefore the difference between the two values is 100%, that is the larger value is twice the smaller value. This illustrates that while the Manning equation may be reasonably accurate, it may be difficult to estimate the values of the variables, so the answer obtained is nothing more than an estimate.

SELF TEST QUESTION 8.1

A trapezoidal channel has a base width of 8.3 m and sides that rise 1 m vertically for every 2 m horizontally. The depth of flow in the channel is 2.7 m, its gradient is 0.001 and Manning's n is $0.035\,\text{s/m}^{1/3}$. Determine the mean velocity and discharge in the channel.

8.2.3 Solving the discharge equations for depth of flow

The calculation of the discharge in a channel at a known depth is simple. The reverse calculation, to obtain the normal depth, D, corresponding to a known discharge, is more difficult. This is because the depth appears in both the area, A, and the hydraulic radius, R, as shown in Example 8.2. However, with **wide rectangular channels** ($B \gg D$) this difficulty can be avoided by assuming that $R = D$. The justification is that:

$$R = \frac{A}{P} = \frac{BD}{(B+2D)} \approx \frac{BD}{B} \approx D \tag{8.11}$$

By assuming that $R = D$ the discharge equation can be solved easily. This only applies to wide, rectangular channels, and the depth obtained may be inaccurate (Example 8.5a). For other channels a trial-and-error procedure is used to obtain D (Example 8.5b). Alternatively, for many channels nomograms or charts have been devised that relate roughness, depth and discharge or velocity (see Chow, 1981; French, 1986; Hydraulics Research, 1990).

EXAMPLE 8.5

For the rectangular channel in Example 8.4, taking the n value as $0.040\,\text{s/m}^{1/3}$, calculate the depth of flow, D, that corresponds to a discharge of $2.83\,\text{m}^3/\text{s}$ by (a) assuming that the river can be considered as a wide rectangular channel, (b) by trial and error.

Figure 8.6 Trapezoidal flood relief channel for the River Exe at Exeter. The channel capacity is 450 m³/s. The sides of the channel are grassed while the base has a concrete lining. This adds to the difficulty of estimating the n value of the channel, and hence its capacity

(a) **Assuming a wide rectangular channel with $R = D$**

$Q = A(1/n)R^{2/3}S_O^{1/2}$ and $A = BD = 4.6D$

$2.83 = 4.6D(1/0.040)D^{2/3} \times 0.0025^{1/2}$

$2.83 = 115D^{5/3} \times 0.05$

$D^{5/3} = 0.492$

$D = (0.492)^{3/5}$

$D = 0.65\text{m}$

(b) **By trial and error**

$Q = A(1/n)R^{2/3}S_O^{1/2}$ where $B = 4.6\text{m}$, $n = 0.040\,\text{s/m}^{1/3}$ and $S_O = 0.0025$. Thus:

$2.83 = 4.6D(1/0.040)[4.6D/(4.6 + 2D)]^{2/3}0.0025^{1/2}$

$2.83 = 5.75D[4.6D/(4.6 + 2D)]^{2/3}$

$0.492 = D[4.6D/(4.6 + 2D)]^{2/3}$

This equation has to be solved by trial and error. Guess a value of D and substitute in the equation above. The value of D is correct when the right-hand side (RHS) of the equation equals the left, that is 0.492 in this case. A table helps facilitate a solution:

Try $D = 0.65\,\text{m}$,   RHS = 0.413

      0.70 m   0.462

      0.75 m   0.513

Now RHS > 0.492 so D lies between 0.70 m and 0.75 m. Changing D by 0.05 m caused a change in the RHS of $(0.513 - 0.462) = 0.051$. So to increase RHS by 0.030, from 0.462 to 0.492, would need a corresponding increase in D of $(0.030/0.051) \times 0.05 = 0.029$ m. Therefore $D = 0.729$ m.

Check:

$A = 4.6 \times 0.729 = 3.353 \text{m}^2, P = 4.6 + 2 \times 0.729 = 6.058 \text{m}$

$R = A/P = 3.353/6.058 = 0.553 \text{m}$

$Q = (A/n)R^{2/3}S_0^{1/2} = (3.353/0.040) \times 0.553^{2/3} \times 0.0025^{1/2}$

$\quad = 83.825 \times 0.674 \times 0.050 = 2.825 \text{m}^3/\text{s} \quad \text{OK}$

Note that in this instance a more accurate estimate is obtained by trial and error than by assuming that the river approximates to a wide rectangular channel (with $B = 4.6$ m and $D = 0.7$ m the river is not really wide enough for that). This approximation does give a rough idea of the first depth to use in a trial-and-error solution though.

8.3 ▶ Channel proportions for maximum discharge or velocity

 66 If an engineer has to design a channel like that in Fig. 8.6, it is generally desirable to make the cross-sectional area of the channel as small as possible to minimise excavation and construction costs. Thus the question arises, what proportions should the channel have to give the maximum discharge for a given cross-sectional area of flow? Trapezoidal, square, rectangular, wide and shallow, narrow and deep, or what? Any ideas? 99

 66 No, how can this be evaluated? 99

Think of the Chezy equation, where the force producing motion (the weight of the water acting down the channel) exactly balances the force resisting motion (friction around the wetted perimeter). Does this give you a clue?

Yes. If we can design the channel so that the resisting force is as small as possible, we will maximise the discharge.

Correct. And how can we minimise the resistance force?

By making the wetted perimeter as short as possible.

Yes, that minimises the area over which the resistance forces act and maximises the flow. Now do you know which geometrical shape has the smallest perimeter for the area it encloses?

A circle.

That is right. Try this for yourself. A circle of 1 m radius has an area of 3.142 m² and a perimeter of 6.284 m. The square that has an area of 3.142 m² has sides 1.7725 m long giving it a perimeter of 7.090 m, which is longer than the circle. A rectangle 2.000 m wide by 1.571 m high also has an area of 3.142 m² but a perimeter of 7.142 m. Thus the circle has the smallest perimeter for its area, so the optimum channel shape would be a semicircle. This would minimise P and maximise Q. However, it is very impractical to excavate or construct semicircular channels, so in many situations a more pragmatic (REMEMBER!!!) solution must be adopted. The engineering compromise is explored below.

8.3.1 Rectangular channel

Remember that to obtain the minimum value of something mathematically we differentiate with respect to the controlling variable and equate to zero. Consider a rectangular channel of width, B, and depth, D. The wetted perimeter, $P = B + 2D$ and the cross-sectional area of flow, $A = BD$ so $B = A/D$. Hence P can be written as:

$$P = (A/D) + 2D \quad \text{or} \quad P = AD^{-1} + 2D$$

Therefore, for a given value of A, surface roughness and channel slope, P will be a minimum when $dP/dD = 0$, thus:

$$dP/dD = -AD^{-2} + 2 = 0$$
$$(A/D^2) = 2 \quad \text{or} \quad A = 2D^2$$

But $A = BD$ so

$$BD = 2D^2 \quad \text{giving}$$

$$\mathbf{B = 2D} \tag{8.12}$$

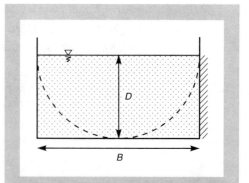

Figure 8.7 Optimum hydraulic section for a rectangular channel

So for maximum discharge the width, B, should be twice the depth, D. This is the shape of a rectangular channel into which a semicircle would fit, as shown in Fig. 8.7. Under these conditions the hydraulic radius is $D/2$. So the rectangular channel that would maximise the discharge for a given cross-sectional area, surface roughness and channel slope is the one that has the proportions that most closely resemble a semicircle, which is the logical result.

8.3.2 Trapezoidal channel

The same arguments that were applied to a rectangular channel can also be applied to a trapezoidal one. In this case let the sides slope at an angle of 1 vertically to S horizontally. Thus if the depth of flow is D then by simple geometry the horizontal width of the side slopes on each side of the channel is SD. The width of the water surface, B_S, is:

$$\mathbf{B_S = B + 2SD} \tag{8.13}$$

$$A = \frac{1}{2}(B + [B + 2SD])D$$

$$\mathbf{A = (B + SD)D} \tag{8.14}$$

Thus $B = AD^{-1} - SD$ (1)

By geometry the wetted length of each side slope is $([SD]^2 + D^2)^{1/2}$ or $D(S^2 + 1)^{1/2}$. Thus:

$$\mathbf{P = B + 2D(S^2 + 1)^{1/2}} \tag{8.15}$$

Substituting the expression for B in equation (1) into equation (8.15) gives:

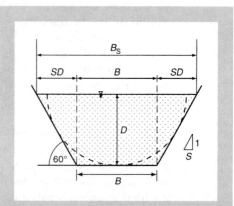

Figure 8.8 Optimum hydraulic section for a trapezoidal section

$$P = AD^{-1} - SD + 2D(S^2 + 1)^{1/2}$$

For a given value of A, surface roughness and channel slope, the discharge will be a maximum when $dP/dD = 0$, that is:

$$dP/dD = -AD^{-2} - S + 2(S^2 + 1)^{1/2} = 0 \qquad (2)$$

Substituting $A = (B + SD)D$ from equation (8.14) into equation (2) and rearranging gives:

$$\frac{(B + SD)D}{D^2} + S = 2(S^2 + 1)^{1/2}$$

Thus the optimum hydraulic section is when:

$$(B + 2SD) = 2D(S^2 + 1)^{1/2} \qquad (8.16)$$

Thus the most efficient hydraulic section is when the top width $(B + 2SD)$ is twice the wetted slope length $D(S^2 + 1)^{1/2}$. This is the property of one half of a hexagon. Under these conditions the hydraulic radius $R = D/2$, as it did for the optimum rectangular channel. Again the optimum hydraulic section is the one that most closely resembles a semicircle.

8.3.3 Circular channel

Surface water and foul sewers are often designed to operate part full under gravity, so the Chezy and Manning equations are applicable. The analysis to obtain the optimum flow condition in a circular conduit has to differ from those above because the shape of the conduit is fixed. The question is therefore at what angle, θ, and depth of flow, D, is the maximum velocity obtained in the circular conduit? If it assumed that the surface roughness and slope of the conduit are fixed, then the maximum velocity will be obtained when $R = A/P$ has a maximum value, that is when P is at a minimum. This can be expressed as:

$$\frac{d(A/P)}{d\theta} = 0$$

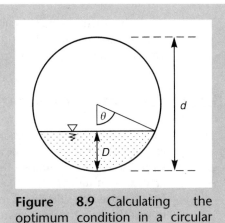

Figure 8.9 Calculating the optimum condition in a circular conduit

The solution to this problem involves obtaining expressions for the area, A, and wetted perimeter, P, corresponding to any angle θ and then differentiating as shown above. This is a lengthy process (see Douglas *et al.*, 1995) so only the result will be quoted here. This is:

$$D = 0.81d \qquad (8.17)$$

Thus the maximum mean velocity (V) occurs in a pipe when it is flowing at a depth equivalent to 0.81 of its diameter (d). After this, any additional increase in depth near the crown of the pipe results in a much longer wetted perimeter (and hence resistance to flow) but only a small increase in the area of flow. Thus pipes are often designed to run not quite full. In fact the maximum discharge (as opposed to velocity) occurs at a proportional depth of about 0.95d and is about 1.09 times the full bore discharge (and coincidentally about 1.1

Table 8.3 Geometric characteristics of best hydraulic sections [*after French, 1986*]

Shape of section	Area of flow, A	Wetted perimeter, P	Hydraulic radius, $R = A/P$	Surface width, B_S	Hydraulic mean depth $D_M = A/B_S$
Rectangle: half of a square	$2D^2$	$4D$	$0.500D$	$2D$	D
Trapezoid: half of a hexagon	$1.732D^2$	$3.463D$	$0.500D$	$2.309D$	$0.750D$
Triangle: half of a square	D^2	$2.83D$	$0.354D$	$2D$	$0.50D$
Semicircle	$0.500\pi D^2$ $0.125\pi d^2$	πD $0.500\pi d$	$0.500D$ $0.250d$	$2D$ $1.000d$	$0.250\pi D$ $0.125\pi d$
Circle: depth of flow 0.80d	$0.215\pi d^2$	$0.705\pi d$	$0.304d$	$0.800d$	$0.269\pi d$
Circle: depth of flow 0.95d	$0.245\pi d^2$	$0.856\pi d$	$0.286d$	$0.436d$	$0.562\pi d$

Note: D = depth of flow; d = diameter of circular conduit.

times the discharge that occurs at a proportional depth of $0.81d$). If the depth was greater than $0.95d$, then it is likely that the pipe would cease to operate as an open channel in which the flow is caused by gravity.

8.3.4 Practical considerations relating to the optimum flow

Table 8.3 provides a useful and more user-friendly summary of the above analysis. It is worth emphasising that the analysis above is of a rather theoretical nature and is solely concerned with finding the optimum shape for a channel when other factors such as the hydraulic gradient and roughness are kept constant. The answers obtained are not always practical. For instance, the side slopes in an unlined channel must be determined from a considera-tion of the stability of the material forming the embankments. If the objective is to increase the capacity of a river channel, then increasing the hydraulic gradient by removing bends and meanders (that is effectively shortening the length of flow between two points) is a more practical proposition than re-shaping the channel. Similarly, removing weeds, debris and sand bars from the channel may effectively reduce the Manning n value, so increasing discharge.

EXAMPLE 8.6

A rectangular, concrete lined channel is to be constructed to carry floodwater. The slope of the ground surface is 1 in 500. The design discharge is $10\,\text{m}^3/\text{s}$. (a) Calculate the proportions of the rectangular channel that will minimise excavation and result in the optimum hydraulic section. (b) If the cross-sectional area of flow is kept the same as in part (a) but for safety reasons the depth of flow in the channel is limited to $1.00\,\text{m}$, what will be the discharge now?

(a) For concrete, Manning's $n = 0.015 \text{s/m}^{1/3}$ (from Table 8.1).

From Table 8.3, for the optimum hydraulic section $A = 2D^2$ and $R = 0.5D$. Thus:

$$Q = (A/n)R^{2/3}S_0^{1/2}$$
$$10 = (2D^2/0.015) \times (0.5D)^{2/3} \times (1/500)^{1/2}$$
$$0.15 = 2D^2 \times 0.63D^{2/3} \times 0.0447$$
$$0.15 = 1.26D^{8/3} \times 0.0447$$
$$D^{8/3} = 2.663$$
$$D = 1.444 \text{m}$$

For the optimum section $B = 2D = 2.888 \text{m}$.

Thus the cross-sectional area of flow is 2.89 m wide by 1.44 m deep. Due allowance should be made for inaccuracies (such as estimating n) and waves on the water surface by adding a sensible freeboard to the depth.

(b) If $D = 1.000 \text{m}$ and $A = 1.444 \times 2.888 = 4.170 \text{m}^2$, then the width of channel now required is:

$$B = 4.170/1.000 = 4.170 \text{m}.$$
$$P = 4.170 + 2 \times 1.000 = 6.170 \text{m}$$

Using the Manning equation:

$$Q = (4.170/0.015) \times (4.170/6.170)^{2/3} \times (1/500)^{1/2}$$
$$Q = 278 \times 0.770 \times 0.0447$$
$$Q = 9.57 \text{m}^3/\text{s}$$

This is a reduction of 4.3% compared to the optimum hydraulic section.

EXAMPLE 8.7

If the conditions were as described in Example 8.6 but a trapezoidal channel was used instead of a rectangular one: (a) what would be the dimensions of the optimum hydraulic section? (b) would the cross-sectional area of flow be more or less than that for the equivalent rectangular channel?

(a) From Table 8.3, for the optimum hydraulic section $A = 1.73D^2$ and $R = 0.5D$. Thus:

$$Q = (A/n)R^{2/3}S_0^{1/2}$$
$$10 = (1.73D^2/0.015) \times (0.5D)^{2/3} \times (1/500)^{1/2}$$
$$0.15 = 1.73D^2 \times 0.63D^{2/3} \times 0.0447$$
$$0.15 = 1.09D^{8/3} \times 0.0447$$
$$D^{8/3} = 3.079$$
$$D = 1.525 \text{m}$$

For the optimum section, surface width $B_s = 2.31D = 2.31 \times 1.525 = 3.523 \text{m}$.
$A = 1.73D^2 = 1.73 \times 1.525^2 = 4.023 \text{m}^2$.
But $A = \frac{1}{2}(B + B_s)D$ so $4.023 = \frac{1}{2}(B + 3.523)1.525$ giving bottom width $B = 1.753 \text{m}$. The sides of the channel slope outwards at an angle of 30° to the vertical (as in Fig. 8.8).

(b) For the rectangular channel in Example 8.6 the cross-sectional area of flow, $A = 4.170 \text{m}^2$. For the trapezoidal channel above, $A = 4.023 \text{m}^2$. Therefore the trapezoidal channel has a cross-sectional area about 4% smaller than the equivalent rectangular channel.

EXAMPLE 8.8

As an alternative to a rectangular or trapezoidal channel, it has been suggested that the flow could be taken by a 2.40 m diameter concrete pipe flowing under the optimum discharge conditions (say at a depth 0.95 of the diameter). Investigate the feasibility of this proposal.

From Table 8.1, Manning's n for a concrete pipe $= 0.013\,\text{s/m}^{1/3}$.

From Table 8.3, area of flow $A = 0.245\pi d^2 = 0.245\pi \times 2.4^2 = 4.433\,\text{m}^2$.

$R = 0.286d = 0.286 \times 2.4 = 0.686\,\text{m}$.

$Q = (A/n)R^{2/3}S_0^{1/2}$

$Q = (4.433/0.013) \times (0.686)^{2/3} \times (1/500)^{1/2} = 341 \times 0.778 \times 0.0447$

$Q = 11.859\,\text{m}^3/\text{s}$

Thus the proposal to use a precast 2.40 m diameter pipe is feasible. It can carry more than the required 10 m³/s, and may be a more economical option than constructing a concrete lined channel *in situ*. In fact a smaller diameter pipe would suffice. It is worth noting that in practice there is a limited range of available pipe sizes and that a relatively small change in diameter results in a large change in discharge. In this case, repeating the calculation above reveals the following relationship between diameter and discharge:

2.40 m diameter	11.86 m³/s
2.25 m diameter	9.99 m³/s
2.10 m diameter	8.31 m³/s

SELF TEST QUESTION 8.2

An open channel has the cross-section shown in Fig. 8.10. The curved parts of the invert (bottom of the channel) have a radius of 1.0 m and form a quarter of a circle, while the remainder of the sides of the channel are vertical. The channel is concrete lined with Manning's $n = 0.017\,\text{s/m}^{1/3}$. The bed gradient is 1 in 600. Calculate the discharge in the channel when the depth of flow is 1.0 m, 2.0 m and 3.0 m.

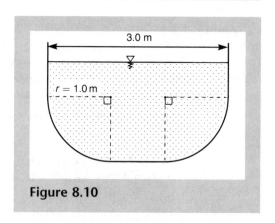

Figure 8.10

8.4 Compound channels and the composite Manning's n

So far it has been assumed that there is only one main channel, be it rectangular, trapezoidal or circular. However, natural rivers in flood usually comprise a main channel with a floodplain on each side. This is called a compound or two stage channel. Within a compound channel it is unlikely that the roughness will be uniform around the entire wetted

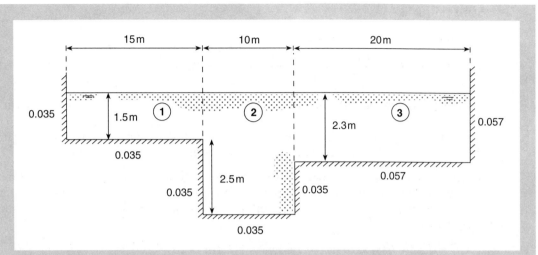

Figure 8.11 A compound channel with three subsections, each having a constant *n* value (shown adjacent to the boundary). The vertical scale is enlarged for clarity

perimeter. Thus it is necessary to deal with both the compound nature of the channel and the variable roughness. Although there are simple techniques that can be employed, it should be appreciated that the answers obtained are often relatively inaccurate. This can easily lead to the over-design or under-design of flood alleviation works, for example.

The compound channel in Fig. 8.11 has been split into three subsections, each subsection having the same Manning's *n* value. Subsection 1 is the left floodplain, and this has the same *n* as the main channel (subsection 2). The right floodplain is subsection 3. The assumption is that each subsection can be analysed separately, then:

Total discharge = Σsubsection discharges (8.18)

i.e. $Q = Q_1 + Q_2 + Q_3$

where the subscripts refer to the subsections. When conducting the calculations, only the real perimeters are assigned a roughness and included in the calculation of P and R, the imaginary vertical dividing lines between subsections (shown dashed) are not. It is assumed that the main channel and floodplains all have the same longitudinal bed slope, $S_O = 1$ in 600 in this example. Proceeding on this basis the calculations for each of the subsections in the diagram are as follows.

Subsection 1

$A_1 = 1.5 \times 15.0 = 22.5\,\text{m}^2$, $P_1 = 1.5 + 15.0 = 16.5\,\text{m}$, $R_1 = A_1/P_1 = 22.5/16.5 = 1.364\,\text{m}$.

$Q_1 = \dfrac{A_1}{n_1} R_1^{2/3} S_O^{1/2} = \dfrac{22.5}{0.035} \times 1.364^{2/3} \times \left(\dfrac{1}{600}\right)^{1/2} = 32.3\,\text{m}^3/\text{s}$.

Subsection 2

$A_2 = 4.0 \times 10.0 = 40.0\,\text{m}^2$, $P_2 = 2.5 + 10.0 + 1.7 = 14.2\,\text{m}$, $R_2 = A_2/P_2 = 40.0/14.2 = 2.817\,\text{m}$.

$Q_2 = \dfrac{A_2}{n_2} R_2^{2/3} S_O^{1/2} = \dfrac{40.0}{0.035} \times 2.817^{2/3} \times \left(\dfrac{1}{600}\right)^{1/2} = 93.1\,\text{m}^3/\text{s}$.

Subsection 3

$A_3 = 2.3 \times 20.0 = 46.0\,\text{m}^2, P_3 = 2.3 + 20.0 = 22.3\,\text{m}, R_3 = A_3/P_3 = 46.0/22.3 = 2.063\,\text{m}.$

$Q_3 = \dfrac{A_3}{n_3} R_3^{2/3} S_O^{1/2} = \dfrac{46.0}{0.057} \times 2.063^{2/3} \times \left(\dfrac{1}{600}\right)^{1/2} = 53.4\,\text{m}^3/\text{s}.$

Total discharge through the main channel and floodplains $Q = 32.3 + 93.1 + 53.4 = 178.8\,\text{m}^3/\text{s}.$

The apparent simplicity of the above calculations masks the difficulty of analysing the flow through compound channels. Particular problems arise due to the following:

- With variable compound channels there is more than one method of analysis and more than one possible answer. Treating subsections 1 and 2 as one channel (since they have the same roughness) gives $A_{1+2} = 62.5\,\text{m}^2$, $P_{1+2} = 30.7\,\text{m}$, $R_{1+2} = 2.036\,\text{m}$ and the combined discharge as:

$Q_{1+2} = (62.5/0.035) \times 2.036^{2/3} \times (1/600)^{1/2} = 117.1\,\text{m}^3/\text{s}$

whereas the previous (and probably more accurate) value was $125.4\,\text{m}^3/\text{s}$. Generally it is advisable to insert a dividing vertical where the channel changes significantly in shape or roughness. Since answers can vary, care must be used in deciding how to undertake an analysis (see Self Test Question 8.3).

- The difficulty of selecting appropriate n values was discussed earlier, but applies especially to compound channels (see Table 8.2). When flow first occurs over the floodplains the depth will be very small so their effective roughness is much increased. Consequently n may decrease with stage up to bank-full level, then increase as the flow goes over-bank, and then decrease again as the depth increases on the floodplains. This varies from one location to another.

- There is a complex interaction between the flow in the main channel and that on the floodplains that can lead to large head losses. As a result the discharge can be over-estimated in extreme cases by as much as the bank-full discharge. For a more accurate method of analysis see Ackers (1992; 1993).

- Just because there is water on the floodplain it does not mean that there is flow through the entire cross-section. With wide floodplains in particular, some water may be trapped and stationary, so including it in the calculations produces a gross overestimate of the discharge.

- The Froude number in compound channels is difficult to calculate and can be more or less meaningless (see French, 1986).

- In compound channels α may be around 2.0, or even larger near obstructions, so it cannot be automatically assumed to be 1.0 or left out of equations. With the above example it is possible to calculate α using equation (4.26) which, with the current notation, can be written as $\alpha = \Sigma\left(V_i^3 A_i\right)/V^3 A$ where

$V_1 = Q_1/A_1 = 32.3/22.5 = 1.436\,\text{m/s}$ and $V_1^3 A_1 = 66.6$

$V_2 = Q_2/A_2 = 93.1/40.0 = 2.328\,\text{m/s}$ and $V_2^3 A_2 = 504.7$

$V_3 = Q_3/A_3 = 53.4/46.0 = 1.161\,\text{m/s}$ and $V_3^3 A_3 = 72.0$

$V = Q/A = 178.8/108.5 = 1.648\,\text{m/s}$ and $V^3 A = 485.6$

thus $\alpha = (66.6 + 504.7 + 72.0)/485.6 = 1.33.$

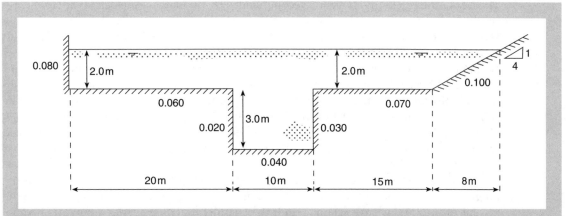

Figure 8.12 A compound channel where n (shown adjacent to the boundary) is different for each part of the perimeter

Despite the problems, this method of analysis can be used successfully to interpolate between known stages and discharges. Initially, for a known stage, this often means adjusting the boundary n values until the calculated discharge agrees with that actually measured. When the method is applied to rivers for which there is no known stage–discharge data, then the results should be used cautiously.

Figure 8.12 illustrates a situation involving a compound channel where the calculations are complicated by the fact that every part of the perimeter has a different n value. This may be encountered in both natural and laboratory channels; in the latter case the channel bottom is often roughened artificially to produce Froude numbers similar to those experienced in rivers. In such cases we need to calculate the average (composite) roughness of the entire cross-section and then use this in the Manning equation to obtain the total discharge.

This average roughness (n_{AV}) cannot be a straight-forward mathematical average of the n values since the various parts of the subsection have different lengths and the effect of the roughness is non-linear. However, the following equations can be used to determine the average roughness of a cross-section that has a total overall wetted perimeter and hydraulic radius of P and R, but which comprises N different lengths, any one of which is denoted by P_i, R_i and n_i.

$$n_{AV} = \frac{PR^{5/3}}{\sum\limits_{i=1}^{N}\left(\dfrac{P_i R_i^{5/3}}{n_i}\right)} \tag{8.19}$$

$$n_{AV} = \left[\frac{\sum\limits_{i=1}^{N}\left(P_i n_i^{3/2}\right)}{P}\right]^{2/3} \tag{8.20}$$

$$n_{AV} = \left[\frac{\sum\limits_{i=1}^{N}\left(P_i n_i^{2}\right)}{P}\right]^{1/2} \tag{8.21}$$

The first of the equations above was recommended by Hydraulics Research (1988) for British rivers. It assumes that the total discharge through the section equals the sum of the sub-section discharges, as in equation (8.18). However, with vertical boundaries this can result in part of the perimeter being omitted since $A_i = 0$ and hence $R_i = 0$, so one of the other equations is easier to use. Equation (8.20) assumes the average velocity in the subsections equals the average velocity in the whole cross-section. Equation (8.21) is used in some commercial software; it assumes that the total force resisting motion is equal to the sum of the subsection resisting forces. It has the merit of being simple, so it is used in Example 8.9. Note that the assumptions inherent in these equations mean that they will often yield different values of n_{AV}, particularly if there is a large variation in n over the cross-section. This is illustrated by Self Test Question 8.3 below.

Later in this chapter we will discuss gradually varying flow and how to calculate the longitudinal elevation of the water surface along a channel. These calculations require the n value of the various cross-sections, so if the channel roughness varies around its perimeter it will be necessary to use one of the above equations and to substitute n_{AV} for n.

EXAMPLE 8.9

A compound channel has the cross-section shown in Fig. 8.12. The values next to the perimeter boundaries are the Manning roughness values ($ns/m^{1/3}$). Use equation (8.21) to calculate n_{AV}.

The calculations are conducted in the table below and the total values substituted into the equation.

Perimeter	n_i ($s/m^{1/3}$)	P_i (m)	$P_i n_i^2$
1	0.080	2.000	0.0128
2	0.060	20.000	0.0720
3	0.020	3.000	0.0012
4	0.040	10.000	0.0160
5	0.030	3.000	0.0027
6	0.070	15.000	0.0735
7	0.100	8.246	0.0825
Total		61.246	0.2607

$$n_{AV} = \left[\frac{\sum_{i=1}^{N}(P_i n_i^2)}{P} \right]^{1/2} = \left[\frac{0.2607}{61.246} \right]^{1/2} = 0.065\,s/m^{1/3}$$

SELF TEST QUESTION 8.3

A cross-section of a compound channel is shown in Fig. 8.13. The values next to the perimeter boundaries are the Manning roughness values ($ns/m^{1/3}$). The channel slope $S_O = 1$ in 500. (a) Use equations (8.18), (8.19), (8.20), and (8.21) to calculate the total discharge through the

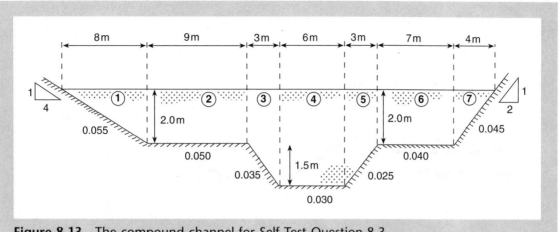

Figure 8.13 The compound channel for Self Test Question 8.3

section. (b) What is the percentage difference between the answers, and how can these be explained? (c) Calculate the value of α for the section.

8.5 Environmentally acceptable channels

With the increased environmental awareness of the late 20th century it became necessary to design not just for functionality but also environmental acceptability. This meant that straight, simple, trapezoidal channels that are relatively easy to analyse were no longer considered appropriate. Instead, meandering compound channels that afford a wider range of habitat for plants and animals were promoted (Fig. 8.14). These have the desired effect of looking natural, but the analysis required for their design is more complex since vegetation and trees may be deliberately included within a curving channel. As Self Test Question 8.3 shows, when calculating the discharge in a compound channel it is possible to obtain significantly different values. This means the engineer must be more diligent in ensuring the accuracy of the design calculations. Frequently specialised computer software is used, but the same difficulties exist and it is incumbent upon the engineer to ensure it is used appropriately and that the answers are valid.

Some useful guides to the design of environmentally acceptable channels and flood alleviation works were provided by Water Space Amenity Commision (1983), Hydraulics Research (1988), RSPB *et al.* (1994) and Brookes and Shields (1996).

8.6 Specific energy and critical depth

In some calculations it is more convenient to use the specific energy of the flow instead of the total energy. The **specific energy**, E, at a particular cross-section of a channel is the energy calculated above the bottom of the channel, thus:

$$E = D + \alpha V^2/2g \tag{8.22}$$

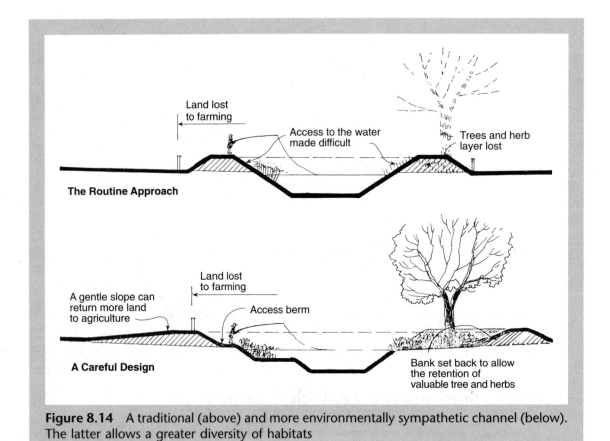

Land lost to farming

Access to the water made difficult

Trees and herb layer lost

The Routine Approach

Land lost to farming

A gentle slope can return more land to agriculture

Access berm

A Careful Design

Bank set back to allow the retention of valuable tree and herbs

Figure 8.14 A traditional (above) and more environmentally sympathetic channel (below). The latter allows a greater diversity of habitats

This expression is similar to equation (8.1) for the total energy head, but with the omission of the 'z' term because specific energy (head) is calculated above the bed of the channel not above an arbitrary datum. Consequently for a particular discharge in a channel with uniform flow where D, α and V are constant it follows that ***the specific energy, E, must be the same at all cross-sections along the channel***. This is an important distinction between specific energy and total energy. Total energy must always decrease along the channel in the direction of flow because of the fall in the elevation of the bed (that is z), as shown in Fig. 8.2.

REMEMBER!!!

Since $V = Q/A$, equation (8.22) can also be conveniently expressed as:

$$E = D + \alpha Q^2/2gA^2 \tag{8.23}$$

In the following text the energy coefficient, α, will normally be taken as 1.0 and will be omitted (but see section 4.8.1).

8.6.1 The concept of two alternate depths of flow in a channel

In section 8.1.2 the existence of two types of flow was introduced: subcritical flow and supercritical flow. **Subcritical flow** is the most common in nature and is relatively deep and slow moving. **Supercritical flow** is less common and is characterised by a very fast,

Table 8.4 Variation of specific energy and Froude number with depth ($Q = 1.0\,\text{m}^3/\text{s}$, $B = 1.0\,\text{m}$)

Depth, D (m)	Area of flow, A (m²)	$V = Q/A$ (m/s)	$V^2/2g$ (m)	$E = D + V^2/2g$ (m)	$F = V/(gD)^{1/2}$
5.000	5.000	0.200	0.002	5.002	0.03
1.000	1.000	1.000	0.051	1.051	0.32
0.500	0.500	2.000	0.204	0.704	0.90
0.467	0.467	2.140	0.234	0.701	1.00
0.300	0.300	3.333	0.566	0.866	1.94
0.253	0.253	3.953	0.796	1.049	2.51
0.200	0.200	5.000	1.274	1.474	3.57
0.102	0.102	9.804	4.899	5.001	9.80

relatively shallow flow. *However, both may occur in the same channel at the same discharge.*

How is this posssible?

To answer this question, let us consider the following hypothetical situation. A rectangular channel is 1.0 m wide. It carries a flow, Q, of $1.0\,\text{m}^3/\text{s}$ (with $\alpha = 1.0$). We will assume various depths, D, ranging from 5.0 m to about 0.1 m and then investigate the characteristics of the flow by calculating the variation of the specific energy, E, and the Froude number, F. We will then draw a graph of D against E.

The calculations are shown in Table 8.4, and the data are plotted in Fig. 8.15. Table 8.4 illustrates several interesting points. Firstly, for the discharge of $1.0\,\text{m}^3/\text{s}$, the specific energy is virtually the same (5.002 m) when the depth of flow is either 5.000 m or 0.102 m. Or, putting this statement the other way around, for a particular specific energy (5.002 m) two **alternate depths** of flow are possible, 5.000 m and 0.102 m. The larger is the subcritical depth of flow, and the smaller the supercritical depth. There is nothing unique about this pair of values: depths of 1.000 m and 0.253 m also have virtually the same specific energy, but this time it is 1.050 m. You can calculate other similar pairs of values if you want to.

Secondly, the specific energy initially decreases with decreasing depth, then increases as the depth continues to decrease, as shown in Fig. 8.15. Thus the specific energy has a minimum value of 0.701 m that corresponds to a depth of 0.467 m. Prove this for yourself by repeating the calculations in the table for depths on either side of 0.467 m.

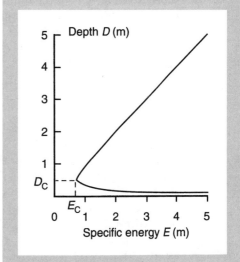

Figure 8.15 Depth–specific energy curve for Table 8.4

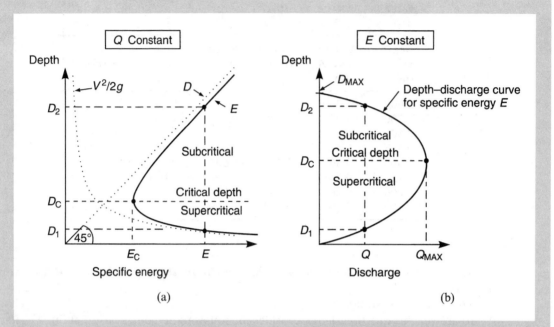

Figure 8.16 (a) A general depth–specific energy graph (Q constant) showing the components: static energy, D, and kinetic energy, $V^2/2g$. (b) The depth–discharge curve (E constant)

Thirdly, it is apparent from the table that the minimum specific energy corresponds to a Froude number of 1.00, which represents the 'dividing line' between subcritical and supercritical flow. Thus the minimum specific energy for a particular discharge occurs at the **critical depth**, D_C. For other discharges, there would be a different minimum value of specific energy and a different critical depth. Note that the critical depth does not have to be found by trial and error, it can be calculated more directly using equations like (8.26) or (8.32).

Figure 8.16a represents a more general form of Fig. 8.15 and shows the two components of specific energy, the static energy, D, and the kinetic energy, $V^2/2g$. The specific energy curve, E, is obtained by adding the horizontal values of D and $V^2/2g$ on the graph. A family of such curves can be drawn, each representing a different discharge.

To a large extent Fig. 8.16a summarises much of what has been discussed above. With steady, uniform flow in a channel of constant width, roughness and bed slope (S_O) then the depth (D) and mean velocity (V) are constant so the specific energy, $E = D + V^2/2g$, is constant along the channel. For this and any other value of E (except E_C which corresponds to critical depth) there are two alternate depths of flow: D_1 and D_2. If S_O exceeds the bed slope required for flow at the critical depth (i.e. S_C) then the flow is supercritical and the depth will be D_1. If $S_O < S_C$ the flow will be subcritical and the depth will be D_2. Of course, this assumes that the physical characteristics of the channel do not vary. If they did, then the D–E curve in Fig. 8.16a maps the change in depth that would result. For example, suppose the bed slope of the channel changes from $S_O < S_C$ to $S_O > S_C$ as in Fig. 8.17b. Initially the flow is at the normal depth calculated from the Manning equation with $D_N = D_2$. As the flow approaches the increase in bed slope it accelerates and passes through critical depth, which is represented by a movement from D_2 to D_C on the curve in Fig 8.16a. Once the flow is established on the steeper gradient it has the supercritical depth D_1. This is represented by the movement from D_C to D_1 on the curve. Note that Q is still constant, but the non-uniform channel condition causes E to vary, as in Table 8.4. Also, the movement on the curve from D_C to D_1 requires an increase in energy, which is provided by the increase in bed slope.

The diagram also indicates another difference between subcritical and supercritical flow. If there is a loss of energy in subcritical flow so that E decreases, then the D–E curve indicates that the depth reduces. For instance, the water surface is drawn-down as flow passes through a bridge opening or over a weir. In supercritical flow, the diagram indicates that the opposite happens: a decrease in E results in an increase in depth (Fig. 8.17c). Thus subcritical and supercritical flow behave differently, and it is well for the engineer to know which type is being dealt with.

Figure 8.16b provides another means of visualising the Q–D–E relationship. This time E is held constant, and the D–Q relationship is plotted. Again, the parts of the diagram representing subcritical and supercritical flow are separated by the critical depth line. Again, for the discharge Q, the alternate depths are D_1 and D_2. However, the important thing to note is that for a particular value of E, the maximum discharge (i.e. Q_{MAX}) is obtained at the

---

Box 8.1 ▶**Remember**

REMEMBER!!!

1. Table 8.4 shows that critical depth corresponds to F = 1, which is the 'border line' between subcritical and supercritical flow. At critical depth $V_c^2/2g = 0.5D_C$ so $E_C = 1.5D_C$ (see section 8.9.1).

2. Figures 8.16a and 8.16b show that for every value of specific energy other than E_C, which corresponds to the critical depth condition, there are two possible depths of flow, one smaller than the critical depth (supercritical) and one larger than the critical depth (subcritical). These **alternate depths** are marked on the diagrams as D_1 and D_2.

3. The minimum specific energy for a particular discharge occurs at the critical depth, D_C (Fig. 8.16a). Put another way, the critical depth condition represents the minimum specific energy at which a particular discharge can be maintained. Therefore, *the critical depth could be said to represent the most efficient flow condition and to give the maximum possible discharge for a given specific energy* (Fig. 8.16b).

4. The depth–specific energy curve in Fig. 8.16a shows that:
   in subcritical flow a loss of energy results in a *decrease* in the depth of flow
   in supercritical flow a loss of energy results in an *increase* in the depth of flow.
   This is illustrated by Fig. 8.17c which shows what happens in subcritical and supercritical flow when there is a rise in bed level. The same effect is apparent if the width of the channel is reduced, or a loss of energy is induced by some other means.

5. At or near the critical depth, a small change in specific energy results in a relatively large change in depth of flow. Thus the flow tends to be unstable and constantly changing. Normally when designing a channel a stable depth–discharge relationship is required, so this part of the curve should be avoided.

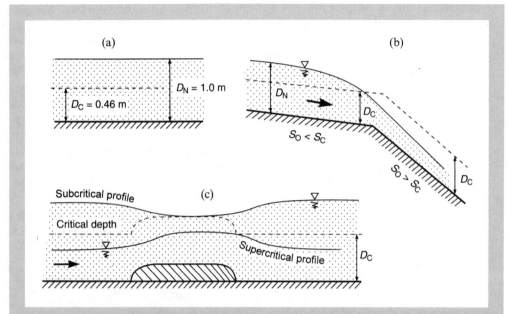

Figure 8.17 Use of the critical depth line to indicate the type of flow (a) in a uniform channel (b) in a channel which gets steeper, and (c) where there is a locally raised bed. The critical depth line is dashed, the actual water surface is shown by a solid line. If the actual water surface falls below the critical depth line, this indicates supercritical flow

critical depth D_C. Thus the critical flow condition is often considered to be the most efficient, although not necessarily the most desirable since a small change in E can result in significant fluctuations in depth (Fig. 8.16a). Also, the high velocities associated with critical and supercritical flow can cause serious erosion of the channel, if it is not reinforced.

Box 8.1 gives a summary of the important points to remember in connection with Fig. 8.16.

8.6.2 Using critical depth

For the particular channel and discharge in Table 8.4 it was found that the critical depth, D_C, was 0.467 m. **How do we use this value?**

Well, in the example above the critical depth, D_C, was 0.467 m. This can be represented by a dashed line drawn above the base of the channel, as in Fig. 8.17a. ***This does not signify that the flow is actually at the critical depth***, it merely acts as a reference level. If the actual depth of flow in the channel is 1.0 m, this too can be represented as a line on the longitudinal section (Fig. 8.17a). Since the actual depth is ***above*** the critical depth line (1.0 m > 0.467 m) this automatically tells us that the flow is subcritical. Similarly, if the flow had been supercritical, then the line representing the actual depth would have plotted below the critical depth line. For example, in Table 8.4 the alternate depth to 1.0 m is 0.253 m (for the same specific energy) and since this is lower than the critical depth line of 0.467 m this automatically indicates supercritical flow (0.253 m < 0.467 m). ***Thus whether the flow is sub-***

critical or supercritical can be determined simply by comparing the actual depth with the critical depth, D_C. The same logic applies to the critical veloc- ity, which is the mean velocity of flow at the critical depth.

Another important point to remember is that the critical depth depends upon the discharge and the geometry of the channel. This can be illustrated by borrowing (from later in the chapter) an equation for calculating the critical depth in a rectangular channel. This is:

$$D_C = (Q^2/gB^2)^{1/3} \tag{8.32}$$

The equation illustrates clearly that the *critical depth increases with increasing discharge*. However, for a constant Q the *critical depth increases with decreasing width*. A good example of this is a bridge waterway opening that is narrower than the channel (in plan) so that the flow has to contract to pass through it. With subcritical flow, when water passes through the opening the velocity increases because of the reduced cross-sectional area, so the elevation of the water surface falls as the flow accelerates and loses energy (Fig. 8.18a). On the other hand, the decreased width of flow means that the critical depth, D_C, is larger in the bridge opening than in the other parts of the channel (equation (8.32)). This increases the likelihood of the water surface passing through the critical depth, with the result that critical or supercritical flow conditions occur either in the bridge waterway opening or just downstream. The significance of this is explained below in section 8.6.3.

If the flow approaching a channel constriction such as a bridge opening is supercritical to begin with, then the effect of the reduction in width is almost the opposite of that experienced in subcritical flow (Fig. 8.18b). In this case the bridge will cause a reduction in

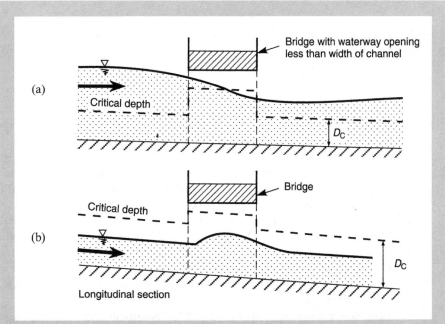

Figure 8.18 Longitudinal section showing the effect of a width constriction caused by a narrow bridge waterway opening on (a) subcritical and (b) supercritical flow

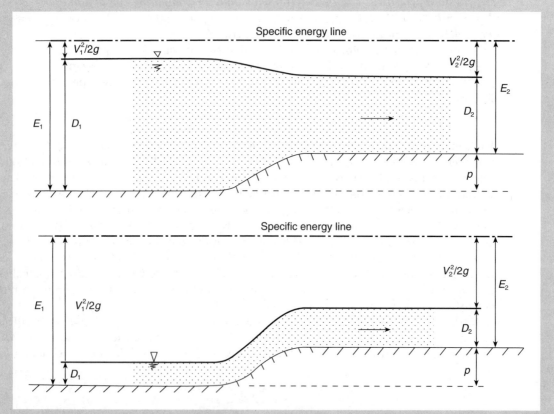

Figure 8.19 Flow over a raised bed hump where the flow is intitially (a) subcritical and (b) supercritical

velocity and a loss of energy which, because of the shape of the specific energy curve (Fig. 8.16a), results in an increase in depth. Thus the actual water surface rises and may approach or cut the critical depth line so that the flow becomes subcritical.

Another example of a variation in channel geometry affecting the flow is a sudden rise in bed level (Fig. 8.19). Because specific energy (E) is calculated above bed level, an increase in bed level of p results in E being reduced by this amount. In subcritical flow (diagram a) the initial specific energy is E_1 and the depth D_1. On the raised portion the equivalent values are E_2 and D_2 where $E_2 = E_1 - p$. As p increases, the depth of flow D_2 decreases until the condition becomes critical with $D_2 = D_C$ and $E_2 = E_C$. If p is increased further, the depth remains critical but the upstream water level increases in order to retain critical energy on the raised portion. In other words, the flow backs up until $E_1 - p = E_C$. Note that the raised bed results in a reduced cross-sectional area of flow, consequently $D_2 < D_1$ and $V_2 > V_1$ so the water surface is drawn-down.

If the flow approaching the raised portion is initially supercritical, the specific energy is again reduced with $E_2 = E_1 - p$. Figure 8.16b shows that in supercritical flow, as E decreases D increases, hence $D_2 > D_1$ and $V_2 < V_1$. As p increases, D_2 increases until conditions become critical (i.e. E_C, D_C). If p is increased again, then the upstream flow becomes subcritical so that critical conditions are regained at section 2.

Sometimes either a reduction in the width of the channel (like the throated flume in section 9.6) or an increase in the elevation of the bed (like the broad crested weir in section 9.5) may be used to induce deliberately critical flow.

Why would we want to do that?

One reason is that the critical depth represents the maximum possible discharge for a given specific energy, so it could be said to be the most efficient flow condition. Sometimes the presence of supercritical flow in a bridge opening can be used to optimise the discharge for the upstream water level (see the next section). Another reason is that it is relatively easy to calculate the critical depth (D_C) and the critical velocity (V_C), so if we can make the flow pass through the critical depth the discharge can be obtained easily from $Q = BD_CV_C$. Such devices are often called critical depth meters, an example of which is the broad crested weir. Thus we are using critical depth to our advantage.

8.6.3 Understanding the practical difference between subcritical and supercritical flow

Perhaps the best way to describe this is in terms of a laboratory experiment that you may be able to try for yourself. The experiment utilises a model sluice gate (see Fig. 1.18) in a rectangular channel which is supplied with water at a constant rate by a pump. At the downstream end of the channel is a tail-gate which can be raised or lowered to control the depth of flow.

First, let us consider what happens with a large opening under the sluice gate and subcritical flow throughout. The water profile initially is as shown by the solid line in Fig. 8.20a. If the tail-gate is raised, this will increase the depth of water between the tail-gate and the

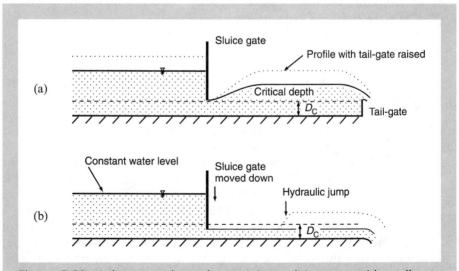

Figure 8.20 Laboratory channel containing a sluice gate, with a tail-gate at the end to control the water depth: (a) with subcritical flow, and (b) with supercritical flow. The solid line shows the initial water surface profile, and the dotted line the profile when the tail-gate is raised

sluice gate, which will in turn raise the water level upstream of the sluice gate. This is shown by the dotted line in Fig. 8.20a. Thus the downstream condition (that is the tail-gate) controls the depth of flow upstream. This is sometimes called the 'backwater' condition, which is the most common condition in rivers and man-made channels. To calculate the profile of the water surface in the channel, you have to start at the downstream control point and work back upstream.

Now, if we drop the tail-gate completely and then progressively lower the sluice gate and/or increase the discharge, the velocity of flow under the gate will get larger and larger, and eventually supercritical flow will occur immediately downstream of the sluice gate. In a relatively short, smooth laboratory channel at a reasonable slope the flow will remain supercritical to the end, as shown by the solid line in Fig. 8.20b. However, if the tail-gate is gradually raised, eventually a sharp increase in depth will occur just upstream of it. This is because the tail-gate is obstructing the flow, forcing it to become subcritical again. The transition from supercritical to subcritical flow takes the form of a steep wave, which after a moment or two will remain stationary. This is called a **standing wave** or **hydraulic jump**. If the tail-gate is raised even further, the hydraulic jump will move progressively upstream towards the sluice gate. An intermediate position is shown by the dotted line in Fig. 8.20b. However, the most significant point here is that provided the hydraulic jump has not reached the sluice gate, the depth of water upstream of the sluice gate will remain constant and will be totally unaffected by what is happening downstream of it. Thus with supercritical flow the control is provided by the structure itself and the upstream conditions. *This is the complete opposite of what happened in subcritical flow where the downstream conditions provided the control.* Consequently when analysing a hydraulic problem, one of the first steps must be to determine whether the flow is critical or supercritical anywhere in the length of channel concerned. Getting this first step wrong (REMEMBER!!!) can result in a totally invalid analysis.

❝ When the flow is supercritical, why is the depth upstream of the sluice gate unaffected by the conditions in the downstream part of the channel?
I do not understand this. Explain please. ❞

To answer this, let us consider more closely what the Froude number actually represents. We will approach the problem by means of an analogy which, although not exact, does illustrate most of the concepts involved. The analogy is between subsonic and supersonic aeroplanes and the subcritical or supercritical flow of water in an open channel.

Probably just about everyone knows that subsonic aeroplanes fly below the speed of sound, while supersonic aircraft fly above the speed of sound. The Mach number (Ma) is a dimensionless ratio that can be thought of as the ratio of the aircraft's velocity to the speed of sound in still air. If Ma < 1 the aircraft is flying subsonically, if Ma > 1 it is supersonic.

Now consider this. If a slow flying subsonic aeroplane drops a bomb, some time later the bomber crew will hear the noise of the explosion as they are overtaken by the sound (pressure) wave travelling at the speed of sound. However, if a supersonic aircraft flying through stationary air drops a bomb while continuing to fly in a straight line, the crew of the bomber will not hear the sound of the explosion. Because they are travelling faster than the speed of sound, the pressure wave that represents the noise of the bomb exploding can never catch up to them. We can use the idea of relative velocity to reverse the motion, so the air is now moving supersonically while the bomber is stationary some distance directly upstream from the point where the bomb explodes. Since this is the same situation as before but with the motion reversed, it follows that the bomber crew still cannot hear the explosion. This is because the air is travelling downstream faster than the sound wave can travel upstream, so the noise of the explosion is swept away downstream and does not reach anyone upstream. ***Thus at supersonic speeds a disturbance cannot travel upstream***. However, the disturbance can move with the flow, so if the stationary bomber was downstream of the explosion the noise could be heard.

With water in an open channel, subcritical flow occurs when F < 1 and supercritical flow when F > 1. The Froude number is the dimensionless ratio, $F = V/(gD)^{1/2}$. Here V is the mean velocity of the water, and $(gD)^{1/2}$ is the velocity of propagation of a gravity (pressure) wave in water. Thus the Froude number is similar to the Mach number, only it relates to water not air (see section 8.1.2).

Now suppose water flows down an open channel with a velocity, V, and someone drops a large boulder into the channel, causing gravity water waves to spread out from the source of the disturbance with a velocity $(gD)^{1/2}$. In subcritical flow $(gD)^{1/2} > V$ so a gravity wave can move upstream and affect the water level there (the bomber hears the explosion), as well as affecting the downstream conditions. In supercritical flow, $(gD)^{1/2} < V$, that is the wave speed is less than that of the water so the wave cannot travel upstream: all the ripples are swept downstream. Thus the conditions downstream are affected, but not those upstream (the bomber cannot hear the explosion). So, ***in supercritical flow a disturbance cannot propagate upstream***. This confirms the observations made in Fig. 8.20.

REMEMBER!!!

As explained earlier, when analysing a problem one of the first steps must be to determine whether the flow is subcritical or supercritical since this determines where the control point is and what equations should be used. Additionally, with supercritical flow the design of curved or non-prismatic channels is not straightforward since complex cross waves may occur which interfere with each other and form a disturbance pattern (see Chow, 1981). Thus it pays to be sure what type of flow is to be dealt with. One easy way to do this is to compare the actual depth of flow in the channel with the critical depth, so the ability to calculate the critical depth for any particular channel and discharge is vitally important.

8.7 ◆ Calculation of the critical flow conditions in any channel

❝ So how do we calculate the critical depth and critical velocity? Do we have to use a trial-and-error procedure as in Table 8.4? ❞

No, it can be calculated from an ***appropriate*** equation. The important point is that some equations can be applied to channels of any shape, including those with an irregular cross-section, while other equations have been specially derived for (say) rectangular channels. Take care not to apply a particular equation to the wrong channel shape. We will start with the general equation for critical depth that can be applied to any channel.

8.7.1 Critical depth

Consider the irregularly shaped channel in Fig. 8.21. Assuming $\alpha = 1.0$, then for any depth of flow, D, and discharge, Q, the specific energy is given by equation (8.23) as:

$$E = \frac{Q^2}{2gA^2} + D \qquad (8.23)$$

Now for a particular value of Q, when the flow is at critical depth the specific energy is at a minimum (Fig. 8.16a). This is expressed mathematically as $dE/dD = 0$. So if equation (8.23) is differentiated with respect to D and equated to zero, we obtain an expression that relates to the critical depth condition. To differentiate $Q^2/2gA^2$ we have to use the chain rule since it contains the variable A not D. For this particular problem the chain rule can be expressed as:

$$\frac{dE}{dD} = \frac{dE}{dA} \times \frac{dA}{dD} \qquad (8.24)$$

So the differentiation of equation (8.23) becomes:

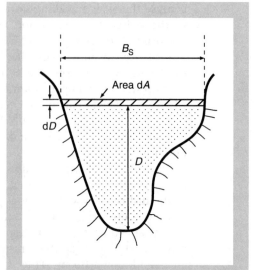

Figure 8.21 An irregular channel with no specific shape and cross-sectional area, A, when the discharge is Q

$$\frac{dE}{dD} = \frac{Q^2}{2g} \times \frac{1}{A^2} \frac{dE}{dA} \times \frac{dA}{dD} + D\frac{dE}{dD}$$

$$0 = \frac{Q^2}{2g} \times \frac{(-2)}{A^3} \times \frac{dA}{dD} + 1 \qquad (1)$$

Now if dD represents an infinitely thin strip of the cross-sectional area at the water surface, then from Fig. 8.21 it is apparent that a change in depth dD will cause a change in area $dA = B_S dD$. Thus $dA/dD = B_S$. Making this substitution in equation (1) above and rearranging gives:

$$\frac{Q^2 B_{SC}}{gA_C^3} = 1 \qquad (8.25)$$

This the general equation that must be satisfied, irrespective of the shape of the channel, if the flow is to be **at the critical depth**. To emphasise the point, B_S and A in equation (8.25) have been written with a subscript 'c' to indicate that they are the surface breadth, B_{SC}, and cross-sectional area, A_C, of the channel that correspond to the critical depth, D_C, at the particular discharge, Q. If it is necessary to include α in equation (8.25), then $\alpha Q^2 B_{SC}/gA_C^3 = 1$.

Of course, with an irregular channel like that in Fig. 8.21, the flow does not have a constant depth. It varies from zero at the sides to a maximum somewhere near the middle. Consequently it is convenient to introduce (from equation (8.3)) the **hydraulic mean critical depth**, $D_{MC} = A_C/B_{SC}$ to represent the average depth of flow in the channel. Thus equation (8.25) becomes:

$$\frac{Q^2}{gA_C^2} \times \frac{B_{SC}}{A_C} = 1$$

$$D_{MC} = Q^2/gA_C^2 \qquad (8.26)$$

With irregular natural channels that have no specific shape, equations (8.25) and (8.26) cannot be solved directly for the critical depth because it is impossible to evaluate A_C when D_{MC} is unknown. A trial-and-error solution must be adopted, although in many cases charts or tables have been devised to give an answer. However, equations (8.25) and (8.26) can be used to confirm the existence (or otherwise) of critical conditions, as shown in Example 8.10, while with prismatic channels that have a simple geometrical shape they can be solved more easily.

A useful way to remember these relationships is simply to say that with critical flow F = 1, i.e. $V_C/(gD_{MC})^{1/2} = 1$. Squaring both sides gives $V_C^2/(gD_{MC}) = 1$ where, as above, $D_{MC} = A_C/B_{SC}$, so substituting gives $V_C^2 B_{SC}/gA_C = 1$. Since $V_C = Q/A_C$ then this again becomes equation (8.25): $Q^2 B_{SC}/gA_C^3 = 1$. This time the equation was obtained using little more than the Froude number. It is often written as:

$$Q = \sqrt{\frac{gA_C^3}{B_{SC}}}$$

(8.27)

8.7.2 Critical velocity

The mean velocity of the flow at the critical depth is called the **critical velocity**, V_C. This can be calculated using equation (8.27).

From the continuity equation, $V_C = \dfrac{Q}{A_C} = \sqrt{\dfrac{gA_C^{\,3}}{A_C^{\,2}B_{SC}}} = \sqrt{\dfrac{gA_C}{B_{SC}}} = \sqrt{gD_{MC}}$

$$V_C = (gD_{MC})^{1/2}$$

(8.28)

By rearranging this equation we get:

$$V_C \big/ (gD_{MC})^{1/2} = 1$$

(8.29)

This is the Froude number (equation (8.4)) written for the special case of flow at the critical depth. Under any other circumstances the right-hand side of equation (8.29) would not equal 1, but some other value. As indicated in section 8.1.2, a Froude number less than 1.0 represents subcritical flow, and a number greater than 1.0 supercritical flow.

EXAMPLE 8.10

Table 8.4 showed the variation of area, velocity and specific energy in a 1.0 m wide channel when carrying a discharge of 1.0 m³/s. Confirm the existence of critical flow conditions at a depth of 0.467 m (and at no other depth) using equations (8.25) to (8.28). Use the following data extracted from Table 8.4:

Depth, D	Area of flow, A	V = Q/A
5.000 m	5.000 m²	0.200 m/s
0.467	0.467	2.140
0.200	0.200	5.000

Equation (8.25) will be applied first. For critical flow the value obtained must be 1.00.
When the depth D = 5.000 m, then $Q^2 B/gA^3 = (1.0^2 \times 1.0)/(9.81 \times 5.000^3) = 0.00082$
When the depth D = 0.467 m, then $Q^2 B/gA^3 = (1.0^2 \times 1.0)/(9.81 \times 0.467^3) = 1.00$
When the depth D = 0.200 m, then $Q^2 B/gA^3 = (1.0^2 \times 1.0)/(9.81 \times 0.2^3) = 12.74$
Thus the three depths correspond to subcritical, critical and supercritical flow.

Now applying equations (8.26) and (8.28):

$$D_{MC} = Q^2/gA_c^2 = 1.0^2/(9.81 \times 0.467^2) = 0.467 m$$
$$V_C = (gD_{MC})^{1/2} = (9.81 \times 0.467)^{1/2} = 2.140 m/s$$

This, again, confirms the values calculated in Table 8.4.

EXAMPLE 8.11

Water flows down a channel of triangular cross-section as shown. The maximum depth of flow on the centre line is 1.23 m when the discharge is 2.144 m³/s. It is thought that this represents the critical flow condition. Confirm whether or not this is so, and if it is calculate the mean critical depth and the critical velocity in the channel.

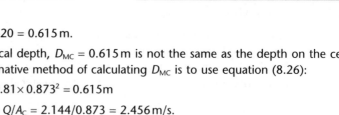

Figure 8.22

$$B_S = 2 \times \tan 30° \times 1.23 = 1.420 m.$$
$$A = 1/2 \times 1.420 \times 1.230 = 0.873 m^2$$

Using equation (8.25):

$$Q^2 B_S/gA^3 = (2.144^2 \times 1.420)/(9.81 \times 0.873^3) = 1$$

Therefore the flow is critical, and B_S and A above are the critical values B_{SC} and A_C.

The mean critical depth,
$$D_{MC} = A_C/B_{SC} = 0.873/1.420 = 0.615 m.$$

Note that the mean critical depth, $D_{MC} = 0.615 m$ is not the same as the depth on the centreline (1.23 m). An alternative method of calculating D_{MC} is to use equation (8.26):

$$D_{MC} = Q^2/gA_C^2 = 2.144^2/9.81 \times 0.873^2 = 0.615 m$$

The critical velocity, $V_C = Q/A_C = 2.144/0.873 = 2.456 m/s$.
Note that $V_C = (gD_{MC})^{1/2} = (9.81 \times 0.615)^{1/2} = 2.456 m/s$ gives exactly the same result.

EXAMPLE 8.12

An irregularly shaped channel is known to be operating at the critical depth. Under these conditions the area of the flow has been measured on site as 3.7 m² and the width of the water surface is 5.3 m. Calculate the discharge.

There are at least two ways of doing this:

Either
$$Q = \sqrt{\frac{gA_c^3}{B_{SC}}}$$
$$= \sqrt{\frac{9.81 \times 3.7^3}{5.3}}$$
$$Q = 9.683 m^3/s$$

or
$$D_{MC} = A_C/B_{SC} = 3.7/5.3 = 0.698 m$$
$$V_C = (gD_{MC})^{1/2}$$
$$= (9.81 \times 0.698)^{1/2} = 2.617 m/s$$
$$Q = A_C V_C = 3.7 \times 2.617$$
$$= 9.683 m^3/s$$

EXAMPLE 8.13

A channel of no definite shape has a mean depth of flow, D_M, of 0.81 m and the mean velocity, V, is 2.97 m/s. Is the flow subcritical or supercritical?

$$F = V/(gD_M)^{1/2} = 2.97/(9.81 \times 0.81)^{1/2} = 1.05$$

The flow is weakly supercritical and is very close to the critical depth condition (F = 1.00). However, in irregular channels it should be appreciated that this calculation may not be very reliable, partly as a result of the variation in velocity across the area of flow. Remember that α has been omitted in the above analysis.

8.7.3 Critical slope

The bed slope of a channel that results in uniform flow at the critical depth is called the **critical slope**, S_C. An expression for the critical slope can be obtained from the Manning equation:

$$V = (1/n)R^{2/3}S_O^{1/2} \tag{8.8}$$

by giving the variables the values corresponding to critical flow conditions. These are critical velocity V_C, mean critical depth, D_{MC}, hydraulic radius R_C, and bed slope, S_C. Thus:

$$V_C = (1/n)R_C^{2/3}S_C^{1/2}$$
$$S_C^{1/2} = V_C n/R_C^{2/3}$$
$$S_C = V_C^2 n^2/R_C^{4/3} \tag{8.30}$$

where $V_C^2 = gD_{MC}$ from equation (8.28). If the actual bed slope $S_O < S_C$ then this is called a **mild slope** and the flow must be subcritical. If the actual bed slope $S_O > S_C$ this is referred to as a **steep slope** and the flow is supercritical. If $S_O = S_C$ then the flow is critical. This provides another means of identifying the type of flow occurring in an open channel.

EXAMPLE 8.14

For the same triangular channel and conditions as in Example 8.11, determine the bed slope that would just maintain the flow at the critical depth. Assume the channel is lined with an n value of 0.015 s/m$^{1/3}$.

Using equation (8.30) with $V_C^2 = gD_{MC}$ then: $S_C = gD_{MC}n^2/R_C^{4/3}$
Now $R_C = A_C/P_C$
    where the wetted perimeter at the critical depth, $P_C = 2 \times$ wetted side slope.
    The wetted side slope = 1.230/cos 30° = 1.230/0.866 = 1.420 m
    so $P_C = 2 \times 1.420 = 2.840$ m
$R_C = A_C/P_C = 0.873/2.840 = 0.307$ m.
Therefore, $S_C = 9.81 \times 0.615 \times 0.015^2/0.307^{4/3}$
           = 0.00655 or 1 in 153

A good habit to get into in practice is to check that all the calculated values agree with each other. For instance, use the calculated values to check that they give the discharge stipulated in the original equation, thus:

$$Q = (A_C/n)R_C^{2/3}S_C^{1/2}$$
$$= (0.873/0.015) \times 0.307^{2/3} \times 0.00655^{1/2}$$
$$= 2.143 \text{m}^3/\text{s} \quad \text{OK}$$

8.8 ▶ Calculation of the critical flow in a trapezoidal channel

Equations (8.25) and (8.26), which can be used to calculate the critical flow conditions in a channel of any shape, include the hydraulic mean critical depth, D_{MC}. With a trapezoidal channel it must be remembered that D_{MC} is not the same as D_C, the actual physical depth of water on the centreline of the channel when the flow is critical because the depth varies across the width of a trapezoidal channel. Thus D_C must be used to calculate A_C and B_{SC}, and then $D_{MC} = A_C/B_{SC}$. However, in most situations we do not know the value of D_C, so we cannot obtain a direct solution to equation (8.25) or (8.26). Instead we must use a trial-and-error procedure. This is quite straightforward, as shown below and in Example 8.15.

8.8.1 Critical depth

On the centreline of the trapezoidal channel in Fig. 8.23 the depth is the critical depth, D_C. The side slopes of the channel are $1:S$, and the bottom width is B. The width of the water surface, B_{SC}, is:

$$B_{SC} = (B + 2SD_C) \tag{8.13}$$

$$A_C = (B + SD_C)D_C \tag{8.14}$$

$$D_{MC} = A_C/B_{SC}$$
$$= (B + SD_C)D_C/(B + 2SD_C) \tag{1}$$

If the discharge in the channel is Q, then the critical velocity, V_C, is given by:

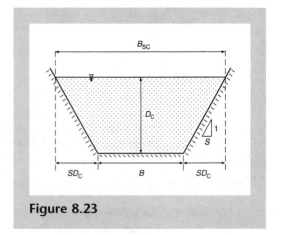

Figure 8.23

$$V_C = Q/A_C$$
$$= Q/(B + SD_C)D_C \tag{2}$$

Now equation (8.28) is $V_C = (gD_{MC})^{1/2}$ which can be rearranged to give $D_{MC} = V_C^2/g$. Substituting for V_C from equation (2) gives:

$$D_{MC} = Q^2/g[(B + SD_C)D_C]^2 \tag{3}$$

Now D_{MC} is defined by both equations (1) and (3). Equating the two expressions gives:

$$\frac{(B + SD_C)D_C}{(B + 2SD_C)} = \frac{Q^2}{g[(B + SD_C)D_C]^2}$$

$$[(B + SD_C)D_C]^3 = (Q^2/g)(B + 2SD_C) \tag{8.31}$$

Example 8.15 shows how this is solved by trial and error. Study it, then try Self Test Question 8.4.

8.8.2 Critical velocity and critical slope

Equations (8.28) and (8.30) are unchanged when applied to a trapezoidal channel.

EXAMPLE 8.15

A trapezoidal channel with a bottom width, B, of 5.0 m and side slopes of 1 in 2 carries a flow of 13.0 m³/s. When the conditions are critical, calculate the actual depth of flow on the channel centreline, the mean critical depth, and the critical velocity.

$$[(B + SD_C)D_C]^3 = (Q^2/g)(B + 2SD_C) \tag{8.31}$$

$$[(5.0 + 2D_C)D_C]^3 = (13.0^2/9.81)(5.0 + 2 \times 2D_C)$$

$$[(5.0 + 2D_C)D_C]^3 = 17.227(5.0 + 4D_C)$$

Now solving by trial and error. Guess a value of D_C and see if the left-hand (LH) side of the equation has the same numerical value as the right-hand (RH) side. A table helps to obtain a rapid solution.

Try D_C =	LH =	RH =
0.9 m	229.2	148.2
0.8	147.2	141.3
0.79	140.5	140.6
0.791	141.1	140.6

Therefore the actual centreline depth when the flow is critical is $D_C = 0.79$ m

The mean critical depth in the channel, $D_{MC} = A_C/B_{SC}$ where:

$$A_C = (B + SD_C)D_C = (5.0 + 2 \times 0.79)0.79 = 5.198 \text{ m}^2$$
$$B_{SC} = (B + 2SD_C) = (5.0 + 2 \times 2 \times 0.79) = 8.16 \text{ m}$$

Thus $D_{MC} = 5.198/8.16 = 0.637$ m
(Note that this is not the same as the centreline depth, $D_C = 0.79$ m.)

The critical velocity, $V_C = (gD_{MC})^{1/2} = (9.81 \times 0.637)^{1/2} = 2.50$ m/s

The accuracy of the answers above can be checked using any of the other equations.
Check: $V_C = Q/A_C = 13.0/5.198 = 2.50$ m/s   OK
Check: $D_{MC} = Q^2/gA_C^2 = 13.0^2/9.81 \times 5.198^2 = 0.637$ m   OK by equation (8.26).
Check: $Q^2B_{SC}/gA_C^3 = 13.0^2 \times 8.16/9.81 \times 5.198^3 = 1.00$ OK by equation (8.25).

SELF TEST QUESTION 8.4

Water flows down a trapezoidal channel at the rate of 35 m³/s. The bottom width of the channel is 9 m and the side slopes are 1 in 4. When the flow in the channel is critical, calculate:

(a) the actual critical depth, D_C, occurring in the channel,
(b) the mean critical depth, D_{MC}, and
(c) the critical velocity, V_C.

8.9 Calculation of the critical flow in a rectangular channel

The equations relating to the critical depth condition in a rectangular channel can be obtained by adapting the general equations derived above. In the case of a rectangular channel this results in a simplification of some equations because:

(a) the depth of flow in the channel is constant across its width, so $D = D_M$ and $D_C = D_{MC}$;

(b) the width, B, of the channel is constant at all depths of flow and $B = B_S = B_{SC}$;

(c) the area of flow, A, is easily calculated as $A = BD$.

8.9.1 Critical depth and critical velocity

Starting with equation (8.26): $D_{MC} = Q^2/gA_C^2$
For the critical depth condition, $A_C = BD_C$. Substitution into the equation above gives:

$$D_C = Q^2/gB^2D_C^2 \text{ and hence } D_C^3 = Q^2/gB^2$$

$$\boldsymbol{D_C = (Q^2/gB^2)^{1/3}} \tag{8.32}$$

If the discharge per metre width of the channel, $q = Q/B$, is used then equation (8.32) becomes:

$$\boldsymbol{D_C = (q^2/g)^{1/3}} \tag{8.33}$$

The general expression (equation (8.28)) for critical velocity, $V_C = (gD_{MC})^{1/2}$ also applies to a rectangular channel and can be written as $V_C = (gD_C)^{1/2}$ since the depth is constant across the channel. Furthermore, the specific energy in the channel at critical depth is:

$$E_C = V_C^2/2g + D_C \text{ where } V_C = (gD_C)^{1/2}$$
$$\text{thus } E_C = (gD_C)/2g + D_C$$
$$= D_C/2 + D_C$$

$$\boldsymbol{E_C = \frac{3}{2}D_C} \tag{8.34}$$

$$\text{or } \boldsymbol{D_C = \frac{2}{3}E_C} \tag{8.35}$$

The fact that the critical depth is two-thirds of the corresponding critical specific energy, E_C, in a rectangular channel provides another method of calculating D_C. Note, however, that this relationship only applies to the critical condition which represents the minimum on the specific energy curve in Fig. 8.16a. At other depths of flow, calculating two-thirds of the specific energy will not give the critical depth (see Example 8.16).

8.9.2 Critical slope

The calculation of the critical slope is exactly the same as before, with D_{MC} becoming D_C because of the constant depth across a rectangular channel.

EXAMPLE 8.16

Water flows down a rectangular channel that is 4.32 m wide at a depth of 2.34 m with a corresponding velocity of 0.97 m/s. Calculate the critical depth in the channel and determine whether the flow is subcritical or supercritical.

Correct solution

$D_C = (Q^2/gB^2)^{1/3}$

$Q = AV = 4.32 \times 2.34 \times 0.97 = 9.806\,m^3/s$

$D_C = (9.806^2/9.81 \times 4.32^2)^{1/3}$

$D_C = 0.807\,m$ (CORRECT)

Incorrect solution

$E = V^2/2g + D$

$E = (0.97^2/2 \times 9.81) + 2.34$

$E = 2.388\,m$

$D_C = \frac{2}{3}E_C = \frac{2}{3} \times 2.388 = 1.592\,m$

The flow in the channel is subcritical because the actual depth of flow (2.34 m) is greater than the critical depth (0.81 m). No other calculation is needed to prove the flow is subcritical, although this can be confirmed by calculating the Froude number:

$$F = V/(gD)^{1/2} = 0.97/(9.81 \times 2.34)^{1/2} = 0.20 \quad (<1.0 \text{ so subcritical})$$

The incorrect solution is wrong because $D_C = \frac{2}{3}E_C$ only applies to critical flow, and the conditions are not critical in this example. Take care not to mis-apply equations!

EXAMPLE 8.17

A rectangular channel is being designed to carry a flow of 12.5 m³/s. At one particular point the channel must pass under a busy railway line, and there is a problem with the vertical clearance. Because of economic considerations the width of the channel must be kept to a minimum, and because of physical restrictions a width of 3.0 m would be preferred. The channel is to be concrete lined ($n = 0.016\,s/m^{1/3}$). Calculate the dimensions and slope of a channel that will carry the required discharge most efficiently.

The greatest discharge for a given specific energy occurs at the critical depth, and this can be regarded as the optimum discharge condition (see Fig. 8.16 and Box 8.1). Although there are disadvantages associated with designing a channel to operate at critical depth, over a short distance this approach may yield a reasonable solution to the problem.

Assuming a rectangular cross-section:

$D_C = (Q^2/gB^2)^{1/3} = (12.5^2/9.81 \times 3.0^2)^{1/3} = 1.210\,m$

$V_C = (gD_C)^{1/2} = (9.81 \times 1.210)^{1/2} = 3.445\,m/s$

$A_C = 1.210 \times 3.0 = 3.630\,m^2$

$P_C = 3.0 + 2 \times 1.210 = 5.420\,m$

$R_C = A_C/P_C = 3.630/5.420 = 0.670\,m$

$S_C = gD_C n^2/R_C^{4/3} = 9.81 \times 1.210 \times 0.016^2/0.670^{4/3} = 0.00518 \text{ or } 1 \text{ in } 193$

Thus a 3.000 m wide channel operating at a depth of 1.210 m and having a slope of 1 in 193 would just carry the design discharge of 12.5 m³/s.

Check: $Q = (A/n)R^{2/3}S_0^{1/2} = (3.63/0.016) \times 0.67^{2/3} \times 0.00518^{1/2} = 12.50\,m^3/s$ OK

Other calculations would have to be conducted to determine the length of channel required to accelerate the flow from whatever depth (presumable subcritical) it had upstream of the railway line to the critical condition, and to determine what happens to the flow afterwards (see flow transitions).

Box 8.2 **Summary**

Remember that the type of flow can be determined by comparing any of the following variables:

	subcritical	critical	supercritical
Actual depth, D, with critical depth, D_C:	$D > D_C$	$D = D_C$	$D < D_C$
Actual flow area, A, with critical area, A_C:	$A > A_C$	$A = A_C$	$A < A_C$
Actual velocity, V, with critical velocity, V_C:	$V < V_C$	$V = V_C$	$V > V_C$
Actual Froude number with critical value:	$F < 1.0$	$F = 1.0$	$F > 1.0$
Actual bed slope, S_O, with critical slope, S_C:	$S_O < S_C$	$S_O = S_C$	$S_O > S_C$

8.10 Flow transitions

The flow in most natural open channels, and indeed most man-made channels, is subcritical. However, we have already seen briefly how the flow can become supercritical, and how it can revert back to subcritical again. The ability to predict what form these flow transitions will take is important to the hydraulic engineer, and can be an integral part of the design of dam spillways, weirs and other hydraulic structures. A knowledge of flow transitions is also needed to calculate the profile of the water surface.

First we will consider the transition from subcritical to supercritical flow, and then the more complex case of the transition from supercritical to subcritical.

8.10.1 The transition from subcritical to supercritical flow

For the flow to change from subcritical to supercritical the velocity of the flow has to increase. This can be caused by an increase in the bed slope (Fig. 8.17b), or a reduction in the area of flow (Fig. 8.18a). As can be seen from the diagrams, this is a relatively simple transition where the water surface gradually decreases in elevation with respect to the base of the channel until the flow attains the supercritical depth. As a rough guide, when there is a sudden increase in the bed slope the flow passes through critical depth at or near the change in gradient. Similarly, at a free overfall (like a vertical cliff) then the flow passes through critical depth (D_C) about 3 to $10D_C$ from the edge of the fall.

8.10.2 The transition from supercritical to subcritical and the hydraulic jump

Situations where the flow changes from supercritical to subcritical include the conditions downstream of a sluice gate (Fig. 8.20b) and where water flows over a steep dam spillway into a stilling basin (Fig. 8.24a). In all such cases a fast, shallow flow has to change to a slow, deep flow. Unlike the transition described above, this cannot happen smoothly and gradually. Instead there is a sudden increase in depth in the form of a **hydraulic jump** or **standing wave**. This zone of **rapidly varying flow** consists of highly turbulent water that

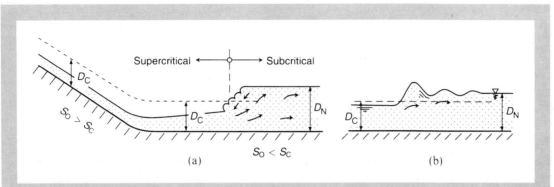

Figure 8.24 (a) A strong hydraulic jump at the toe of a dam spillway, and (b) an undular jump [*after Webber (1971)*]

froths and boils. The water on the wave front tends to have a motion that is directed downwards and back upstream, while the underlying flow is expanding upwards in a downstream direction. Hence there is great turbulence, a lot of air entrainment, and a considerable loss of energy. Downstream of the jump the water surface quickly becomes calm, and the flow continues smoothly at the subcritical normal depth.

It should be noted that the type of jump shown in Fig. 8.24a occurs when the upstream flow has a Froude number of 2 or more. At lower F values the jump tends to be either **undular** in nature (Fig. 8.24b) or weakly developed so that it is not always easy to determine by visual observation that there is a flow transition taking place. This is particularly true in natural rivers where the flow can be turbulent and the water surface undulates anyway.

So why does the jump form?

To answer this let us start at the upstream end of Fig. 8.24a and follow what happens to the flow. Initially the slope is steep ($S_O > S_C$) and the flow supercritical with a relatively high velocity. However, this velocity cannot be maintained when the flow enters the lower part of the channel that has a mild channel slope ($S_O < S_C$). The high velocity flow slows as a result of frictional resistance and the reduced bed gradient, the depth increases to maintain continuity of flow, and there is a loss of specific energy (ΔE in Fig. 8.25a). Note that in this part of the depth–specific energy curve a loss of energy results in an increase in depth (Fig. 8.25b). Thus the water surface rises gradually towards the critical depth line, but before it gets there the hydraulic jump occurs.

Why?

Well, the flow is ultimately going to become subcritical. We have seen how two alternate depths of flow can occur in the same channel at the same discharge, and how the specific energy varies with the depth of flow (section 8.6.1 and Table 8.4). So we know that the flow has to attain a point somewhere on the subcritical part of the depth–specific energy curve. However, if we follow the depth–specific energy curve in Fig. 8.25b from a starting point of supercritical flow with a depth D_1 and specific energy E_1 around towards critical depth, this describes what is happening to the flow as the depth gradually rises prior to the jump. Now we come to the problem. If the depth gradually increased all the way to the critical value, D_C, any further increase in depth could only occur if there was an increase in specific energy. This would require additional energy to be added to the flow, which cannot happen. Thus it is impossible for the flow to change smoothly from supercritical to subcritical, that is from

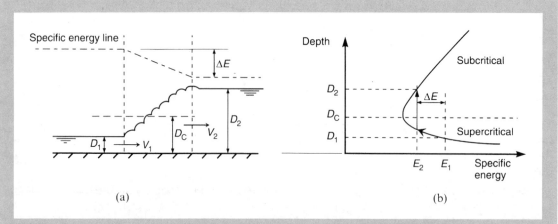

Figure 8.25 Initial depth D_1, sequent depth D_2 and energy loss ΔE: (a) at the hydraulic jump, and (b) on the depth–specific energy diagram [*after Webber (1971)*]

D_1 to D_2. (When the flow transition is from subcritical to supercritical as in Fig. 8.17b the increase in the channel slope provides the additional energy requirement.)

Having ruled out the possibility of a smooth transition from supercritical to subcritical, what happens is that the depth initially increases gradually from D_1 towards D_C, then before it attains this depth the hydraulic jump occurs at energy level E_2 with the flow switching to a depth D_2 on the upper, subcritical part of the depth–specific energy curve (Fig. 8.25b). As a result of the loss of energy, ΔE, in the jump, the **sequent** or **conjugate depth** of flow after the jump, D_2, is less than the alternate depth vertically above E_1 that would otherwise have occurred. This energy loss means that the hydraulic jump cannot be analysed simply with either the depth–specific energy curve or the energy equation. However, it can be analysed using the momentum equation.

For supercritical flow in a horizontal rectangular channel the application of the momentum equation gives a quadratic equation that enables the depths before and after a hydraulic jump to be calculated. If D_1 is the initial depth at which a jump will start and D_2 is the sequent depth after the jump:

$$D_1 = \frac{D_2}{2}\left(\sqrt{1+8F_2^2} - 1\right) \tag{8.36}$$

$$D_2 = \frac{D_1}{2}\left(\sqrt{1+8F_1^2} - 1\right) \tag{8.37}$$

where F_1 and F_2 are the Froude numbers of the flow before and after the jump. From this it can be shown that the energy loss at the jump is:

$$\Delta E = (D_2 - D_1)^3 / 4D_1 D_2 \tag{8.38}$$

The relative height of the jump, h_j/E_1, was given by Chow (1981) as:

$$\frac{h_j}{E_1} = \frac{\sqrt{1+8F_1^2} - 3}{F_1^2 + 2} \tag{8.39}$$

where h_j is the height of the jump, $(D_2 - D_1)$. The length of the jump, L_j, can be determined from experimental data. L_j varies with F_1, from about $L_j = 4D_2$ when $F_1 = 1.7$, to a maximum of $L_j = 6.15D_2$ when $F_1 = 7$, after which the length decreases (Table 8.5).

Table 8.5 Variation of the length of a hydraulic jump L_J with F_1

F_1	< 1.7	1.7	2.0	2.5	3.0	4.0	5.0	7.0	14.0	20.0
L_J	Undular jump	$4.0D_2$	$4.4D_2$	$4.8D_2$	$5.3D_2$	$5.8D_2$	$6.0D_2$	$6.2D_2$	$6.0D_2$	$5.5D_2$

Note: L_J is expressed in terms of the sequent depth D_2.

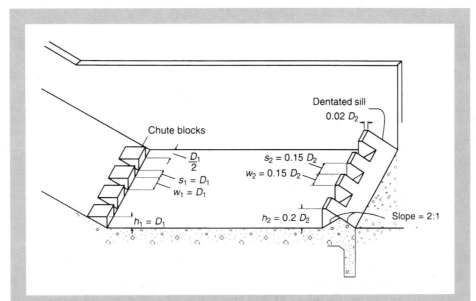

Figure 8.26 A typical stilling basin, the USBR basin II, suitable for $F_1 > 4.5$. The chute blocks lift the jet off the floor and aid jump development. The sill further reduces the length of jump and reduces scour. The diagram shows the height (*h*), width (*w*) and spacing (*s*) of the blocks [*Courtesy of United States Department of the Interior – Bureau of Reclamation*]

The nature of the hydraulic jump gives it considerable merit as an energy dissipator. In situations like the bottom of a dam spillway, it provides a good mean of reducing the large kinetic energy of the flow to a level that will not scour the downstream river channel. Consequently stilling basins are often designed to induce a hydraulic jump by obstructing the flow with blocks or something similar (Figs 8.26 and 9.4). This also ensures that the jump occurs in a controlled manner at a position determined by the designer, rather than allowing the jump to form naturally and having to reinforce a much longer length of channel. The most effective hydraulic jumps occur in the range $F_1 = 4.5$ to 9, with an acceptable performance up to $F_1 = 13$. After this suitable stilling basins become much more expensive and perform less well.

EXAMPLE 8.18

Water flows down a steep concrete lined rectangular channel 5.0 m wide at a depth of 0.65 m when the discharge is 19.0 m³/s. At the bottom of the slope the channel becomes horizontal, but is otherwise unchanged. Determine whether a hydraulic jump will form, the energy loss, and the approximate dimensions of the jump.

Considering the flow in the steep upstream channel, $D_1 = 0.65\,$m.
The upstream velocity, $V_1 = Q/A_1 = 19.0/(5.0 \times 0.65) = 5.846\,$m/s.
Upstream Froude number, $F_1 = V_1/(gD_1)^{1/2} = 5.846/(9.81 \times 0.65)^{1/2} = 2.32$
Therefore the flow is supercritical in the upstream channel. It is probable that a weak jump would form in the horizontal part of the channel. Assuming that the initial height of the jump is $D_1 = 0.65\,$m then equation (8.37) is:

$$D_2 = \frac{D_1}{2}\left(\sqrt{1+8F_1^2} - 1\right)$$

$$D_2 = \frac{0.65}{2}\left(\sqrt{1+8 \times 2.32^2} - 1\right) = 1.83\text{m}$$

Therefore the downstream depth would be $D_2 = 1.83\,$m, giving

$$V_2 = Q/BD_2 = 19.0/(5.0 \times 1.83) = 2.077\text{m/s}$$

Thus the height of the jump is $(1.83 - 0.65) = 1.18\,$m.

$$F_2 = V_2/(gD_2)^{1/2} = 2.077/(9.81 \times 1.83)^{1/2} = 0.49$$

The energy loss is defined by equation (8.38):

$$\Delta E = (D_2 - D_1)^3 / 4D_1 D_2$$
$$= (1.83 - 0.65)^3 / 4 \times 0.65 \times 1.83$$
$$= 0.35\text{m}$$

From Table 8.5 the length of the jump would be about $4.6 \times D_2 = 4.6 \times 1.83 = 8.4\,$m. Alternatively, the height of the jump can be calculated from equation (8.39):

$$h_J = \frac{E_1\left(\sqrt{1+8F_1^2} - 3\right)}{F_1^2 + 2} = \frac{E_1\left(\sqrt{1+8 \times 2.32^2} - 3\right)}{2.32^2 + 2} = 0.493E_1$$

Now $E_1 = D_1 + V_1^2/2g = 0.65 + 5.846^2/19.62 = 2.392\,$m.
So $h_J = 0.493E_1 = 0.493 \times 2.392 = 1.18\,$m.

Thus both equations give the height of the jump as about 1.18 m.
See Examples 8.21 and 8.22 which also illustrate the use of the equations.

◀8.11▶ Gradually varying non-uniform flow

As described in section 8.1.1, with gradually varying flow the depth (and hence the velocity) varies longitudinally along the channel. As a result the bed, the water surface and the total energy line are no longer parallel (Fig. 8.2b) so it is necessary to use the slope of the energy line in calculations. This line shows the rate at which energy head is lost overcoming frictional forces, and so is referred to as the friction gradient and symbolised by S_F. The importance of using S_F (instead of the bed slope, S_O) in the Manning equation is illustrated clearly by Example 8.19. Make sure you appreciate the significance of the example before continuing!

EXAMPLE 8.19

The Exwick flood relief channel is shown in Fig. 8.6, and Fig. 8.27 is a simplified longitudinal section. Assume the part of the channel that is full of water is trapezoidal in cross-section with a base width of 26.0 m and side slopes of 1:2 (vertical:horizontal). The average depth of flow is 1.2 m. The channel is concrete lined ($n = 0.015\,\text{s/m}^{1/3}$) and has a longitudinal bed slope of 1 in 1600. Calculate the discharge.

Incorrect solution assuming uniform flow and using S_O

Width of water surface $= (2 \times 1.2) + 26.0 + (2 \times 1.2) = 30.8\,\text{m}$

Cross-sectional area of flow $A = \frac{1}{2}(26.0 + 30.8) \times 1.2 = 34.08\,\text{m}$

Length of side slopes $= (1.2^2 + 2.4^2)^{1/2} = 2.68\,\text{m}$

Length of wetted perimeter $P = (2 \times 2.68) + 26.0 = 31.36\,\text{m}$

Hydraulic radius $R = A/P = 34.08/31.36 = 1.09\,\text{m}$

Using the Manning equation, $Q = \dfrac{A}{n}R^{2/3}S_O^{1/2} = \dfrac{34.08}{0.015} \times 1.09^{2/3} \times (1/1600)^{1/2} = 60.15\,\text{m}^3/\text{s}$

Thus the discharge in the channel is calculated at 60.15 m³/s when it is really zero!

Correct solution assuming non-uniform flow and using S_F

Assuming there is no flow into the channel at the upstream end, the water in the channel is trapped between the inflow and outflow weirs. Consequently the longitudinal gradient of the water surface is zero, because it is horizontal. With zero flow down the channel $V^2/2g = 0$ so the total energy line is also horizontal and $S_F = 0$. The discharge calculated from the Manning equation is now:

$$Q = \dfrac{A}{n}R^{2/3}S_F^{1/2} = 0\,\text{m}^3/\text{s}$$

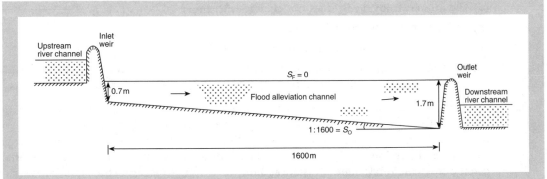

Figure 8.27 Diagrammatic longitudinal section of part of the Exeter flood relief channel. Note that the water is trapped between the inlet and outlet weirs, so $Q = 0$. Thus $S_F = 0$ must be used in the Manning equation, not $S_O = 1/1600$

Thus the assumption that $S_F = S_O$ must never be made (unless it can be justified), otherwise totally incorrect answers will be obtained.

Gradually varying non-uniform flow is common. Indeed, how many natural river channels meet the requirements of uniform flow where successive cross-sections must have exactly the same width, area, depth of flow, roughness, gradient and velocity? Only man-made channels like that in Fig. 8.1 come close; most natural rivers vary from section to section. However, gradually varied flow may occur in any type of channel if there is a change in its characteristics, such as a change in bed slope. Generally of most interest to engineers is the situation where gradually varying flow occurs as a result of an obstruction that raises the water surface above the uniform flow normal depth line (e.g. Fig. 8.32). Obstructions take many forms, such as fallen trees, bridge abutments and piers, weirs, sluice gates and dams. All of these may cause the water to 'back up', that is to rise above the normal depth upstream of the obstruction. Often the effect of an obstruction may extend for a surprisingly large distance, perhaps many kilometres. For example, if a river with a longitudinal gradient of 1 in 1000 has a barrage constructed across it that raises the water level by 5 m, then the ponded water extends upstream for 5 km. If a dam is constructed that raises the water level by 50 m, the stored water will extend 50 km. In fact, the distances may be considerably larger as a result of increased water levels in the transition between the flowing river and the ponded water. Therefore, engineers must not build anything that obstructs a river channel without first calculating by how much the water level will be raised at various distances upstream. For instance, suppose there is a proposal to build a new bridge in a city centre. If at high discharges the structure forms an obstruction, it is vitally important that the water level is not raised so much that river-side properties, shops and factories are flooded. If they are, the cost could be £100 million or more depending upon the circumstances, with the designer being sued for damages (see section 13.4). Consequently engineers must be able to calculate the **backwater curve**, that is the longitudinal profile of the water surface upstream of the obstruction. These curves do not include rapidly varying transitions, such as those described in section 8.10 or where the flow passes over a weir crest as in section 9.5.

Example 8.19 illustrated that when analysing gradually varying flow with the Manning equation you must use the friction gradient (i.e. S_F, the slope of the total energy line). In such an analysis it is necessary to make a number of assumptions:

(a) the flow is steady, i.e. it does not change with time;

(b) the streamlines are virtually parallel;

(c) the friction losses are the same as in uniform flow;

(d) the channel slope is relatively small, so that the depth of flow measured perpendicular to the bed is essentially the same as the depth measured in a vertical plane.

Clearly, gradually varying flow is slightly more complex than uniform flow. In the latter case, once the flow depth for a particular discharge has been calculated at one cross-section, the same depth occurs at all other sections. With gradually varying flow this is no longer the case: each section is different. With natural rivers a field survey can be used to establish the cross-sectional shape at distances along its length, to measure the elevation of water surface at a known discharge, and subsequently to calculate $\alpha V^2/2g$ and the elevation of the total energy line at a particular discharge. The change in elevation of this line between successive cross-sections gives the friction slope, S_F. For different discharges, such as floods, either the survey has to be repeated (difficult!) or the water surface profile calculated, as described later. However, at any cross-section of a channel (subscript i) the Manning equation can be used to calculate S_{Fi}, a fact which is used later to obtain the elevation of the water surface. The Manning equation is:

$$V_i = \frac{1}{n_i} R_i^{2/3} S_{Fi}^{1/2} \quad \text{so} \quad S_{Fi}^{1/2} = \frac{V_i n_i}{R_i^{2/3}}$$

$$S_{Fi} = \frac{V_i^2 n_i^2}{R_i^{4/3}} \tag{8.40}$$

8.11.1 Classification of gradually varying water surface profiles

With gradually varying flow the water surface is not parallel to the bed. The gradually varying flow equation gives the change in depth (D) with distance (L) along the channel:

$$\frac{dD}{dL} = \frac{S_O - S_F}{1 - F^2} \tag{8.41}$$

The derivation of this equation can be found in Appendix 1 as Proof 8.2. The equation shows that the rate of change of depth depends upon two things: the slope of the friction gradient (S_F) relative to the bed slope (S_O), and the Froude number (F). Note the following:

■ +ve values of dD/dL indicate increasing depths.

■ −ve values of dD/dL indicate decreasing depths.

■ With uniform flow $S_O = S_F$ so $dD/dL = 0$, that is the water surface is parallel to the bed.

■ With critical flow F = 1 so $dD/dL = \infty$, that is the water surface is perpendicular to the bed. This is not possible, although the hydraulic jump in Fig. 8.25a has a steep gradient (but not infinity) so it can be argued that the equation gives some indication of what is happening.

The gradually varied flow equation can be used to define and sketch twelve standard types of water surface profile, which are denoted by a combination of a letter and a number (e.g. M1, S3). There are five letters, which indicate the bed slope of the channel (S_O) relative to the critical slope (S_C):

M – mild bed slope with $S_O < S_C$

S – steep bed slope with $S_O > S_C$

C – critical bed slope with $S_O = S_C$

H – horizontal bed

A – adverse bed slope, i.e. sloping upwards in the direction of flow

Adverse bed slopes may sometimes be necessary in man-made channels, or they may occur naturally in rivers. Bridges often have a scour hole immediately downstream where the high velocity in the bridge waterway has eroded the bed, resulting in an adverse gradient for a short distance.

The number which is attached to the letter indicates in which of three vertical zones the surface profile is located. In most situations the normal depth (D_N) is larger than the critical depth (D_C), so drawing the normal depth line (N.D.L.), critical depth line (C.D.L.) and the channel bed gives the three zones, 1, 2, 3, shown in Fig. 8.28. These three numbers are used in combination with the letters above to identify the twelve standard surface water

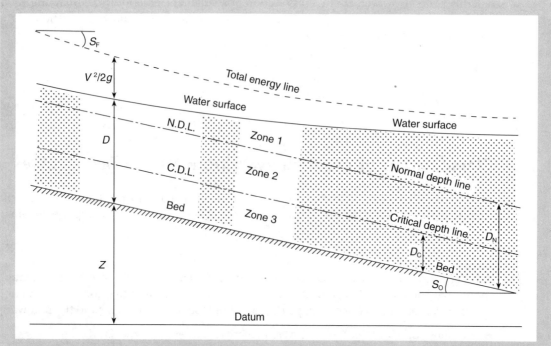

Figure 8.28 The use of the normal depth line, critical depth line and channel bed to denote the three zones which are used in conjuction with the channel slope to identify the water surface profile. Here $D > D_N > D_C$ so the water surface is in zone 1 and we have an M1 backwater curve (see Table 8.6)

profiles shown in Fig. 8.29 and Table 8.6. Zone 1 is above the normal depth line where sub-critical flow occurs, and all of the curves are of the backwater type. Zone 2 is the middle zone, where only drawdown curves are experienced. Zone 3 is closest to the bed where super-critical flow occurs, and the associated curves represent an increase in water level. Each of the three zones (1–3) have their own distinctive curves which are valid only within the limits of that zone. In the diagram, the surface profiles are shown dashed where the conditions required for gradually varying flow are violated, such as near a hydraulic jump. Each of the zone boundaries is approached in a characteristic way:

- Above normal depth ($D > D_N$), the backwater curve tends to become asymptotic to a horizontal water surface line, such as a reservoir surface; i.e. the gap between the two lines gradually decreases to zero as the distance travelled increases.
- The normal depth line is approached asymptotically.
- The critical depth line is approached at right angles.
- The channel bed is approached at right angles.

8.11.2 Control points

The section above described the standard water surface profiles and indicated where they are likely to occur, but for engineers it is essential that the actual elevation of the water

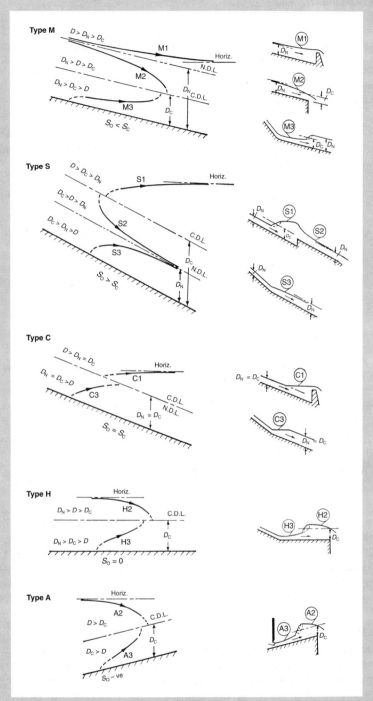

Figure 8.29 Diagrammatic illustration of the twelve types of water surface profile and where they may be typically found. The profile is denoted by the combination of the bed slope initial letter (Mild, Steep, Critical, Horizontal and Adverse) and the number of the zone in which the water surface occurs (1, 2, 3 as in Fig. 8.28). For clarity, the horizontal length is much compressed [*after Webber, 1971; reproduced by permission of Routledge*]

Table 8.6 Classification of gradually varied flow surface water profiles

Channel slope	Symbol	Depth and type of flow	Type of profile	$\frac{dD}{dL}$	Notes, typical location, causes
MILD $0 < S_O < S_C$	M1	$D > D_N > D_C$ Subcritical	Backwater	+ve	Very common: backwater from an obstruction raises upstream water level.
	M2	$D_N > D > D_C$ Subcritical	Drawdown	−ve	Drawdown curve: e.g. approach to a free overfall.
	M3	$D_N > D_C > D$ Supercritical	Backwater	+ve	Downstream of sluice gate, etc.; change from steep to mild slope.
STEEP $S_O > S_C > 0$	S1	$D > D_C > D_N$ Subcritical	Backwater	+ve	Profile starts with a hydraulic jump and ends at an obstruction or control structure.
	S2	$D_C > D > D_N$ Supercritical	Drawdown	−ve	Change from mild to steep slope; entrance to a steep channel.
	S3	$D_C > D_N > D$ Supercritical	Backwater	+ve	Downstream of sluice gate; change from steep to less steep slope.
CRITICAL $S_O = S_C$	C1	$D > D_N = D_C$ Subcritical	Backwater	+ve	Backwater from an obstruction; ends at obstruction.
	C2	$D_C = D = D_N$ Critical, uniform	Parallel to bed	0	Uniform flow at the critical depth in a channel.
	C3	$D_N = D_C > D$ Supercritical	Backwater	+ve	Downstream of sluice gate; change from steep to less steep slope.
HORIZONTAL $S_O = 0$	H2	$D_N > D > D_C$ Subcritical	Drawdown	−ve	Drawdown in approach to free overfall.
	H3	$D_N > D_C > D$ Supercritical	Backwater	+ve	Downstream of sluice gate; change from sloping to horizontal channel.
ADVERSE $S_O = $ −ve	A2	$D > D_C$ Subcritical	Drawdown	−ve	Drawdown to an overfall, fall over a weir. Uncommon, usually short in length.
	A3	$D_C > D$ Supercritical	Backwater	+ve	Downstream of sluice gate. Uncommon, usually short in length.

Note: a subcritical backwater profile starts at an obstruction such as a sluice gate and continues **upstream** until the water level decreases back to the normal depth. A supercritical backwater starts at the sluice gate and continues **downstream** until the water level increases to the normal depth.

surface can be calculated, for reasons already explained. This can only be accomplished if there is a control point with a known stage–discharge relationship from which to start. Some examples of control points are listed below and indicated in Fig. 8.30 by CP:

■ Control structures such as weirs (sections 5.5 and 9.5) and sluice gates (section 9.2) have rating curves which relate head to discharge. Dams may also provide a control, where there is a relationship between reservoir level and discharge over a spillway.

■ Where there is an increase in bed slope causing the flow to pass through critical depth near the change in gradient (Fig. 8.17b).

■ At the entry to a long uniform channel. After an initial entry head loss, depending upon the channel slope the flow will attain the normal depth or pass through critical near the

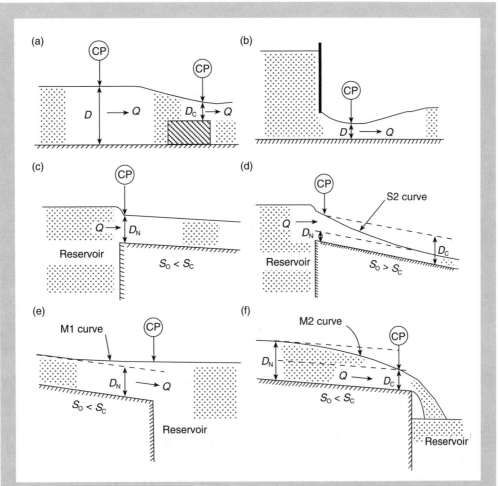

Figure 8.30 Diagrammatic examples of control points (CP) where there is a known relationship between head and discharge: (a) weir crest; (b) sluice gate vena contracta; (c and d) flow from a reservoir into an open channel having mild and steep slopes respectively, after allowing for the entry head loss; (e) flow from a mild open channel into a reservoir with a fixed water level (e.g. controlled by a spillway or overflow); (f) a mild open channel ending with a vertical drop into a reservoir causing the water surface to pass through critical depth just upstream of the fall

entrance. Although we are dealing with a channel, the loss can be estimated from Table 6.4 and may typically be $0.5V^2/2g$.

■ Where a channel discharges into a large reservoir that has a known water surface level which is above that in the approach channel.

■ At a free overfall, where the water level will pass through critical depth (D_C) around 3–10D_C upstream of the drop.

Box 8.3 ▶ **Summary so far**

(REMEMBER!!!)

When starting to calculate the profile of the water surface in gradually varied flow, the steps involved are as follows.

Step 1 Draw lines representing normal depth, critical depth and the channel bed.

Step 2 Insert the appropriate channel control points at entry, change of slope, exit, overfall, or where there is critical depth or a control structure.

Step 3 Between the control points, sketch the appropriate surface profile from Fig. 8.29.

Step 4 Calculate the actual elevation of the water surface along the channel, working upstream with subcritical flow and downstream with supercritical flow. An easy introduction to this is provided by the standard step method.

Often critical depth makes a good control (e.g. see equation (8.32)), so this appears several times above under different circumstances. Control points may be identified at one or more points along a channel, after which the water surface profile can be sketched between them and then established by calculation. In subcritical flow the control point is downstream, and the calculations proceed in an upstream direction, as in Example 8.20. In supercritical flow the control point is upstream, and the calculations proceed in a downstream direction, as in Example 8.21. The reason for this was explained in section 8.6.3.

8.11.3 Standard step method of profile evaluation for river channels

The standard step (or step by step) method is so called because the water surface elevation is calculated at known distances along the channel. This technique can be used with either prismatic channels or natural rivers that have a varying cross-section. With uniform, prismatic channels a regular interval such as 20 m may be adopted, and it is convenient to work with the specific energy measured from the bed, as described later in section 8.11.4. Although prismatic channels are the most convenient for tutorial and exam questions, in practice it is probably river channels that have to be analysed most frequently. All rivers vary in width, depth and roughness so usually cross-sections are located and surveyed where significant changes occur. By this means the channel is split into reaches that have relatively uniform characteristics. The uneven bed makes water depth meaningless, so the water surface profile is defined by its elevation (Z) above a horizontal datum, as in Fig. 8.31. The diagram shows the energy heads at two sections (represented by the subscripts 1 and 2) which are a distance ΔL apart. Assuming the channel gradient is relatively small, ΔL can be taken as either the plan or slope length. It is also assumed that cross-sections have been surveyed at intervals along the channel so the roughness, flow area and hydraulic radius at any stage can be determined. Equating the total head at the two sections and allowing for the friction loss in the channel:

$$Z_1 + \alpha_1 V_1^2/2g = Z_2 + \alpha_2 V_2^2/2g + \overline{S_F}\Delta L \tag{8.42}$$

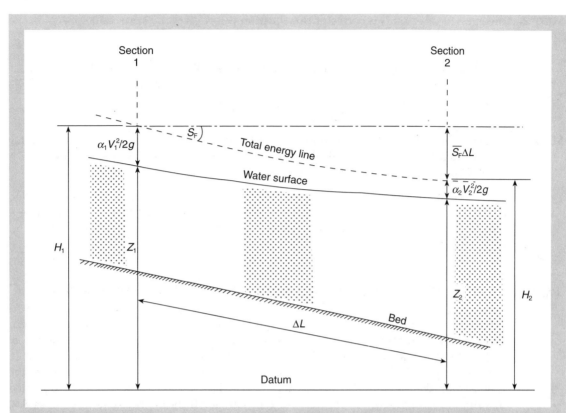

Figure 8.31 In non-prismatic channels, water depth is meaningless so the water surface profile is defined by its elevation above datum. The total heads H_1 and H_2 are used in the standard step method of profile evaluation

where $\overline{S_F}$ is the average of the friction gradients S_{F1} and S_{F2} at the ends of the reach. Thus if $S_{F1} = V_1^2 n_1^2 / R_1^{4/3}$ and $S_{F2} = V_2^2 n_2^2 / R_2^{4/3}$ as in equation (8.40) then:

$$\overline{S_F} = \frac{S_{F1} + S_{F2}}{2} \tag{8.43}$$

This is valid only if ΔL has a relatively short length and the friction gradient (which may be curved) can be approximated by a straight line.

Suppose we are working from section 1, where the values of α_1, n_1, Z_1, V_1 and S_{F1} are known, to section 2 with the objective of determining the value of Z_2. We can devise a simple iterative procedure from the equation above. Note that an iterative procedure is needed because the velocity, V_2, is included in step 1 below but this also determines the friction loss in step 2. All the procedure does is make sure that the value of Z_2 guessed in step 1 is consistent with the fall in the friction gradient between sections 1 and 2 as calulated in step 2. This is illustrated diagramatically by Fig. 8.31.

Step 1 guess Z_2 so we can calculate the total head H_2:

$$H_2 = Z_2 + \alpha_2 V_2^2 / 2g \tag{8.44}$$

Step 2   the guessed value of Z_2 is correct if the equation below gives the same value of H_2.

$$H_2 = H_1 - \overline{S_F}\Delta L \tag{8.45}$$

Note that the sign in this equation is –ve when working downstream since the total head must fall (as in Fig. 8.31), but +ve when working upstream since the total head must increase. REMEMBER!!!

Step 3   compare the values of H_2 obtained from steps 1 and 2. If they are different, adjust the value of Z_2 in step 1 and repeat the calculations. If they agree the guessed value of H_2 in step 1 is correct, so move to the next section and repeat the procedure.

The procedure is the same whether working upstream or downstream, but remember to adjust the sign in step 2 according to whether the total head must increase or decrease. A table or spreadsheet can be used to conduct the calculations so that they become a matter of routine, as in Example 8.20.

EXAMPLE 8.20   STANDARD STEP METHOD USING TOTAL ENERGY HEAD IN A RIVER

A broad crested rectangular weir forms a control point in a river channel (Fig. 8.32). At a distance of 10 m upstream of the weir the rectangular channel has an average bed elevation of 48.895 m above ordnance datum (mOD), a width of 10 m and the depth of flow is 1.49 m when the discharge is 19 m³/s. Upstream of the control point the width, area and roughness of the channel vary along its length. Cross-sections of the channel have been surveyed and plotted at 10 m intervals upstream of the weir. For the water levels shown in column 3 of Table 8.7, the appropriate values A, n and R are shown in columns 4, 7 and 8. Assume $\alpha = 1.0$ and the bed slope of the channel at the weir is 1 in 400. It is thought that normal depth in the channel is about 1.3 m. There is some concern that the weir may cause the river to overtop its

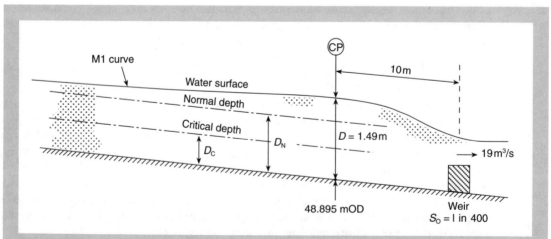

Figure 8.32   In Example 8.20 there is an M1 backwater curve upstream of the weir. CP is the control point. Typically gauging stations record the stage upstream of the weir, not at the weir

Table 8.7 Calculations for Example 8.20 – standard step method with an irregular channel

$Q = 19.000\,\text{m}^3/\text{s}$; Z, A and n are from survey data.
In column 11, H_2 is the value for the current row; H_1 is the value in column 6 of the previous row.

Col. 1 Chainage m	2 ΔL m	3 Z mOD	4 A m²	5 V m/s	6 $H =$ $Z + V^2/2g$ Eqn (8.44) mOD	7 n s/m$^{1/3}$	8 R m	9 S_F Eqn (8.40)	10 $\overline{S_F}$ Eqn (8.43)	11 $H_2 = H_1 + \overline{S_F}\Delta L$ Eqn (8.45) mOD
0 (weir)										
10 (CP)	10	50.385	15.111	1.257	50.466	0.035	1.144	0.00162	—	—
20	10	50.411	16.350	1.162	50.480	0.035	1.158	0.00136	0.00149	50.480
30	10	50.419	15.833	1.200	50.492	0.030	1.149	0.00108	0.00122	50.492
40	10	50.427	15.558	1.221	50.503	0.030	1.147	0.00112	0.00110	50.503
50	10	50.444	16.156	1.176	50.514	0.030	1.152	0.00103	0.00107	50.514
60	10	50.464	17.576	1.081	50.524	0.030	1.180	0.00084	0.00094	50.524
70	10	50.480	19.201	0.990	50.530	0.025	1.263	0.00045	0.00065	50.530
80	10	50.497	22.174	0.857	50.534	0.025	1.341	0.00031	0.00038	50.534
90	10	50.498	20.700	0.918	50.541	0.040	1.279	0.00097	0.00064	50.541
100	10	50.499	18.676	1.017	50.552	0.040	1.183	0.00132	0.00115	50.552

banks in the 100 m length of channel nearest to the weir, particularly at chainage 80 m where the bank dips to 50.7 mOD. Calculate the elevation of the water surface and determine if this is a problem.

Assuming uniform flow, check that at the weir the normal depth $D_N = 1.3\,\text{m}$ as stated. Assuming a rectangular channel, $A = 1.3 \times 10 = 13.00\,\text{m}^2$, $P = 1.30 + 10 + 1.30 = 12.60\,\text{m}$ and $R = A/P = 13.00/12.60 = 1.03\,\text{m}$. Using Manning:

$$Q = \frac{A}{n}R^{2/3}S_0^{1/2} = \frac{13.00}{0.035} \times 1.03^{2/3} \times (1/400)^{1/2} = 18.94\,\text{m}^3/\text{s}$$

Thus with uniform flow near the weir the depth would be about 1.30 m when the discharge is 19.00 m³/s. The critical depth in the channel is given by equation (8.32):

$$D_C = (Q^2/gB^2)^{1/3} = (19.00^2/9.81 \times 10.00^2)^{1/3} = 0.72\,\text{m}$$

Since the actual depth upstream of the weir $D = 1.49\,\text{m}$ then $D > D_N > D_C$, so the channel slope is mild and we have a M1 backwater curve with the water surface above normal depth (Table 8.6 and Fig. 8.29). The calculations to obtain the elevation of the water surface (Z) with distance upstream of the weir are shown in Table 8.7 (where in column 11, H_2 is the value for the current row; H_1 is the value in column 6 of the previous row). The first two rows of calculation are given below in full.

Chainage 10 m (control point)

Elevation of water surface $Z_1 = 48.895 + 1.490 = 50.385\,\text{mOD}$ (column 3) when $Q = 19.000\,\text{m}^3/\text{s}$.

With $A_1 = 15.111\,\text{m}^2$ (from survey data), $V_1 = 19.000/15.111 = 1.257\,\text{m/s}$ (column 5).

Total energy head $H_1 = Z_1 + V_1^2/2g = 50.385 + (1.257^2/19.62) = 50.466\,\text{m}$ (column 6).

With $n_1 = 0.035\,\text{s/m}^{1/3}$ and $R_1 = 1.144\,\text{m}$ (columns 7 and 8, from survey data), from equation (8.40) we get $S_{F1} = V_1^2 n_1^2/R_1^{4/3} = (1.257^2 \times 0.035^2)/1.144^{4/3} = 0.00162$ (column 9).

Since this is the control point (CP) the calculations for this row stop here.

Chainage 20 m

Guess that the elevation of the water surface $Z_2 = 50.411$ m and see if this checks out. With $A_2 = 16.350$ m² (from survey data), $V_2 = 19.000/16.350 = 1.162$ m/s.

Step 1 from equation (8.44), $H_2 = Z_2 + V_2^2/2g = 50.411 + (1.162^2/19.62) = 50.480$ m (column 6).

Step 2 with $n_2 = 0.035$ s/m$^{1/3}$ and $R_2 = 1.158$ m (columns 7 and 8), from equation (8.40) we get $S_{F2} = V_2^2 n_2^2/R_2^{4/3} = (1.162^2 \times 0.035^2)/1.158^{4/3} = 0.00136$ (column 9).

From equation (8.43), $\overline{S_F} = (S_{F1} + S_{F2})/2 = (0.00162 + 0.00136)/2 = 0.00149$ (column 10).
From equation (8.45), $H_2 = H_1 + \overline{S_F} \, \Delta L = 50.466 + (0.00149 \times 10) = 50.480$ m (column 11).

Step 3 compare the values of H_2 from steps 1 and 2 (i.e. compare columns 6 and 11 within the same row). They are the same, so the guessed value of $Z_2 = 50.411$ m is correct. Note that when using a calculator, under appropriate circumstances values of H_2 which agree to within 10 mm may be regarded as satisfactory, but the advent of spreadsheets makes agreement to the nearest mm extremely easy to obtain quickly.

The remainder of the example proceeds as above with the elevation of the water surface (Z) being recorded in column 3. At chainage 80 m it is apparent that $Z = 50.5$ mOD while the bank elevation is 50.7 mOD, so the water level is 0.2 m below the top of the bank. This is a rather small freeboard given that the water surface is likely to undulate and the calculations may not be entirely accurate (e.g. due to errors in determining n, A, etc.; the assumption of a uniform channel within a reach, and other approximations) so some overtopping is probable unless the bank level is raised.

This example illustrates the general method and highlights the following:

- With non-prismatic channels n, A and R vary and have to be found from cross-sectional data obtained from a survey of the channel.

- The solution gives no information regarding flow depths, which are meaningless in river channels, only the elevation of the water surface above datum, Z.

- With $\Delta L = 10$ m, the change in Z between rows is relatively small, M1 curves being relatively gentle and approaching the normal depth line asymptotically, touching it at infinity. For practical purposes the backwater curve is often assumed to end when it is within 10 mm of the normal depth.

- In this example we are working back upstream so Z and H must increase, so the sign in front of $\overline{S_F} \, \Delta L$ in equation (8.45) has to be +ve (Fig. 8.31 illustrates working downstream when it would have to be negative to give a decrease in Z_2 and H_2).

- The friction gradient, which represents the loss of energy head, is small where V and n are small.

- Over the 90 m covered by the calculations the fall in the total head line is $(50.552 - 50.466) = 0.086$ m while the fall in the water surface is $(50.499 - 50.385) = 0.114$ m, which represent gradients of 1 in 1047 and 1 in 789, confirming that the flow is non-uniform.

- The values in Table 8.7 are derived from a spreadsheet which makes it easy to conduct iterative calculations to the nearest mm, although such precision is not justified. When repeating the calculations within a row, the right side of the table (i.e. from column 7) changes relatively little, so Z in column 3 can be adjusted accordingly.

- Example 9.7 uses the same data to estimate the depth upstream of the weir.

8.11.4 Standard step method of profile evaluation for prismatic channels

The procedure for prismatic channels is basically the same as that above for rivers, but with two changes. The first is that we will use the specific energy measured as a head $E\mathrm{m}$ above the sloping bed in Fig. 8.33, instead of the total energy head $H\mathrm{m}$ measured above the horizontal datum in Fig. 8.31. The second is that because the bed is the datum and has a slope S_O, the slope of the specific energy line S_F must be measured relative to this, hence the term $(S_\mathrm{O} - \overline{S_\mathrm{F}})$ in the equations below. Referring to Fig. 8.33, with the bed level at section 2 as the datum, equating the total heads at sections 1 and 2:

$$S_\mathrm{O}\Delta L + D_1 + \alpha_1 V_1^2/2g = D_2 + \alpha_2 V_1^2/2g + \overline{S_\mathrm{F}}\,\Delta L \tag{8.46}$$

$$S_\mathrm{O}\Delta L + E_1 = E_2 + \overline{S_\mathrm{F}}\Delta L \tag{8.47}$$

or $$E_2 = E_1 + \left(S_\mathrm{O} - \overline{S_\mathrm{F}}\right)\Delta L \tag{8.48}$$

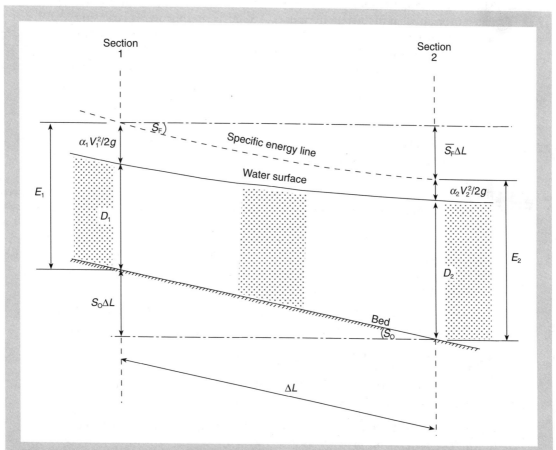

Figure 8.33 The specific energy heads E_1 and E_2 used with the standard step method of profile evaluation in prismatic channels

where $\overline{S_F}$ is again the average friction gradient from equation (8.43). Thus if the conditions at section 1 are known (i.e. α_1, n_1, D_1, V_1) we can use an iterative process to determine the elevation (D_2) of the water surface at section 2. Again, we have to use iteration because the value of D_2 determines V_2 which affects the value of both E_2 and $\overline{S_F}$. All we need to do is ensure that the guessed value of D_2, and hence E_2, in step 1 is consistent with the relative fall in the friction gradient between sections 1 and 2 as calculated in step 2.

Step 1 guess D_2 and calculate:

$$E_2 = D_2 + \alpha_2 V_2^2/2g \tag{8.49}$$

Step 2 the guessed value of D_2 is correct if the equation below gives the same value of E_2:

$$E_2 = E_1 + \left(S_O - \overline{S_F}\right)\Delta L \tag{8.50}$$

Again, the values of S_{F1}, S_{F2} and $\overline{S_F}$ are evaluated using equations (8.40) and (8.43).

Step 3 compare the values of E_2 obtained from steps 1 and 2. If they are different, adjust the value of D_2 in step 1 and repeat the calculations. If they agree the guessed value of D_2 in step 1 is correct, so move to the next section and repeat the procedure.

The procedure is the same whether working upstream or downstream, but care must be taken to ensure that the sign in front of $(S_O - \overline{S_F})\Delta L$ is appropriate. This depends upon whether the depth of flow is increasing or decreasing and the starting point on the depth–specific energy curve (see Table 8.4 and Fig. 8.16a). A table or spreadsheet can be used to conduct the calculations so that they become a matter of routine, as in Example 8.21. (REMEMBER!!!)

EXAMPLE 8.21 STANDARD STEP METHOD USING SPECIFIC ENERGY IN A PRISMATIC CHANNEL

The jet emerging from an underflow vertical sluice gate (similar to that in Fig. 1.18) contracts to a vena contracta just downstream of the gate where the depth is $D = 0.85\,\text{m}$. Downstream of the sluice gate where the flow is unaffected the corresponding normal depth is 2.60 m (Fig. 8.34). The channel has a uniformly rectangular cross-section 7.00 m wide, $n = 0.030\,\text{s/m}^{1/3}$ and a survey has revealed that its bed slope is 1 in 200. Assume $\alpha = 1.00$. Using the standard step specific energy method, determine the surface water profile and the location of the hydraulic jump.

Assuming uniform flow in the downstream channel, with $D_N = 2.60\,\text{m}$ calculate the discharge.
$A = 2.60 \times 7.00 = 18.20\text{m}^2$, $P = 2.60 + 7.00 + 2.60 = 12.20\text{m}$, $R = A/P = 18.20/12.20 = 1.49\text{m}$.

$$Q = \frac{A}{n}R^{2/3}S_O^{1/2} = \frac{18.20}{0.03} \times 1.49^{2/3} \times (1/200)^{1/2} = 55.96\text{m}^3/\text{s}.$$

Take the discharge corresponding to $D_N = 2.60\,\text{m}$ as $55.96\,\text{m}^3/\text{s}$.
The mean flow velocity $V_N = 55.96/18.20 = 3.07\,\text{m/s}$.

The critical depth D_C for a rectangular channel can be obtained from equation (8.32):
$$D_C = (Q^2/gB^2)^{1/3} = (55.96^2/9.81 \times 7.00^2)^{1/3} = 1.87\text{m}.$$

Thus at the vena contracta $D < D_C < D_N$ so the flow is supercritical.

The critical channel slope can be obtained from equation (8.28) and (8.30):

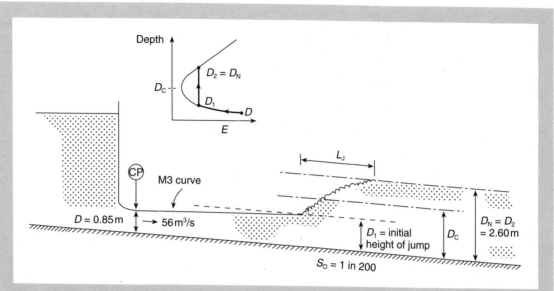

Figure 8.34 In Example 8.21 the supercritical flow after the sluice gate returns to subcritical via a hydraulic jump of length L_J

$V_C = (gD_{MC})^{1/2} = (9.81 \times 1.87)^{1/2} = 4.28 \, \text{m/s}.$

$A_C = 1.87 \times 7.00 = 13.09 \, \text{m}^2, P_C = 1.87 + 7.00 + 1.87 = 10.74 \, \text{m}^2, R_C = A_C/P_C = 13.09/10.74 = 1.22 \, \text{m}.$

$S_C = V_C^2 n^2/R_C^{4/3} = 4.28^2 \times 0.03^2/1.22^{4/3} = 0.0126 \text{ or } 1 \text{ in } 79.$

Thus $S_O < S_C$ so from Table 8.6 and Fig. 8.29 it is apparent that we have a mild bed slope, which when combined with supercritical flow will result in an M3 backwater curve where the depth recovers downstream of the sluice gate and regains normal depth via a hydraulic jump. In order to determine where the jump will occur we need to use equation (8.36) to obtain the intitial depth D_1 at which the jump will start. The sequent depth after the jump $D_2 = D_N$ in this instance. With $D_N = 2.60 \, \text{m}$ and $V_N = 3.07 \, \text{m/s}$ we get $F_N = F_2 = V_N/(gD_N)^{1/2} = 3.07/(9.81 \times 2.60)^{1/2} = 0.61$ so:

$$D_1 = \frac{D_2}{2}\left(\sqrt{1 + 8F_2^2} - 1\right) = \frac{2.60}{2}\left(\sqrt{1 + 8 \times 0.61^2} - 1\right) = 1.29 \, \text{m}$$

Thus we can sketch the outline of the M3 surface water profile: the depth will be 0.85 m just downstream of the sluice gate (control point 1), and will gradually recover with distance downstream to 1.29 m, at which depth a hydraulic jump will commence that will return the depth to normal at 2.60 m (control point 2). The remaining part of the question involves calculating the actual elevation of the water surface downstream of the gate with the objective of determining exactly where the water surface passess through a depth of 1.29 m. The calculations, based upon specific energy head and equations (8.49) and (8.50), are shown in full in Table 8.8, the first two rows of which are explained below.

Chainage 0 m (vena contracta control point)

$D = 0.850 \, \text{m}$ and $B = 7.000 \, \text{m}$ so $A = 0.850 \times 7.000 = 5.950 \, \text{m}^2$ (column 4).
$Q = 55.960 \, \text{m}^3/\text{s}$ so $V = 55.960/5.950 = 9.405 \, \text{m/s}$ (column 5).

Table 8.8 Calculations for Example 8.21 – standard step method with a prismatic channel

$Q = 55.960 \, \text{m}^3/\text{s}$, $n = 0.030 \, \text{s}/\text{m}^{1/3}$, $B = 7.000 \, \text{m}$, $S_O = 1$ in $200 = 0.005$.
In column 11, E_2 is the value for the current row; E_1 is the value in column 6 of the previous row.

Col. 1 Chainage m	2 ΔL m	3 D m	4 A m^2	5 V m/s	6 $E = D + V^2/2g$ Eqn (8.49) m	7 n $\text{s}/\text{m}^{1/3}$	8 R m	9 S_F Eqn (8.40)	10 $\overline{S_F}$ Eqn (8.43)	11 $E_2 = E_1 + (S_O - \overline{S_F})\Delta L$ Eqn (8.50) m
0 (CP)	0	0.850	5.950	9.405	5.358	0.030	0.684	0.132	—	—
5	5	0.917	6.419	8.718	4.791	0.030	0.727	0.105	0.118	4.791
10	5	0.985	6.895	8.116	4.342	0.030	0.769	0.084	0.094	4.343
15	5	1.054	7.378	7.585	3.986	0.030	0.810	0.069	0.076	3.985
20	5	1.125	7.875	7.106	3.699	0.030	0.851	0.056	0.062	3.699
25	5	1.198	8.386	6.673	3.468	0.030	0.893	0.047	0.051	3.466
30	5	1.275	8.925	6.270	3.279	0.030	0.935	0.039	0.043	3.279
35	5	1.356	9.492	5.895	3.127	0.030	0.977	0.032	0.035	3.126

Specific energy head $E = D + V^2/2g = 0.850 + (9.405^2/19.62) = 5.358 \, \text{m}$ (column 6).
$P = 0.850 + 7.000 + 0.850 = 8.700 \, \text{m}$, $A = 5.950 \, \text{m}^2$ so $R = A/P = 5.950/8.700 = 0.684 \, \text{m}$ (column 8).
From equation (8.40), $S_F = V^2 n^2/R^{4/3} = 9.405^2 \times 0.030^2/0.684^{4/3} = 0.132$ (column 9).
This is the control point, so in this row the calculation stops here.

Chainage 5 m

Guess that the depth $D_2 = 0.917 \, \text{m}$ and see if this checks out.
With $A_2 = 0.917 \times 7.000 = 6.419 \, \text{m}^2$, $V_2 = 55.960/6.419 = 8.718 \, \text{m/s}$ (column 5).

Step 1 from equation (8.49), $E_2 = D_2 + V_2^2/2g = 0.917 + (8.718^2/19.62) = 4.791 \, \text{m}$ (column 6).

Step 2 $P = 0.917 + 7.000 + 0.917 = 8.834 \, \text{m}$, so $R = A/P = 6.419/8.834 = 0.727 \, \text{m}$ (column 8).

With $n_2 = 0.030 \, \text{s}/\text{m}^{1/3}$, equation (8.40) gives $S_{F2} = V_2^2 n_2^2/R_2^{4/3} = 8.718^2 \times 0.030^2/0.727^{4/3} = 0.105$ (column 9).
From equation (8.43), $\overline{S_F} = (S_{F1} + S_{F2})/2 = (0.132 + 0.105)/2 = 0.118$ (column 10).
$S_O = 1$ in $200 = 0.005$.
From equation (8.50), $E_2 = E_1 + (S_O - \overline{S_F})\Delta L = 5.358 + (0.005 - 0.118) \times 5 = 4.791 \, \text{m}$ (column 11).

Step 3 compare the values of E_2 from steps 1 and 2 (i.e. compare columns 6 and 11 within the same row). They are the same, so the guessed value of $D_2 = 0.917 \, \text{m}$ is correct.

The calculations for the remaining chainages are summarised in Table 8.8. This is based on the results from a spreadsheet that used more significant figures than above, hence there are one or two insignificant discrepancies due to rounding errors. The table shows that the depth of 1.29 m required to initiate the hydraulic jump occurs between chainages 30 m and 35 m. Interpolation can be used to find where.

required change in depth after chainage 30 m = $(1.290 - 1.275) = 0.015 \, \text{m}$
change in depth between chainages 30 m and 35 m = $(1.356 - 1.275) = 0.081 \, \text{m}$
distance required for 0.015 m change in depth = $(0.015/0.081) \times 5 = 0.926 \, \text{m}$
thus depth of 1.290 m occurs at chainage 30.926 m.

The length of the jump can be obtained from Table 8.5 and Fig. 8.25a. If F_1 is the Froude number corresponding to the initial condition where $D_1 = 1.29$ m then $A_1 = 1.29 \times 7 = 9.03$ m² and $V_1 = 55.96/9.03 = 6.20$ m/s. Hence $F_1 = V_1/(gD_1)^{1/2} = 6.20/(9.81 \times 1.29)^{1/2} = 1.74$. From the table this gives $L_J = 4.05D_2 = 4.05 \times 2.60 = 10.53$ m. Thus the jump would end, and normal depth would start, at about chainage $30.93 + 10.53 = 41.46$ m.

This example illustrates the following:

- Using specific energy it is possible to see the variation in depth. However this technique is restricted to uniform, prismatic channels.

- Since the flow is initially supercritical, as the depth in the channel increases towards the critical depth the specific energy must fall, as indicated by the *Depth–E* curve in Fig. 8.34. In this example, this happens automatically since the value of $(S_O - \overline{S_F})\Delta L = (0.005 - 0.118) \times 5 = -0.565$ m. When applying equation (8.50), think of the position on the depth–specific energy curve and determine whether the required change in depth will result in an increase or decrease in E.

- The above is made easier by using Fig. 8.29 to sketch the profile of the water surface between control points **before** starting the calculations.

- In Table 8.8 only the final, correct depth is recorded in column 3. To prove to yourself that other depths are wrong, pick one row of the table and change the value of D significantly; you will find that your new value will not satisfy the equations in steps 1 and 2. Remember, you must also change A, R etc. as well.

- Example 9.2 uses the same data to calculate the depth upstream of the sluice gate and the force acting on it.

SELF TEST QUESTION 8.5

A prismatic, rectangular channel is 6 m wide, but otherwise similar to that in Fig. 8.1. The channel has a bed slope of 1 in 800 and $n = 0.017$ s/m$^{1/3}$. Assume $\alpha = 1.0$. It is designed to carry 35 m³/s at a normal depth of 2.34 m. The channel terminates in a vertical drop so that the flow falls freely into a lower reservoir (Fig. 8.35). Assume the flow passes through critical depth at the drop. Calculate the elevation of the water surface in the channel using specific energy head and the standard step method, and so determine: (a) the distance at which the water surface is within 10 mm of the normal depth; (b) the change in the elevation of the water surface in the first 100 m; (c) the change in the elevation of the water surface in the last 100 m of your calculations. (Hint: take $\Delta L = 100$ m since theoretically the backwater curve reaches normal depth at infinity. A good way to learn the method and save time is to create a spreadsheet, e.g. using Excel or similar. If you are using a calculator there is little to be gained by completing all of the calculations, so stop after 4 or 5 rows and refer to the solution in Appendix 2.)

8.11.5 Direct step method of profile evaluation for prismatic channels

With the direct step method, the distances (ΔL) required for the water depth to change by a fixed amount (ΔD) are calculated. This method of determining the water surface profile along a channel is preferred by many, since it affords a direct solution without the need to guess values and undertake several iterations. However, this is a marginal advantage when

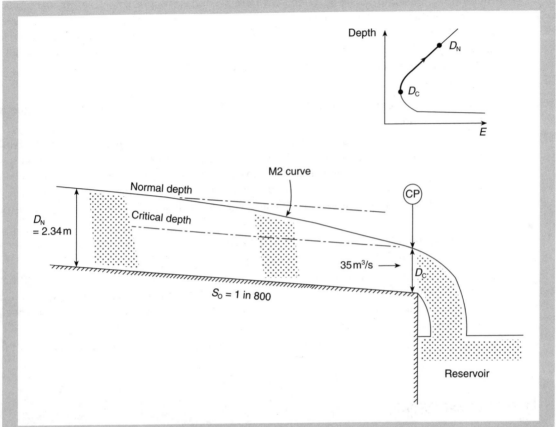

Figure 8.35 In Self Test Question 8.5 the flow from a mildly sloping channel enters the reservoir via a free fall

a spreadsheet is being used, since an iteration can be completed in a few seconds, and there are a few disadvatages associated with the direct step technique. One is that if the difference in depth at the control points at the ends of the surface water profile is divided into a number of equal intervals, then ΔD is likely to have a rather arbitrary value. For instance, in Example 8.8 the depth had to change from 0.85 m at the control point to 1.29 m at the start of the hydraulic jump, so if this difference of 0.44 m was split into five steps then ΔD = 0.088 m. Additionally, the distance corresponding to each 0.088 m interval would be determined from the calculations and cannot be preselected.

The direct step method is based on the gradually varying flow equation introduced earlier:

$$\frac{\mathrm{d}D}{\mathrm{d}L} = \frac{S_O - S_F}{1 - F^2} \tag{8.41}$$

Writing this equation in the form of finite differences (ΔD, ΔL) and rearranging it to solve for the distance (ΔL) required for a change in depth (ΔD) gives:

$$\Delta L = \Delta D \times \frac{(1 - F^2)_{AV}}{(S_O - S_F)_{AV}} \tag{8.51}$$

where $(1 - F^2)_{AV}$ and $(S_O - S_F)_{AV}$ are the average of the values at the two ends of the reach. Since S_O is known and F and S_F can be calculated, by giving ΔD a suitable value the equation can be solved to find the distance ΔL. Solving for successive intervals allows the surface water profile to be determined. The calculations can be conducted in a computer spreadsheet or in a table, as in Example 8.22. The calculations for the first two rows of the table are given in full below.

EXAMPLE 8.22   DIRECT STEP METHOD APPLIED TO A PRISMATIC CHANNEL

In Self Test Question 8.5, the 6 m wide rectangular channel ended in a vertical drop into a reservoir. An alternative arrangement is to be investigated in which the channel would end in a ramp or spillway with a slope of 1 in 10. Assume that this part of the channel is long enough for the flow to become established at the normal depth before it encounters any backwater from the reservoir. At the end of the channel the depth will be constant at 4.755 m as dictated by the water surface level in the reservoir (Fig. 8.36). Take $\alpha = 1.0$ and $n = 0.017\,\text{s/m}^{1/3}$. Determine the elevation of the water surface profile in the channel.

As in Example 8.5, determine the normal depth in the channel by trial and error:

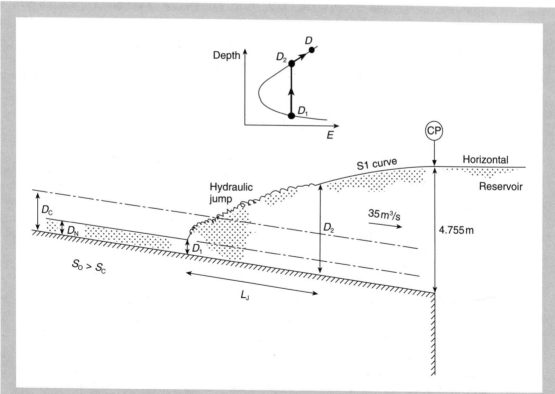

Figure 8.36   In Example 8.22 the steeply sloping channel enters a reservoir with a higher water level

$$Q = \frac{A}{n} R^{2/3} S_O^{1/2}$$

$$35.0 = \frac{6 \times D_N}{0.017} \left(\frac{6 \times D_N}{6 + 2D_N} \right)^{2/3} \left(\frac{1}{10} \right)^{1/2}$$

$$0.3136 = D_N \left(\frac{6 \times D_N}{6 + 2D_N} \right)^{2/3}$$

Guess D_N and see if the right hand side (RHS) of the equation = 0.3136.
Try D_N = 0.500 m, RHS = 0.2842.
Try D_N = 0.532 m, RHS = 0.3133.
Take normal depth D_N = 0.532 m.

Critical depth $D_C = (Q^2/gB^2)^{1/3} = (35^2/9.81 \times 6^2)^{1/3} = 1.514$ m.
$A_C = 6 \times 1.514 = 9.084$ m^2.
Critical velocity $V_C = Q/A_C = 35/9.084 = 3.853$ m/s.
$P_C = B + 2D_C = 6 + (2 \times 1.514) = 9.028$ m and $R = A_C/P_C = 9.084/9.028 = 1.006$ m.
Critical slope $S_C = V_C^2 n^2/R_C^{4/3} = 3.853^2 \times 0.017^2/1.006^{4/3} = 0.00426$ or 1 in 235.
Thus $S_O > S_C > 0$ so we are dealing with a steep slope and an S curve.
Assume that the reservoir level forms a control point where at the end of the channel D = 4.755 m.
Thus $D > D_C > D_N$ which Table 8.6 defines as an S1 curve that starts with a hydraulic jump.
Referring to Fig. 8.36, the water will flow down the ramp at the initial depth $D_1 = D_N$, undergo a hydraulic jump to the sequent depth D_2, then an S1 curve will form a transition to the reservoir surface level.

With D_1 = 0.532 m then V_1 = 35/(6 × 0.532) = 10.965 m/s.
$F_1 = V_1/(gD_1)^{1/2} = 10.965/(9.81 \times 0.532)^{1/2} = 4.800$.
Although the channel is not horizontal, assume that the sequent depth of the hydraulic jump (D_2) is:

$$D_2 = \frac{D_1}{2}(\sqrt{1 + 8F_1^2} - 1) = \frac{0.532}{2}(\sqrt{1 + 8 \times 4.800^2} - 1) = 3.355 \text{m}$$

Approximately, from Table 8.5, the length of the jump L_J = 5.96D_2 = 5.96 × 3.355 = 20 m.
Thus the S1 backwater curve will form a transition between the depth D_2 = 3.355 m at the end of the jump and the reservoir level that corresponds to a depth D = 4.755 m at the end of the channel. This means the depth changes by (4.755 − 3.355) = 1.400 m. The S1 curve will probably not be very long, so starting at the reservoir take ΔD = 0.200 m and calculate the distances (ΔL) required for successive 0.200 m changes in depth. The calculations are summarised in Table 8.9 with the first two rows given in full below.

D = 4.755 m, control point at the reservoir (chainage 0)

Depth D = 4.755 m, so A = 4.755 × 6.000 = 28.530 m^2 and V = 35.000/28.530 = 1.227 m/s.
$F = V/(gD)^{1/2} = 1.227/(9.81 \times 4.755)^{1/2} = 0.180$ (column 4).
$(1 - F^2) = (1 - 0.180^2) = 0.968$ (column 5).
P = 6.000 + (2 × 4.755) = 15.510 m and $R = A/P$ = 28.530/15.510 = 1.839 m (column 7).
From equation (8.40), $S_F = V^2 n^2/R^{4/3} = 1.227^2 \times 0.017^2/1.839^{4/3} = 0.000193$ (column 8).
S_O = 1 in 10 or 0.100 thus $(S_O - S_F) = (0.100 - 0.000193) = 0.0998$ (column 9).
This is the control point, so in this row the calculations stop here.

Table 8.9 Calculations for Example 8.22 – direct step method with a prismatic channel
$Q = 35.000\,\text{m}^3/\text{s}$, $n = 0.017\,\text{s/m}^{1/3}$, $B = 6.000\,\text{m}$, $S_O = 1$ in $10 = 0.100$.

Col. 1 D m	2 A m²	3 V m/s	4 F	5 $(1 - F^2)$	6 $(1 - F^2)_{AV}$	7 R m	8 S_F Eqn (8.40)	9 $(S_O - S_F)$	10 $(S_O - S_F)_{AV}$	11 ΔL Eqn (8.51) m	12 Chainage $\Sigma(\Delta L)$ m
4.755	28.530	1.227	0.180	0.968	—	1.839	0.000193	0.0998	—	—	0 (CP)
4.555	27.330	1.281	0.192	0.963	0.966	1.809	0.000215	0.0998	0.0998	1.936	1.936
4.355	26.130	1.339	0.205	0.958	0.961	1.776	0.000241	0.0998	0.0998	1.926	3.862
4.155	24.930	1.404	0.220	0.952	0.955	1.742	0.000272	0.0997	0.0998	1.914	5.776
3.955	23.730	1.475	0.237	0.944	0.948	1.706	0.000308	0.0997	0.0997	1.902	7.678
3.755	22.530	1.553	0.256	0.934	0.939	1.668	0.000352	0.0996	0.0997	1.884	9.562
3.555	21.330	1.641	0.278	0.923	0.929	1.627	0.000407	0.0996	0.0996	1.865	11.427
3.355	20.130	1.739	0.303	0.908	0.916	1.584	0.000473	0.0995	0.0996	1.839	13.266

$D = 4.555\,\text{m}$, first 0.200 m step

Depth is reduced by 0.200 m to $D = 4.555\,\text{m}$, so $A = 4.555 \times 6.000 = 27.330\,\text{m}^2$ (column 2).
$V = 35.000/27.330 = 1.281\,\text{m/s}$ (column 3).
$F = V/(gD)^{1/2} = 1.281/(9.81 \times 4.555)^{1/2} = 0.192$ (column 4).
$(1 - F^2) = (1 - 0.192^2) = 0.963$ (column 5).
For the reach, $(1 - F^2)_{AV} = (0.968 + 0.963)/2 = 0.966$ (column 6).
$P = 6.000 + (2 \times 4.555) = 15.110\,\text{m}$ and $R = A/P = 27.330/15.110 = 1.809\,\text{m}$ (column 7).
From equation (8.40), $S_F = V^2n^2/R^{4/3} = 1.281^2 \times 0.017^2/1.809^{4/3} = 0.000215$ (column 8).
$S_O = 1$ in 10 or 0.100, thus $(S_O - S_F) = (0.100 - 0.000215) = 0.0998$ (column 9).
$(S_O - S_F)_{AV} = (0.0998 + 0.0998)/2 = 0.0998$ (column 10).

From equation (8.51), $\Delta L = \Delta D \times \dfrac{(1 - F^2)_{AV}}{(S_O - S_F)_{AV}} = 0.200 \times \dfrac{0.966}{0.0998} = 1.936\,\text{m}$ (column 11).

Thus chainage at which $D = 4.555\,\text{m}$ is 1.936 m (column 12).

The remainder of the calculations are conducted in similar fashion. By adding the values of ΔL it is apparent that a depth of 3.355 m (i.e. the depth at the end of the hydraulic jump) is reached at chainage 13.266 m, so the beginning of the jump would be at about chainage 33.3 m.

This example illustrates the following:

- The direct step method allows the depth and surface profile to be evaluated without having to guess values or use an iterative procedure.

- Dividing the required change in depth into equal steps of ΔD means that the intermediate water depths may have 'odd' values, e.g. 4.155 m rather than 4.100 m or 4.150 m.

- Because the calculations yield the distances (ΔL) required for the water level to change by a predetermined amount (ΔD) there is no control over the chainages obtained, which have rather 'odd' values like 1.936 m.

- The method can only be used in a prismatic channel.

- The example shows how a hydraulic jump occurs before a S1 curve.

8.12 Surge waves in open channels

The surge waves shown in Fig. 8.37 are an example of rapidly varying unsteady flow. They can arise by various means, but a good illustration is when a sluice gate is lowered or raised. The water depths corresponding to the initial position of the sluice gate are shown dashed, and those afterwards by the solid lines. In diagram (a) the gate is lowered, causing an increase in depth upstream and a reduction in depth downstream. In diagram (b) the gate

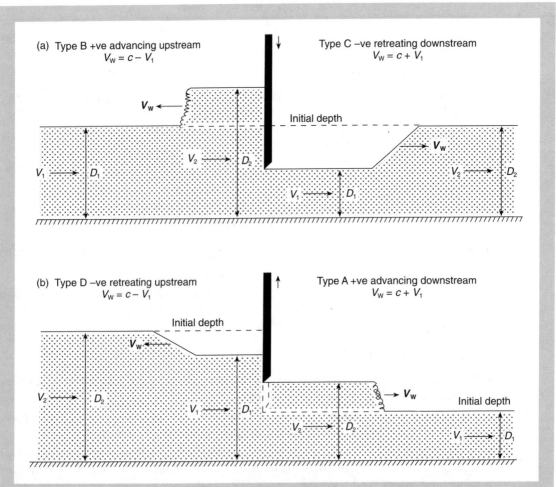

Figure 8.37 Surge waves produced by (a) lowering and (b) raising a sluice gate. The dashed line shows the initial depth, the solid line the depth after the change. Positive waves moving towards a smaller depth are steep fronted, while negative waves have sloping fronts. Note D_1 is always less than D_2, and that this is unsteady flow with a change in discharge so $A_1V_1 \neq A_2V_2$

is raised, causing a reduction in depth upstream and an increase in depth downstream. Positive surge waves are those which move in the direction of a smaller water depth and negative surges those which move towards greater depth. Positive surges often have steep wave fronts, but negative waves have sloping fronts since, as will be seen later, the leading edge of the front retreats quicker than the back edge. Chow (1981) also designated the four types of surge wave in the figure as A, B, C and D, and this can sometimes be a convenient means of classification. Frequently of greatest interest is type B, since this is the condition most likely to overtop the riverbanks. It is also similar to a tidal bore, where the incoming tide is funnelled by an estuary towards the mouth of a river creating a surge wave. The bore of the River Severn is the best known example in Britain, having a wave height of up to 1.5 m and a velocity of up to 6 m/s (Webber, 1971).

With a surge wave the flow is unsteady because the conditions change with time. For example, anyone standing on a bridge watching the Severn bore will see a surge wave approach and pass underneath, thus the depth and discharge at this point change with time. We also have to distinguish between the velocity or celerity of the surge wave (c) measured relative to the initial flow velocity V_1, and the absolute wave velocity (V_W) relative to a stationary observer. As an illustration, consider the wave front in Fig. 8.38a. Suppose this is a standing wave or hydraulic jump with $V_W = 0$ so that the wave remains opposite the same point on the riverbank. If $V_1 = 3$ m/s then $c = 3$ m/s in the opposite direction so $V_W = c - V_1 = 0$. Despite remaining in the same position the wave is actually moving forwards at $c = 3$ m/s, which is one reason a hydraulic jump is so turbulent. Note that with surge wave types A and C that move in the same direction as V_1 then $V_W = c + V_1$.

In order to use the steady flow equations to analyse the moving surge wave we need to bring the wave front to rest by superimposing an equal and opposite velocity (V_W) on the system, and then calculate absolute velocities relative to the stationary front. For example, consider two cars travelling in opposite directions that collide head on. If one is travelling at 40 mph and the other at 60 mph, the same impact would occur if one car was stationary and the other hit it with an absolute velocity of 100 mph. Similarly, in Fig. 8.38b the wave front is stationary as seen by a stationary observer (i.e. $V_W = 0$) but the approach flow now has the absolute velocity $V_1 + V_W$. Likewise the downstream velocity becomes $V_2 + V_W$. This means that conditions are steady relative to the control volume and we can apply the momentum equation in the direction of motion, as in section 4.5.2. In doing so we will assume the channel bed is horizontal so the weight of water is eliminated, that the channel has a uniform, rectangular cross-section of area $A = BD$, and that friction is negligible over the short length involved.

$$\Sigma F = \rho Q(V_2 - V_1) \tag{4.9}$$
$$P_1 A_1 - P_2 A_2 = \rho A_1 (V_1 + V_W)(V_2 - V_1)$$

Assuming a hydrostatic pressure distribution, $P_1 = (0 + \rho g D_1)/2 = \rho g D_1/2$. For the time being it is convenient to write this as $P_1 = \rho g \bar{y}_1$ where $\bar{y}_1 = D_1/2$ is the centroidal depth of the area A_1. Similarly $P_2 = \rho g \bar{y}_2$ so:

$$\rho g \bar{y}_1 A_1 - \rho g \bar{y}_2 A_2 = \rho A_1 (V_1 + V_W)(V_2 - V_1)$$

It is fairly obvious that $\bar{y}_1 A_1 < \bar{y}_2 A_2$ and that $V_2 < V_1$ so that our normal sign convention of Chapter 4 results in a negative quantity on each side of the equation. Rearranging and cancelling the ρ's gives.

$$g(\bar{y}_2 A_2 - \bar{y}_1 A_1) = A_1 (V_1 + V_W)(V_1 + V_2) \tag{8.52}$$

Now apply the continuity equation to the control volume:

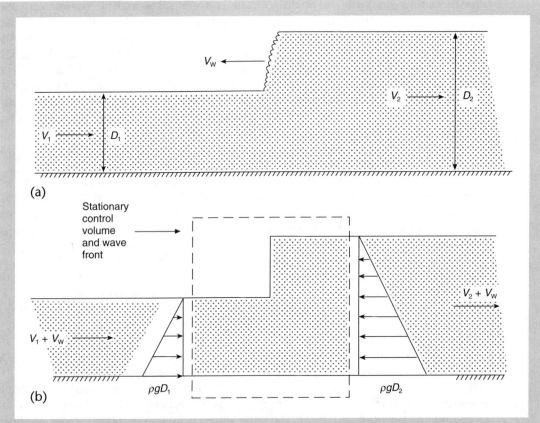

Figure 8.38 (a) A positive surge wave advancing from right to left with absolute velocity V_W. (b) The equivalent stationary wave front obtained by superimposing V_W from left to right, i.e. adding V_W to both V_1 and V_2. The absolute velocities remain the same, but the steady flow equations can now be applied to the stationary control volume and surge wave

$$(V_1 + V_W)A_1 = (V_2 + V_W)A_2 \tag{8.53}$$

$$V_2 = \frac{A_1}{A_2}(V_1 + V_W) - V_W \tag{8.54}$$

Substituting this expression for V_2 into equation (8.52) gives:

$$g(\bar{y}_2 A_2 - \bar{y}_1 A_1) = A_1(V_1 + V_W)\left(V_1 - \left[\frac{A_1}{A_2}(V_1 + V_W) - V_W\right]\right)$$

$$g(\bar{y}_2 A_2 - \bar{y}_1 A_1) = A_1(V_1 + V_W)\left(V_1 + V_W - \left[\frac{A_1}{A_2}(V_1 + V_W)\right]\right)$$

$$g(\bar{y}_2 A_2 - \bar{y}_1 A_1) = A_1(V_1 + V_W)^2\left(1 - \frac{A_1}{A_2}\right)$$

$$(V_1 + V_W) = \sqrt{\frac{g(\bar{y}_2 A_2 - \bar{y}_1 A_1)}{A_1(1 - A_1/A_2)}} \tag{8.55}$$

or $\quad V_W = \sqrt{\dfrac{g(\bar{y}_2 A_2 - \bar{y}_1 A_1)}{A_1(1 - A_1/A_2)}} - V_1$ (8.56)

$(V_1 + V_W) = c$ is the celerity of the surge wave relative to the initial flow velocity, and $V_W = c - V_1$ is the absolute velocity relative to the riverbanks. For a rectangular channel $\bar{y}_1 = D_1/2$, $\bar{y}_2 = D_2/2$, $A_1 = BD_1$ and $A_2 = BD_2$ where B is the channel width. Before making this substitution above it is helpful if equation (8.55) is rearranged slightly and it is remembered that $(D_2^2 - D_1^2) = (D_2 - D_1)(D_2 + D_1)$.

$(V_1 + V_W) = \sqrt{\dfrac{gA_2(\bar{y}_2 A_2 - \bar{y}_1 A_1)}{A_1(A_2 - A_1)}} = \sqrt{\dfrac{gBD_2([BD_2^2/2])}{BD_1(BD_2 - BD_1)}} = \sqrt{\dfrac{gD_2(D_2 - D_1)(D_2 + D_1)}{2D_1(D_2 - D_1)}}$

$(V_1 + V_W) = \sqrt{\dfrac{gD_2(D_2 + D_1)}{2D_1}}$ (8.57)

For a small wave $D_2 \to D_1$ and hence its celerity is:

$(V_1 + V_W) = \sqrt{gD_1}$ (8.58)

and the absolute velocity of the wave as it moves upstream as seen by a stationary observer is:

$V_W = \sqrt{gD_1} - V_1$ (8.59)

Remember that the equations above are for a surge travelling upstream. For a surge wave travelling downstream, equation (8.56), and similar, would be V_W (REMEMBER!!!) $= c + V_1$. Thus relative to the banks the wave moves faster because it is assisted by the normal river flow. Note V_1 is always the velocity corresponding to the smaller depth D_1 as in Fig. 8.37. In either case if $V_1 = 0$ then $\sqrt{gD_1}$ is the velocity of a small gravity wave in still water and the ratio of $V_1/\sqrt{gD_1}$ is the Froude number (F). For a standing wave or hydraulic jump that is stationary with respect to the riverbanks then $V_W = 0$ and $V_1 = \sqrt{gD_1}$. As stated earlier in the chapter, in supercritical flow a small disturbance cannot propagate upstream, it being swept away downstream. However, an exception is that a very large disturbance can move upstream, but it will transform the flow to subcritical in the process. One drastic example is almost completely closing a sluice gate located in a channel that initially experiences supercritical flow.

Note that if a hydraulic jump is analysed as a stationary surge then with $V_W = 0$ in equation (8.57) squaring both sides gives:

$V_1^2 = \dfrac{gD_2(D_2 + D_1)}{2D_1}$ or $\dfrac{2V_1^2 D_1}{g} = D_2^2 + D_2 D_1$

Now squaring the Froude number gives $F_1^2 = V_1^2/gD_1$ so $V_1^2/g = F_1^2 D_1$. Substituting above gives:

$2F_1^2 D_1^2 = D_2^2 + D_2 D_1$

Rearranging and dividing through by D_1^2:

$\left(\dfrac{D_2}{D_1}\right)^2 + \left(\dfrac{D_2}{D_1}\right) - 2F_1^2 = 0$

This is a quadratic equation whose solution is:

$\dfrac{D_2}{D_1} = \dfrac{1}{2}\left(\sqrt{1 + 8F_1^2} - 1\right)$ (8.37)

This is equation (8.37) in section 8.10.2.

EXAMPLE 8.23

A rectangular channel 3.0 m wide has a normal depth of 1.4 m while it discharges 9.956 m³/s of water to the turbines of a hydroelectric scheme. A sudden failure of part of the electrical transmision system cuts the water demand to 3.500 m³/s, the flow reduction being achieved by quickly dropping a vertical sluice gate. This causes a positive surge wave to travel upstream as in Fig. 8.38a. Determine the initial depth (D_2) and absolute velocity (V_W) of the wave, ignoring friction and the slope of the channel.

For the first condition $V_1 = Q_1/A_1 = 9.956/(3.0 \times 1.4) = 2.370$ m/s.
In the steady condition that exists after the surge wave has passed the continuity equation is $Q_2 = A_2V_2$ so $3.50 = 3.0 \times D_2V_2$ giving $V_2 = 3.50/3.0D_2$ or:

$$V_2 = 1.167/D_2 \qquad (1)$$

For the surge wave, equation (8.53) can be applied and simplified since $A_1 = BD_1$ and $A_2 = BD_2$.

$$(V_1 + V_W)D_1 = (V_2 + V_W)D_2$$
$$(2.370 + V_W)1.4 = \left(\frac{1.167}{D_2} + V_W\right)D_2$$
$$3.318 + 1.4V_W = 1.167 + V_W D_2$$
$$V_W(D_2 - 1.4) = 2.151$$
$$V_W = 2.151/(D_2 - 1.4) \qquad (2)$$

Substitute the above expression for V_W into equation (8.57) written as:

$$V_W = \sqrt{\frac{gD_2(D_2 + D_1)}{2D_1}} - V_1$$
$$\frac{2.151}{(D_2 - 1.4)} = \sqrt{\frac{9.81 \times D_2(D_2 + 1.4)}{2 \times 1.4}} - 2.370$$
$$\frac{2.151}{(D_2 - 1.4)} = \sqrt{3.504D_2(D_2 + 1.4)} - 2.370$$

This equation must be solved by trial and error. When $D_2 = 2.162$ m the left and right side yield the same value:

left side $= 2.151/(2.162 - 1.4) = 2.823$
right side $= [3.504 \times 2.162 (2.162 + 1.4)]^{1/2} - 2.370 = 2.825$ so OK

From equation (1): $V_2 = 1.167/2.162 = 0.540$ m/s.
From equation (2): $V_W = 2.151/(2.162 - 1.4) = 2.823$ m/s relative to the banks.
Thus after the closure of the gate, the depth will be 2.162 m and the absolute wave velocity 2.823 m/s.
Check: $Q_2 = A_2V_2 = 2.162 \times 3.0 \times 0.540 = 3.502$ m³/s OK.
Note that because of the reduction in discharge $Q_1 \neq Q_2$ and $A_1V_1 \neq A_2V_2$.

Summary

1. When gravitational and resistance forces balance, uniform flow occurs where the velocity and depth are constant and the water surface is parallel to the bed. This scenario is described by the Manning equation (8.8): $V = (1/n)R^{2/3}S_O^{1/2}$ where V is the mean velocity (m/s), n Manning's roughness factor (s/m$^{1/3}$), R the hydraulic radius (m) and S_O the bed slope (e.g. 1/200). The value of n depends not just on the bed material, but also channel alignment, irregularity and variation (Table 8.2). Discharge (m^3/s) is: $Q = (A/n)R^{2/3}S_O^{1/2}$ where A is the cross-sectional area of flow (m^2). With compound channels and where n is not uniform around the perimeter, the total discharge equals that in the subsections (equation (8.18) and section 8.4).

2. The specific energy (Em) of the flow is that calculated using the channel bed as a datum, so $E = D + \alpha V^2/2g$ where D is the depth (m) and α allows for variations in velocity over the cross-section. Specific energy can be used to show (e.g. in Table 8.4) that for any Q there are usually two alternate depths of flow: one supercritical and less than the critical depth ($D_1 < D_C$ in Fig. 8.16) and one subcritical and greater than the critical depth ($D_2 > D_C$). The exception is critical depth itself, which represents the minimum E at which Q is possible. Subcritical and supercritical flow are different in character and it is important to understand this (see sections 8.6.2 and 8.6.3).

3. For the simplest case of a rectangular channel of width B, equation (8.32) gives the critical depth (D_C) as: $D_C = (Q^2/gB^2)^{1/3}$ which corresponds to $D_C = \frac{2}{3}E_C$ where E_C is the specific energy with critical flow. The corresponding critical velocity is $V_C = (gD_C)^{1/2}$. The critical bed slope at which the flow is critical can be found by substituting V_C, D_C, etc. into the Manning equation and solving for S_C.

4. Flow cannot change smoothly from supercritical to subcritical: in Fig. 8.16 the transition from D_1 to D_C is OK, but then an increase in energy is needed to get from D_C to D_2, which is not possible. Thus the change occurs suddenly as a hydraulic jump. The initial supercritical depth at which a jump will form and the sequent depth after the jump are given by equations (8.36) and (8.37). It is important that the jump occurs predictably on a reinforced bed to prevent damage to the channel and its surroundings.

5. Gradually varying non-uniform is common, and means that the water surface is not parallel to the bed so that the depth of flow and velocity vary along the length of the channel. Consequently the slope of the total energy line (i.e. friction gradient, S_F) should be used in the Manning equation. At any cross-section i, equation (8.40) gives $S_{Fi} = V_i^2 n_i^2/R_i^{4/3}$ and the average friction gradient over a length (ΔL) of the channel as the average value at the two ends: $\overline{S_F} = (S_{F1} + S_{F2})/2$.

6. With gradually varying non-uniform flow the elevation of the water surface is often unknown and has to be calculated. However, the profile may have the standard shapes shown in Fig. 8.29 and Table 8.6. The profiles are denoted by an initial letter (Mild, Steep, Critical, Horizontal or Adverse) and number (1 = above normal depth line, 2 = between critical depth and normal depth lines, 3 = between the bed and critical depth line). Calculation of the profile starts at a control point with a known depth–discharge relationship (e.g. where critcal depth occurs or at a structure) and proceeds in an upstream direction with subcritical flow and downstream with supercritical flow. The calculations are iterative and use either total head (H) or specific energy head (E). Knowing D_1 and V_1 at the control point, the

standard step method consists of little more than guessing D_2 (and thus V_2) at the next cross-section then ensuring that the head at section 2 (e.g. $E_2 = D_2 + V_2^2/2g$) is consistent with that obtained using the equations in point 5 above (i.e. $E_2 = E_1 + (S_O - \overline{S_F})\Delta L$), remembering to include the reference bed slope (S_O) when using specific energy.

7. The direct step method uses the gradually varying flow equation (8.41) to calculate the distances (ΔL) required for the water depth (ΔD) to change by a fixed amount. Equation (8.51) is:

$$\Delta L = \Delta D \times \frac{(1-F^2)_{AV}}{(S_O - S_F)_{AV}}$$

8. Surge waves in open channels are an example of unsteady flow. They can be analysed using steady flow equations provided the wave front is brought to rest by superimposing on the system an equal and opposite absolute wave velocity V_W measured relative to the banks. When the surge wave moves in the same direction as the river flow $V_W = c + V_1$, and when it moves in the opposite direction $V_W = c - V_1$ where c is the velocity or celerity of the wave relative to the velocity V_1 occurring at the smaller depth of flow. With a uniform rectangular channel, for small waves $c = \sqrt{gD_1}$ and for larger waves:

$$c = \sqrt{\frac{gD_2(D_2 + D_1)}{2D_1}}$$

Revision questions

8.1 (a) When is the flow in a pipe considered to be open channel flow, and when is it not? Explain the distinction. (b) What is meant by the total energy of the flow in an open channel? (c) Define specific energy. (d) Describe the major differences between total energy and specific energy. (e) Define what is meant by the terms wetted perimeter, hydraulic radius, hydraulic mean depth, normal depth, subcritical flow, supercritical flow, critical depth, hydraulic mean critical depth, critical velocity and critical slope.

8.2 Water flows down a trapezoidal open channel that has a bottom width B of 4.0 m and a water surface width B_S of 17.2 m when the depth of flow on the channel centreline is 3.3 m. (a) Without using any charts or tables, calculate the value of the wetted perimeter, the hydraulic radius and the hydraulic mean depth. (b) If the optimum hydraulic section for a trapezoidal channel is half of a hexagon, prove that the top width of the channel is 2.309D where D is the centreline depth. (c) If D = 3.3 m, what should be the dimensions and side slopes of the channel to obtain the maximum discharge?

[(a) 18.758 m; 1.865 m; 2.034 m. (c) B = 3.81 m, B_S = 7.62 m, 30° to vertical]

8.3 Water flows down a half-full circular pipeline of diameter 1.4 m. The pipeline is laid at a gradient of 1 in 250. (a) If the Chezy coefficient, C, is 55 $m^{1/2}$/s, what is the discharge? (b) Using equation (8.7), what value of Manning's n corresponds to $C = 55 m^{1/2}$/s? (c) By considering the units of the quantities involved, show that equation (8.7) is a valid relationship. (d) Calculate the discharge in the pipeline using the Manning equation and the result of (b), assuming everything else is the same.

[1.584 m^3/s; n = 0.0153 $s/m^{1/3}$; 1.580 m^3/s]

8.4 (a) Water flows down a rectangular channel that has a width of 20 m, a Manning n value of 0.032 $s/m^{1/3}$ and a slope of 1 in 100. Calculate the discharge in the channel when the depth of flow is 0.5 m, 1.0 m, 2.0 m and 4.0 m. Plot a graph of depth against discharge. Is a straight line obtained? (b) For the channel in (a) calculate the depth corresponding to a discharge of 212.5 m^3/s.

[(a) 19.06 m^3/s; 58.65 m^3/s; 175.72 m^3/s; 503.38 m^3/s; no. (b) 2.261 m]

8.5 (a) Explain why it is essential for an engineer to know whether the flow is subcritical or supercritical when designing (i) an open channel, and (ii) a bridge that will have its abutments in the channel. (b) Summarise the behavioural differences between subcritical and supercritical flow. How do these two types of flow behave differently? (c) If the flow at a particular cross-section of an open channel is supercritical, will it be affected by the downstream conditions? If the answer is 'no', explain why this is the case.

8.6 At a cross-section of a channel that has an irregular shape the water surface width is 2.561 m and the area of flow is 1.340 m^2 when the discharge is 3.036 m^3/s. Determine (a) whether or not the flow is at the critical depth; (b) the hydraulic mean critical depth; (c) the critical velocity.

[Yes; 0.523 m; 2.266 m/s]

8.7 A trapezoidal channel that carries a discharge of 24.6 m^3/s has a bottom width, B, of 12.5 m and side slopes of 1:1. For the critical flow condition calculate (a) the actual water depth on the channel centreline, D_C; (b) the hydraulic mean critical depth, D_{MC}; (c) the critical velocity, V_C.

[0.719 m; 0.682 m; 2.588 m/s]

8.8 (a) Explain how it is possible for the flow at a particular cross-section in a channel to occur at two significantly different (alternate) depths of flow at the same discharge. (b) Water flows down a rectangular channel 4.0 m wide at a depth of 1.7 m. The discharge is 15.0 m^3/s. (i) Determine whether the flow is subcritical or supercritical. (ii) Calculate the second (alternate) depth of flow that could occur in the channel at the same discharge. (iii) Determine the critical slope of the channel assuming that it is lined with concrete with a Manning n of 0.012 s/m$^{1/3}$. [Subcritical; 0.785 m; 1 in 406]

8.9 (a) Describe what is meant by a 'hydraulic jump'. (b) Explain where, how and why a hydraulic jump forms. (c) A rectangular channel 10 m wide forms part of a dam spillway. The discharge is 36.5 m^3/s when the depth of flow is 0.43 m. At the foot of the spillway the channel is almost horizontal, with a hydraulic jump. Calculate (i) the depth of flow after the jump; (ii) the height of the hydraulic jump; (iii) its length; (iv) the energy loss at the jump.

[2.31 m; 1.88 m; 13.45 m; 1.67 m]

8.10 (a) What are the essential differences between uniform flow and gradually varying non-uniform flow? (b) Sketch and describe the conditions required for an M1, M3 and S1 surface profile. Give examples of where you might find such curves. (c) Repeat the calculations for Self Test Question 8.5 up to chainage 400 m, but this time using a standard step $\Delta L = 40$ m. Is there a significant difference in the calculated depths? If so, how do you think the accuracy of the calculated depth is affected by the value selected for ΔL? Why does ΔL have an effect?

[(c) About 0.045 m and 0.023 m difference at chainage 200 and 400 m respectively]

8.11 A weir built to supply water to an old paper mill raises the water level in the river for some distance upstream. To reduce flooding it is proposed that the weir should be removed, but the effect of this needs to be determined first. Just upstream of the weir the depth is 3.722 m when the discharge is 15.333 m^3/s. The river channel is rectangular and prismatic, 6.5 m wide, $n = 0.035$ s/m$^{1/3}$, the bed slope is 1 in 200, and α can be taken as 1.00. (a) Estimate the normal depth in the channel. (b) Determine by how much the depth of flow is increased by the weir at the weir. (c) Using the standard step method, obtain the elevation of the water surface upstream of the weir. (d) At approximately what distance does the backwater curve regain normal depth? (e) If houses and commercial properties line both banks, comment upon the possible repercussions of removing the weir.

[(a) 1.250 m; (b) 2.472 m; (d) 900 m]

8.12 Repeat question 8.11 above using the direct step method. Assume that the required change in water level is (3.722 − 1.250) = 2.472 m divided into eight steps of $\Delta D = 0.309$ m. (a) Compare the chainages at which the upstream depth in the two questions first becomes 1.250 m. Is there a significant difference? (b) Why might the answers in part (a) be expected to differ? What factors affect the accuracy of the analysis?

[(a) roughly 190 m difference]

9

Hydraulic structures

Hydraulic structures include dams, which store water for water supply, and sluice gates which are used to control the discharge in rivers and to alleviate flooding. Bridges and culverts, which carry roads and railways over rivers, are very numerous examples of hydraulic structures; few roads are constructed without them. Knowledge and skill are needed to design a hydraulically efficient bridge or culvert that has a waterway of an appropriate size, does not cause upstream flooding, and is unlikely to be damaged by floods. Concrete weirs are used to measure the discharge of a river. They are designed to operate with critical depth on the crest, and illustrate the use of the principles outlined in Chapter 8. Thus some of the questions answered in this chapter include:

What are the main types of dam, and what conditions suit each type?

How is a simple concrete gravity dam designed?

What are the most common causes of dam failure?

What types of dam spillway are suitable for the main types of dam?

How can the head–discharge relationship of a sluice gate be determined?

How can bridge waterways and culverts be analysed and designed?

How can a weir or flume be designed to measure the flow in a river?

 **9.1 Dams**

Dams may be constructed to store water for domestic and industrial use, for irrigation, to generate hydro-electricity or to prevent flooding. Some of these topics are discussed in later chapters. Dam design and construction is a highly specialised branch of civil engineering, necessarily so since a failure could cost thousands of lives. The type of dam constructed in a particular location must be determined after considering all relevant factors such as the local geology, shape of the valley, environmental concerns, climate, and the availability of local materials, expertise, manpower and plant. However, there are only really three basic types of dam: gravity, arch, and buttress, which are further classified according to the material from which they are constructed: concrete, masonry, rock, earth etc. (see Box 9.1). Briefly the characteristics of these dams are as follows.

Gravity dams

- Essentially, gravity dams are relatively simple. They are usually straight or slightly curved in plan and rely on their own weight to resist the hydrostatic force that is trying to overturn the dam and push it forward. Gravity dams can be constructed using concrete (Figs 9.1a and 9.2), masonry, or embankments of rock or earth (earth here meaning any clay–silt–sand–gravel–rock mixture). There is a wide variety of embankment dams, Fig. 9.1b illustrating only one generalised type.

- Earth or rock embankment dams form about 83% of the world's large dams (i.e. >15 m high), and concrete gravity dams another 11% (Novak *et al.*, 1990). Typical heights of large gravity dams are in the range 50–150 m. The concrete Grand Coulee dam in Washington is 168 m high with a 122 m wide base.

- Concrete gravity dams may be built almost anywhere, but when over 20 m high require a relatively strong rock foundation to cope with the compressive stress generated by their weight. Such dams are now relatively expensive since they require large amounts of concrete. Problems arise with the heat generated as the concrete sets and with shrinkage cracks.

- Since 1982 roller-compacted concrete dams have become quite common. They can be 60% less expensive than a concrete gravity dam, the saving arising from using a dry, low cement mix incorporating fly ash and then effectively using earth moving machinery and roller compactors to build an embankment dam.

- A rock or earth fill embankment dam is relatively wide at its base and so can be built on a relatively weak foundation. They are ideal for wide shallow valleys. Provided a suitable material is available locally, rock and earth dams can be economical to construct using large earth-moving equipment. They are rendered water-tight via an impermeable core or a membrane on the upstream face.

- Care must be taken to ensure that earth and rockfill dams are not destroyed by erosion as a result of water spilling over the embankment. Consequently the spillway is usually not part of the dam itself. Settlement and internal erosion are other common causes of failure.

Arch dams

- Arch dams are almost always constructed from reinforced concrete, and use perhaps only 20% of the concrete required for a gravity dam. An arch dam is rather like an arch bridge lying on its side with the crown upstream (Figs 9.1c and 9.3). The strength of the arch is used to transmit the hydrostatic force to the foundations, which must consist of rock strong enough to withstand the high loads (see Table 9.1). Arch dams require a narrow, steep sided valley or gorge. Typically the length of the dam's crest is limited to about 10 times its height.

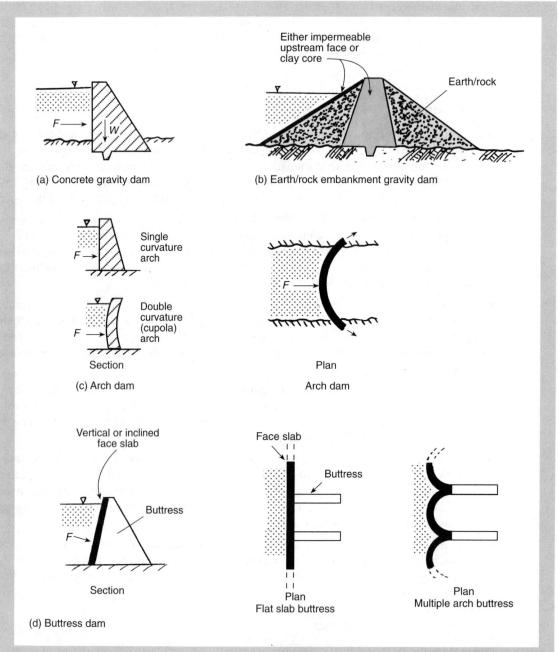

Figure 9.1 Major dam types. There are many different varieties, so those shown are just a generalised example

Figure 9.2 Cow Green Dam on the river Tees. Because the geology changes, unusually the far side of the dam is an earth embankment while the near side is a concrete gravity dam. The overflow spillway is in the centre of the photo, with a downstream stilling basin. The compensation flow is discharging from the pipe centre left [*Photo © L. Hamill*]

Table 9.1 Allowable compressive stresses for foundation materials
[After Linsley *et al.*, *Water Resources Engineering*, 1992; reproduced with permission of The McGraw-Hill Companies]

Material	Allowable stress (10^3 N/m^2)*
Granite	4000–6000
Limestone	2500–5000
Sandstone	2500–4000
Gravel	250–500
Sand	150–400
Firm clay	250–300
Soft clay	50–100

* 10^3 N/m^2 = kN/m^2.

- Arch dams comprise only about 4% of the world's large dams. A typical height for a large arch dam is 70–250 m. The infamous Vaiont Dam that survived the disaster in Table 9.3 is 266 m high but has a maximum thickness of only 22 m (Linsley *et al.*, 1992). This is 98 m higher but 100 m narrower than the Grand Coulee Dam mentioned above.

- Arch dams are technically complex, particularly thin double curvature (cupola) dams which are curved both in plan and section. Most of the hydrostatic force is resisted by

Figure 9.3 Monticello Dam, California, is a medium thick arch dam having a height of 93 m and a crest length of 312 m. Its crest width is 3.7 m, its base width 30.5 m. It contains about 249 000 m³ of concrete [*Photo reproduced with permission of US Bureau of Reclamation*]

the arching action between the abutments, and the remainder by cantilever action at the base. If an abutment fails, as at Malpasset where the left abutment moved by as much as 2 m, then the consequences can be disastrous (see Table 9.3).

Buttress dams

- These are a hybrid of an arch and concrete gravity dam. The flat slab variety consists of a continuous upstream face slab, either vertical or angled to increase stability, with downstream buttresses to provide strength and support (Figs 9.1d and 9.4). The multiple arch type consists of a series of arches, and is used when a valley is too wide for a single arch dam.

- Buttress dams form about 2% of the world's large dams, typical heights being 30–90 m with a flat upstream slab and 40–220 m with a multiple arch.

- Buttress dams use only about 60% of the concrete required for a gravity dam, but may not be any cheaper because of the need to reinforce the concrete and to use more complex formwork. However, their reduced weight means they can be built on weaker foundations. The gaps between the buttresses also result in a smaller uplift (see below).

The design of a dam would be undertaken using sophisticated computer software, but a brief introduction to some aspects is given below.

Figure 9.4. Wimbleball Dam, Somerset. This is a concrete buttress dam. Note the overflow spillway (top left) leading to a series of cascades parallel to the dam. The flat-V Crump weir measures the discharge emerging from the stilling basin. The concrete chevron is to initiate a controlled hydraulic jump [*Photo © L. Hamill*]

9.1.1 Elementary concrete gravity dam design

The design of a dam is a complex process, but some of the basic principles can be illustrated using a simple rectangular concrete gravity dam. The two unavoidable forces acting on the dam are the hydrostatic force (F) and the weight of the dam (W). If the line of the resultant of these two forces (R) lies outside the downstream toe of the base then the dam would tip or overturn about this point. To prevent tension cracks developing at the upstream face and potentially damaging water penetration of the concrete, the 'middle-third rule' states that R should pass through the middle-third of the dam's base. Normally a factor of safety (about 1.3 to 4.0 according to circumstances) would be adopted so that R lies well inside the middle-third, but Fig. 9.5a shows the limiting condition where R passes through the very edge of the middle-third. Example 9.1 illustrates how the minimum base width is calculated. To increase stability, generally the width of a dam increases towards its base (Fig. 9.5b). With this more complex cross-section the middle third-rule can be applied at different levels: R_1, R_2 and R_3 should all fall within the middle-third of their base. The calculations are slightly more difficult, but the procedure is similar.

The uplift force on the base of the dam was ignored in Example 9.1; it can significantly reduce the effective weight (W) of the dam. Uplift occurs because water under pres-

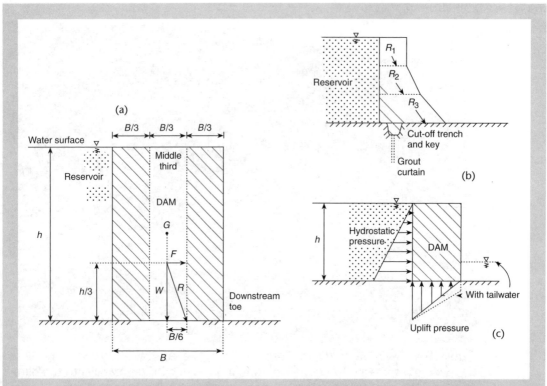

Figure 9.5 (a) To satisfy the 'middle-third rule' the resultant of *F* and *W* must pass through the base of the dam within the middle-third (i.e. *B*/3). This can be used to calculate the minimum width, *B*. (b) If the dam width varies, the middle-third rule can be applied at various levels. (c) The analysis in part (a) ignores the uplift pressure on the base, which effectively reduces *W*

sure exerts a force at right angles to any surface it comes into contact with, including upwards (see section 1.7). If the water underneath the dam at the upstream face is at the full hydrostatic pressure (*h*) while at the downstream face the pressure is atmospheric, the pressure distribution is as shown by the solid line in Fig. 9.5c; a high tailwater increases the pressure on the base (dashed line). Usually the uplift pressure can be reduced to around 33–66% of the maximum upstream value by constructing a relatively impermeable grout curtain or cut-off underneath the upstream part of the dam. As the water seeps through the curtain there is a large head loss, so reducing the head and uplift pressure on the base, which can be further reduced by drains drilled between the curtain and downstream toe. These measures also reduce water seepage under the dam. Seepage from the high to low pressure areas can cause erosion of the foundation.

To help prevent a gravity dam failing through sliding, it is usually keyed into the foundation to increase resistance. A sliding failure can occur at foundation level, or at any level where the net horizontal force exceeds the shear resistance. We are assuming here that the net horizontal force is due to hydrostatic pressure, but in practice horizontal earthquake forces, ice forces and wave forces also have to be evaluated. For a horizontal plane, the shear resistance is the product of the vertical force (weight) acting on the shear plane and the coefficient of friction (*f*) between its two surfaces (Table 9.2).

Table 9.2 Representative coefficients of friction for foundation materials
[After US Department of the Interior, Bureau of Reclamation, 1987]

Material*	f
Sound rock, clean and irregular surface	0.8
Rock, some jointing and laminations	0.7
Gravel and coarse sand	0.4
Sand	0.3
Shale	0.3

* For silt and clay, testing is required.

EXAMPLE 9.1

A concrete gravity dam of rectangular cross-section is to be constructed (Fig. 9.5a). The density of the concrete is 2350 kg/m^3 and that of the water 1000 kg/m^3. When the reservoir is full the maximum depth of water equals the dam height of 30 m. By considering a 1 m length and using the middle-third rule, determine the minimum dam width.

The hydrostatic force due to the water $F = \rho g h_G A$
$$= 1000 \times 9.81 \times (30/2) \times 3 \times 1$$
$$= 4414.50 \times 10^3 \, N$$

Assume the width of the dam base $= B$, as shown in Fig. 9.5a.
The weight of the dam W = weight density of concrete × volume of dam per m length
$$= 2350 \times 9.81 \times (30 \times B \times 1)$$
$$= 691.61B \times 10^3 \, N$$

By similar triangles (or $\tan \theta =$) it is apparent that:
$$\frac{W}{F} = \frac{10}{(B/6)}$$

so
$$\frac{691.61B \times 10^3}{4414.50 \times 10^3} = \frac{60}{B}$$

$$691.61B^2 = 264\,870$$

$$B = 19.57 \, m$$

Note: often B is about $\frac{2}{3}$ of the water depth (h), so this is a good starting point for calculations.

SELF TEST QUESTION 9.1

Repeat the calculations above using masonry of density 2700 kg/m^3 instead of concrete. By how much does this change the minimum thickness of the dam?

Table 9.3 Some notable dam incidents
[Based on Binnie, 1981; Novak *et al.*, 1990; Kiersch, 1964; New Civil Engineer, 1979a,b,c; Vischer and Hager, 1998]

Dam	Location	Height & type	Failure date	Cause	Deaths
Dale Dyke (or Bradfield)	Near Sheffield, England	29 m earthfill	1864	Poor design and construction; internal erosion caused dam breach.	238
Malpasset	France	66 m concrete arch	1959	Foundation failure initiated sudden dam collapse.	421
Vaiont	Italy	266 m concrete arch	1963	240×10^6 m^3 landslip into reservoir. 110 m splash wave overtopped dam. The dam survived.	2043
Machhu II	India	26 m earthfill	1979	Catastrophic flood, inadequate spillway design and operation, overtopping caused breach.	>2000

9.1.2 Dam failures

A sobering thought is that a dam is 200 times more likely to fail than a nuclear power-station. There are many ways a dam can fail: cracking, overturning about the toe, sliding and material failure being possibilities that require calculation. Causes of failure can include overtopping and erosion of earth embankments as a consequence of floods that exceed the capacity of the dam spillway (overflow), earthquakes, foundation failure, and poor design and construction. Failures of large dams are quite frequent with about 1100 having been recorded, some 515 occurring between 1950 and 1975 (Widmann, 1984). Of the latter, about 30% were attributable to operational facilities, especially to inadequate spillways; 27% were due to unexpected structural behaviour, partly caused by excessively optimistic load assumptions; 20% were caused by underseepage, for instance by increased uplift; 11% were due to inadequate material properties, and the remaining 12% were due to a variety of causes. Singh (1996) found much the same thing, with earth embankment dams having the highest probability of failure: after 1900 almost half of the failures were due to overtopping. Of these overtopping failures, 41% were due to the spillway being underdesigned and 21% due to problems with spillway gate operation (Schnitter, 1993; Vischer and Hager, 1998). Thus to minimise the possibility of a failure an accurate estimate of the design flood is needed, an adequate spillway must be provided, the dam has to be designed correctly, construction must be to a good standard and, when complete, the dam must be operated well and inspected regularly. Some of these topics arise later.

Earth dams tend to fail slowly. Sometimes the deterioration of a dam and its impending failure are discovered early enough for the reservoir to be emptied, so that the incident passes virtually unnoticed. However, concrete dams can fail instantaneously, releasing a huge, fast moving floodwave into the valley downstream. Tragically, in a few instances catastrophic failures have resulted in a large loss of life (Table 9.3). One specialised branch of hydraulics concerns dambreak waves (e.g. see Vischer and Hager, 1998).

See for yourself – dam types and cracking dams

Although the longevity of internet sites is hard to predict, here are a couple that you might like to investigate:

1. **www.usbr.gov/cdams/** is the US Bureau of Reclamation site and contains photos, technical and hydrological data of a large number of dams. It illustrates nicely the major dam and spillway types described above. Among the hydrological data it lists details of the probable maximum flood (e.g. rainfall on snow). This highlights the significance of the topics in Chapters 12 and 13.

2. **http://simscience.org** illustrates how dams can crack and the consequences. You can draw a dam, indicate where you want the cracks, and then watch what happens.

9.1.3 Dam spillways and drawoffs

The function of the spillway is to allow floods to pass safely over, around or through the dam and to discharge them to the downstream river channel without causing damage. Consequently there is usually a stilling basin downstream of the spillway (Fig. 8.26). As indicated above, poor spillway design and operational problems account for around 30% of all dam failures, so clearly this aspect of the design is crucial (see Table 13.3 for the flood return period). There are many types of spillway: overflow, chute, side-overflow, shaft, crest gates and syphon. Which is used in any location will depend upon the type of dam (e.g. concrete or earth), its size, the topography of the area and operational considerations.

An **overflow spillway** is simply part of the dam that is designed to allow water to flow over it. This is quite common with concrete and masonry dams (Fig. 9.2). With earth or rock dams, which would erode and possibly fail if water flowed over them, either a special concrete spillway section must be constructed or an alternative such as a shaft spillway adopted. The ideal overflow spillway should guide the water over the crest as smoothly as possible, and in cross-section should resemble the underside of the aerated nappe in Fig. 5.11. If the water lifts from the spillway a vacuum can form resulting in cavitation damage (see section 11.8). The discharge over the spillway can be related to the head (H) over the crest using a weir equation like equation (9.20), where C has a value between 1.7 and 2.3 depending upon H and the spillway geometry (Linsley *et al.*, 1992). Sometimes overflow spillways end in a ski jump that throws the water upwards so that it lands far enough downstream not to cause scour at the base of the dam.

A **chute spillway** is a steep concrete channel to take water from reservoir level down to river level. Where the topography permits they are often built on natural ground at the end of an earth or a rock fill dam, since it is undesirable to locate the spillway on the dam itself, but chutes may be adopted with any type of dam. The flow in the chute is supercritical, the high velocity resulting in the entrainment of air and bulking (i.e. expansion vertically) of the water. Either to avoid building on earth or rock dams or for other reasons such as topography, sometimes the water may enter a shute spillway via a side-overflow weir located perpendicular to the upstream face of the dam. Figures 9.6 and 9.7 show the side-overflow weir at Kielder Dam, and the steeply inclined chute down which water flows

Figure 9.6 The side overflow weir at Kielder Dam leads to the chute spillway in Fig. 9.7. The 52 m high earth embankment dam and entrance to the chute is at the top of the photo [*Photo © L. Hamill*]

at speeds of up to 31 m/s (70 mph). In narrow valleys, a side overflow weir may be used to allow a chute spillway to be constructed along the downstream face of the dam. A similar arrangement, but using a series of steps instead of a chute, is shown at the top left of Fig. 9.4.

A bellmouth overflow and **shaft spillway** consist essentially of a funnel entrance of fixed elevation and a vertical shaft down which the overflow from the reservoir falls (Fig. 9.8). The vertical shaft leads to a gradual 90° bend and a horizontal tunnel that passes around or under the dam and hence to the downstream river channel. These spillways can be used with any type of dam, being particularly useful where space is restricted and with earth or rock dams that require a separate spillway structure. They are also called drop inlets and morning glory spillways. Provided the shaft diameter is large enough to take the flow, the discharge over the bellmouth lip can be calculated by assuming a straight weir of the same length. At larger discharges the tunnel itself may provide the flow control, as in pipeflow. Problems can arise due to cavitation, while the bellmouth must be provided with some sort of screen to prevent logs or other debris becoming wedged in the shaft.

Crest gates can be used to raise the water level in a reservoir above the height of the dam and to control the level. Flat sluice gates, like the one in Fig. 1.18, slide vertically upwards on rollers running in grooves, but big gates (e.g. 15 m wide) will be subject to large forces and difficult to lift. The **radial** or **Tainter gate** is a better and more widely used option (Fig. 9.9). Being part of a circle that is pivoted at its centre, the hydrostatic force acts perpendicular to the curved gate and passes through the pivot, so theoreti-

Figure 9.7 The chute spillway at Kielder Dam leading to the stilling basin and River North Tyne. Water velocities can reach 70 miles/h (31 m/s) (*Photo © L. Hamill*)

cally there is no turning moment (other than the weight of the gate). Friction occurs mostly at the pivot and is relatively small compared to a flat gate, so radial gates can be lifted with a fairly modest motor driven winch. With all gates it is essential that there is some means of lifting them in an emergency, such as a power failure, or they should be designed to be overtopped. Dams have failed in the past as a result of gates that could not be opened.

Drum gates are sometimes used with long dams. In cross-section the gate resembles one-quarter of a cylindrical oil-drum that is attached at its centre to the dam crest by a hinge (Fig. 9.10). When not in use the gate fits into a recess in the crest (the large recess makes these gates unsuitable for small dams). By allowing water into the recess the gate is forced out into the raised position. Automatic operation is possible.

Roller (or **rolling**) **gates** consist of a steel cylinder that has gear teeth at the ends. The gears engage an inclined rack attached to piers on the dam crest. When closed the horizontal cylinder sits on the crest. The gate is raised by pulling on a hoisting cable, with the result that the cylinder rolls up the rack. The water in the reservoir then discharges underneath the gate (see Fig. 9.12e). Roller gates are suitable for long spans (e.g. 45 m) provided the variation in water level is moderate (e.g. 6 m).

Syphons spillways are useful where automatic action is desirable, the required discharge is not large, space is restricted, it is necessary to increase the capacity of an existing

Figure 9.8 The bellmouth entrance to the shaft spillway at Balderhead Reservoir, and the drawoff/access tower [*Photo © L. Hamill*]

Figure 9.9 A radial gate on the crest of Dundreggan Dam, Scotland. Being part of a circle, the force due to hydrostatic pressure passes through the pivot located at the centre of curvature, so there is no turning moment on the gate, making it easy to control [*Photo © L. Hamill*]

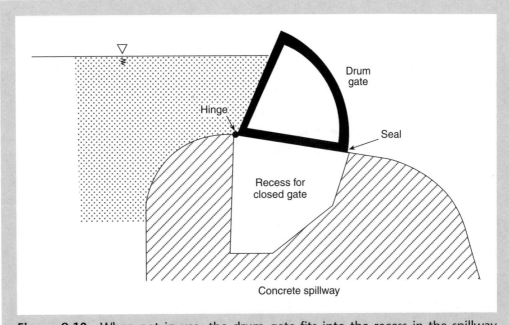

Figure 9.10 When not in use, the drum gate fits into the recess in the spillway crest. It can be raised by allowing water into the recess

spillway, or where it is necessary to keep the reservoir level within a narrow range. To increase the range, syphons may be set with crests at different levels. Syphons are not suitable where they are likely to be blocked by ice in winter. Design guidelines are vague, so they are often the subject of model tests.

Figure 9.11a shows a dam crest with a syphon spillway (these can be retro-fitted if the existing spillway is inadequate). With air in the syphon, as the reservoir surface rises to just above crest level the syphon operates like a conventional overflow spillway or weir. Usually the syphon exit is submerged to prevent air entering from downstream so, as the water level continues to rise and the flow over the crest increases, the air at the crest becomes entrained in the flow and is removed. The 'primed' condition is reached when the air is exhausted, allowing the syphon to run full at maximum capacity; the corresponding reservoir level is marked by a dashed chain line in the diagram. As the reservoir level falls, the water level drops below the upper lip of the inlet allowing air to enter, breaking the syphon and stopping the discharge. Thus the syphon operates over a restricted range of reservoir levels.

The top of the syphon lies above the hydraulic grade line, so when it is running full there is a sub-atmospheric pressure at the crest. Effectively, atmospheric pressure acting on the reservoir surface forces water up the syphon. Once primed, the discharge can be calculated using the small orifice equation (5.12), where C_D has a value of about 0.9 and H is the vertical difference between the reservoir level and tailwater level (Fig. 9.11a). Since the discharge depends upon H, which may vary relatively little, syphons tend to give a fairly constant discharge regardless of reservoir level. As with all syphons, flow will cease if the pressure at the crest falls below about −7.5 m of water (or 2.8 m of water absolute pressure). Thus 7.5 m is the maximum elevation of the syphon crest above the hydraulic grade line (i.e. total head line minus the velocity head).

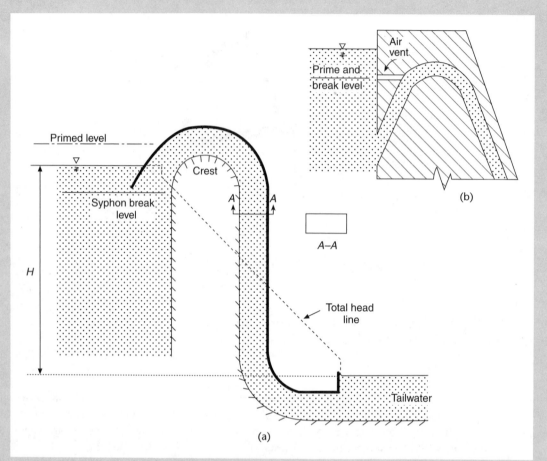

Figure 9.11 (a) Primed syphon spillway on a dam crest. (b) A syphon spillway set at a lower level. Syphons may be set at several different levels. The syphon will not prime unless water in the reservoir covers the air vent

Syphons can also be built into a dam at various levels. In Fig. 9.11b the syphon operates as described above, except that an air vent is incorporated near crest level so that the syphon will run in the primed condition only when water in the reservoir covers the vent and prevents air from entering. This further restricts the range of head and ensures an almost constant discharge from the syphon.

Although not part of the spillway (except with the shaft type) reservoirs also need a **drawoff tower** constructed within the reservoir to allow water to be abstracted, either for water supply or as compensation water to provide some flow in the river downstream. Drawoff towers often have several drawoff points at different levels, thus enabling water from near the surface of the full reservoir to be selected, or from lower down.

9.2 Sluice gates and other control gates

Sluice gates like the one in Fig. 1.18 are used to control the flow in rivers and man-made open channels. They are sometimes referred to as underflow gates, since the flow passes

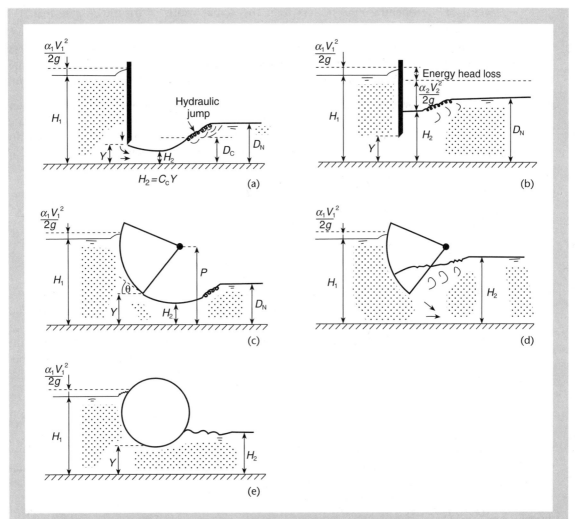

Figure 9.12 Underflow gates. (a) Free vertical sluice gate. (b) Drowned vertical sluice gate. (c) Free radial gate. (d) Drowned radial gate. (e) Roller gate

under the bottom edge of the gate, as shown diagrammatically in Fig. 9.12 (a and b). Once calibrated, either by measurements in the field or model tests, they can also be used to measure the discharge. There are similarities between the discharge through an orifice and under a sluice gate, but also important differences:

■ With a small orifice there is a contraction from all sides, whereas with a sluice gate there is no contraction from the flat, horizontal channel bed. Close to the bed the water can pass through the opening without significant deviation. However, the water passing down the upstream face of the gate and through the top of the opening experiences a larger than normal contraction. This offsets the suppressed bottom contraction so that overall the amount of contraction is similar to an orifice. Depending upon the relative width of the opening to the channel, there will also be some degree of side contraction in the same way as for an orifice or sharp crested rectangular weir, hence the effective width of the opening may be less than the actual width of the sluice gate (e.g. see equation (5.23)).

■ Unlike flow through a small orifice where atmospheric pressure exists at the vena contracta of the jet, with the sluice gate the flow is along the channel bottom so that the pressure distribution at a vertical section through the jet is approximately hydrostatic (i.e. $P = \rho gh$).

The analysis of flow under a sluice gate was introduced in Example 4.11. This was based on the energy equation applied to cross-section 1 upstream of the gate and section 2 at the vena contracta, as in Fig. 9.12a. Taking the head (depth) at section 1 to be H_1, as for an orifice, and the downstream depth as H_2, then with no loss of energy head:

$$H_1 + \alpha_1 V_1^2/2g = H_2 + \alpha_2 V_2^2/2g$$

so $\quad \alpha_2 V_2^2/2g = H_1 + \alpha_1 V_1^2/2g - H_2$

$$V_2 = \left[\frac{2g}{\alpha_2}\left(H_1 + \frac{\alpha_1 V_1^2}{2g} - H_2\right)\right]^{1/2}$$

The theoretical discharge under the sluice gate is $Q_T = a_O V_2$ where a_O is the area of the opening; $a_O = bY$ where b is the width of the gate across the channel and Y the height of the opening. The actual discharge can be obtained by introducing the coefficient of discharge $C_D = C_C \times C_V$. The value of the coefficient of contraction (C_C) depends upon the shape of the gate and its relative height from the bed, but often has a value of around 0.6. The coefficient of velocity (C_V) has a value just less than unity. Thus the actual discharge $Q_A = C_D a_O V_2$ is given by:

$$Q_A = C_D a_O \left[\frac{2g}{\alpha_2}\left(H_1 + \frac{\alpha_1 V_1^2}{2g} - H_2\right)\right]^{1/2} \tag{9.1}$$

In free flow it is often assumed that $H_2 = C_C Y$. If an overall sluice gate coefficient (C) is introduced that incorporates C_D, α_2, the velocity of approach and the depth H_2 then equation (9.1) can be conveniently simplified to:

$$Q_A = C a_O \sqrt{2gH_1} \tag{9.2}$$

When the depth $H_2 \leq$ the critical depth (D_C) the flow is free (i.e. the jet can discharge freely into the downstream channel) and the value of C for a sharp edged sluice gate is between 0.5 and 0.6, as shown in Fig. 9.13. Under these conditions the value of C depends largely upon the relative height of the opening H_1/Y. For example, if $H_1/Y = 4$ then the free flow line gives $C = 0.54$.

If the normal depth of flow (D_N) in the channel some distance downstream of the sluice gate is relatively high compared to the height of the opening (Y) then it will submerge the jet and affect the discharge, as with a drowned orifice. One test for submergence is to assume that $D_N = D_2$, the depth after a hydraulic jump. The sequent depth (D_1) required to initiate the jump can be obtained from equation (8.36). If the actual depth at the vena contracta $H_2 > D_1$ then a jump cannot form and submergence is likely. Alternatively, it is possible to start with $D_1 = H_2$ and to use equation (8.37) to calculate the sequent depth D_2. The submerged condition will occur if $D_N > D_2$. This is illustrated in Example 9.2.

The submerged condition is more complex and requires two variables to obtain C from Fig. 9.13. These are D_N/Y (used instead of H_1/Y) and the Froude number in the opening $F_O = V_O/(gY)^{1/2}$, the values of which are printed within the diagram. For example, if $D_N/Y = 4$

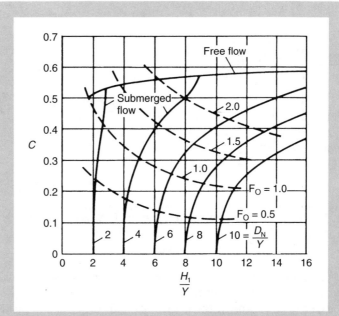

Figure 9.13 Coefficient of discharge, C, for a vertical sluice gate. For free flow, use the appropriate value of H_1/Y and the top free flow line to obtain C. For submerged flow, use D_N/Y and F_O (printed within the diagram) to find the point of intersection of the two lines and then C

and $F_O = 1.5$ then $C = 0.41$, considerably less than for the free flow situation (to compensate for **not** using the differential head $H_D = H_1 - H_2$ in equation (9.2)).

Equation (9.2) can also be applied to the other underflow radial gates in Fig. 9.12. For the radial gate in (c) and (d) all of the following affect the value of C: the upstream head H_1, the radius of the gate R, the height of the opening Y and the height of the pivot P. In the free flow condition the value of C is about 0.50–0.59 when $P/R = 0.1$; 0.58–0.63 when $P/R = 0.5$; 0.68–0.76 when $P/R = 0.9$ (see Lewin, 1995; Roberson et al., 1998). As a radial gate is opened and closed the angle θ at which the flow hits the bottom edge of the gate changes, which affects the contraction of the jet. In the free flow condition, with θ in degrees, a frequently quoted empirical expression for C_C is:

$$C_C = 1 - 0.75(\theta/90) + 0.36(\theta/90)^2 \tag{9.3}$$

In the submerged condition the value of C must be obtained from graphs similar to Fig. 9.13 (e.g. see Chow, 1981; Roberson et al., 1998).

EXAMPLE 9.2

An underflow vertical sluice gate discharges freely 56 m³/s into a rectangular channel 7.00 m wide. The gate, which is the same width as the channel, is set at a height of 1.40 m above the bed and the depth at the vena contracta is 0.85 m (Fig. 9.14a). The energy loss in the converging flow at

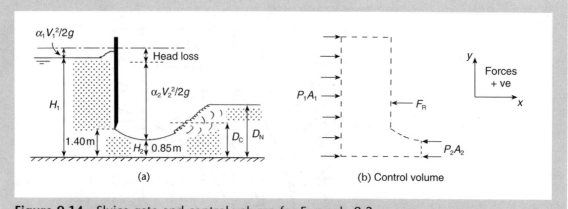

Figure 9.14 Sluice gate and control volume for Example 9.2

a sluice gate is quite small, so take the head loss between sections 1 and 2 as $0.05V_2^2/2g$. The normal depth in the downstream channel $D_N = 2.60$ m. (a) Calculate the depth upstream of the gate (take the energy coefficients α as 1.05). (b) Confirm that the gate is able to discharge freely. Does a hydraulic jump occur downstream of the gate and, if so, where? (c) Use the momentum equation to calculate the force on the gate (assume the momentum coefficient $\beta = 1.00$).

(a) Apply the energy equation between sections 1 and 2:

$$H_1 + \alpha_1 V_1^2/2g = H_2 + \alpha_2 V_2^2/2g + 0.05V_2^2/2g$$

For the first iteration, assume $V_1 = 0$. With $H_2 = 0.85$ m then $V_2 = 56.00/(0.85 \times 7.00) = 9.41$ m/s.

$$H_1 + 0 = 0.85 + (1.05 \times 9.41^2/19.62) + (0.05 \times 9.41^2/19.62)$$

$$H_1 = 0.85 + 4.74 + 0.23$$

$$H_1 = 5.82 \text{ m}$$

For the second iteration $V_1 = 56.00/(5.82 \times 7.00) = 1.37$ m/s and $\alpha_1 V_1^2/2g = 0.10$ m, so:

$$H_1 + 0.10 = 0.85 + 4.74 + 0.23$$

$$H_1 = 5.72 \text{ m}$$

For the third iteration $V_1 = 56.00/(5.72 \times 7.00) = 1.40$ m/s and $\alpha_1 V_1^2/2g = 0.11$ m, so $H_1 = 5.71$ m.
The fourth iteration gives $V_1 = 56.00/(5.71 \times 7.00) = 1.40$ m/s, so $H_1 = 5.71$ m.

This answer can be checked by substituting the values into equation (9.2) and solving for H_1:

$$Q_A = Ca_0\sqrt{2gH_1}$$

With $H_1 = 5.71$ m and $Y = 1.40$ m then $H_1/Y = 4.08$ so for free flow Fig. 9.13 gives $C = 0.54$.

$$56.00 = 0.54 \times (1.40 \times 7.00)[19.62H_1]^{1/2}$$

$$H_1 = 5.71 \text{ m}$$

Note that the exact agreement between the two answers is due largely to luck: the answer obtained from the energy equations depends upon the value assumed for the energy head loss and particularly the value of α_2, as illustrated by Example 4.11 and Self Test Question 4.4.

(b) From equation (8.32) the critical depth is:

$$D_C = (Q^2/gB^2)^{1/3} = (56.00^2/9.81 \times 7.00^2)^{1/3} = 1.87\text{m}$$

This indicates that the flow at the vena contracta is supercritical (0.85 m < 1.87 m) so the jet is likely to discharge freely. However, this can be confirmed using equation (8.36), with the depth after a hydraulic jump has occurred $D_2 = D_N = 2.60$ m. With $V_2 = 56.00/(2.60 \times 7.00) = 3.08$ m/s and $F_2 = V_2/(gD^2)^{1/2} = 3.08/(9.81 \times 2.60)^{1/2} = 0.61$ then the sequent depth to 2.60 m, i.e. the initial depth D_1 required for the jump to form, is:

$$D_1 = \frac{D_2}{2}\left(\sqrt{1+8F_2^2} - 1\right) = \frac{2.60}{2}\left(\sqrt{1+8 \times 0.61^2} - 1\right) = 1.29\text{m}$$

The depth at the vena contracta is $H_2 = 0.85$ m so $H_2 < D_1$ which means the jump can form and the gate is not submerged (if $H_2 > D_1$ then a jump cannot form, so the gate would probably be submerged). The remainder of the question is answered in Example 8.21, which shows that the depth downstream of the vena contracta gradually increases from 0.85 m to 1.29 m over a distance of 30.93 m, which is where the hydraulic jump will start. Note that an alternative solution is to say that $D_1 = H_2 = 0.85$ m which gives $V_1 = 9.41$ m/s and $F_1 = 3.26$, when equation (8.37) gives the sequent depth to 0.85 m as $D_2 = 3.52$ m. Since $D_N < D_2$ the gate is not submerged.

(c) Take a control volume between sections 1 and 2 as in Fig. 9.14b, then considering the external forces acting horizontally:

$$P_1 A_1 - F_R - P_2 A_2 = \rho Q(V_2 - V_1)$$

Assuming a hydrostatic pressure distribution at sections 1 and 2 then the average pressures are $P_1 = \rho g H_1/2$ and $P_2 = \rho g H_2/2$ thus:

$$0.5\rho g H_1 A_1 - F_R - 0.5\rho g H_2 A_2 = \rho Q(V_2 - V_1)$$

Taking the values from part (a):

$$0.5 \times 1000 \times 9.81 \times 5.71 \times (5.71 \times 7.00) - F_R - 0.5 \times 1000 \times 9.81 \times 0.85 \times (0.85 \times 7.00)$$
$$= 1000 \times 56.00(9.41 - 1.40)$$

$$1119.46 \times 10^3 - F_R - 24.81 \times 10^3 = 448.56 \times 10^3$$

$$F_R = 646.09 \times 10^3\,\text{N}$$

9.3 ▶ Flow around bridge piers and through bridge waterways

A common hydraulic problem encountered by engineers is that of flow through a bridge. Although the procedures for the structural design of bridges are well established, those relating to hydraulic design are relatively vague. Additionally, **scour** around piers and abutment causes many problems. Scour is the erosion of the river bed by flowing water. Particular difficulties are encountered in predicting the depth of scour, and in providing adequate scour protection. Floods, scour and movement of the (REMEMBER!!!)

Figure 9.15 In February 1989 part of the railway viaduct over the River Ness collapsed, probably as a result of scour [*reproduced by permission of John Paul Photography, Inverness*]

foundations are the most common cause of bridge failure (Hamill, 1999). In any particular year, hundreds of bridges around the world may be destroyed. Many involve loss of life when vehicles drive or fall off the collapsed bridge into the river. Fortunately this was not the case when part of the viaduct in Fig. 9.15 collapsed, probably as a result of scour.

When a bridge is built across a river, for reasons of economy it is frequently necessary to have piers and/or abutments in the channel. These obstruct the normal flow of water and cause an increase in upstream water level called the **afflux**. The variation of the afflux with upstream distance is the **backwater curve** (see section 8.11). The greater the obstruction, the greater the afflux. Thus the smaller the distance between the abutments and the larger the number of piers, the greater the afflux and the risk of upstream flooding and flood damage.

The afflux is basically caused by the energy head loss as the flow passes through the bridge opening. This loss arises from the need for the water on the upstream floodplain to contract through the opening and then, crucially, to expand back onto the floodplain. The energy head loss in expanding flow is generally quite large (see Box 5.2 and section 6.6.1). To overcome this head loss and to maintain continuity of flow there has to be an increase in upstream water level (i.e. the afflux) to force the water through the obstruction. Usually the greatest afflux occurs between one and two spans upstream, that is between b and $2b$ where b is the opening width.

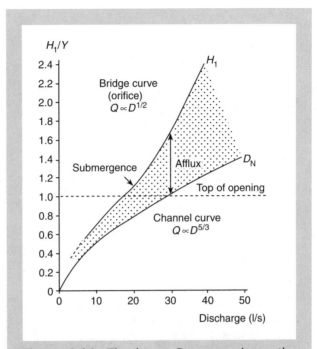

Figure 9.16 The lower D_N curve shows the head–discharge relationship of an open channel, while the upper curve shows a typical relationship for a bridge located in the channel. The vertical difference between the two lines represents the afflux (or backwater), which increases rapidly after the waterway opening of height Y becomes submerged at $H_1/Y \geq 1.1$. The data are obtained from a 250 mm wide by 125 mm high rectangular bridge opening placed in a laboratory channel at a slope of 1/200

The afflux increases rapidly if the deck of the bridge comes into contact with the water, causing the bridge waterway opening to become submerged so that the flow resembles that through an orifice (free or submerged depending upon whether or not the downstream water level is also above the top of the opening). The expansion is now three dimensional and friction losses, which are normally relatively small, increase significantly. This is illustrated by Fig. 9.16 which shows a typical stage–discharge curve for a submerged bridge opening. The curve is initially concave downwards since in channel flow $Q \propto D^{5/3}$. After submergence, in orifice flow, it is concave upwards since $Q \propto D^{1/2}$ (with the area of the opening constant). It might be imagined that it would be possible to avoid the opening becoming submerged, but this is not always possible either because of cost or site conditions. For example, in Fig. 8.6 the bridge deck and railway line could not be raised because of the proximity to Exeter Station, consequently the deck and abutments have been rounded and slim piers constructed to optimise hydraulic efficiency.

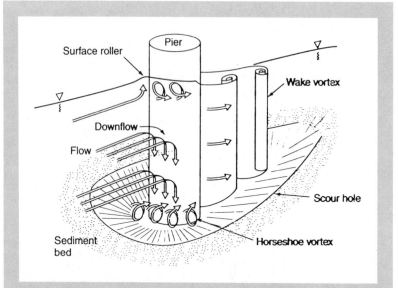

Figure 9.17 The flow pattern and scour hole at a cylindrical bridge pier. The principal causes of scour are the downflow, horseshoe and wake vortices. [*after Melville, 1988. Reproduced by permission of Technomic Publishing Company Inc., Lancaster, PA, USA*]

9.3.1 Scour at bridge piers and abutments

A highway bridge would typically be designed for floods that have a return period of between 1 in 30 and 1 in 150 years, depending upon its importance and location. As shown in Example 13.5, the probability that the design flood will be exceeded during the life of a bridge is higher than might be expected. Thus the designer must consider the consequences of this happening. If there is nothing of value on the upstream floodplains, then the increased afflux may not be a problem. However, from the structural point of view, if a larger flow is forced through the waterway opening this may significantly increase the velocity in the vicinity of the piers and abutments, causing a larger depth of scour than designed for and foundation failure.

The mechanism of scour around a bridge pier is illustrated in Fig. 9.17. When the approaching flow hits the pier it is forced downwards, the resulting horseshoe and wake vortices being the principal causes of local bed erosion. The squarer the pier the deeper the erosion, as indicated by Table 9.4. This table provides a very quick means of estimating the potential scour depth at a pier; for more accurate calculations there are many equations that can be used (e.g. see Hamill, 1999). Note that part (b) of the table indicates that the scour depth increases if the flow approaches the pier at an angle instead of head on. The table assumes that the bed material is easily eroded (Table 9.5). However, similar scour depths can be achieved in more competent material, but over a much longer period of time. Protection against scour can be provided by various means, such as using foundations that extend below the scour depth, employing rip-rap (large angular stones) to protect the piers and abutments, using sacrificial piles to protect

Table 9.4 Typical scour depths d_{SP}

[After Neill, 1973. Reproduced by permission of University of Totonto Press]

Part (a) relates to different pier shapes when pointing directly into the flow, while part (b) is a multiplication factor to be used when the flow hits the pier at an angle. The effect of any abutments may be additional.

(a)

Pier shape in plan	Pier shape in profile	Suggested allowance for local scour
		$d_{SP} = 1.5\,b_P$
	Ditto	Ditto
	Ditto	$d_{SP} = 2.0\,b_P$
	Ditto	$d_{SP} = 1.2\,b_P$
		$d_{SP} = 1.0\,b_P$
Ditto		$d_{SP} = 2.0\,b_P$

(b) **Multiplying factors for local scour at skewed piers***
(to be applied to local scour allowances of part a).

Angle of attack	Length-to-width ratio of pier in plan		
	4	8	12
0°	1.0	1.0	1.0
15°	1.5	2.0	2.5
30°	2.0	2.5	3.5
45°	2.5	3.5	4.5

* The table is intended to indicate the approximate range only. Design depths for severely skewed piers, where the use of these is unavoidable, should preferably be determined by means of special model tests. The values quoted are based approximately on graphs by Laursen (1962).

Table 9.5 Typical clear water erosion threshold velocities, V_e

If water velocities exceed V_e then erosion is probable. Note that for water carrying a lot of silt or sediment V_e may be up to 50% higher.

Material	V_e m/s
Silt	0.2
Fine sand	0.3–0.5
Coarse sand	0.6–0.8
Alluvial silt (non-colloidal)	0.6
Gravel	1.0
Stiff clay	1.1
Coarse gravel	1.2
Pebbles/stones (>60 mm diameter)	3.0

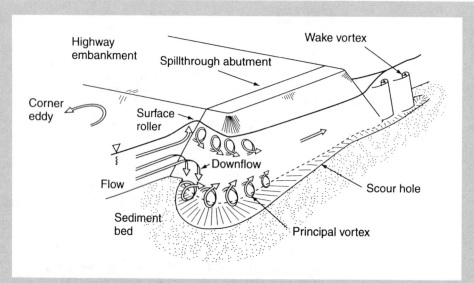

Figure 9.18 The flow pattern at a sloping abutment. The downflow and principal vortex are the chief causes of scour, as in Fig. 9.19 [*after Melville, 1988. Reproduced by permission of Technomic Publishing Company Inc., Lancaster, PA, USA*]

the piers, concreting the bed, and using electronic instruments to sound an alarm if significant scour occurs. The first option is the best, the others rather an admission of a poor initial design.

The cause of scour at an abutment is shown in Fig. 9.18. Note that the principal vortex is chiefly responsible for the scour hole, and that the potential for scour increases with hydraulically inefficient square edged abutments that result in high local velocities. Comparing Figs 9.18 and 9.19, it is easy to see the damage caused by the principal vortex, mostly (it would appear) to the filled ground adjacent to the abutment foundation.

Figure 9.19 Scour at the upstream abutment of Nether Bridge, near Launceston, Cornwall. The action of the downflow and principal vortex are apparent [*Photo © L. Hamill*]

Scour is not an easy problem to assess: laboratory model studies suffer from scale effects, while field studies during flood have been very difficult to conduct accurately, although modern electronic equipment such as fathometers and ground penetrating radar is making this easier. However, scour assessment is not an area of hydraulics where great accuracy should be expected. Additional complications arise when debris becomes snagged on piers or wedged across a bridge opening, significantly altering the anticipated flow velocities. Other difficulties include evaluating the effect of exposed footings and piles on scour depths. Useful guides to the subject and the equations available to estimate scour depth have been provided by Jones (1984), Melville (1988), Highways Agency (1994), Richardson and Davis (1995) and Hamill (1999).

9.3.2 Afflux resulting from bridge piers

With bridges that consist of many spans the discharge (Q m^3/s) and upstream afflux ($H_1^\star$) are controlled by the number, shape and thickness of the piers (Fig. 9.20). Many investigators have studied this, with one of the earliest and simplest equations being credited to d'Aubuisson. The equation requires a trial and error solution because V_1 depends upon $H_1^\star$.

$$Q = K_A b D_N (2gH_1^\star + V_1^2)^{1/2} \tag{9.4}$$

where K_A is a dimensionless coefficient having a value between 0.96 and 1.31 depending upon the shape of the piers and the proportion of the channel width they occupy (Table

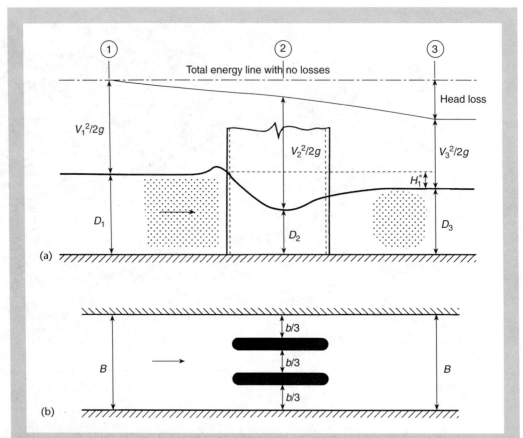

Figure 9.20 Definition of the afflux, H_1^*, caused by bridge piers located in a river. If the approach velocity (V_1) is large and/or the piers cause a significant obstruction then, as the water accelerates between the piers, critical or super-critical flow may occur at section 2, and possibly in the downstream channel (see Fig. 9.21) [*after L. Hamill, Bridge Hydraulics, 1999. Reproduced by permission of Routledge*]

9.6), b is the total width of the openings between the piers (m), D_N is the normal depth (= D_3 m) downstream of the bridge and V_1 is the approach velocity (m/s) some distance upstream of the bridge. In the table, m is the dimensionless **channel contraction ratio**, and is defined as $m = (1 - b/B)$ where B is the total width of the river channel. Thus if a bridge opening is 20 m wide (= b) and is located in a channel 30 m wide (= B) then $m = (1 - 20/30)$ = 0.33. Since the flow in a compound channel is not evenly distributed throughout its cross-section, m can be defined more accurately using the discharge ratio $m = (1 - q/Q)$ where q is the flow in m³/s that can pass through the bridge opening without having to deviate (e.g. move off the floodplain) and Q m³/s is the total discharge. This ratio should be adopted whenever possible (see section 8.4 and Hamill, 1993).

Another much quoted equation for the afflux arising from piers is:

$$H_1^* = K_Y D_N F_N^2 (K_Y + 5F_N^2 - 0.6)(m + 15m^4) \tag{9.5}$$

Table 9.6 d'Aubuisson and Yarnell bridge pier coefficients for afflux

[After Yarnell, 1934]

Type of pier	d'Aubuission coefficient, K_A Channel contraction ratio, m					Yarnell, K_Y Channel contraction ratio 0.12–0.50
	0.10	0.20	0.30	0.40	0.50	
Square nose and tail	0.96	1.02	1.02	1.00	0.97	1.25
Semicircular nose and tail	0.99	1.13	1.20	1.26	1.31	0.90
90° triangular nose and tail	—	—	—	—	—	1.05
Twin cylinder pier without diaphragm	—	—	—	—	—	1.05
Twin cylinder pier with diaphragm	—	—	—	—	—	0.95
Lens shaped nose and tail	1.00	1.14	1.22	—	—	0.90

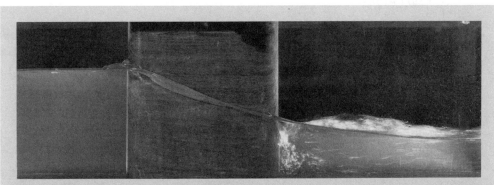

Figure 9.21 Flow past a round nosed model bridge pier, illustrating the reduction in depth and the possible occurrence of supercritical flow [*reproduced by permission of TecQuipment International Ltd, Nottingham*]

where K_Y is Yarnell's pier coefficient (Table 9.6), D_N is the downstream normal depth (= D_3 m) and F_N the corresponding Froude number. This equation can be solved relatively easily.

The equations above relate to the performance of piers in a flow which is subcritical. However, the reduced cross-sectional area of flow between the piers can sometimes result in a significant drawdown of the water surface, high velocities and supercritical flow (Fig. 9.21). The behaviour of supercritical flow would be very different, but it is relatively rare, so for this condition either consult a more specialised text (e.g. Chow, 1981) or obtain a simple solution using the equation below. This assumes that critical depth occurs at section 2 somewhere between the piers, with supercritical flow and a hydraulic jump further downstream. Based on a consideration of the energy head loss, the specific energy head (E_1)

at section 1, where maximum afflux occurs, can be estimated from the conditions at section 2 thus:

$$E_1 = C_P V_2^2/2g + E_2 \qquad (9.6)$$

where C_P depends on pier shape and has a value of about 0.35 for square ended piers and 0.18 for rounded ends. The upstream depth D_1 can be estimated from E_1, then the afflux $H_1^* = D_1 - D_N$ where D_N is the normal depth in the channel without the piers. The values of C_P are for piers with a length:width ratio of 4:1 pointing directly into the flow; longer piers with rounded noses increase H_1^* by as much as 5% (7:1) or 10% (13:1). Longer square nosed piers result in a slight decrease in H_1^*. A skew of 20° increases H_1^* by a factor of 2.3, since the pier has an effective width approximately 2.3 times as large when hit by the flow at this angle (Henderson, 1966).

EXAMPLE 9.3

A rectangular river channel 67.5 m wide is to be completely spanned by a new bridge that will have five equally spaced piers. Each of the piers is 2.0 m wide, 16 m long and has a square nose and tail. The abutments are not in the channel and so do not affect the flow. During a 1 in 100 year flood the discharge is 850 m³/s and the normal depth 4.4 m. Estimate (a) the afflux and (b) the suggested allowance for scour assuming the flow hits the piers at an angle of up to 15°.

(a) $D_N = 4.40$ m, $A_N = 4.40 \times 67.50 = 297.00$ m², $V_N = 850/297.00 = 2.86$ m/s, and $F_N = V_N/(gD_N)^{1/2} = 2.86/(9.81 \times 4.40)^{1/2} = 0.44$. This is subcritical so use Yarnell's equation for pier afflux:

$$H_1^* = K_Y D_N F_N^2 (K_Y + 5F_N^2 - 0.6)(m + 15\,m^4) \qquad (9.5)$$

Take $K_Y = 1.25$ (Table 9.6) and the channel contraction ratio $m = (1 - b/B)$.
With $b = 67.5 - (5 \times 2.0) = 57.5$ m and $B = 67.5$ m then $m = (1 - 57.5/67.5) = 0.15$.
Thus $H_1^* = 1.25 \times 4.40 \times 0.44^2 (1.25 + 5 \times 0.44^2 - 0.6)(0.15 + 15 \times 0.15^4)$
$= 1.06 \times 1.62 \times 0.16$
$H_1^* = 0.27$ m

(b) The length/width ratio is $16/2 = 8$. With square noses and tails the suggested scour allowance in Table 9.4a is $2.0b_P$, but with a 15° angle of attack the multiplying factor in part (b) of the table is 2.0 so the potential scour depth is $4.0b_P = 4.0 \times 2.0 = 8.0$ m.

SELF TEST QUESTION 9.2

The afflux in Example 9.3 is of some concern since the bridge is to be constructed in a town centre that is already subjected to periodic flooding. A contraction ratio of about 0.11 is easily possible, and a more efficient pier shape could be used. Repeat the calculations to see what reduction in afflux can be achieved.

9.3.3 Afflux arising from abutments (and piers)

Here it is assumed that the primary constriction of the flow is caused by the abutments, the piers being of secondary importance. This problem has many variables, not least the geometry of the abutments, so it is not possible to present a single equation that can be used. Several investigations of the problem have been undertaken, and these resulted in useful, but lengthy, guides to the analysis and design of bridge openings (Matthai, 1967; Bradley, 1978). Readers should consult more specialised texts for details, such as Chow (1981), French (1986) or Hamill (1999). However, in many cases the afflux in channel flow is relatively small, particularly when the opening width (b) is similar to the channel width (B). The most troublesome condition is when the opening becomes submerged, as described earlier. If the water level is above the top of the opening upstream, but below the top downstream, the flow is analogous to the discharge underneath a free sluice gate, or through a free orifice with the total head measured from the centre of the opening (Fig. 9.22a). Bradley (1978) represented this type of flow by:

$$Q = C_d a_O \left[2g \left(H_1 + \frac{\alpha_1 V_1^2}{2g} - \frac{Y}{2} \right) \right]^{1/2} \tag{9.7}$$

where Q is the discharge (m³/s), C_d is a dimensionless coefficient of discharge with a value of about 0.4 to 0.6 depending upon the degree of submergence, a_O is the total cross-sectional area (m²) of the waterway openings flowing full (i.e. not counting pier areas), H_1 is the upstream depth (m), V_1 is the corresponding mean flow velocity, α_1 is the dimensionless energy coefficient and Y is the height of the opening above bed level. Generally the opening does not drown completely until $H_1 > 1.1Y$ when rectangular openings typically have a C_d of about 0.40. This can be a useful basis for preliminary calculations, as in Example 9.4. When $H_1 = 1.3Y$ then C_d is about 0.48, rising to a fairly constant value of 0.50 when $H_1 > 1.6Y$. The afflux can be estimated as $H_1^* = H_1 - D_N$ where D_N is the normal depth in the channel without the bridge.

Figure 9.22b shows the water level above the top of the opening at both the upstream and downstream face of the bridge. This is analogous to flow through a drowned orifice.

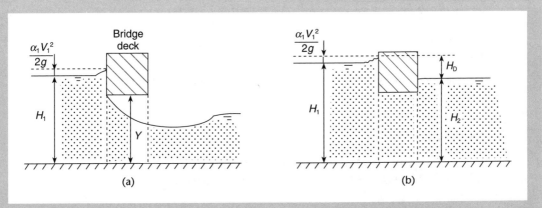

Figure 9.22 (a) A bridge waterway opening submerged at the upstream face, which is analogous to a sluice gate or free orifice. (b) The opening is submerged upstream and downstream making it analogous to a drowned orifice

This condition is most likely to occur when the downstream channel has a relatively low bed or friction gradient, a poor alignment, is overgrown, or is affected by the backwater from another obstruction. It can be represented by:

$$Q = C_d a_O (2gH_D)^{1/2} \tag{9.8}$$

and

$$H_D = \left(H_1 + \frac{\alpha_1 V_1^2}{2g} - H_2 \right) \tag{9.9}$$

where H_D is the differential head (m) across the structure and H_2 is the downstream depth (m). Often the velocity head is negligible and can be ignored, so equation (9.9) becomes $H_D = H_1 - H_2$. Here C_d has a value between 0.65 and 0.90 with 0.80 being typical (Hamill, 1999).

In either the free or drowned condition, rounding the entrance to the bridge opening can significantly increase C_d (as with any orifice). This can result either in an increased Q for a given upstream water level, or a reduced afflux and upstream water level for a given Q. See Hamill (1997) for details.

EXAMPLE 9.4

A two span bridge is to be built across a river that is 16.4 m wide. It is estimated that $\alpha_1 = 1.3$. The bridge will have two rectangular openings each 5.0 m wide and 2.5 m high. The channel downstream of the bridge is straight and clear, so it is thought that the bridge would behave like a free sluice gate if the opening should become submerged. The bridge is to be designed for a 1 in 100 year flood, which is estimated to be 52 m³/s. Will the bridge pass its design flood without becoming submerged?

Assume that the opening becomes submerged when $H_1 = 1.1Y = 1.1 \times 2.5 = 2.75$ m and that in this condition $C_d = 0.40$ (see text above). The discharge at which this occurs is:

$$Q = C_d a_O [2g(H_1 + \alpha_1 V_1^2/2g - Y/2)]^{1/2} \tag{9.7}$$

where $a_O = 2 (5.0 \times 2.5) = 25.00$ m² and $Y/2 = 2.5/2 = 1.25$ m.

First iteration

Assume the velocity head is negligible so $Q = 0.40 \times 25.00 [19.62 (2.75 - 1.25)]^{1/2} = 54.25$ m³/s.

Second iteration

$V_1 = 54.25/(16.40 \times 2.75) = 1.20$ m/s and $\alpha_1 V_1^2/2g = (1.30 \times 1.20^2)/19.62 = 0.10$ m. $Q = 0.40 \times 25.00 [19.62 (2.75 + 0.10 - 1.25)]^{1/2} = 56.03$ m³/s.

Third iteration

$V_1 = 56.03/(16.40 \times 2.75) = 1.24$ m/s and $\alpha_1 V_1^2/2g = (1.30 \times 1.24^2)/19.62 = 0.10$ m. $Q = 56.03$ m³/s as above.

This answer is only a rough estimate but since $56\,\text{m}^3/\text{s} > 52\,\text{m}^3/\text{s}$ it can be tentatively concluded that the bridge would be able to pass its design flood without the openings becoming drowned. However, it is normal practice to allow at least 0.6 m clearance between the design water level and the top of the opening to allow the passage of debris. It is also possible that the openings could be partially blocked by debris, reducing a_O. The exact value of C_d is also unknown. Thus it may be wise to slightly increase the height and/or width of the openings, depending upon the geometry of the upstream channel.

9.4 Culverts

Culverts are very common, often being constructed to allow rivers to pass under highways or railway embankments. They have also been used to carry watercourses under built-up areas, and many towns flood because the culvert's capacity is insufficient to carry large flood flows. Culverts vary in length from tens to hundreds of metres. Depending upon the size and importance of the culvert, the inlet may be a special structure designed to allow water to enter smoothly, or simply a pipe protruding into the upstream channel (Fig. 9.23). The essential components of a culvert are the inlet, the barrel and the exit. The depth at the inlet is (H_1), and the water surface is called the headwater level, as shown in Fig. 9.24. The barrel (of length L and height Y) can be formed from circular pipes, rectangular concrete box sections, oval corrugated metal sections or *in situ* concrete. The depth of flow in the barrel is D_B. The bed slope of the barrel (S_B) may equal the natural slope of the stream (S_O), or may be steeper to eliminate or reduce any potential problem with siltation and debris accumulation. Depending upon S_B and other factors, the flow in the barrel can be subcritical ($D_C < D_B < Y$), supercritical ($D_B < D_C < Y$) or the barrel can flow full ($D_B = Y$). Large culverts usually have an exit structure designed to return the flow to the river channel smoothly, without erosion. The depth at exit is H_2 and the water surface is called the tailwater level.

Unlike bridge openings, culverts can be long enough to be treated as open channels during low flows. Whatever the discharge, their length means that the friction head loss can be significant and must be included in any calculations. During floods, when the inlet (and possibly the outlet) is submerged, the flow in the culvert can be analysed in the same way as the flow between two reservoirs (as in section 6.2). Thus some of the factors affecting flow through a culvert are length, roughness, inlet geometry, inlet conditions, barrel slope, size, and tailwater (exit) conditions. This makes an accurate analysis of culverts difficult unless standard types that have characteristics proven by experience or model tests are used. A badly designed culvert can lead to upstream flooding, and property damage both upstream and downstream. As with a bridge, when a culvert is placed in a natural stream channel, flood flows must be funnelled off the upstream floodplain and through the relatively narrow barrel. The contraction and subsequent expansion of the flow combined with friction means that there is a loss of energy head. To compensate, and to maintain continuity, there is an increase in upstream water level to force the water through the culvert. This increase in level is the afflux (backwater). A hydraulically inefficient culvert will cause a much larger afflux than an efficient, well designed one. The afflux increases rapidly when the barrel becomes submerged (see Fig. 9.16), so when designing a culvert it is essential to know what type of flow is occurring.

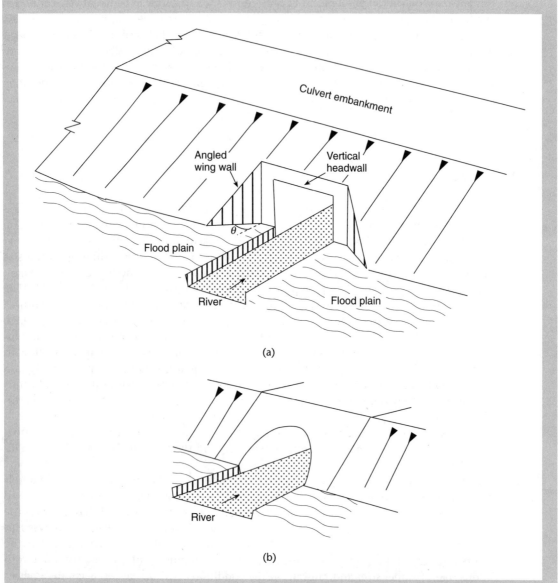

Figure 9.23 Types of culvert. (a) An embankment with a concrete culvert incorporating a vertical headwall and wingwalls at an angle θ. (b) An arch culvert with the headwall flush to the embankment

9.4.1 Types of flow in culverts

There are many ways to classify the flow through culverts. Several or all of the flow conditions shown in Table 9.7 and Fig. 9.25 may be experienced at one culvert as the discharge and water level increases during a flood, and then falls. There are many different flow classifications and design guidelines, such as those given by Chow (1981), American Iron and

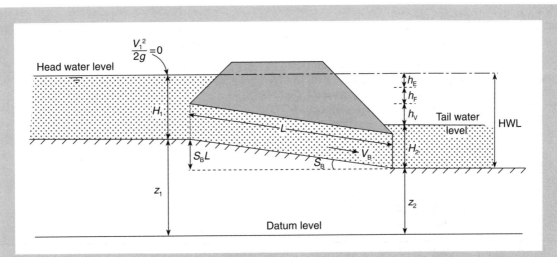

Figure 9.24 A culvert with the headwater and tailwater level above the entrance and exit respectively. The culvert barrel has a length L, height Y, flow velocity V_B and bed slope S_B. The entrance, friction and exit (velocity) head losses are denoted by h_E, h_F and h_V respectively

Steel Institute (1984), French (1986) and CIRIA (1997), but in broad terms the types of flow above can be listed in four categories:

- channel flow with unsubmerged inlet (types 1, 3 and 4);
- submerged inlet, but barrel only part full (type 2);
- submerged inlet, barrel full, but free discharge at the outlet (type 5);
- submerged inlet and outlet (type 6).

Culverts are often described as operating under inlet control or outlet control. The **control point** (CP) has the lowest discharge capacity. Thus the term 'inlet control' means simply that the inlet has a lower discharge capacity than either the culvert barrel or the outlet, so the inlet is limiting the discharge through the culvert (flow types 1 and 2). Inlet control tends to occur with larger culverts on relatively steep gradients where generally the outlet will flow freely and will not submerge. Often the control is where the velocity increases and critical depth occurs. Remember that D_C increases inside a width constriction (see Fig. 8.18 and equation (8.32)) so if the culvert is narrower than the upstream channel it is possible that the flow will pass through critical depth near the entrance establishing inlet control. With critical flow established, the characteristics of the channel downstream of the control do not affect the flow upstream. Thus an example of inlet control is when the flow passes through critical depth as it enters the barrel, and then for the remainder of the culvert the flow is supercritical so that the partially-full barrel can cope comfortably with any discharge that passes through the inlet. A variation on type 1 and 2 flow is that a hydraulic jump occurs in the barrel and the flow returns to subcritical before the exit.

When operating in inlet control the headwater depth is determined by the flow rate ($Q\,\mathrm{m^3/s}$), the cross-sectional area of the barrel ($A_B\,\mathrm{m^2}$) and the shape (i.e. hydraulic efficiency)

Table 9.7 Types of culvert flow

Type		Inlet	Barrel	Outlet	Method of analysis or comments
I N L E T C O N T R O L	1	Unsubmerged. Inlet control with critical depth at inlet.	Part full. Supercritical flow. $D_B < D_C < Y$ $S_B > S_C$	Unsubmerged. $H_2 < D_C$	As an open channel.
	2	Submerged. $H_1 > 1.2$–$1.5Y$ Inlet (orifice) control. Contracting flow passes through critical just inside barrel.	Part full. Supercritical flow over at least part of its length. $D_B < D_C < Y$	Unsubmerged. $H_2 < D_C$ May become type 5 or 6 flow if tailwater rises rapidly.	Inlet as orifice, barrel as open channel flow.
O U T L E T C O N T R O L	3	Unsubmerged. Subcritical flow.	Part full. Subcritical flow. $D_C < D_B < Y$ $S_B < S_C$	Unsubmerged. Outlet control with critical depth at exit. $H_2 < D_C$	As open channel.
	4	Unsubmerged. Subcritical flow.	Part full. Subcritical flow. $D_C < D_B < Y$ $S_B < S_C$	Unsubmerged. Outlet control; control point in channel downstream of exit. Subcritical flow. $D_C < H_2 < Y$	As open channel. Normal design condition for new culverts.
	5	Submerged. $H_1 > 1.2$–$1.5Y$	Flowing full along or all or part of its length.	Unsubmerged. Outlet control with control point at or downstream of the exit. $D_C < H_2 < Y$	Flow from a reservoir through a pipeline: see Example 6.3.
	6	Submerged. $H_1 > 1.2$–$1.5Y$	Full. $D_B = Y$	Submerged $H_2 > Y$ Outlet control with control point in downstream river channel.	As pipeflow between reservoirs: see Examples 6.2 and 6.4.

of the inlet. In this condition the length and roughness of the barrel and the outlet conditions do not matter. When Q is small and the entrance unsubmerged the culvert behaves like an open channel, and can be analysed as such.

Culverts operating with inlet control are also sometimes described as being hydraulically short or hydraulically long. This has nothing to do with their actual length, i.e. L in Fig. 9.24, but can be explained as follows. With the inlet submerged, the flow contracts as it enters the culvert so that the barrel is initially running part full, and then gradually expands again as friction slows the flow. The culvert is hydraulically short if the flow exits the culvert before having expanded to fill the barrel; such a culvert will never flow full like a pipe. The culvert is hydraulically long if the barrel is full when the exit is reached.

Generally culverts operating in outlet control will be found where the bed slope is relatively flat. In outlet control, it is the outlet which restricts the discharge through

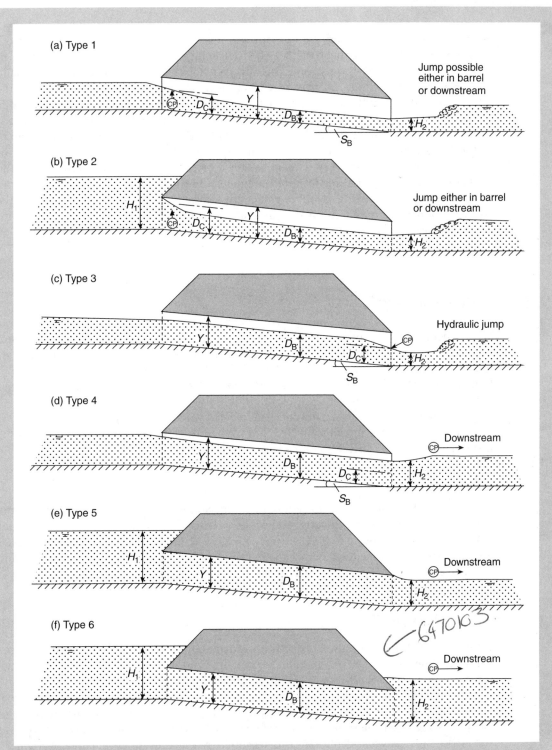

Figure 9.25 Illustration of the culvert flow types listed in Table 9.7. The control point is marked (CP). Types 1 (unsubmerged) and 2 (submerged) are under inlet control. Types 3 and 4 (unsubmerged) and types 5 and 6 (submerged) are all under outlet control

the culvert. The critical factors are the tailwater level in the outlet channel (H_2), and the slope, roughness and length of the barrel. If H_2 is high as a result of an obstruction in the river channel further downstream, this affects the discharge through the barrel. Usually the barrel will tend to run full over part or all of its length, and at the design discharge will probably have a fully submerged inlet and quite possibly a submerged outlet. In this condition an increase in tailwater level will produce a corresponding increase in head-water level in order to maintain the required differential head. This condition is unde-sirable for several reasons: it results in a relatively large head loss and increases the risk of upstream flooding and property damage (as for a bridge, Fig. 9.16); in severe storms the level and quantity of water upstream of the inlet could threaten the safety of the embank-ment; blockage by floating debris significantly exacerbates these problems; extensive downstream damage may result if the barrel at exit is flowing full and blasting water into the downstream channel. In the latter case an impact stilling basin (or similar) may be advisable.

9.4.2 Culvert analysis

When operating as an open channel, the techniques in Chapter 8 relating to uniform and non-uniform flow can be used to determine the profile of the water surface in the culvert (see also Chow, 1981, Chapter 10). If the flow is subcritical, then the downstream condi-tions would determine the water level upstream. If the flow passes through critical, then the conditions downstream of this point do not matter.

The entrance to a culvert may become submerged when the headwater depth $H_1 = 1.2Y$, but this cannot be guaranteed until $H_1 > 1.5Y$. When the inlet is submerged but the exit is free (type 2 flow) this resembles a free small orifice so:

$$Q = C_d A_B [2gH_1]^{1/2} \qquad (9.10)$$

where C_d is a dimensionless coefficient of discharge and A_B is the cross-sectional area of the barrel (m²). For pipe or box culverts set flush in a vertical headwall, the value of C_d is about 0.44 when $H_1/Y = 1.4$ rising to 0.51 when $H_1/Y = 2.0$ and 0.59 when $H_1/Y = 5.0$. These values are applicable to pipes with or without wingwalls. For box culverts with 45° wingwalls and a square edged soffit, C_d is about 0.44 when $H_1/Y = 1.3$ rising to 0.53 when $H_1/Y = 2.0$ and 0.62 when $H_1/Y = 5.0$ (French, 1986). Alternatively:

$$Q = C_D A_B [2g(H_1 - Y/2)]^{1/2} \qquad (9.11)$$

where $(H_1 - Y/2)$ is the headwater depth measured above the centre of the orifice. For cir-cular and pipe-arch culverts C_D has a value ranging from 0.62 for square edged inlet struc-tures to 1.0 for well rounded ones (ARMCO, undated).

If the inlet is submerged and the barrel is full (type 5) or the outlet is submerged (type 6) then the flow can be analysed using the methods for reservoir–pipeline problems in Chapter 6, i.e. discharge to the atmosphere or flow between two reservoirs respectively. In such cases the headwater depth (H_1) is determined by the tailwater level (H_2) and the head loss through the culvert as shown in Fig. 9.24. They can be analysed by applying the energy equation to an upstream and downstream section:

$$z_1 + H_1 + V_1^2/2g = z_2 + H_2 + h_E + h_F + h_V \qquad (9.12)$$

If S_B is the bed slope of the culvert barrel then $(z_1 - z_2) = S_B L$. The entrance head loss is $h_E = K_E V_2^2/2g$ (Table 6.4). The friction head loss in the barrel (subscript B) may be taken as $h_F = S_F L$ where $S_F = V_B^2 n_B^2/R_B^{4/3}$ from equation (8.40), L is its length, V_B is the velocity in the barrel, n_B is its average Manning's n, and R_B is its hydraulic radius. Because of the static tail-water submerging the outlet the kinetic energy of the water emerging from the culvert is lost, so the outlet head loss is $h_V = V_2^2/2g$. If the outlet is not submerged then $V_2^2/2g$ would be included in equation (9.12) anyway as the standard velocity head term. Thus equation (9.12) becomes:

$$S_B L + H_1 + V_1^2/2g = H_2 + K_E V_2^2/2g + V_2^2 n_B^2 L/R_B^{4/3} + V_2^2/2g \qquad (9.13)$$

If the approach velocity head $V_1^2/2g = 0$ due to the ponding of water upstream, then $S_B L + H_1$ is the headwater level (HWL) above the mean bed level at the culvert outlet (Fig. 9.24). If $V_2 = V_B$, then equation (9.13) becomes:

$$HWL = H_2 + V_B^2/2g + K_E V_B^2/2g + V_B^2 n_B^2 L/R_B^{4/3} \qquad (9.14)$$

With a bevelled ring entrance, $K_E = 0.25$; if the barrel projects from the fill with no head-walls, K_E can be as high as 0.90, but typically if the end of the barrel matches the fill slope or has square edged wingwalls or headwalls then $K_E = 0.50$ so:

$$HWL = H_2 + 1.5 V_B^2/2g + V_B^2 n_B^2 L/R_B^{4/3} \qquad (9.15)$$

so **headwater elevation above datum** $= z_2 + HWL$ $\qquad (9.16)$

and $H_1 = HWL - S_B L$ $\qquad (9.17)$

These equations can be applied in both submerged and open channel flow. However, if the tailwater level (H_2) is below the top of the culvert outlet, when calculating the head-water level the total head loss should be added to the larger of H_2 or $0.5(D_C + Y)$ where D_C is the critical depth at the flow rate in question. Note that the headwater elevation can be minimised by using a rounded entrance, a good alignment, a smooth barrel and effective exit.

9.4.3 Design objectives for new culverts

Sometimes it is necessary to recalculate the discharge capacity of an existing culvert to determine if replacement is necessary. Obviously under these circumstances the characteristics of the culvert are predetermined. However, when a new structure is proposed the designer has many options, so the following desirable characteristics should be kept in mind:

(a) The culvert alignment should be chosen to avoid the river having to bend sharply at entry and exit. Construction costs are reduced by minimising culvert length, channel realignment and stream diversion; it may not be possible to minimise all three, so a compromise may be needed according to the geometry of the site.

(b) Whenever possible the shape and capacity of the culvert should match that of the natural channel, and careful consideration should be given to whether or not the entrance should be allowed to submerge. Many culverts under highways are relatively large, perhaps 3 m high, so if they submerge (say $H_1 > 3.6$ m) there may be water to a depth of 3 m on the floodplain. It is quite possible to have a floodplain 60 m wide and extending 150 m upstream, so this amounts to $3 \times 60 \times 150 = 27\,000\,m^3$ of stored water. Remember that when there is more than $25\,000\,m^3$ of water above the level of adjacent ground then the UK Reservoirs Act (1975) applies, so the culvert and embankment would have to be considered as a dam (see Table 13.3). In practice this can result in complications and additional expense, and may mean that to avoid potential failure the embankment has to be designed to be overtopped safely by extreme floods.

(c) If necessary, entrance screens should be provided to collect debris and prevent blockage of the culvert. After allowing for freeboard and the passage of any debris, the capacity of the culvert should exceed the design flow. If the area around the culvert is to be developed (e.g. for housing), any future increase in run-off should be allowed for. At many sites height restrictions mean that the culvert will become submerged during high flows so, as with bridges, a significant increase in upstream depth should be anticipated and allowed for (Fig. 9.16).

(d) The culvert barrel should be straight and smooth with no bends or changes of section. In the early stages of design a barrel gradient (S_B) equal to the natural watercourse (S_O) may be assumed, although it can be different when there is a good reason, e.g. made steeper to achieve a self-cleansing velocity and avoid problems with the accumulation of sediment and debris. Sometimes the overall gradient of the stream can be increased by realigning it to shorten its length (culvert construction is easier on dry ground anyway, and eliminates the need for stream diversion). However, there is no point making the culvert so steep that its outlet is buried in the downstream channel. At the design discharge, ideally the minimum velocity should be 0.75 m/s, and the maximum 2.0 m/s to avoid scour in the downstream channel. Culverts with two or more barrels set at different heights are sometimes employed, the lowest barrel conveying low flows and the others coming into use as the discharge increases. Although the accumulation of sediment is to be avoided, it is sometimes good practice to have the invert below normal bed level, say a quarter of the diameter of a large pipe or a minimum of 0.15 m for box culverts. This allows an environmentally friendly bed similar to the stream to be provided, or facilitates some natural siltation (allow for this when assigning n_B). It also allows for any future erosion or regrading of the channel: the culvert bed should not be higher than that of the adjacent channel.

(e) There should be suitable, safe access for maintenance. Pipe diameters less than 0.45 m should be avoided as they are prone to blockage, while long motorway culverts of less than 1.0 m diameter are difficult to inspect and maintain. Concrete pipes are available up to 2.4 m diameter (between 1.2 m and 2.4 m they increase in 0.15 m increments). Precast concrete box sections are available from 1.0 m wide and 0.6 m high to 6.0 m × 3.6 m (from 1.2 m both width and height increase in 0.3 m increments). Corrugated metal culverts may have a span up to 12 m and a rise of 9 m. Clearly these larger sizes allow easy access.

(f) The environmental impact should be as small as possible. Visual appearance and the needs of wildlife should be considered (e.g. fish and wildlife migration).

9.4.4 Design calculations

Opinions vary regarding the ideal design flow condition in a culvert. One school of thought is that the culvert should match as closely as possible the slope and bankful shape of the natural stream channel, in which case the culvert should be designed for type 3 flow with outlet control, the barrel running part full and subcritical conditions throughout (CIRIA,

1997). However, this tends to produce a rather large culvert and assumes the flow in the potentially smooth culvert is the same as in the naturally rough channel, which need not be the case. Others have advised that the culvert should be designed for inlet control and modest submergence of the inlet (ARMCO, undated). This avoids an oversized culvert barrel, usually eliminates excessive upstream flooding as a result of high tailwater levels, and minimises potential downstream damage caused by a full barrel discharging into the channel. A compromise is to design the culvert to operate unsubmerged during a 1 in 10 year flood, but to pass a 1 in 100 year flood with an acceptable amount of upstream submergence.

To design a new culvert the information requirements are:

- The design return period and discharge, Q (see section 13.3.2). The return period is often 1 in 100 years for urban areas, but something less extreme for other areas.
- The tailwater level corresponding to Q. This can be based on the depth obtained from the Manning equation (uniform flow, section 8.2), the calculated surface profile (gradually varying flow, section 8.11) or field observations.
- The maximum permissible headwater elevation that will not cause unacceptable upstream flooding. After deducting some amount for freeboard (say 0.3 m), this level must at least equal the tailwater level plus head losses.
- The number, type, roughness, length, slope and invert level of the barrels.
- The cross-sectional area of the barrels, A_B. As a first estimate CIRIA (1997) suggested that $A_B = A_{TW} + (B_S \times F_B)$ where A_{TW} is the cross-sectional area of flow between the banks of the natural channel at the design tailwater level (i.e. ignore the floodplains), B_S is the corresponding width of the water surface and F_B is the design freeboard in the culvert barrel. With free flow, F_B should be at least 0.3 m for small culverts and 0.6 m for large culverts.

EXAMPLE 9.5

A single barrel, rectangular culvert has to be designed for a river that has a 1 in 10 year flood flow of 12.90 m³/s and a 1 in 100 year discharge of 23.00 m³/s. The maximum permissible upstream flood level is 78.60 m above Ordnance Datum (mOD). The length of the culvert barrel is 45 m and its design freeboard is 0.60 m. The mean bed level at the outlet is 74.80 mOD, and 60 m upstream it is 74.90 mOD. During a 1 in 10 year flood the downstream channel has a tailwater depth of 2.10 m and a surface width (B_S) of 5.40 m. During a 100 year flood the tailwater depth is 3.06 m. The river has a bed of gravel and some stones averaging around 120 mm diameter. Determine a suitable size and slope for the culvert.

The approach adopted will be to make the culvert match as closely as possible the size, shape and slope of the natural channel during the 1 in 10 year flood and to design the culvert for type 3 subcritical channel flow. Then a check of what happens during the 1 in 100 year event will be undertaken.

1 in 10 year flood, $Q = 12.90$ m³/s

Say maximum permissible upstream flood level = 78.60 − 0.30 m freeboard = 78.30 mOD.
Natural slope of stream, $S_O = (74.90 − 74.80)/60 = 1$ in 600.

Make slope of culvert barrel $S_B = S_O = 1$ in 600.

Length of culvert = 45 m.

Say invert level of entrance = $74.80 + 45 \times (1/600) = 74.88$ mOD.

Say width of culvert barrel = bankful width $B_S = 5.40$ m.

Assume depth of flow in culvert barrel, $D_B = 2.10$ m.

Velocity of flow in barrel, $V_B = 12.90/(5.40 \times 2.10) = 1.14$ m/s ($0.75 < 1.14 < 2.00$ m/s, so OK).

Make culvert height Y = tailwater depth + culvert freeboard = $2.10 + 0.60 = 2.70$ m.

Use a rectangular concrete box section 5.40 m wide × 2.70 m high.

From equation (8.32), critical depth in the rectangular culvert is:

$D_C = (Q^2/gB^2)^{1/3} = (12.90^2/9.81 \times 5.4^2)^{1/3} = 0.83$ m.

Check tailwater level: $0.5(D_C + Y) = 0.5(0.83 + 2.70) = 1.77$ m (i.e. < 2.10 m), so use 2.10 m.

Allow for the possibility of giving the culvert a bed of gravel and stones or for the natural transport of such material into the culvert, so say the composite bed/concrete roughness of the barrel is $n_B = 0.030\,\text{s/m}^{1/3}$.

$$R_B = A_B/P_B = (2.10 \times 5.40)/(5.40 + 2 \times 2.10) = 1.18 \text{ m}.$$

Estimate the headwater level from equation (9.15):

$$HWL = H_2 + 1.5V_B^2/2g + V_B^2 n_B^2 L/R_B^{4/3}$$

$$= 2.10 + \frac{1.5 \times 1.14^2}{19.62} + \frac{1.14^2 \times 0.03^2 \times 45}{1.18^{4/3}}$$

$$= 2.10 + 0.10 + 0.04$$

$$= 2.24 \text{m}$$

Therefore headwater elevation = $74.80 + 2.24 = 77.04$ mOD (< 78.30 mOD maximum, so OK).

This method tends to overdesign, and a smaller culvert could be possible. It also assumes that the depth of flow in the culvert is the same as in downstream channel. Check the actual flow depth in the barrel (assuming uniform flow) using the Manning equation:

$$Q = \frac{A_B}{n_B} R_B^{2/3} S_B^{1/2}$$

$$12.90 = \frac{5.40D_B}{0.03} \left[\frac{5.40D_B}{5.40 + 2D_B} \right]^{2/3} \left(\frac{1}{600} \right)^{1/2}$$

$$1.76 = D_B \left[\frac{5.40D_B}{5.40 + 2D_B} \right]^{2/3}$$

By trial and error $D_B = 1.71$ m (note that this would give a higher velocity and higher losses). Thus $D_B = 1.71$ m > D_C (0.83 m) so type 4 subcritical flow occurs, probably with a depth D_B between 1.71 and 2.10 m. It is unlikely that the culvert's inlet will become submerged (see below).

1 in 100 year flood, $Q = 23.00 \text{ m}^3/s$

The tailwater depth = 3.06 m > 2.70 m height of the barrel (Y), so it is possible the barrel will be full with type 6 flow and outlet control (i.e. the control is in the downstream channel). With a full barrel $V_B = 23.00/(5.40 \times 2.70) = 1.58$ m/s and $R_B = (5.40 \times 2.70)/(2 \times 5.40 + 2 \times 2.70) = 0.90$ m.

$$HWL = H_2 + 1.5V_B^2/2g + V_B^2 n_B^2 L/R_B^{4/3}$$
$$= 3.06 + \frac{1.5 \times 1.58^2}{19.62} + \frac{1.58^2 \times 0.03^2 \times 45}{0.90^{4/3}}$$
$$= 3.06 + 0.19 + 0.12$$
$$= 3.37\,\text{m}$$

Therefore headwater elevation = 74.80 + 3.37 = 78.17 mOD (< 78.30 mOD maximum, so OK). From equation (9.17), $H_1 = HWL - S_B L = 3.37 - (1/600) \times 45 = 3.23\,\text{m}$.

The upstream depth ratio $H_1/Y = 3.23/2.70 = 1.20$, so it is just possible that the entrance may submerge. Check the discharge capacity of the entrance using an orifice equation. Assuming the entrance to the box culvert has 45° wingwalls and a square top edge, for equation (9.10) the value of C_d is about 0.44 when $H_1/Y = 1.3$ so assuming submergence has occurred:

$$Q = C_D A_B (2gH_1)^{1/2} = 0.44 \times 5.4 \times 2.7 [19.62 \times 3.23]^{1/2} = 51.07\,\text{m}^3/\text{s}$$

Alternatively, using equation (9.11) as a cross-check:

$$Q = C_d A_B [2g(H_1 - Y/2)]^{1/2} = 0.62 \times 5.4 \times 2.7 [19.62 \times (3.23 - 2.70/2)]^{1/2} = 54.90\,\text{m}^3/\text{s}$$

Thus both equations indicate that at the values of H_1/Y required for submergence the capacity of the inlet far exceeds the actual discharge, so the culvert is not operating under inlet control and must be in outlet control. Note if a higher headwater level and inlet control was acceptable then the culvert could be made smaller.

9.5 Broad crested and Crump weirs

Broad crested weirs are robust structures that are generally constructed from reinforced concrete and which usually span the full width of the channel. They are used to measure the discharge of rivers, and are much more suited for this purpose than the relatively flimsy sharp crested weirs. Additionally, by virtue of being a critical depth meter, the broad crested weir has the advantage that it operates effectively with higher downstream water levels than a sharp crested weir.

Mostly rectangular broad crested weirs will be considered below, although there are a variety of possible shapes: triangular, trapezoidal and round crested all being quite common. If a standard shape is used then there is a large body of literature available relating to their design, operation, calibration and coefficient of discharge (see BS 3680). However, if a unique design is adopted, then it will have to be calibrated either in the field by river gauging or by means of a scaled-down model in the laboratory (see Chapter 10).

9.5.1 Head–discharge relationship

A rectangular broad crested weir is shown in Fig. 9.26. When the length, L, of the crest is greater than about three times the upstream head, the weir is broad enough for the flow to pass through critical depth somewhere near to its downstream edge. Consequently this makes the calculation of the discharge relatively straightforward. Applying the continuity equation to the section on the weir crest where the flow is at critical depth gives: $Q = A_C V_C$.

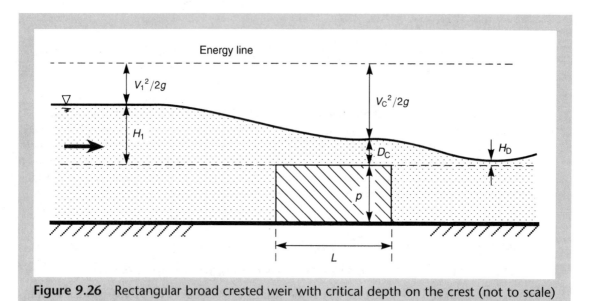

Figure 9.26 Rectangular broad crested weir with critical depth on the crest (not to scale)

Now assuming that the breadth of the weir (b) spans the full width (B) of the channel and that the cross-sectional area of flow is rectangular, then:

$$A_C = bD_C \text{ and } V_C = (gD_C)^{1/2}$$
$$Q = bD_C(gD_C)^{1/2}$$
$$\boldsymbol{Q = \sqrt{g}\,bD_C^{3/2}} \tag{9.18}$$

The same result can be obtained by rearranging equation (8.32) with $B = b$. However, equation (9.18) does not provide a very practical means of calculating Q. it is much easier to use a stilling well located in a gauging hut just upstream of the weir to measure the head of water, H_1, above the crest than to attempt to measure the critical depth on the crest itself. In order to eliminate D_C from the equation, we can use the fact that in a rectangular channel $D_C = \frac{2}{3}E_C$ (equation (8.35)). Using the weir crest as the datum level, and assuming no loss of energy, the specific energy at an upstream section (subscript 1, Fig. 9.26) equals that at the critical section:

$$V_1^2/2g + H_1 = V_C^2/2g + D_C = E_C$$
$$E_C = H_1 + V_1^2/2g \text{ and } D_C = \frac{2}{3}E_C \text{ so } D_C = \frac{2}{3}(H_1 + V_1^2/2g)$$

Substituting this expression for D_C into equation (9.18) gives:

$$Q = \sqrt{g}\,b\left[\frac{2}{3}(H_1 + V_1^2/2g)\right]^{3/2} = (9.81)^{1/2}\left(\frac{2}{3}\right)^{3/2}b\left(H_1 + V_1^2/2g\right)^{3/2}$$

$$\boldsymbol{Q = 1.705\,b\left(H_1 + V_1^2/2g\right)^{3/2}} \tag{9.19}$$

The term $V_1^2/2g$ in the above equation is the velocity head of the approaching flow. As with the rectangular sharp crested weir, the problem arises that the velocity of approach, V_1, cannot be calculated until Q is known, and Q cannot be calculated until V_1 is known. A way around this involving an iterative procedure was described in Chapter 5, but in practice it is often found that the velocity head is so small as to be negligible. Alternatively, a coefficient of discharge, C, can be introduced into the equation to allow for the velocity of approach, non-parallel streamlines over the crest, and energy losses. C varies between about 1.4 and 2.1 according to the shape of the weir and the discharge, but frequently has a value of about 1.6. Thus:

$$Q = CbH_1^{3/2} \tag{9.20}$$

The broad crested weir will cease to operate according to the above equations if a backwater from further downstream causes the weir to submerge. Equations (9.19) and (9.20) can be applied until the head of water above the crest on the downstream side of the weir, H_D, exceeds the critical depth on the crest. This is often expressed as the **submergence ratio**, H_D/H_1. The weir will operate satisfactorily up to a submergence ratio or **modular limit** of about 0.66, that is when $H_D = 0.66H_1$. For sharp crested weirs the head–discharge relationship becomes inaccurate at a submergence ratio of around 0.22, so the broad crested type has a wider operating range. Once the weir has submerged, the downstream water level must also be measured and the discharge calculated using a combination of weir and orifice equations. However, this requires the evaluation of two coefficients of discharge, which means that the weir must be calibrated by river gauging during high flows. This can be accomplished using a propeller type velocity (current) meter.

A Crump weir has a triangular cross-section with (generally) upstream and downstream slopes of 1:2 and 1:5 respectively and a horizontal crest. Critical or supercritical flow occurs on the downstream slope. The depth of flow is measured at tapping points upstream of the weir and (sometimes) just downstream of the crest, the latter being used in calculations when the weir is submerged. The advantage of a Crump weir is a wider range of measurement, a more predictable performance when submerged, smaller head losses and less afflux. Twort *et al.* (1994) stated that it was perhaps the most successful of all weirs with a simple head discharge relationship of approximately $Q = 1.96bH_1^{3/2}\,\text{m}^3/\text{s}$ up to a submergence ratio of about 0.75. Reasonable results can be obtained up to a submergence ratio of about 0.90 by using the downstream crest tappings, although these are prone to blockage by silt and sediment. For accuracy a sharp crest is important, so sometimes it is formed by a metal strip, and there should be a depth of at least 60 mm over the crest. This can be achieved either by a using a compound weir that has several crests at different levels, or the flat-V form shown in Fig. 9.4. The latter gives greater accuaracy and sensitivity at low flows in much the same way as the triangular sharp crested weir in section 5.5.3.

9.5.2 Minimum height of a broad crested weir

A common mistake made by many students in design classes is to calculate the head that will occur over a weir at a particular discharge without considering at all the height of weir required to obtain critical depth on the crest. For example, suppose the depth of flow approaching the weir is 2 m. If the height, p, of the weir crest above the bottom of the

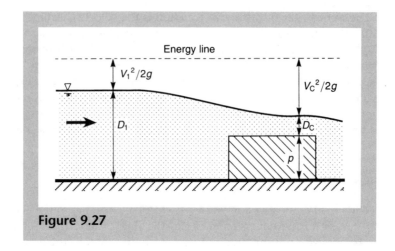

Figure 9.27

channel is only 50 mm, the weir is so low that the flow would be totally unaffected by it and certainly would not be induced to pass through critical depth. Equally ridiculously, if the weir is 4 m high it would behave as a small dam and would raise the upstream water level very considerably and cause quite serious flooding.

So how can we work out the optimum height for the weir? What height will give supercritical flow without unduly raising the upstream water level?

The answer is obtained by applying the energy equation to two sections (Fig. 9.27), one some distance upstream of the weir (subscript 1) and the second on the weir crest where critical depth occurs (subscript c). In this case the bottom of the channel is used as the datum level. Assuming that the channel is horizontal over this relatively short distance, that both cross-sectional areas of flow are rectangular, and that there is no loss of energy, then:

$$V_1^2/2g + D_1 = V_C^2/2g + D_C + p \qquad (9.21)$$

where $V_1 = Q/A_1$, $D_C = (Q^2/gB^2)^{1/3}$, and $V_C = (gD_C)^{1/2}$. This is usually sufficient to enable equation (9.21) to be solved for p when Q and D_1 are known (Example 9.6). Alternatively, the depth, D_1, upstream of the weir can be calculated if Q and p are known (Example 9.7). When calculating the 'ideal' height of weir, it must be appreciated that it is only ideal for the design discharge. The weir cannot adjust its height to suit the flow, so at low flows it may be too high, and at high flows it may be too low. Consequently 'V' shaped concrete weirs are often used, or compound crump weirs that have crests set at different levels.

EXAMPLE 9.6

Water flows along a rectangular channel at a depth of 1.3 m when the discharge is 8.74 m³/s. The channel width (B) is 5.5 m, the same as the weir (b). Ignoring energy losses, what is the minimum height (p) of a rectangular broad crested weir if it is to function with critical depth on the crest?

$V_1 = Q/A = 8.74/(1.3 \times 5.5) = 1.222 \, \text{m/s}.$

$D_C = (Q^2/gB^2)^{1/3} = (8.74^2/9.81 \times 5.5^2)^{1/3} = 0.636 \, \text{m}$

$V_C = (gD_C)^{1/2} = (9.81 \times 0.636)^{1/2} = 2.498 \, \text{m/s}$

Substituting these values into equation (9.21) and then solving for p gives:

$$1.222^2/19.62 + 1.300 = 2.498^2/19.62 + 0.636 + p$$
$$0.076 + 1.300 = 0.318 + 0.636 + p$$
$$p = 0.42\,m$$

The weir should have a height of 0.42 m measured from bed level.

EXAMPLE 9.7

Water flows over a broad crested weir 0.5 m high that completely spans a rectangular channel 10.0 m wide ($b = B$). When the discharge is 19.0 m³/s, estimate the depth of flow upstream of the weir. Assume no loss of energy and that critical depth occurs on the weir crest.

$$D_C = (Q^2/gB^2)^{1/3} = (19.0^2/9.81 \times 10.0^2)^{1/3} = 0.717\,m$$
$$V_C = (gD_C)^{1/2} = (9.81 \times 0.717)^{1/2} = 2.652\,m/s$$

Substitution of these values into equation (9.21) and the fact that $V_1 = Q/BD_1$ gives:

$$V_1^2/2g + D_1 = 2.652^2/19.62 + 0.717 + 0.500$$
$$Q^2/2gB^2D_1^2 + D_1 = 0.358 + 0.717 + 0.500$$
$$[19.0^2/(19.62 \times 10.0^2 \times D_1^2)] + D_1 = 1.575$$
$$[0.184/D_1^2] + D_1 = 1.575$$

D_1 has to be found by trial and error but it is often possible to make a reasonably accurate first guess because the upstream velocity head is usually small (see the previous example). So to begin with, guess $D_1 = 1.55$ m and evaluate the left-hand (LH) side of the equation, then adjust D_1 until the LH and the right-hand (RH) sides agree.

Try $D_1 =$	LH =	RH =
1.55 m	1.627	1.575
1.50	1.582	1.575
1.48	1.564	1.575
1.49	1.573	1.575

The water upstream of the weir is approximately 1.49 m deep.

9.6 Throated flumes

With small open channels a **throated flume** may prove a better alternative than a weir, and they have been used successfully to measure relatively large flows. The throated flume is basically a width constriction that, when seen in plan, has a shape similar to a Venturi meter. Thus it has a rounded (bell-mouth) converging section, a parallel throat, and a diverging section (Figs 9.28 and 9.29). Typically it is constructed from concrete, although other materials can be used. Advantages of the flume include:

(a) The obstacle to the flow is relatively small so there is little afflux or backwater (that is increase in the upstream water level), which is an asset where the channel has little freeboard, or has a very small slope.

Figure 9.28 A throated flume on Devonport Leat, Devon, looking upstream. Note the acceleration of the water through the throat with the flow passing through critical depth, then the small hydraulic jump downsteam of the throat. The advantage of flumes like this is that they do not cause an increase in water level upstream. This flume has a flat bed without a hump

(b) It is easily constructed and very robust, since there is little to damage.

(c) Easy maintenance, since there is unlikely to be any siltation, and there is little to trap floating debris. Consequently flumes are often used in sewage treatment works.

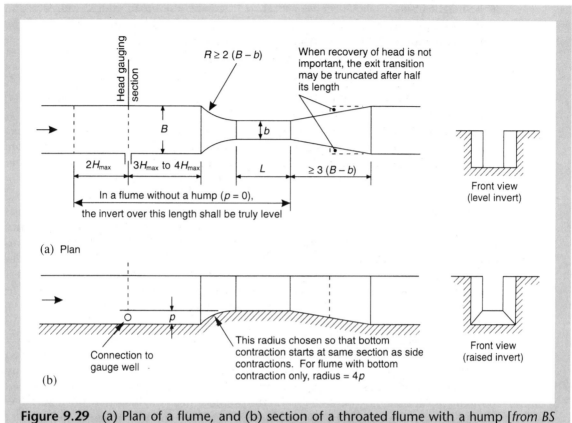

Figure 9.29 (a) Plan of a flume, and (b) section of a throated flume with a hump [*from BS 3680, courtesy of BSI*]

(d) Like a Venturi meter, there is little loss of energy when water flows through a flume, much less than with a weir.

9.6.1 Head–discharge relationship

The throated flume is essentially a width constriction like that discussed in section 8.6.2. Under normal operating conditions, throated flumes are designed so that the flow passes through critical depth in the throat, with a weak hydraulic jump forming in the diverging section (Fig. 9.29). Thus they are sometimes called '**standing wave flumes**' as well as '**Venturi flumes**'. Being a critical depth meter, these flumes have the advantage that all of the equations applicable to critical flow in a rectangular channel can be applied to derive the head–discharge equation. The procedure is the same as for the broad crested weir, and the result identical:

$$Q = 1.705\, b_{\mathrm{C}}(D_1 + V_1^2/2g)^{3/2} \tag{9.22}$$

where b_{C} is the width of the throat where critical flow occurs (which is not the same as the full channel width, B) and D_1 is the depth of water above the flat bed of the flume (instead of the height of water above the weir crest, H_1). As with the broad crested weir, a coefficient of discharge should be intoduced to allow for energy losses, and the fact that the velocity of approach, V_1, is often assumed to be negligible, so:

$$Q = Cb_C D_1^{3/2} \qquad (9.23)$$

where C has a value of about 1.65, slightly higher than the coefficient for a broad crested weir (see BS3680).

It is often difficult to design a throated flume that will operate satisfactorily over a range of discharges with critical depth in the throat and a standing wave in the diverging portion. The principal problem is usually submergence of the flume due to a high downstream water level, D_D. When this happens the head–discharge relationship becomes invalid. The submergence ratio is defined as D_D/D_1, with submergence of the flume occurring when the ratio is about 0.75, that is when $D_D = 0.75D_1$. The maximum permissible value of this ratio is called the modular limit. To ensure that the flow is induced to pass through critical depth, the throat must be made narrow enough to provide a strong control over the flow. Unfortunately, this can lead to a large afflux at higher discharges. As with the broad crested weir, the proportions of the structure are fixed and are 'ideal' for only a limited range of discharge. This problem can be alleviated by designing a flume with a streamlined hump. In the converging part of the flume the channel bed (invert) gradually rises, becomes flat in the throat, and then gradually falls to the original level in the diverging portion. The hump aids the formation of critical flow (see section 8.6.2 and Fig. 8.19) and results in greater accuracy, although construction is more difficult. Normally a streamlined hump is employed in addition to a contraction in width, but it is possible to use a hump by itself without any narrowing of the channel. The upstream and downstream depths (D_1 and D_D) are measured above the top of the hump (Fig. 9.29b) and the submergence ratio is again defined as D_D/D_1. The equations above can still be applied, but may require some adjustment to the value of C. If the hump is made more pronounced so that it is triangular with a 1:2 upstream slope and a 1:5 downstream slope and the throated sides are omitted, then it becomes a Crump weir. These weirs have a modular limit of about 0.8.

A shortened flume without any curves is favoured in the USA, and is often referred to as a Parshall flume (US Bureau of Reclamation, 1967).

9.6.2 Design of throated flumes

The design of these flumes requires some degree of compromise between ensuring that the throat is narrow enough to control the flow and prevent submergence, but not so narrow as to cause excessive afflux. The equations and principles presented earlier can be used as the basis of a simple design procedure, as in Example 9.8.

EXAMPLE 9.8

A throated flume is to be built on a uniform man-made rectangular channel like that in Fig. 9.28. The flow in the channel is maintained at about 0.3 m³/s with a normal depth of around 0.35 m. The freeboard of the channel is very limited, so the afflux should not exceed 0.2 m otherwise overtopping and scouring of the banks will result. Determine a suitable throat width for a flat bed flume. Assume a modular limit of 0.75 and a coefficient C of 1.65.

Assume the normal depth of flow exists at all sections in the unconstricted channel, so $D_N = D_D = 0.35\,m$. The flume will submerge at the modular limit when $D_D = 0.75D_1$. Thus the minimum depth upstream must be $D_1 = D_D/0.75 = 0.35/0.75 = 0.467\,m$. Putting $D_1 = 0.467\,m$ into equation (9.23):

$$0.3 = 1.65b_C 0.467^{3/2} \quad \text{which gives } b_C = 0.570\,m$$

Therefore the **maximum** throat width that will induce critical flow = 0.57 m.
Alternatively, if the upstream depth, $D_1 = D_N + 0.2\,m$ (the maximum allowable afflux) then $D_1 = 0.35 + 0.20 = 0.55\,m$. Putting $D_1 = 0.55\,m$ in equation (9.23) now gives:

$$0.3 = 1.65b_C 0.55^{3/2} \quad \text{and hence } b_C = 0.446\,m$$

Therefore to limit the afflux to 0.2 m the **minimum** throat width is 0.45 m.
Since the repercussions of exceeding the maximum afflux sound quite severe, a throat width closer to the maximum than the minimum value would be sensible, say about 0.53 m.

Summary
1. The type of dam constructed in any location depends upon many factors, but earth embankments are the most common. They are particularly suited to wide shallow valleys and locations where the foundations are relatively weak. Conversely arch dams must have strong rock foundations and are suited to steep sided valleys. Buttress dams are a concrete hybrid between gravity and arch types.

2. A common cause of failure of earth dams is overtopping and scour, often the result of the spillway having an inadequate capacity. Generally earth dams have a spillway that is not part of the dam crest (e.g. a shaft or side overflow-chute spillway). Concrete dams are not susceptible to scour, and so may employ an overflow spillway, crest gates or syphons.

3. Vertical lift sluice gates can be used on dam crests, or in open channels to control the flow of water. In free flow their hydraulic performance is summarised by equation (9.2): $Q_A = Ca_O(2gH_1)^{1/2}$ where C is an overall sluice gate coefficient, a_O is the area of the opening and H_1 is the upstream head. As illustrated by Fig. 9.13, the variation of C is complex, particularly when the opening is drowned by the downstream tailwater level. Equation (9.2) is applicable to all underflow radial gates, but with a different value of C.

4. The downflow and horseshoe vortex at a bridge pier or the downflow and principal vortex at a bridge abutment can deeply scour the bed, and may cause the bridge to collapse. Square corners increase scour depths, rounded streamlined shapes minimise it. Bridge piers and abutments form an obstruction to the flow, the resulting increase in upstream water level being called the afflux. This can be evaluated using the equations in the text.

5. There are many types of flow in culverts (Fig. 9.25 and Table 9.7), but one important consideration is whether the control point with the lowest discharge capacity is located at the inlet or the outlet. With inlet control it is the contraction of the flow into the barrel that limits the discharge. When the inlet (but not the outlet) is submerged it is effectively a free

orifice, so equation (9.10) is: $Q = C_dA_B[2gH_1]^{1/2}$. With relatively flat bed slopes, outlet control is common and it is the outlet or tailwater depth in the downstream channel that controls the depth and discharge in the culvert. With the inlet (and possibly the outlet) submerged and the barrel flowing full, the problem can be analysed in the same way as the discharge from a reservoir in Chapter 6.

6. Broad crested weirs are robust concrete structures that are designed to operate with critical depth (D_C) on their crest. Critical flow provides known relationships that enable Q to be obtained, i.e. $Q = A_CV_C$ where $A_C = bD_C$ and $V_C = (gD_C)^{1/2}$. This gives $Q = (g)^{1/2}bD_C^{3/2}$ where b is the weir length. It is preferable to measure the depth upstream of the weir (H_1), so equating specific energy at the upstream and critical sections results in equation (9.20): $Q = CbH_1^{3/2}$ where C is a coefficient of discharge that includes g. Note that there is a minimum weir height needed to obtain critical flow

on the crest (see section 9.5.2). Throated flumes use a narrowing of the channel (and possibly a streamlined bed hump) to induce critical flow in the throat. They can be analysed in the same way as the broad crested weir and result in basically the same equations. Flat bed flumes have the advantage of causing minimal obstruction and little or no afflux.

7. If the depth downstream of a weir (H_D) or flume (D_D) becomes too large it can submerge the measuring section and prevent critical flow occurring, so it will no longer be effective. The degree of submergence is indicated by the submergence or modular ratio, i.e. H_D/H_1 or D_D/D_1 where H_1 and D_1 are the upstream depths. The maximum downstream depth at which the weir or flume will operate satisfactorily is the modular limit, which is 0.66 for a broad crested weir (i.e. $H_D = 0.66H_1$). The modular limit is about 0.75–0.90 for a Crump weir, 0.75 for a throated flume and 0.22 for a sharp crested weir.

Revision questions

9.1 List the principal advantages and disadvantages of gravity, arch and buttress dams, and give an indication of where each type may be used. What type of spillway is typically used with each of these dams?

9.2 Look at Tables 9.3 and 13.3, which list some notable dam incidents and spillway design standards, and the associated text. What message does this give to engineers and politicians with respect to dam safety?

9.3 A 4.0 m wide vertical sluice gate is positioned in a horizontal, rectangular channel of the same width. The gate operates freely and must pass a discharge of 15.0 m³/s without the upstream head exceeding a value of 3.5 m. (a) Use equation (9.2) and Fig. 9.13 to determine the height at which the gate should be set to give an upstream depth of 3.5 m. (b) What is the approximate depth of the jet

at the vena contracta? (c) Assuming an energy head loss through the gate of $0.05V_2^2/2g$ and that $\alpha_1 = \alpha_2 = 1.05$, check the answer from part (b) using the energy equation. (d) If the normal depth in the channel is 1.95 m, confirm that the gate is actually discharging freely.

[(a) 0.84 m; (b) about 0.50 m]

9.4 A vertical underflow sluice gate 5.5 m wide discharges 18.2 m³/s into a rectangular channel of the same width. The gate is set 0.90 m above the bed (Y), the downstream depth at the gate is 1.3 m and the normal depth in the channel is $D_N = 2.4$ m. (a) Confirm that the gate is operating in the submerged condition. (b) Use equation (9.2) and Fig. 9.13 to determine the upstream depth.

[(b) 3.40 m]

9.5 A bridge has nine round nosed masonry piers each 2.5 m thick. They are equally spaced in a rectangular river channel that is 127 m wide. The normal depth in the channel is 2.1 m when the discharge is 530 m³/s. (a) Use Yarnell's equation to calculate the afflux caused by the piers. (b) Check the answer in (a) using the d'Aubuisson equation. (c) If the piers have a length to width ratio of 4, what is the potential scour depth at the piers if the flow hits the piers at an angle of 0° and 30°?

[(a) 0.088 m; (b) 0.055 m; (c) 3.75 m and 7.50 m]

9.6 The design discharge for a culvert is 13.8 m³/s when the normal depth in the rectangular river channel is 1.9 m. The channel has a bankful width of 4.2 m, a slope of 1 in 300 and a Manning's n of about 0.040 s/m$^{1/3}$. The maximum permissible upstream depth is 2.2 m (i.e. 2.5 m less 0.3 m freeboard). The culvert will have a length of 35 m. Design a single barrel, rectangular culvert that will operate with outlet control, following the general procedure and example given in the text. The site is environmentally sensitive so allow for the invert being 0.15 m below existing bed level, thus enabling a layer of stones and gravel to be provided on the channel bottom.

[Use a culvert about 4.20 m wide by 2.70 m high at a slope of 1 in 300]

9.7 (a) With respect to a broad crested weir, what is meant by the 'submergence ratio' and why is it important? (b) Water flows along a rectangular channel in which there is a broad crested weir with a horizontal crest. The channel is 9.0 m wide, and upstream of the weir the depth of flow measured from the channel bed is 1.1 m when the discharge is 8.24 m³/s. Ignoring any loss of energy, what is the minimum height of the weir that will allow it to function with critical depth on the crest? (c) If a broad crested weir has a coefficient of discharge, C, of 1.65, and if it completely spans a 17.4 m wide rectangular channel, what would be the head over the weir when the discharge is 6.8 m³/s?

[(b) 0.475 m; (c) 0.383 m]

9.8 (a) List the advantages and disadvantages of a throated flume compared to a broad crested weir when used to measure the discharge in an open channel. (b) A flat bed throated flume is to be constructed at a position in a rectangular open channel where the normal depth is 0.55 m, the channel width is 7.5 m, the channel slope is 1/250 and the channel has a Manning n value of 0.035 s/m$^{1/3}$. Assuming that the flume has a typical coefficient of discharge of 1.65, what is the maximum throat width that will still induce critical flow in the flume?

[4.41 m assuming a modular ratio of 0.75]

10

Dimensional analysis and hydraulic models

This chapter explores the difference between units and dimensions. It then shows how the analysis of dimensions can be used to derive the equations that govern hydraulic phenomena. In some cases it is possible to obtain dimensionless groupings of variables, such as the Reynolds and Froude numbers, that have a particular hydraulic significance. Since such groupings are dimensionless, they do not change with the size or scale of the hydraulic system concerned. This leads to the concept of hydraulic models, where scaled-down versions of a system are used to predict the performance of the real thing. Examples include the analysis of the head-discharge characteristics of unusually shaped weirs, as mentioned in the last chapter, and the determination of the equations and performance characteristics of pumps and turbines, as described in the next chapter. Thus dimensional analysis is a powerful and useful tool that can be used to investigate and obtain solutions to real problems. The questions answered include:

What is the difference between a unit and a dimension?

What is dimensional homogeneity, and why is it important?

How can a hydraulic model be used to predict the performance of the real thing?

What is meant by hydraulic similarity?

What are scale effects?

Why do we sometimes use distorted models?

How do you go about undertaking a hydraulic model investigation?

10.1 Units and dimensions

❝ What is the difference between a unit and a dimension? ❞

Well, a metre is a unit. It is a unit of length. Length is a fundamental dimension, which could also be expressed in other units such as mm, inches, feet, yards, km, or miles. So a dimension can be expressed in many different units. **Now you try to name some other fundamental dimensions.**

Length (L) has already been mentioned. How about area?

No, an area is a product of two lengths, that is $L \times L$, which is the same fundamental dimension squared. You are looking for something that cannot be broken down into any component parts.

How about mass? M must be a fundamental dimension, it cannot be broken down. And time, T. I cannot think of any more.

You are right. There are only three fundamental dimensions: M, L and T. Everything else can be expressed in terms of M, L and T. When it comes to dimensional analysis you need to know, for example, that force has the fundamental dimensions MLT^{-2}. The same argument applies to pressure, density, energy and so on.

How on earth can anyone remember the dimensions of these things?

OK, there are two ways to tackle this. The first approach is suitable for very simple quantities like area and volume where the dimensions are similar to the units, that is L^2 replaces m² and L^3 replaces m³. The second approach is to break down the more complex quantities into constituent parts. Table 10.1 shows how we can start with the simplest quantities and work up to the more complicated ones like force by remembering how the quantity is defined.

REMEMBER!!!

Table 10.1 does not include all of the quantities encountered in this text, only the most

Table 10.1 Quantities, units and fundamental dimensions

Quantity	Definition	Units	Dimensions
Length	—	**m**	*L*
Area	Length × Length	m²	L^2
Volume	Length × Length × Length	m³	L^3
Time	—	**s**	*T*
Discharge	Volume/Time	m³/s	L^3T^{-1}
Velocity	Length/Time	m/s	LT^{-1}
Acceleration	Velocity/Time	m/s²	LT^{-2}
Mass	—	**kg**	*M*
Force	Mass × Acceleration	N	MLT^{-2}
Work or energy	Force × Distance	Nm or J	ML^2T^{-2}
Power	Work/Time	J/s or W	ML^2T^{-3}
Pressure	Force/Area	N/m²	$ML^{-1}T^{-2}$
Mass density	Mass/Volume	kg/m³	ML^{-3}
Weight density	Force/Volume	N/m³	$ML^{-2}T^{-2}$

common. However, it illustrates how the dimensions of a quantity can be determined by a logical progression from the simple to the more complex.

 ## 10.2 Dimensional homogeneity

The concept of dimensional homogeneity was introduced in the Introduction to this book. There it was stated that 'For dimensional homogeneity, both sides of an equation must have the same units'. Now that we understand the difference between dimensions and units, it would be better to say that 'for dimensional homogeneity, both sides of an equation must have the same fundamental dimensions'. In the Introduction this concept was illustrated by substituting units into the Bernoulli equation to show that all of the terms could be reduced to metres. The same result can be obtained by substituting dimensions (instead of units) into the equation, thus:

$$(V^2/2g) + (P/\rho g) + z = \text{constant}$$
$$\frac{(LT^{-1})^2}{LT^{-2}} + \frac{ML^{-1}T^{-2}}{ML^{-3}LT^{-2}} + L = \text{constant}$$
$$L + L + L = \text{constant}$$

Therefore the constant must also have the dimension L, and is in fact referred to as the total head and expressed in metres. Sections 8.2.1 and 8.2.2 employ the idea of dimensional homogeneity to find the units of Chezy's C and Manning's n. Up to now units rather than dimensions have been substituted into the equations because this is easier, but the principle is the same. If you find the manipulation of the powers confusing, see Box 10.1 below.

Dimensional analysis can also be used to formulate equations. Take, for instance, Einstein's famous equation $E = Mc^2$, where E is energy, M is mass and c is the velocity of light. Could some other form of the equation have been correct, say $E = Mc$ or $E = Mc^3$? Dimensional homogeneity provides the answer.

$$E = Mc^2$$
$$ML^2T^{-2} = M(LT^{-1})^2$$
$$= ML^2T^{-2}$$

Box 10.1 **Laws of indices**

$1/a = a^{-1}$

$a^b \times a^c = a^{b+c}$ e.g. $a^2 \times a^3 = a^5$ $\qquad a^2 \times a^{-4} = a^{-2}$ or $1/a^2$

$a^b/a^c = a^{b-c}$ e.g. $a^4/a^2 = a^2$ $\qquad a^2/a^3 = a^{-1}$ or $1/a$ $\qquad a^2/a^{-3} = a^5$

$(a^b)^c = a^{b \times c}$ e.g. $(a^2)^3 = a^6$ $\qquad (a^{-2})^3 = a^{-6}$

$a^{b+c} = a^b \times a^c$ e.g. $a^{2+b} = a^2 \times a^b$

$a^{b-c} = a^b/a^c$ e.g. $a^{2-b} = a^2 \times a^{-b} = a^2/a^b$

Thus the dimensions of the left side of the equation only equal those on the right when the power of c is 2; any other power does not satisfy the requirements of dimensional homogeneity so such an equation would be meaningless, even if it balanced numerically. This concept is developed further below.

SELF TEST QUESTION 10.1

The equation for the discharge (Q) over a sharp crested rectangular weir is:

$$Q = 0.667 C_D L(2g)^{1/2} H^{3/2}$$

where L is the length of the weir, g the acceleration due to gravity, and H the head over the weir. Show that dimensional homogeneity is satisfied only if the coefficient of discharge (C_D) is dimensionless.

◀10.3▶ Dimensional analysis using the Rayleigh method

Dimensional analysis by the Rayleigh (or indicial) technique is little more than an extension of the principle of dimensional homogeneity outlined above, that is the dimensions on one side of the equation must equal those on the other side. The steps in the procedure are:

1. For a particular problem, list the variables that must be taken into consideration. For instance, if we want to investigate the discharge (Q) from a pump, then we could guess that this would be influenced by the speed of rotation of the impeller (N), the diameter of the impeller (D), the increase in pressure (P) of the liquid imparted by the pump and the density (ρ) of the liquid being pumped.

2. Incorporate the variables into a power equation, thus:

$$Q = KN^a D^b P^c \rho^d \qquad (1)$$

where K is a dimensionless constant. The powers a, b, c and d are unknown and must be determined by dimensional analysis.

3. List the units and fundamental dimensions of the quantities in the power equation:

Q	m³/s	$L^3 T^{-1}$
N	revolutions/s	T^{-1}
D	m	L
P	N/m²	$ML^{-1}T^{-2}$
ρ	kg/m³	ML^{-3}

Note that a revolution is dimensionless: think of it as a distance divided by the length of the circumference of a circle. Substituting the fundamental dimensions into equation (1):

$$L^3 T^{-1} = K[T^{-1}]^a L^b [ML^{-1}T^{-2}]^c [ML^{-3}]^d \qquad (2)$$

4. To satisfy the requirement of dimensional homogeneity, the power of each dimension (*MLT*) must be the same on each side of the equation. So taking each fundamental dimension in turn and equating the powers on the left and right sides of equation (2):

M: $0 = c + d$ (3)

L: $3 = b - c - 3d$ (4)

T: $-1 = -a - 2c$ (5)

5. Solve equation (3), (4) and (5) for the values of the unknown powers. Since there are four unknowns and only three equations, solve a, b and d in terms of c, thus:

from equation (3): $d = -c$

from equation (4): $3 = b - c - 3d$ where $d = -c$

$3 = b - c + 3c$

$b = 3 - 2c$

from equation (5): $a = 1 - 2c$

6. Substitute these powers back into the original power equation (1):

$Q = KN^{1-2c}D^{3-2c}P^c\rho^{-c}$

or $Q = KNN^{-2c}D^3D^{-2c}P^c\rho^{-c}$

7. Group the variables so that those with known powers and those with the unknown power c are together:

$Q = KND^3[P/N^2D^2\rho]^c$ (6)

This equation relates discharge to the other variables. Alternatively, equation (6) can be written as:

$(Q/ND^3) = f[P/N^2D^2\rho]$ (7)

where $f[\ ]$ means a function of the term in the square bracket. Note that the resulting equation is dimensionally correct, but not numerically correct. The value of any numerical constants like K must be found by experiment.

That completes the dimensional analysis, but if we develop equation (7) further using the fact that $P = \rho gH$, where H is the increase in head imparted by the pump, then:

$(Q/ND^3) = f[gH/N^2D^2]$ (10.1)

Equation (10.1) gives the head–discharge relationship of a pump. Furthermore, it allows the performance of similar pumps of different size to be compared. This is possible because of the nature of the groupings in equation (10.1), that is:

$(Q/ND^3) = (L^3T^{-1}/T^{-1}L^3) = 1 = \text{dimensionless}$ (10.2)

$[gH/N^2D^2] = (LT^{-2}L/T^{-2}L^2) = 1 = \text{dimensionless}$ (10.3)

Thus the left- and right-hand sides of the equation are dimensionless. This satisfies the requirement of dimensional homogeneity, but also means that theoretically the head–discharge relationship is valid regardless of scale (although in practice there may be some scale effects that should be taken into consideration). This means that, for economy, a model or scaled-down pump can be used to predict the performance of a full-size prototype. Thus if subscripts A and B represent two similar pumps of different size then it follows that:

$$(Q/ND^3)_A = (Q/ND^3)_B \qquad (10.4)$$

and $[gH/N^2D^2]_A = [gH/N^2D^2]_B \qquad (10.5)$

These equations can be used quantitatively, as shown in section 11.6. By simple inspection it is apparent from equation (10.2) or (10.4), for example, that if the diameter of the pump impeller is increased then the speed of rotation can be reduced for the same discharge.

Other applications of the Rayleigh method of dimensional analysis are given in Examples 10.1 and 10.2. The former is a simple example that illustrates how the equation for discharge over a broad crested weir can be obtained by dimensional analysis, as an alternative to the method of derivation in section 9.5.1. The second example is more complex and concerns the power developed by a turbine. This will be referred to in Chapter 11. Work through the examples, remembering the advice in Box 10.2, and then try Self Test Question 10.2.

EXAMPLE 10.1

Assuming that for a broad concrete weir the discharge per metre length of crest (q) depends upon gravity (g) and the head (H) over the crest, find the form of the head–discharge equation.

discharge/m length	q	m³s⁻¹/m	L^3T^{-1}/L or L^2T^{-1}
gravity	g	m/s²	LT^{-2}
head	H	m	L

Let $q = Kg^aH^b$ (1) (K is a constant)

$L^2T^{-1} = K[LT^{-2}]^aL^b$ (2)

equating powers of T: $-1 = -2a$ (3)

$$a = \tfrac{1}{2}$$

equating powers of L: $2 = a + b$ (4)

Thus $2 = \tfrac{1}{2} + b$

$$b = \tfrac{3}{2}$$

Box 10.2   **Remember**

1. When undertaking a dimensional analysis try to formulate the problem so that it contains all three of the dimensions M, L and T, so that three equations are obtained. Failure to do this will result in more unknown powers.

2. When there are more than three variables there is no easy way to know which powers to solve for. This is a matter of experience and trial and error. Often more than one solution to the same problem is possible. Generally the best solution is the one that yields some recognisable dimensionless grouping, such as the Froude number, Reynolds number or Mach number.

Substituting into equation (1) gives: $q = Kg^{1/2}H^{3/2}$

Now if $q = Q/b$ where Q is the total discharge over the total length, b, of the weir then:

$$Q = Kbg^{1/2}H^{3/2}$$

If $g^{1/2}$ is incorporated in the weir coefficient, C, then: $Q = CbH^{3/2}$ (as equation (9.20)).

EXAMPLE 10.2

The power developed by a turbine may be assumed to depend upon the following:

speed of runner rotation	N	revs/s	T^{-1}
diameter of the runner	D	m	L
liquid pressure on entry	P	N/m²	$ML^{-1}T^{-2}$
liquid density	ρ	kg/m³	ML^{-3}
(Power	Pow	J/s	ML^2T^{-3})

Using dimensional analysis obtain an equation relating Pow to the other variables, and arrange this in a form that would enable the performance of similar turbines of different size to be investigated.

Assume $Pow = KN^aD^bP^c\rho^d$ (1)

$$ML^2T^{-3} = K[T^{-1}]^a L^b [ML^{-1}T^{-2}]^c [ML^{-3}]^d \quad (2)$$

Equating powers:

M: $1 = c + d$ (3)
L: $2 = b - c - 3d$ (4)
T: $-3 = -a - 2c$ (5)

Since there are three equations and four unknowns, express a, b and d in terms of c, thus:

from equation (3): $d = 1 - c$
from equation (5): $a = 3 - 2c$
from equation (4): $2 = b - c - 3d$
$$2 = b - c - 3(1-c)$$
$$2 = b - c - 3 + 3c$$
$$b = 5 - 2c$$

Substituting these powers back into equation (1):

$$Pow = KN^{3-2c}D^{5-2c}P^c\rho^{1-c}$$
$$Pow = KN^3D^5\rho[P/N^2D^2\rho]^c$$
$$(Pow/\rho N^3D^5) = f[P/N^2D^2\rho]$$

If $P = \rho gH$ then:

$$(Pow/\rho N^3D^5) = f[gH/N^2D^2]$$

Note:

$$(Pow/\rho N^3D^5) = \left(ML^2T^{-3}/ML^{-3}[T^{-1}]^3L^5\right) = (ML^2T^{-3}/ML^2T^{-3})$$
$$= 1 = \text{dimensionless}$$

$$[gH/N^2D^2] = \left[LT^{-2}L/(T^{-1})^2L^2\right] = 1 = \text{dimensionless}$$

Thus if two similar turbines A and B are to be compared, this can be done as follows:

$$(Pow/\rho N^3 D^5)_A = (Pow/\rho N^3 D^5)_B \text{ and}$$
$$[gH/N^2 D^2]_A = [gH/N^2 D^2]_B$$

SELF TEST QUESTION 10.2

The force (F) exerted on a vane by a water jet depends upon the density of the liquid (ρ), and the area (A) and mean velocity (V) of the jet. Using the Rayleigh method of dimensional analysis, find the form of the equation for the force.

10.4 Dimensional analysis using the Buckingham Π theorem

When encountered for the first time without any explanation, the Buckingham Π theorem appears to be just about meaningless to most students. Do not worry, it is not as bad as it appears. First let us look at the theorem, then at what it means.

The theorem states that in a physical problem which involves n quantities (variables) and which contains m fundamental dimensions, the quantities can be arranged into $(n - m)$ independent dimensionless groups. If $Z_1, Z_2, \ldots, Z_n$ are the quantities involved that are essential to the solution (such as diameter, velocity, pressure, etc.) then some functional relationship exists such that:

$$f'(Z_1, Z_2, \ldots, Z_n) = 0 \tag{10.6}$$

If the quantities $Z_1, Z_2, \ldots, Z_n$ are combined to form $(n - m)$ dimensionless groups that are represented by the symbol Π and a subscript (that is $\Pi_1, \Pi_2, \ldots, \Pi_{n-m}$) then an equation exists such that:

$$f''(\Pi_1, \Pi_2, \ldots, \Pi_{n-m}) = 0 \tag{10.7}$$

Now let us see how we can apply the theorem to the same problem that was used to demonstrate the Rayleigh method, that is to determine the form of the equation for the discharge from a pump. The problem will again be broken down into a number of steps.

1. List the total number of quantities involved in the problem:

1	Q	m³/s	$L^3 T^{-1}$
2	N	revolutions/s	T^{-1}
3	D	m	L
4	P	N/m²	$ML^{-1}T^{-2}$
5	ρ	kg/m³	ML^{-3}

Thus $n = 5$.

2. It is assumed that these quantities can be arranged in the form of a functional relationship so that:

$$f'(Q, N, D, P, \rho) = 0 \tag{1}$$

where f' (and f'', f^* and f below) simply means a function of the things in the brackets.

3. The five quantities ($n = 5$) listed above contain all three fundamental dimensions MLT ($m = 3$), and the Buckingham Π theorem tells us that $(n - m) = (5 - 3) = 2$ dimensionless groups can be expected. Thus another way of writing equation (1) is:

$$f''(\Pi_1, \Pi_2) = 0 \qquad (2)$$

4. The overall objective is to find the form of Π_1 and Π_2. To do this we first have to select m of the n (that is 3 of the 5 in this case) quantities to act as **repeating variables**. They are called repeating variables because they appear in all of the equations for Π below, that is they are repeated in each equation. The repeating variables should contain collectively M, L and T, and should not include the quantity whose variation is being investigated (Q in this problem). Density, velocity, length or diameter are often suitable as repeating variables, but to a large extent it is just up to experience, guesswork, and trial and error as to which variables give the best solution (but see Box 10.3). Try N, D and ρ as repeating variables, so the primary (non-repeating) variables are Q and P.

REMEMBER!!!

5. Combine the repeating and non-repeating variables to form an equation for each of the Π's. Each equation contains all of the repeating variables and one of the primary variables. The repeating variables are all raised to an unknown power a, b, c, etc., with a numerical subscript which is the same as that assigned to Π. Thus:

$$\Pi_1 = N^{a_1} D^{b_1} \rho^{c_1} Q \qquad (3)$$

$$\Pi_2 = N^{a_2} D^{b_2} \rho^{c_2} P \qquad (4)$$

6. The next step is to substitute the equivalent fundamental dimensions into equation (3) and solve for the unknown powers of M, L, and T, one at a time, as in the Rayleigh method in section 10.3. Note that the Π's $= M^0 L^0 T^0 = 0$ in the equations below.

$$0 = [T^{-1}]^{a_1} L^{b_1} [ML^{-3}]^{c_1} L^3 T^{-1}$$

equating powers of M: $0 = c_1$ so $c_1 = 0$

$\qquad\qquad L$: $0 = b_1 - 3c_1 + 3$

$\qquad\qquad\qquad$ since $c_1 = 0$, $b_1 = -3$

$\qquad\qquad T$: $0 = -a_1 - 1$ so $a_1 = -1$

Thus $\Pi_1 = N^{-1} D^{-3} \rho^0 Q = (Q/ND^3)$

7. Solve for the unknown powers in equation (4) as in step 6 above:

$$0 = [T^{-1}]^{a_2} L^{b_2} [ML^{-3}]^{c_2} ML^{-1} T^{-2}$$

equating powers of M: $0 = c_2 + 1$ so $c_2 = -1$

$\qquad\qquad L$: $0 = b_2 - 3c_2 - 1$

$\qquad\qquad\qquad 0 = b_2 + 3 - 1$ so $b_2 = -2$

$\qquad\qquad T$: $0 = -a_2 - 2$ so $a_2 = -2$

Thus $\Pi_2 = N^{-2} D^{-2} \rho^{-1} P = (P/N^2 D^2 \rho)$

If $P = \rho gH$, then $\Pi_2 = (gH/N^2 D^2)$

8. Thus equation (2) becomes: $f''\left(\dfrac{Q}{ND^3}, \dfrac{gH}{N^2 D^2}\right) = 0 \qquad (5)$

9. If necessary any of the Π terms can be inverted or raised to some power without affecting their dimensionless status. This enables us to write equation (5) as:

$$(Q/ND^3) = f[gH/N^2 D^2]$$

This is exactly the same as equation (10.1) derived by the Rayleigh method earlier. Now work through Examples 10.3 and 10.4 and Self Test Question 10.3.

EXAMPLE 10.3

According to equation (5.21) the theoretical discharge (Q_T) over a sharp crested rectangular weir is $Q_T = \frac{2}{3}b\,(2g)^{1/2}H^{3/2}$ where b is the width of the weir crest, g is gravity and H is the head above the crest. It might be expected that the liquid's density (ρ) and dynamic viscosity (μ) should be included, since they are important variables. Use the Buckingham Π theorem to investigate whether or not ρ and μ do influence the discharge.

1	discharge	Q_T	m³/s	L^3T^{-1}
2	width	b	m	L
3	gravity	g	m/s²	LT^{-2}
4	head	H	m	L
5	liquid density	ρ	kg/m³	ML^{-3}
6	dynamic viscosity	μ	kg/ms	$ML^{-1}T^{-1}$

The quantities can be written in the form of a functional relationship.

$$f'(Q_T,b,g,H,\rho,\mu) = 0 \qquad (1)$$

In this problem there are 6 ($= n$) quantities involving 3 ($= m$) dimensions so there will be ($n - m$) = 3 dimensionless Π groupings. Thus:

$$f''(\Pi_1,\Pi_2,\Pi_3) = 0 \qquad (2)$$

Let g, H and ρ be the repeating variables, and Q_T, b and μ the non-repeating variables, so:

$$\Pi_1 = g^{a_1}H^{b_1}\rho^{c_1}Q_T \qquad (3)$$

$$\Pi_2 = g^{a_2}H^{b_2}\rho^{c_2}b \qquad (4)$$

$$\Pi_3 = g^{a_3}H^{b_3}\rho^{c_3}\mu \qquad (5)$$

Considering $\Pi_1 = g^{a_1}H^{b_1}\rho^{c_1}Q_T$, substituting the fundamental dimensions gives:

$$0 = [LT^{-2}]^{a_1}L^{b_1}[ML^{-3}]^{c_1}L^3T^{-1}$$

equating powers of M: $0 = c_1$ so $c_1 = 0$

T: $0 = -2a_1 - 1$ so $a_1 = -\frac{1}{2}$

L: $0 = a_1 + b_1 - 3c_1 + 3$

$0 = -\frac{1}{2} + b_1 - 0 + 3$ so $b_1 = -\frac{5}{2}$

Thus $\Pi_1 = g^{-1/2}H^{-5/2}\rho^0 Q_T$ or $(Q_T/g^{1/2}H^{5/2})$

Now considering $\Pi_2 = g^{a_2}H^{b_2}\rho^{c_2}b$ the fundamental dimensions are:

$$0 = [LT^{-2}]^{a_2}L^{b_2}[ML^{-3}]^{c_2}L$$

equating powers of M: $0 = c_2$ so $c_2 = 0$

T: $0 = -2a_2$ so $a_2 = 0$

L: $0 = a_2 + b_2 - 3c_2 + 1$

$0 = 0 + b_2 - 0 + 1$ so $b_2 = -1$

Thus $\Pi_2 = g^0 H^{-1}\rho^0 b$ or (b/H)

Now considering $\Pi_3 = g^{a_3}H^{b_3}\rho^{c_3}\mu$ the fundamental dimensions are:

$$0 = [LT^{-2}]^{a_3} L^{b_3} [ML^{-3}]^{c_3} ML^{-1}T^{-1}$$

equating powers of M: $0 = c_3 + 1$ so $c_3 = -1$
T: $0 = -2a_3 - 1$ so $a_3 = -\frac{1}{2}$
L: $0 = a_3 + b_3 - 3c_3 - 1$
$0 = -\frac{1}{2} + b_3 - 3(-1) - 1$ so $b_3 = -\frac{3}{2}$

Thus $\Pi_3 = g^{-1/2}H^{-3/2}\rho^{-1}\mu$ or $(\mu/\rho g^{1/2}H^{3/2})$

Therefore $f''(\Pi_1, \Pi_2, \Pi_3) = 0$ becomes $f''(Q_T/g^{1/2}H^{5/2}, b/H, \mu/\rho g^{1/2}H^{3/2}) = 0$.
Any of the terms can be rearranged, combined or inverted so:

$$\frac{Q_T}{g^{1/2}H^{5/2}} = f'''\left(\frac{b}{H}, \frac{\mu}{\rho g^{1/2}H^{3/2}}\right) \text{ or}$$

$$Q_T = f\left(bg^{1/2}H^{3/2}, \frac{\rho g^{1/2}H^{3/2}}{\mu}\right)$$

The first term in the brackets is recognisable as the weir discharge equation, while the second indicates that ρ and μ do affect the discharge. Since they do not appear in equation (5.21) their effect has to be included in the coefficient of discharge introduced in equation (5.22). From equation (5.6), $V = (gH)^{1/2}$ so the second term becomes $\rho HV/\mu$ which is the Reynolds number (Re) of equation (4.3) and Table 10.2 with H as the characteristic length. Thus:

$$Q_T = f(bg^{1/2}H^{3/2}, \text{Re})$$

EXAMPLE 10.4

In turbulent flow the head loss when a liquid flows through a smooth pipe is assumed to depend upon the quantities below. Determine the form of the equation using the Buckingham Π theorem.

1	head loss	$h_F = \Delta H/l$	m/m	dimensionless
2	mean velocity	V	m/s	LT^{-1}
3	diameter	D	m	L
4	liquid density	ρ	kg/m^3	ML^{-3}
5	dynamic viscosity	μ	kg/ms	$ML^{-1}T^{-1}$
6	gravity	g	m/s^2	LT^{-2}

The quantities can be written in the form of a functional relationship:

$$f'(h_F, V, D, \rho, \mu, g) = 0 \qquad (1)$$

In this problem there are $6 (= n)$ quantities involving $3 (= m)$ fundamental dimensions, so there will be $(6 - 3) = 3$ dimensionless Π groups. One of these is the head loss h_F. Thus:

$$f''(\Pi_1, \Pi_2, \Pi_3,) = 0 \qquad (2)$$

$$\text{where } \Pi_1 = h_F = \Delta H/l \qquad (3)$$

The number of repeating variables $= m = 3$, and let these be V, D and ρ. The primary variables are μ and g.

$$\Pi_2 = V^{a_2}D^{b_2}\rho^{c_2}\mu \qquad (4)$$

$$\Pi_3 = V^{a_3}D^{b_3}\rho^{c_3}g \qquad (5)$$

Substituting the fundamental dimensions into equation (4) gives:

$$0 = [LT^{-1}]^{a_2} L^{b_2} [ML^{-3}]^{c_2} ML^{-1}T^{-1}$$

equating powers of M: $\quad 0 = c_2 + 1$ so $c_2 = -1$

T: $\quad 0 = -a_2 - 1$ so $a_2 = -1$

L: $\quad 0 = a_2 + b_2 - 3c_2 - 1$

$\quad 0 = -1 + b_2 + 3 - 1$ so $b_2 = -1$

Thus: $\Pi_2 = V^{-1}D^{-1}\rho^{-1}\mu = (\mu/VD\rho)$

$(\mu/VD\rho)$ is the equivalent of (1/Re) where Re is Reynolds number (equation (4.3)).

Now substituting fundamental dimensions into equation (5) gives:

$$0 = [LT^{-1}]^{a_3} L^{b_3} [ML^{-3}]^{c_3} LT^{-2}$$

equating powers of M: $\quad 0 = c_3$ so $c_3 = 0$

T: $\quad 0 = -a_3 - 2$ so $a_3 = -2$

L: $\quad 0 = a_3 + b_3 - 3c_3 + 1$

$\quad 0 = -2 + b_3 - 0 + 1$ so $b_3 = 1$

Thus: $\Pi_3 = V^{-2}D^1\rho^0 g = (gD/V^2)$

Therefore equation (2) can now be written as: $f''\left(\dfrac{\Delta H}{l}, \dfrac{\mu}{VD\rho}, \dfrac{gD}{V^2}\right) = 0$

Writing this as an equation for the head loss: $\dfrac{\Delta H}{l} = f*\left(\dfrac{\mu}{VD\rho}, \dfrac{gD}{V^2}\right)$

Since any of the Π terms can be inverted, and recognising Π_2 as the Reynolds number:

$$\frac{\Delta H}{l} = f\left(Re, \frac{V^2}{gD}\right)$$

This resembles the general form of most equations for head loss in a smooth pipe.

SELF TEST QUESTION 10.3

Using the Buckingham Π method analyse the problem in Example 10.4 again, but this time use ρ, μ and g as the repeating variables and determine the form of the equation obtained.

The dimensionless groupings obtained above lend themselves nicely to the development of hydraulic models. For instance if the equation $(Q/ND^3) = f[gH/N^2D^2]$ is plotted (Fig. 10.1) then the resulting line is a graph of the function f. If the equation was derived in a different form with a different function, $f*$, say, then this would give a different line on the graph. If we want to compare the performance of two similar pumps, A and B, then even if the two pumps are of different size, they will still be represented by the

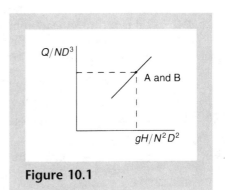

Figure 10.1

same point on the graph. However, this depends upon the similarity of the pumps (see below).

10.5 Hydraulic models and similarity

Many problems can be investigated using mathematical models, which consist of an equation or series of equations that represent the behaviour of the system. For example, even a simple equation like $P = \rho g H$ could be regarded as a mathematical model of the variation of hydrostatic pressure with depth below the surface of a liquid. Unfortunately, not all problems are so easy to analyse. There are situations where complex flow conditions, lack of detailed data, or difficulty determining the form of the equations governing the phenomenon rule out a mathematical solution. Under these circumstances a three-dimensional physical model investigation may be undertaken. Generally the model is a scaled-down version of the real prototype, but exceptionally the model may be larger than the prototype if the real thing is small. Civil Engineers typically use hydraulic models to investigate:

(a) hydraulic structures such as complicated weirs, dam spillways and flow through bridges;

(b) the flow in river channels and the effect of flood relief works;

(c) the flow in estuaries, including tidal currents, the effect of dredging or reclamation works;

(d) the effect of constructing new harbours and marinas on wave heights and wave reflection;

(e) coastal erosion, beach processes, sediment transport and coast protection;

(f) flow in pipes, pipe bends, contractions and expansions;

(g) the performance characteristics of pumps and turbines;

(h) the vibration of multistorey buildings and suspension bridges in strong winds.

Many of the above involve complex three-dimensional flow with a free water surface, and this still frequently presents difficulties with a mathematical modelling approach, despite major advances in computer hardware and software. Consequently there is a continuing need for hydraulic models, although their use has diminished since the 1950s and 1960s. Of course, a physical hydraulic model may be used to verify the effectiveness of a design created within a computer. A dam spillway may work well as a computer model, but before spending millions of pounds on its construction a hydraulic model may be used to confirm the suitability of the design. Since about 23% of dams fail as a result of having an inadequate spillway capacity, with potentially catastrophic consequences for people and property downstream, this is a situation where every precaution must be taken to ensure suitability for purpose. Similarly, wind-tunnel tests on scale models of new cars are the norm; it is impractical to build and test full-size prototypes, particularly during the early stages of development.

10.5.1 Hydraulic similarity

The relationship between model and prototype performance is determined by the laws of hydraulic similarity. Since there are many laws, and models cannot comply with all of them simultaneously, the model will not reflect totally the performance of the prototype. Some error will be incurred, which is referred to as **scale effect** (see section 10.5.4). Fortunately, by careful design of the model, by using a reasonably large scale, and by judicious interpretation of the results (experience helps), scale effects can be minimised. However, hydraulic models should not be viewed as infallible calculators that automatically produce the correct answer to a problem. To design a successful hydraulic model and use it effectively to predict the performance of a prototype requires a knowledge of similarity. There are three types of hydraulic similarity that must be considered.

Geometric similarity: the similarity of shape. The model must physically resemble the prototype, with all of the significant features of the prototype being reproduced to scale in the model. That is any model length, L_M, is related to the equivalent length in the prototype, L_P, via the scale factor, $1 : X$.

Kinematic similarity: the similarity of motion. At similar points at similar times, the model must reproduce to scale the velocity and direction of flow experienced within the prototype.

Dynamic similarity: the similarity of forces. At similar points, the model must reproduce to scale all of the forces experienced within the prototype.

Already it is apparent that there is rather more to building a successful hydraulic model than merely reconstructing the prototype to a different scale. However, some of the topics covered earlier in the chapter help us. The Buckingham Π theorem can be used to combine the variables that govern a hydraulic system into dimensionless groups. Figure 10.1 showed how two geometrically similar pumps of different size can be represented by the same point on a graph, indicating that the ratio of the two dimensionless parameters is the same regardless of size. This must be so for hydraulic similarity. In Box 10.2 it was stated that it pays to keep an eye open for standard dimensionless groups, such as the Reynolds number in Example 10.4. These standard groups not only make the dimensional analysis easier, they

Table 10.2 Forces which may dominate the design of a hydraulic model	
Force and dimensionless parameter	**Situations where the force may be dominant**
Gravity Froude Number, $F = V/(gL)^{1/2}$	Open boundary hydraulics with air/water interface, e.g. flow over a weir or in a river, gravity waves
Viscous drag Reynolds Number, $Re = \rho L V/\mu$	Viscous flow close to boundaries, e.g. drag on a balloon, pipeflow in the transition zone
Pressure Euler Number, $E = V/(2\Delta P/\rho)^{1/2}$	Fully turbulent pressurised flow in enclosed systems, e.g. pumps, pipes (ΔP = difference in pressure)
Surface tension Weber Number, $W = V/(\sigma/\rho L)^{1/2}$	Air/water interface and small head (often unimportant), e.g. low flow over a weir (σ = surface tension, N/m)
Compressibility Mach Number, $M = V/c$	Only significant with compressible fluids and $M > 0.25$, e.g. gas turbine design (c = sonic velocity in the fluid concerned).

also indicate the dominant force(s) governing a particular hydraulic phenomenon. When we design a physical model, in order to minimise scale effects we must make sure that the dominant force is correctly reproduced, at the expense of other, less important forces if necessary (theoretically, for dynamic similarity the ratio of all model and prototype forces should be the same, but this is not always possible).

So how do you know what force will be dominant in a particular situation?

The most important forces and their associated dimensionless parameter are shown in Table 10.2, with typical conditions where each may be dominant. The length, L, represents some significant dimension such as depth of flow or pipe diameter, and V is velocity.

10.5.2 Gravity (Froudian) models

Perhaps the most common type of hydraulic model is that concerned with open boundary hydraulics where the water has a free surface. This includes anything concerned with open channel hydraulics, weirs with a reasonably large head over the crest, dam spillways, stilling basins and models of coastal areas that involve gravity waves. These models may be referred to as Froudian. Because gravitational forces are the most important, the Froude numbers in model and prototype (subscripts M and P) must be the same. Thus $F_M = F_P$ or:

$$\left[V/(gL)^{1/2}\right]_M = \left[V/(gL)^{1/2}\right]_P \text{ which can be written as } \left[V_M/(gL_M)^{1/2}\right] = \left[V_P/(gL_P)^{1/2}\right]$$

where V is the velocity, L is a length and g is the acceleration due to gravity, which is the same in both the model and prototype. Cancelling g and rearranging gives:

$$\frac{V_M}{V_P} = \left(\frac{L_M}{L_P}\right)^{1/2}$$

$$(10.8)$$

For an undistorted model the ratio L_M/L_P is the geometric scale of the model, $1/X$. Therefore:

$$\frac{L_M}{L_P} = \frac{1}{X} \tag{10.9}$$

Combining equations (10.8) and (10.9) gives:

$$\frac{V_M}{V_P} = \frac{1}{X^{1/2}} \tag{10.10}$$

Note that the model velocity is conveniently less than that in the prototype. For example, at a scale of $1:20$ ($X = 20$), a velocity of $1\,\text{m/s}$ in the prototype becomes $0.22\,\text{m/s}$ in the model. Since velocity is a distance divided by time, $V = L/T$ so $T = L/V$. Thus the time scale, T_M/T_P is:

$$\frac{T_M}{T_P} = \frac{L_M}{V_M} \times \frac{V_P}{L_P} \quad \text{which is equivalent to}$$

$$\frac{T_M}{T_P} = \frac{L_M}{L_P} \times \frac{V_P}{V_M} = \frac{1}{X} \times X^{1/2} \, \text{so}$$

$$\frac{T_M}{T_P} = \frac{1}{X^{1/2}} \tag{10.11}$$

Volume is equivalent to L^3. Thus a volume of 1 in the model is equivalent to X^3 in the prototype. Discharge (Q) is volume (Vol) divided by time (T), therefore:

$$\frac{Vol_M}{Vol_P} = \frac{1}{X^3} \quad \text{and} \quad \frac{T_M}{T_P} = \frac{1}{X^{1/2}} \quad \text{so} \quad \frac{Q_M}{Q_P} = \frac{Vol_M}{Vol_P} \times \frac{T_P}{T_M} = \frac{X^{1/2}}{X^3} \quad \text{which gives}$$

$$\frac{Q_M}{Q_P} = \frac{1}{X^{5/2}} \tag{10.12}$$

The Manning equation can be used to determine the relationship between the surface roughness of the model and prototype. The Manning equation is:

$$V = (1/n)R^{2/3}S_O^{1/2} \tag{8.8}$$

where Manning's n is a coefficient of roughness, R is the hydraulic radius (in metres) and S_O is a dimensionless ratio of a vertical distance to horizontal distance. In an undistorted model with vertical and horizontal distances both being to the same scale, S_O is the same in model and prototype and cancels. Using equations (10.9) and (10.10):

$$\frac{V_M}{V_P} = \left(\frac{n_P}{n_M}\right)\left(\frac{L_M}{L_P}\right)^{2/3}\left(\frac{S_{OM}}{S_{OP}}\right)^{1/2} \tag{10.13}$$

$$\frac{1}{X^{1/2}} = \left(\frac{n_P}{n_M}\right)\left(\frac{1}{X}\right)^{2/3}$$

Inverting to get (n_M/n_P) and rearranging gives $(n_M/n_P) = (X^{1/2}/X^{2/3}) = (X^{3/6}/X^{4/6})$ and hence:

$$\frac{n_M}{n_P} = \frac{1}{X^{1/6}} \tag{10.14}$$

This equation can be used to calculate the roughness of surface required in the model. For instance, a prototype concrete channel with $n = 0.017\,\text{s/m}^{1/3}$ when modelled at a scale of $1:20$ (that is $X = 20$) would have to have a roughness of $n_M = n_P/X^{1/6} = 0.017/20^{1/6}$ $= 0.010\,\text{s/m}^{1/3}$. This is about the roughness of glass (Table 8.1). However, it is not this simple to transfer roughness from model to prototype. When a model is constructed, even with the 'correct' surface roughness it is sometimes found that the model is effectively too smooth so that the velocities are too high or that the model does not reproduce the flow patterns in the prototype. Often the solution is to use something like a wire mesh or vertical rods to make the flow more turbulent. These roughness elements are not designed in any way, they are merely added where it would appear they are needed until the model simulates satisfactorily the behaviour of the prototype. It is usually a good idea to calculate the value of Reynolds number at various points in the model to ensure that the flow is turbulent (that is $\text{Re} > 2000$, see Example 10.5), because the characteristics of laminar and turbulent flow are markedly different. The model must have the same flow type as the prototype.

10.5.3 Viscosity (Reynolds) models

Viscosity may become important close to solid boundaries where the fluid velocity is low. A good example is the flow in pipes, where all of the flow is within the boundary layer. In these situations there may be significant viscous forces (whereas at high velocities with $\text{Re} > 4000$ they are not important). Remember that $\text{Re} = \rho L V/\mu$ or $\text{Re} = LV/v$ where v is the kinematic viscosity $= \mu/\rho$, and L is a characteristic dimension such as pipe diameter or the depth of flow in a channel (D). Models dealing with this type of situation must be constructed so that the Reynolds number in model and prototype are similar, that is:

$$V_M L_M/v_M = V_P L_P/v_P \tag{10.15}$$

or $\dfrac{V_M}{V_P} = \dfrac{v_M}{v_P} \times \dfrac{L_P}{L_M}$ and since equation (10.9) gives $L_P/L_M = X$ (the scale factor)

$$\frac{V_M}{V_P} = \frac{v_M}{v_P} \times X \tag{10.16}$$

If the same fluid is used in model and prototype then:

$$V_M = V_P \times X \tag{10.17}$$

This means that if the model scale is $1:20$ (say) then the velocity in the model should be twenty times that in the prototype. In many models viscous forces are not significant so this is ignored, but if they are important then this requirement is impractical. However, by using a different fluid in the model the ratio v_M/v_P in equation (10.16) can be changed and V_M reduced. For example, if the prototype involves water flowing through a pipeline, air might be used in the model. This greatly facilitates the construction and operation of the model.

Equations relating model and prototype volume and discharge according to Reynolds' law can be derived by using equation (10.16) and following a similar procedure to that adopted above for the Froudian model. The same logic can also be applied to any of the other dimensionless parameters listed in Table 10.2. If it is, most of the equations obtained turn out to be different, which leads us to the question of scale effect.

10.5.4 Scale effect – the limitations of hydraulic similarity

If the ratio V_M/V_P is derived on the basis of the Froude, Reynolds, Euler and Weber numbers, and if it is assumed that the ideal model should be able to satisfy all of these equations simultaneously, then this condition is described by:

$$\frac{V_M}{V_P} = \frac{1}{X^{1/2}} = \frac{v_M}{v_P} \times X = \sqrt{\frac{\rho_P}{\rho_M} \times \frac{\Delta P_M}{\Delta P_P}} = \sqrt{\frac{\rho_P}{\rho_M} \times \frac{\sigma_M}{\sigma_P}} \times X^{1/2} \qquad (10.18)$$

The first three terms above are equations (10.10) and (10.16), and the last two are the equivalent expressions for pressure (Euler number) and surface tension (Weber number) that are obtained by following a similar procedure. Thus for the model to have exactly the same behaviour as the prototype, equation (10.18) must be satisfied. However, this is impossible because there is no fluid in existence with the required physical properties, except when $X = 1$, which is when the model is full-size and the same fluid is used. If it is assumed that gravity and viscous forces are generally the most important, then to satisfy F and Re simultaneously requires that:

$$\frac{1}{X^{1/2}} = \frac{v_M}{v_P} \times X$$

or $\quad v_M = \dfrac{v_P}{X^{3/2}}$ $\qquad\qquad (10.19)$

Thus for open boundary hydraulic models with scales of between 1:10 and 1:100 (that is $X = 10$ to 100) the viscosity of the model liquid would have to be considerably less than that in the prototype. Since the prototype liquid is often water, there is no suitable model liquid. Who would want to work in the middle of a lake of petrol? Therefore it is not possible to satisfy F and Re simultaneously. Fortunately viscous forces are usually small and Re is generally large indicating flow in the rough turbulent zone, so provided that the flow in the model is also turbulent with Re not less than 1500 (the actual value is not important) then viscosity can be ignored without causing a large error. If this is the case then a Froudian model is created, but the consequence of ignoring the less predominant effect is to introduce some scale error. Similarly, if modelling a system where viscous forces are dominant and the model is created according to Reynolds' law, there will be a scale effect as a result of ignoring gravity. Thus it is essential that the dominant effect is correctly identified and modelled, and that the limitations and ensuing scale effects are fully appreciated.

10.5.5 Distorted models

When a model needs to be constructed that covers a large area of shallow water, it is often necessary to distort the model so that the vertical scale $(1:Y)$ is larger than the horizontal scale $(1:X)$. This makes the measurement of changes in depth easier and more accurate, reduces the effect of surface tension, and also gives higher Reynolds numbers and a more turbulent flow. Typically distorted models are used to study floods in rivers and river training works, the effect of land reclamation in estuaries, changes in sandbank shape, or beach transportation processes. When changes in bed shape are important, mobile bed models are used. Typically the bed material may be ground coal, sand or wood chips. Note that if waves are to be studied, then the model is usually undistorted. Also, if a detailed study of flow behaviour is required, say flow around a river bend, then a distorted model should not be used since the distortion causes an incorrect representation of detail (although the bulk flow is reproduced well enough).

The degree of distortion varies, but typically the vertical scale $(1:Y)$ may be 5 to 10 times the horizontal scale $(1:X)$. Obviously, distortion requires an alteration to the previously defined scale laws. In a distorted model with $L_M/L_P = 1/Y$ in the vertical direction, the ratio of model to prototype plan area is $1/X^2$, cross-sectional area $1/XY$, and volume $1/X^2Y$. Distorted models are generally used to study problems involving gravity phenomena, so the dimension L in the Froude number should be defined in terms of the vertical direction in which gravity acts. Following the same steps as in section 10.5.2, but with $L_M/L_P = 1/Y$, the equivalent of equation (10.10) for a distorted model is:

$$\frac{V_M}{V_P} = \frac{1}{Y^{1/2}} \tag{10.20}$$

Note that the velocities above refer to the horizontal direction, as is usual with the Froude number. Similarly, the time scale relates to the time required to travel horizontally between two points. So with $L_M/L_P = 1/X$ and V_M/V_P defined by equation (10.20), following the previous procedure, the equivalent of equation (10.11) is:

$$\frac{T_M}{T_P} = \frac{Y^{1/2}}{X} \tag{10.21}$$

Following the same logic that led to equation (10.12), with distortion the model discharge $Q_M = Vol_M/T_M$ with Vol_M being defined by $(Vol_M/Vol_P = 1/X^2Y)$ and T_M by equation (10.21), so:

$$\frac{Q_M}{Q_P} = \frac{Vol_M}{Vol_P} \times \frac{T_P}{T_M} = \frac{1}{X^2Y} \times \frac{X}{Y^{1/2}}$$

$$\frac{Q_M}{Q_P} = \frac{1}{XY^{3/2}} \tag{10.22}$$

With reference to equation (10.13), we now have different vertical and horizontal scales. If it is assumed that the flow is wide relative to its depth (see section 8.2.3) then $R \approx D$, the vertical depth. The bed slope, S_O, must be included, thus:

$$\frac{1}{Y^{1/2}} = \frac{n_P}{n_M} \left(\frac{1}{Y}\right)^{2/3} \left(\frac{X}{Y}\right)^{1/2}$$

or $\quad \dfrac{n_M}{n_P} = \dfrac{X^{1/2}}{Y^{2/3}} \tag{10.23}$

With a vertical scale of about $1:100$ and a horizontal scale of $1:500$ then equation (10.23) indicates that the model roughness should be almost exactly the same as that of the prototype. This relationship cannot be followed exactly. With fixed bed models, roughly finished cement mortar often provides a suitable surface, with additional roughness being provided where necessary by using wire mesh, vertical rods or angular stones.

10.5.6 The steps in the conduct of a hydraulic model investigation

As an illustration, the steps involved in the construction of a river or estuary model are described below. A similar general procedure can be used for other types of model.

1. Carry out a site investigation to obtain topographic and hydrological/hydrographic field data within the area to be covered by the model. This should include, as appropriate, data relating to river discharge, velocity distribution, surface flow patterns, water levels, tide levels and tidal flow.

2. Determine what type of model is required, that is which force is dominant. Decide between undistorted or distorted, and fixed or mobile bed models.

3. Select the model scales, ensuring that the depths are large enough to avoid surface tension effects and to give turbulent flow, but without making the area of the model or the volume of fluid to be pumped excessive.

4. Construct the model, complete with control and measuring equipment. Water levels can be measured with point gauges, velocities with pygmy velocity meters (Fig. 5.19), discharge with orifice meters or sharp crested weirs, and flow patterns with dye or by sprinkling particles (aluminium powder, polystyrene beads) on the water surface.

5. 'Prove' the model by checking that the model can correctly reproduce the observed hydrological/hydrographic data. If it cannot, then modify the model until it can, say by adjusting the bed roughness, or the way in which the water enters and leaves the model.

Figure 10.2 Physical hydraulic model of Goring Weir [*courtesy HR, Wallingford*]

6. Once the model has been proved, use it to obtain data relating to conditions that were not observed in the field. Study the effect of alternative solutions to the problem which necessitated the construction of the model, and determine the optimum solution on the basis of suitability, technical and cost criteria.

EXAMPLE 10.5

Because there are worries about potential flooding, a model investigation is to be undertaken of the afflux that will occur if a bridge is constructed in a city centre. The concrete bridge will have a rectangular opening 4.0 m wide and 2.5 m high. The rectangular river channel is approximately 5.4 m wide with $n = 0.040 \, s/m^{1/3}$ and a bed slope $S_O = 1$ in 380. The largest known flow in the river is $17.4 \, m^3/s$ when the normal depth was 2.2 m. The investigation will be conducted in a rectangular laboratory flume that is 0.45 m wide with glass sides 0.50 m high and a smooth, painted metal bottom. The flume can be set at any slope, has a recirculating water supply that is measured using an orifice meter in the supply pipeline, and has a maximum discharge of $0.100 \, m^3/s$. Briefly outline how the model should be constructed to reproduce the hydraulic performance of the prototype.

At $17.4 \, m^3/s$ the prototype bridge opening is unlikely to submerge (2.2 m < 2.5 m, see section 9.3) so this is a problem involving open channel flow and requiring an undistorted Froudian model. Thus the Froude number in the model and prototype must be the same, i.e. $F_M = F_P$ as described in section 10.5.2. Since the laboratory channel is of fixed width, the scale of the model is:

$$\frac{L_M}{L_P} = \frac{0.45}{5.4} \text{ or } \frac{1}{12}$$

Thus the scale of the model is 1:12 and $X = 12$. This means the bridge opening should be $4.0/12 = 0.333$ m wide and $2.5/12 = 0.208$ m high. The maximum normal depth of flow in the channel will be about $2.2/12 = 0.183$ m. Since $0.183 \, m \ll 0.5 \, m$ there is plenty of freeboard to allow for afflux upstream. From equation (10.12) the maximum model discharge is $Q_M = Q_P/X^{5/2} = 17.4/12^{5/2} = 0.035 \, m^3/s$. The maximum pump capacity is $0.100 \, m^3/s$, so there is plenty of spare capacity and freeboard to study the head–discharge performance of the bridge at higher flows.

From Table 8.1, the concrete bridge would have a surface roughness $n \approx 0.017 \, s/m^{1/3}$, so from equation (10.14) $n_M = n_P/X^{1/6} = 0.017/12^{1/6} = 0.011 \, s/m^{1/3}$. The table indicates that a bridge made from Perspex would have almost the ideal roughness. Similarly the required roughness for the laboratory channel is $n_M = 0.040/12^{1/6} = 0.026 \, s/m^{1/3}$. Since the channel is much smoother than this ($n = 0.010 \, s/m^{1/3}$ for glass) it will have to be roughened. This can be achieved by taping aluminium mesh to the bottom and sides of the channel until the desired V_M and F_M are obtained. With $V_P = 17.4/(5.4 \times 2.2) = 1.465 \, m/s$, equation (10.10) gives $V_M = V_P/X^{1/2} = 1.465/12^{1/2} = 0.423 \, m/s$. Thus the channel should be set with a slope of 1/380 and the roughness adjusted until at a discharge of $0.035 \, m^3/s$ the depth is 0.183 m and the mean velocity is $0.423 \, m/s$ (check: $Q = AV = 0.183 \times 0.45 \times 0.423 = 0.035 \, m^3/s$ as above). When this condition is achieved, the bridge can be introduced to the channel and the increase in upstream water level above normal depth (i.e. the afflux) measured. A similar procedure can be used for other discharges.

Note that $F_M = V_M/(gD_M)^{1/2} = 0.423/(9.81 \times 0.183)^{1/2} = 0.315$ and $F_P = V_P/(gD_P)^{1/2} = 1.465/(9.81 \times 2.2)^{1/2} = 0.315$ as required for hydraulic similarity in a Froudian model.

EXAMPLE 10.6

A town floods periodically when a river bursts its banks. It is proposed to construct an artificial relief channel to carry some of the flood water around the town. The start of this channel is roughly at right-angles to the natural river. A side overflow weir (that is a weir constructed along the riverbank) will be used to control the flow down the relief channel. Ideally the relief channel would not operate until the river carries about $6\,m^3/s$, which represents the bank-full condition with a depth of about 1.5 m and a width of 9.5 m. Since the equations relating to side overflow weirs are not always reliable, it is proposed that a hydraulic model of the entrance structure should be built to determine its optimum position, length and crest height, and to ensure that it will work effectively. Given the data below, outline a suitable model investigation.

Maximum flood flow in the river = $50\,m^3/s$ with a depth of 3.0 m and a width of 17 m.
Maximum proposed discharge in the relief channel = $30\,m^3/s$
Minimum drought flow in river = $0.7\,m^3/s$ with a depth of about 0.4 m and a width of 8 m.

Flow over a weir is controlled by gravity, so a Froudian model is required. This is a hydraulic structure model that does not need to be distorted, nor is a mobile bed required. A typical scale for such investigations would be between 1:20 and 1:50. Try 1:20 for greatest accuracy.

$$\text{At } 1:20 \quad \frac{L_M}{L_P} = \frac{1}{20} \qquad \frac{V_M}{V_P} = \frac{1}{20^{1/2}} = \frac{1}{4.47} \qquad \frac{T_M}{T_P} = \frac{1}{20^{1/2}} = \frac{1}{4.47}$$

$$\frac{Vol_M}{Vol_P} = \frac{1}{20^3} = \frac{1}{8000} \qquad \frac{Q_M}{Q_P} = \frac{1}{20^{5/2}} = \frac{1}{1789} \qquad \frac{n_M}{n_P} = \frac{1}{20^{1/6}} = \frac{1}{1.65}$$

Say that the model extends at least 5× river width upstream and 3× river width downstream of the weir giving the length to be modelled = 8 × 17 = 136 m, that is a model length = 136/20 = 6.8 m + the weir length. Say the width to be modelled is about 6× river width = 6 × 17 = 102 m, that is 102/20 = 5.1 m in the model. This is not excessive, so the area required is OK.

To avoid surface tension effects the minimum head over the model weir should be 6 mm, that is 0.006 × 20 = 0.12 m in the prototype. So results up to 0.12 m of head over the crest are not accurate, which is acceptable.

To avoid large scale effects check that the flow is turbulent in the model by calculating Re = $\rho LV/\mu$. The average velocity in the river during flood = 50/(3 × 17) = 0.98 m/s, so V_M = 0.98/4.47 = 0.22 m/s. The model depth is 3/20 = 0.15 m and hence the Reynolds number, Re = (1000 × 0.15 × 0.22/1.005 × 10^{-3}) = 32 800 > 2000, that is turbulent (see section 4.2.1). Average velocity in the river at low flow = 0.7/(0.4 × 8) = 0.22 m/s, so V_M = 0.22/4.47 = 0.05 m/s. The model depth is 0.4/20 = 0.02 m so Re = (1000 × 0.02 × 0.05/1.005 × 10^{-3}) = 995 < 2000, that is transitional. With Re = 995 the flow is not fully turbulent and there may be some scale effects, but this drought condition is not really of interest.

Check the conditions at the point at which the weir starts operating, that is when the average velocity = 6/(1.5 × 9.5) = 0.42 m/s, so V_M = 0.42/4.47 = 0.09 m/s with a model depth of 1.5/20 = 0.08 m. Thus Re = (1000 × 0.08 × 0.09/1.005 × 10^{-3}) = 7200 > 2000 (fully turbulent). Suppose that the model velocity near the bank is 0.02 m/s at the same depth, then Re = (1000 × 0.08 × 0.02/1.005 × 10^{-3}) = 1600. Re = 1600 is just about OK, so accept that very low flows may not be accurately modelled.

Check the discharges with respect to available pumps and ease of measurement. Maximum discharge in the river during flood = $50\,m^3/s$, so Q_M = 50/1789 = $0.028\,m^3/s$. This is OK.

Discharge at bank-full condition = $6\,m^3/s$, so $Q_M = 6/1789 = 0.003\,m^3/s$ or $3\,l/s$. Small but OK. Maximum discharge to be measured in relief channel = $30\,m^3/s$, so $Q_M = 30/1789 = 0.017\,m^3/s$. The discharge to the channel could be measured using an orifice plate in the pipeline from the pump with the water being admitted to the upstream end of the model via a suitable diffuser, possibly a pipe with holes in the bottom. The discharge at the exit from the river and relief channels could be measured with a sharp crested weir.

If the weir is concrete with $n_P = 0.015$ then $n_M = 0.015/1.65 = 0.009\,s/m^{1/3}$ so perspex or varnished wood is OK. If the river channel has $n_P = 0.035$ then $n_M = 0.035/1.65 = 0.021\,s/m^{1/3}$ so roughened cement mortar is just OK.

Summary

1. There are three fundamental dimensions (M, L and T). Everything else can be broken down into these dimensions. For example, force = MLT^{-2}. In the metric SI system the units of MLT are kg, m and s.

2. Make sure you understand the laws of indices in Box 10.1 and are happy using them, otherwise you will never obtain the correct answers to the problems in this chapter.

3. For dimensional homogeneity both sides of an equation must contain exactly the same dimensions raised to exactly the same powers. This also applies to all of the individual terms of the equation. For example, all three terms in the energy (Bernoulli) equation have the dimension L, so the total must also have the dimension L (i.e. head).

4. Dimensional analysis uses the concept of dimensional homogeneity to determine which variables should be included in equations describing a hydraulic phenomenon, and the powers to which the variables should be raised. There are two methods of dimensional analysis: the Rayleigh (indicial) technique and the Buckingham Π theorem.

5. The Rayleigh method begins by writing a power equation. In Example 10.1 the power equation is $q = Kg^aH^b$ where q, g and H are the variables, K is a constant, and a and b the unknown powers. The method involves sub-stituting into this equation the fundamental dimensions of q, g and h, i.e. $L^2T^{-1} = K[LT^{-2}]^aL^b$. Then, one at a time, the powers of M, L and T are equated and the equations solved for the numerical values of a and b (i.e. $\frac{1}{2}$ and $\frac{3}{2}$ in Example 10.1). Thus $q = Kg^{1/2}H^{3/2}$.

6. The Buckingham Π theorem can be applied to more complex problems. This method starts with a functional relationship. In Example 10.3 (flow over a sharp crested weir) this is $f'(Q_T, b, g, H, \rho, \mu) = 0$. Here there are six variables (quantities) that between them contain all three fundamental dimensions (i.e. MLT) so there will be $(6 - 3) = 3$ equations and three dimensionless Π groupings. The three equations for Π_1, Π_2 and Π_3 all contain the same three repeating variables plus one non-repeating variable. The fundamental dimensions are inserted into each equation in turn and the unknown powers obtained using dimensional homogeneity (effectively using a slightly modified Rayleigh method). Remember that any of the dimensionless groupings obtained can be inverted, combined or raised to some power without affecting their dimensionless status.

7. When selecting the repeating variables for the Buckingham Π method, try to ensure that the combinations adopted will enable standard dimensionless grouping to be obtained. For example, with open channel flow where the

Froude number is important, make sure V, g and D appear in the same equation (D = depth). For pipe flow, make sure that the variables of the Reynold number, ρ, D, V and μ, are together (D = diameter). Often ρ, V and D are suitable as repeating variables. Using different combinations of variables results in different answers, all correct, but some more recognisable than others.

8. Because the Froude number, Reynolds number etc. are dimensionless, their value is not affected by size. Hence a model can be used to reproduce the flow in a full-size prototype. The dominant force must be correctly identified and reproduced, at the expense of the others if necessary. The aim is to achieve geometric similarity (same shape), kinematic similarity (same motion) and dynamic similarity (same forces). For instance, in Example 10.5 both model and prototype have F = 0.315. In practice there is some scale effect because it is not possible for the model to have exactly the same F, Re, E, W and M values as the full-size version (see Table 10.2) because each parameter results in different scale factors (e.g. compare equations (10.10) and (10.16)).

Revision questions

10.1 What are the units and dimensions of the following quantities: (a) shear stress; (b) kinematic viscosity; (c) surface tension; (d) Manning's n; (e) the drag coefficient; (f) hydraulic radius?

10.2 (a) Describe what is meant by dimensional homogeneity. (b) Starting with the Darcy equation (6.12), use dimensional homogeneity to determine the fundamental dimensions (if any) of the pipe friction factor, λ.

10.3 The discharge per metre length of weir crest, q, is thought to depend upon the head, H, over the crest, the height of the crest above the river bed, p, and gravity, g. Find the form of the equation using both the Rayleigh and Buckingham Π methods. Is p significant?
$$[q = f(g^{1/2}H^{3/2}p/H); \text{ yes, } p \text{ affects } C_D]$$

10.4 The discharge through a small orifice, Q, depends upon the area of the orifice, A, the head above the orifice, H, and gravity, g. Derive the discharge equation using both the Rayleigh and Buckingham Π methods, and show that the same equation can be obtained in both cases.
$$[Q = f(Ag^{1/2}H^{1/2}]$$

10.5 Derive an equation for the drag force, F, on a sphere of diameter, D, when it is positioned in a fluid that has a mean flow velocity, V, a density, ρ, and a dynamic viscosity, μ. Again use both the Rayleigh and Buckingham Π methods, and show that the same equation can be obtained using both techniques.
$$[F = f(\rho D^2 V^2, \text{ Re}), \text{ see section 4.9}]$$

10.6 (a) Describe what is meant by similarity. (b) If two pumps are said to be similar, what does this mean? (c) Pumps A and B are similar and can be related by $[Q/ND^3]_A = [Q/ND^3]_B$. Pump A has an impeller diameter of 0.3 m and runs at a speed of 1200 rpm. If pump B has a diameter of 0.6 m, what speed would it have to run at to achieve the same discharge? (d) Alternatively, if pump B in part (c) also ran at 1200 rpm, what would be its discharge compared to pump A?
$$[(c) \text{ } 150 \text{ rpm; (d) } \times 8.0]$$

10.7 (a) What type of hydraulic problems are best investigated using undistorted physical models? (b) Derive fully the equivalent of equations (10.8) to (10.14) for an undistorted model operating according to Reynolds' law.

10.8 (a) What type of hydraulic problems are best investigated using distorted physical models? (b) Derive fully the equivalent of equations (10.8) to (10.14) for a distorted Froudian model.

10.9 The model investigation outlined in Example 10.6 was only just satisfactory in some respects, problems perhaps arising at very low flows due to surface tension effects and the difficulty of ensuring turbulent flow. To ensure that 1:20 is the best scale for the model, repeat the calculations using a scale of 1:15. Tabulate the values of the calculated variables for the two scales, and decide whether it would be better to build the model to a larger scale.

[The larger scale may increase accuracy and reduce the difficulties associated with surface tension and ensuring turbulent flow, but the maximum flood discharge to be pumped is now quite large. The larger scale may be desirable but is not essential]

10.10 A physical hydraulic model is to be built to study the effect of altering the breakwaters at the entrance to a 150 m wide river to allow the navigation of larger ships. However, there is concern that altering the breakwaters will result in changes to the beach levels and offshore sandbanks around the estuary. It is proposed that a distorted model will be constructed. A distortion of about 8 is quite typical, so take the horizontal scale as 1:240 and the vertical scale as 1:30. The model will have a mobile bed of fine sand. It is proposed that the model will include a length of the coastline 2 km to each side of the river, and will cover 2 km of the river upstream of the mouth as well as 2 km of the offshore sea bed. A wave generator is available to reproduce wave effects. A field investigation in the estuary shows that the maximum tidal velocity at the half-tide depth is 1.6 m/s, while the depth of water at the estuary mouth varies from 4.5 m to 8.5 m according to tide. The maximum ebb tidal discharge is 1000 m^3/s with a maximum tidal volume of 9×10^6 m^3. A tide occurs every 12.4 hours and the maximum tidal range is 4.0 m. Calculate (a) the floor space required to construct the model; (b) the maximum model tidal velocity; (c) the range of water depths in the model at the estuary mouth; (d) the model Reynolds number at the half-tide depth at the mouth ($v = 1.14 \times 10^{-6}$ m^2/s); (e) the maximum tidal discharge in the model; (f) the maximum model tidal volume; (g) the model tidal period; and (h) the maximum model tidal range.

[(a) 17.29 m × 16.67 m; (b) 0.29 m/s; (c) 0.15 m to 0.28 m; (d) 56 000; (e) 0.025 m^3/s; (f) 5.21 m^3; (g) 17 mins; (h) 0.13 m]

 11

Turbines and pumps

This chapter starts by considering the difference between positive displacement and rotodynamic machines, the difference between impulse and reaction turbines, and the general definition of efficiency and power. It then uses the momentum equation to obtain the force exerted by a jet of water when it hits a stationary or moving vane, such as the bucket of a Pelton wheel. The Pelton wheel is an impulse turbine suitable for sites with large heads of water. Other types of turbine, such as reaction turbines, are then considered, and their performance discussed. Some pumps can be thought of as turbines operating in reverse, and there are many pumped-storage schemes where water is pumped into storage reservoirs during off-peak periods, then allowed to flow through the same machines (now operating in reverse) to generate electricity when it is required. The performance characteristics of various types of pump are outlined, and what happens when two or more pumps are used in series or in parallel. How to match a pump to a rising main (delivery pipe) is then considered, as well as some common operational problems. This chapter answers such questions as:

How do turbines work?

What sort of turbine is best in a particular location?

How do pumps work?

How do you select the best pump for a particular situation?

How do you obtain the best combination of pumps and rising main?

What are cavitation and surge, and why are they important?

11.1 ▶ Introduction

11.1.1 Turbines and pumps

Turbines are used to generate electricity. They are a key element in hydro-electric schemes. The most familiar schemes may be those where water is stored behind a dam built across a river, then delivered through a pipeline to the turbines. The water drives the turbines. Connected to each turbine is a generator, like the dynamo on a bicycle, that generates the electricity. The basic requirement for these schemes is thus a plentiful supply of water with a large enough head to drive the turbines. Consequently mountainous countries like Wales, Scotland, Norway and Sweden are ideal locations for this type of high head hydro-power. However, advances in turbine design now make low head hydro-power a practical possibility, provided the flow of water is large. Tidal power stations, like that at La Rance, France, employ special turbines to utilise the head of water created across dams or barrages during the tidal cycle.

Pumps are driven by electric motors and are used to raise water over some vertical height. They have been used for centuries to drain mine workings. Other uses are in land drainage schemes. Many low lying areas of Britain, like the Somerset Levels and the Fens, are below sea level (see Fig. 12.3). Water cannot always drain away naturally, it has to be pumped. Other uses of pumps include moving large quantities of water and sewage from source to treatment works, and so on.

A detailed knowledge of the internal workings of turbines and pumps is rarely needed, but it is useful to have a rudimentary knowledge of how they work, how to specify them, and how they fit into Civil Engineering projects. The first thing is to appreciate the basic difference between a turbine and a pump.

Turbines are machines which use an input in the form of a flow of water with a significant head to obtain a mechanical output, that is the rotation of a runner which in turn drives a generator. This is a turbine-generator.

Pumps are machines which use a mechanical input (that is the rotation of a runner powered by a motor) to lift a quantity of water to some particular height. This is a motor-pump.

A motor is basically a generator operating in reverse. Centrifugal pumps are effectively turbines operating in reverse, and some machines are designed to act as both pumps and turbines. For instance, when there is a surplus of cheap electricity available via the national grid (for example during the night) it may be used to pump water from a lower reservoir up to a high level reservoir. Then, at times of high electrical demand water flows back down to the lower reservoir via the turbines to generate electricity and feed it into the grid. One such pumped-storage scheme is that at Foyers, on the shore of Loch Ness (Fig. 11.1). Another is the Dinorwic hydro-power station in Wales, which was the largest in Europe when constructed. At Dinorwic it is necessary to pump water back to the upper reservoir because of the large flow rate (up to $390 \, \text{m}^3/\text{s}$) through the turbines. The reservoir would quickly empty otherwise. Water is fed through tunnels about 10 m in diameter to six reversible Francis pump/turbines which give an average station output of 1681 MW. If there is a sudden demand for electricity, Dinorwic can go from an output of 0 to 1320 MW in 10 seconds ($1 \, \text{MW} = 10^6 \, \text{W}$ or 1000 one-bar electric fires).

11.1.2 Positive displacement and rotodynamic machines

Some of the earliest pumps were the beam engines designed to dewater mine workings. These are a good example of positive displacement machines, which operate on the same principle as the bicycle pump. Essentially they consist of an inlet valve to admit a given quantity of fluid to a cylinder in front of a piston, the piston which is driven through the cylinder expelling the fluid, and an outlet valve. The valves control the flow into and out of the cylinder in synchronisation with the piston and prevent any backflow when the piston returns to its starting position for the next cycle. Because the piston moves backwards and forwards in the cylinder these are called reciprocating pumps. Since the piston

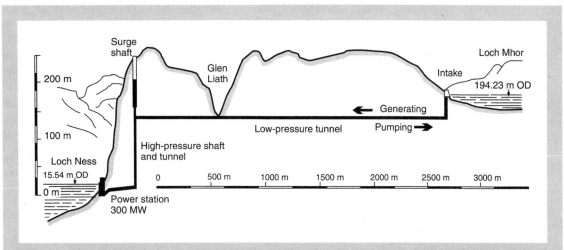

Figure 11.1 The Foyers scheme. Water is pumped up from Loch Ness to Loch Mhor, then flows back through two reversible Francis turbines, each of which is connected to a 150 MW generator. Total discharge pumping is 155 m³/s against 166 m net head, or 190 m³/s with 182 m head generating. The high-pressure tunnels reduce from 7.3 m to 3.0 m diameter at the pump/turbines

travels the same distance each time, the same quantity of fluid is expelled on each stroke irrespective of the head pumped against.

Another type of positive displacement pump is a rotary pump, which uses intermeshing gears, rotors, or lobes which revolve to force liquid around within a closed casing. These pumps are very simple, having no valves. They produce a smooth discharge which is proportional to the speed of rotation. The maximum flow rate is of the order of 0.030 m³/s. They are well suited for pumping viscous liquids and sewage sludges and may be useful for dewatering construction sites, but the discharge is too small for most large Civil Engineering projects. Consequently positive displacement pumps will not be considered further.

This chapter is mainly concerned with rotodynamic machines (the ram pump is entirely different and is considered in section 11.9). An early example of a rotodynamic turbine would be an old fashioned mill waterwheel where a continous flow of water is fed tangentially onto the rotating wheel to drive it and keep it in motion. Thus the chief characteristics of these machines are:

(a) Both the flow of fluid through the machine and the output is continuous. Rotodynamic machines are ideal for situations involving a large flow rate.

(b) All rotodynamic machines have a rotating element, called a runner (turbines) or impeller (pumps). With turbines the fluid often enters the runner tangentially, while with pumps the fluid often leaves the impeller tangentially. This is explained later.

11.1.3 Impulse and reaction turbines

Turbines are often described as being either of the impulse or reaction type. One of the simplest types of turbine to understand is the Pelton wheel, which uses the impact of a high-velocity water jet to turn a runner. The impact takes place in the open atmosphere (or rather in a casing in which the pressure is atmospheric) so pressure does not contribute to the force exerted on the runner. A reaction turbine, on the other hand, derives some of the force from the pressure of the water, in addition to the velocity of the water and the change in direction of the flow. The pressure component distinguishes reaction turbines from impulse turbines. If the Bernoulli equation is applied to the inlet (suffix 1) and outlet (suffix 2) of the turbine assuming no loss of energy and a horizontal streamline, then the energy per unit weight of water, E, transferred to the turbine can be found from:

$$V_1^2/2g + P_1/\rho g = V_2^2/2g + P_2/\rho g + E$$
$$E = (V_1^2 - V_2^2)/2g + (P_1 - P_2)/\rho g \tag{11.1}$$

If $P_1 = P_2$ so that E depends only on the velocity term then we have a pure impulse turbine like the Pelton wheel. If $V_1 = V_2$ so that E depends only on the pressure term, then we have a pure reaction turbine. The much used Francis turbine depends on a combination of velocity and pressure changes.

11.1.4 Efficiency

The **overall efficiency**, ε_T, of a turbine is given by the ratio of the power output to the available power. The available power is the power of the stream of water. Equation (4.22) shows that:

$$\text{Total energy per unit weight (N) of fluid} = z + V^2/2g + P/\rho g = H \tag{4.22}$$

where H is the total head in metres. Thus if we have a stream of liquid with a weight density $\rho g\,\text{N/m}^3$ and a volumetric flow rate $Q\,\text{m}^3/\text{s}$ then the weight of liquid flowing per second is $\rho g Q\,\text{N/s}$. The total energy per second of the stream of liquid, that is the **input power**, is:

$$\rho g Q(z + V^2/2g + P/\rho g)$$
$$\text{or} \quad \rho g Q H\,\text{Nm/s} \tag{11.2}$$

If *Pow* is the **output power** of a turbine, then its overall efficiency, ε_T, is:

$$\varepsilon_T = Pow/\rho g Q H \tag{11.3}$$

where H the total head available, that is the head difference between the inlet of the turbine and the tailwater level of the discharged water after allowing for pipeline head losses between the supply reservoir and the inlet. Turbines can have efficiencies as high as 93% under optimum running conditions.

The definition of the overall efficiency, ε_P, of a pump is the ratio of the fluid power output ($\rho g Q H$) to the mechanical power input (*Pow*) to the machine:

$$\varepsilon_P = \rho g Q H/Pow \tag{11.4}$$

where H is the actual total head difference between the inlet and outlet of the machine. The efficiency of large centrifugal pumps may be as high as 90%, with smaller units and axial flow pumps perhaps having efficiencies nearer 80%.

11.1.5 Synchronous speed

A factor that has to borne in mind when dealing with turbines and pumps is that the choice of the rotational speed, $N\,\text{rpm}$, of the turbine-generator or motor-pump is severely restricted.

This is because, for example, a turbine must produce electricity at a fixed frequency, *fre* (50 Hz in the UK; Hz = cycle/s). If the turbine or motor has p number of poles then:

$$N = (60 \times fre)/p \qquad (11.5)$$

Thus at 50 Hz with 6 poles the synchronous speed would be 500 revolutions per minute. The important point is that a turbine cannot run at any speed it likes as dictated by the discharge and velocity of the stream of water driving it, nor can the speed be adjusted to suit the electrical demand. The required electrical output must be achieved by some other means, usually by adjusting the water supply to the turbine while maintaining a constant rotational speed.

 ## 11.2 Impulse turbines

These are some of the simplest turbines where a jet of water strikes, effectively in the open air, a series of vanes or buckets attached to a runner (or wheel), so turning the runner. This is the hi-tech equivalent of an old fashioned water wheel. The basic principle governing the operation of such machines is the exchange of momentum between the water jet and the vane. The higher the velocity of the jet, the greater its momentum, the greater the impact when it hits the vane, and the greater the force on the vane. As we shall see, the shape of the vane is also crucially important. Before deriving the relevant equations, read Box 11.1 below carefully, and if necessary go back to Chapter 4 and revise the basic principles.

11.2.1 The force exerted on a stationary vane

To begin with we will consider the general case of a jet striking a stationary flat vane, as in Fig. 11.2a. The angle through which the jet is deflected is θ. By considering a control volume the momentum equation can be applied to obtain the components of the resultant force, F_{RX} and F_{RY} acting in the x and y directions (Fig. 11.2b) and thus $F_R = (F_{RX}{}^2 + F_{RY}{}^2)^{1/2}$.

x direction

$$\Sigma F_X = \rho Q(V_{2X} - V_{1X})$$

$$-F_{RX} = \rho Q(V_2 \cos\theta - V_1) \qquad (11.6)$$

y direction

$$\Sigma F_Y = \rho Q(V_{2Y} - V_{1Y})$$

$$F_{RY} = \rho Q(V_2 \sin\theta - 0) \qquad (11.7)$$

Equations (11.6) and (11.7) lead to some interesting results. For example, if θ in Fig. 11.2a is very small so that the jet merely strikes the vane a glancing blow, then both F_{RX} and F_{RY} are very small. This is because the bracket in both equations has a value close to zero. As θ increases towards 90°, so the forces involved increase. This is logical. Far better to be struck a glancing blow on the head by a cricket ball than to be hit squarely between the eyes.

Now consider the flat plate in Fig. 11.2c. In this case the deflection angle $\theta = 90°$. The net force in the y direction is zero, assuming that the jet divides equally on impact with a flow rate of $Q/2$ and velocity V_2 in both directions (gravitational effects being ignored). In the x direction (that is the initial direction of the jet) $\cos\theta = 0$ and equation (11.6) gives:

$$F_{RX} = \rho QV_1 \qquad (11.8)$$

With the hemispherical cup in Fig. 11.2d the deflection angle θ is 180° and $\cos\theta = -1$, so the two velocity terms in equation (11.6) have the same sign. If it assumed that there is

Box 11.1 ▶ **Remember**

By Newton's Laws of Motion there is a force acting on the vane if:

(a) the jet of water changes *velocity*;

(b) the jet of water changes *direction*;

(c) the jet of water changes *velocity and direction*.

Some points to remember when using a control volume are:

(A) only the external forces acting on the control volume are considered;

(B) we use a sign convention with *forces*, that is +ve in the original direction of motion, −ve in the opposite direction;

(C) in a particular direction the momentum equation states that the sum of the external forces is equal to the rate of change of momentum: $\Sigma F = \rho Q(V_2 - V_1)$.

When applying the momentum equation to the impact of a jet it is usual to assume that:

(i) In this specific situation, the surface of the control volume is at atmospheric pressure. Thus there is no net pressure force, *so pressure can be ignored*. This means there are no *PA* terms in the jet impact equations. Compare this with the flow around a pipe bend when the liquid is under pressure, for instance as in Example 4.4.

(ii) The effect of gravity on the jet is negligible, so it can be ignored.

(iii) The weight of water in the control volume can be ignored.

no reduction in velocity as the water flows around the cup so that $V_1 = V_2$, then equation (11.6) becomes:

$$F_{RX} = 2\rho Q V_1 \tag{11.9}$$

Thus the force exerted on the hemispherical cup (in the original direction of the jet) is theoretically twice that exerted on the flat plate. Note that this is only true when θ is exactly 180° and 90° respectively: any other value alters the equations, as shown in Table 11.1.

In Fig. 11.2e the jet enters tangentially along the x axis and exits along the y axis. The velocity in the x direction is initially V_1 and becomes zero, whereas in the y direction the velocity is initially zero and becomes V_1 (assuming no reduction in velocity so that $V_1 = V_2$). Since the mass flow rate (ρQ) is the same, it follows from equations (11.6) and (11.7) that the forces F_{RX} and F_{RY} are numerically equal: the rate of change of momentum is the same along both axes, so the forces are the same. This would, perhaps, appear unlikely from a first inspection of the problem, but it is the logical outcome of applying Newton's Second Law. In fact this arrangement gives the largest force in the y direction, as Table 11.1 shows. Note that the force in the y direction was ignored in Fig. 11.2c because there were two equal jets travelling in opposite directions that cancelled each other out, giving a net force of zero. If, as above, the two jets are not balanced, then the net force is not zero and should be calculated.

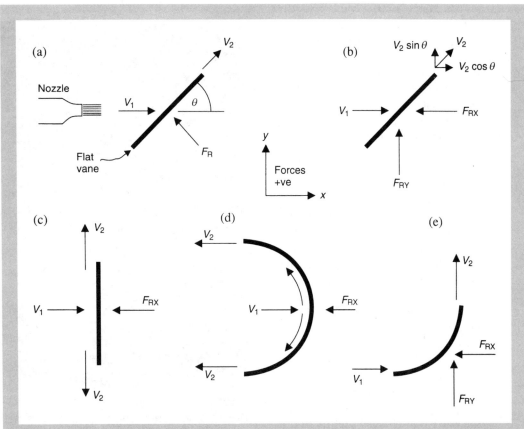

Figure 11.2 (a) A jet of water striking a flat vane with velocity V_1 and leaving parallel to the vane surface with velocity V_2 and (b) the equivalent components of force and velocity in the x and y directions. (c) A jet striking a flat plate at right angles. (d) A jet striking the centre of a hemispherical cup at right angles. (e) Impact on a curved vane with $\theta = 90°$ and the jet striking the vane tangentially

Table 11.1 Illustration of variation of F_{RX} and F_{RY} with deflection angle θ (when $V_1 = V_2$)

θ (degrees)	$\cos\theta$	$F_{RX} = \rho Q(V_2\cos\theta - V_1)$	$\sin\theta$	$F_{RY} = \rho Q(V_2\sin\theta)$
10	0.98	$0.02\rho QV_1$	0.17	$0.17\rho QV_1$
45	0.71	$0.29\rho QV_1$	0.71	$0.71\rho QV_1$
80	0.17	$0.83\rho QV_1$	0.98	$0.98\rho QV_1$
90	0.00	$1.00\rho QV_1$	1.00	$1.00\rho QV_1$
100	−0.17	$1.17\rho QV_1$	0.98	$0.98\rho QV_1$
135	−0.71	$1.71\rho QV_1$	0.71	$0.71\rho QV_1$
170	−0.98	$1.98\rho QV_1$	0.17	$0.17\rho QV_1$
180	−1.00	$2.00\rho QV_1$	0.00	0.00
190	−0.98	$1.98\rho QV_1$	−0.17	$−0.17\rho QV_1$

The shape of the vane and the deflection angle are of vital importance in determining the force exerted on the vane. This in turn determines how much power can be extracted from the jet, and hence the 'efficiency' of the impact type turbine. Obviously, the other variables such as the mass flow rate (ρQ) of the jet and its initial velocity (V_1) are also important, but for any given jet it is the shape of the vane that governs the efficiency. Thus much effort has been devoted to obtaining the optimum vane geometry. The Pelton wheel bucket is one of the most efficient, with the jet striking tangentially a central splitter (fin), then being turned through about 165° and discharged (Fig. 11.3). Ensuring that the discharged water does not interfere with the next bucket on the rotating runner is one of the design considerations. For this reason the deflection angle is not 180° but something less, so that the water is discharged slightly to the side. Having spent its energy, the actual velocity, V_2, of the discharged water is relatively small, as shown by the length of the velocity vector in Fig. 11.4 (Box 11.2). This vector also indicates the actual direction in which the water leaves the vane.

Figure 11.3 A Pelton wheel runner where the jet strikes the central splitter, with half of the water flowing around each side of the buckets. Note the notch in the bucket on the centreline of the jet. As the runner rotates, more than one bucket is hit by the jet at any instant, but the average distance from the nozzle to impact with the buckets remains constant. (*Photo courtesy of Sulzer-Escher Wyss Ltd*)

11.2.2 The force exerted on a single moving vane

In a turbine like the Pelton wheel the jet does not strike one stationary vane, but a number of vanes mounted on the circumference of a rotating runner. Thus the vane is not stationary, as considered previously, but moving in the same direction as the jet when the impact takes place. To analyse this new situation we use the idea of relative velocity. This was explained in Box 11.2, which should be read carefully before continuing. Now consider this question

Do you think that the force exerted on a vane that is moving in the same direction as the jet will be more or less than that exerted on a stationary vane?

The answer is that the force is less, because it is like catching a fast moving cricket ball. You move your hands backwards as you catch the ball to reduce the impact.

Box 11.2 **The concept of relative velocity**

The idea of relative velocity helps with the analysis of many problems, but what do we mean by relative velocity? Well, one much quoted example of relative velocity is if two cars have a head on collision while travelling in exactly opposite directions. If the two cars each have a velocity of 50 mph, then this is the equivalent of one of the cars travelling at 100 mph striking a stationary car head on. Thus the relative velocity is 100 mph. On the other hand, if two cars are travelling in exactly the same direction with speeds of 70 mph and 50 mph, and the faster car runs into the back of the other, the relative velocity is 20 mph. Consequently the impact is the same as that between a stationary car and one travelling at 20 mph. This is an example of the 'bringing to rest technique', where we consider one of the two moving objects to be stationary and measure all velocities relative to the 'stationary' object. We will consider a moving vane to be stationary, and calculate the velocity of the jet relative to the 'stationary' vane.

Remember that velocity has both magnitude and direction, so we have to apply the concept of relative velocity to direction as well. For example, suppose Spike tries to walk slowly down an escalator that is travelling upwards at a greater velocity. Spike would be transported backwards while still walking forwards. Similarly, suppose Spike is riding on a lorry travelling at 70 mph down the motorway. If Spike gently throws a football off the back of the lorry, he would see the football being left behind, apparently going backwards. That is because Spike sees the action relative to himself. A stationary observer standing at the side of the motorway would see Spike throw the football, and then would see the football bounce down the motorway in the same direction as the lorry. This is because the football would initially still have a forward velocity close to 70 mph, despite being thrown off the back of the lorry.

The illustration with the lorry has a relevance to the impact of a jet on a moving vane. The initial velocity of the jet is V_1 and the vane velocity is U. Thus the inlet velocity of the jet relative to a stationary vane is $(V_1 - U)$. When the jet leaves the vane in Fig. 11.4, still with velocity $(V_1 - U)$, it is going backwards relative to the initial jet. But because the bucket is also moving forward with velocity U, the actual direction of the jet, V_2, on leaving the vane may be in a forward direction. This is illustrated by the outlet velocity vectors in Fig. 11.4. Note that as U reduces, V_2 angles around towards the direction of the $(V_1 - U)$ vector below.

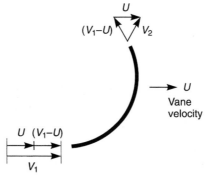

Figure 11.4 Inlet and outlet velocity vectors to a curved vane. The jet velocity is V_1 m/s, and the vane velocity is U m/s. Thus the relative velocity = $(V_1 - U)$. If there is no loss of energy as the water flows over the vane then the relative velocity at exit is still $(V_1 - U)$ but the actual exit velocity is V_2

Box 11.3 **Single moving vanes**

1. If the jet has an initial velocity, V_1, and the vane is moving in the same direction with a velocity, U, then if the vane is considered to be stationary (see Box 11.2) the relative velocity of the jet is $(V_1 - U)$. Thus the fact that the vane is moving away from the jet reduces the effective velocity of the jet compared to an equivalent stationary vane.

2. If the vane is moving away from the nozzle, then the effective mass flow rate is reduced from $\rho A V_1$ to $\rho A(V_1 - U)$ because of the extension of the jet (Fig. 11.5). If $U = V_1$ then the mass per second hitting the vane would be zero.

3. Thus if $U = V_1$ the rate of change of momentum is zero, and no force will be exerted on the vane by the jet. The equations above must be re-derived to take this into account.

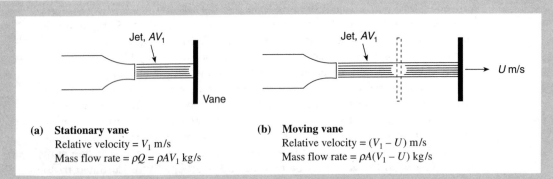

(a) **Stationary vane**
Relative velocity = V_1 m/s
Mass flow rate = $\rho Q = \rho A V_1$ kg/s

(b) **Moving vane**
Relative velocity = $(V_1 - U)$ m/s
Mass flow rate = $\rho A(V_1 - U)$ kg/s

Figure 11.5 The effect of a moving vane on the mass flow rate of a jet that has a velocity V_1 m/s and cross-sectional area of flow of A m²

Remembering the points in Box 11.3, we can now derive the force exerted on the moving vane in Fig. 11.6a. Consider a control volume enclosing the vane. As before, assume that the jet and its surroundings are at atmospheric pressure, that gravitational forces and the weight of the water in the control volume can be ignored, and that the velocity of the water flowing over the vane is constant at $(V_1 - U)$. Applying the momentum equation in the x direction, remembering the sign convention:

$$\Sigma F_X = \text{mass flow rate} \times (V_{2X} - V_{1X})$$

The relative velocity is $(V_1 - U)$ and the mass flow rate is $\rho A(V_1 - U)$

$$-F_{RX} = \rho A(V_1 - U)[(V_1 - U)\cos\theta - (V_1 - U)]$$

$$-F_{RX} = \rho A(V_1 - U)^2[\cos\theta - 1] \tag{11.10}$$

Remember that if $\theta > 90°$ then $\cos\theta$ has a negative value, so the terms in the square brackets are added together and F_{RX} becomes positive. There are no pressure terms because of the assumption of atmospheric pressure. Applying the same procedure to the y direction gives:

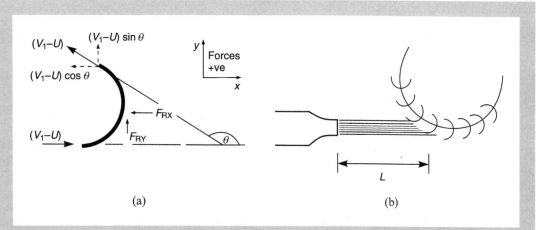

Figure 11.6 (a) The force exerted by a jet of velocity, V_1, on a curved vane moving in the same direction with velocity, U. (b) Impact between a jet and the vanes on a moving runner

$$\Sigma F_Y = \text{mass flow rate} \times (V_{2Y} - V_{1Y})$$
$$F_{RY} = \rho A(V_1 - U)[(V_1 - U)\sin\theta - 0]$$
$$F_{RY} = \rho A(V_1 - U)^2 \sin\theta \tag{11.11}$$

As usual $F_R = (F_{RX}{}^2 + F_{RY}{}^2)^{1/2}$ and its angle to the horizontal is $\tan^{-1}(F_{RY}/F_{RX})$. Examples 11.1 to 11.5 illustrate the use of the above equations. They also show how, by introducing the term, η, we can allow for any reduction in velocity that may occur as the water flows around the vane. However, we still do not have a realistic analysis because we have only considered one moving vane, not a series of similar vanes mounted on a runner as in Figs 11.3 and 11.6b.

11.2.3 The force exerted on a single vane mounted on a runner

This represents more closely the situation encountered with something like a Pelton wheel runner, shown diagrammatically in Fig. 11.6b. We can analyse this situation by adapting the equations above, provided we make the three additional assumptions in Box 11.4.

Although in practice the jet may strike two (or possibly more) vanes simultaneously as the runner rotates, one is nearer to the nozzle and one further away (Figs 11.3 and 11.6b) so the average length of the jet remains L. If the other conditions are as before, with a curved vane of deflection angle, θ, and velocity, U, being hit tangentially by a jet of velocity, V_1, then applying the momentum equation in the initial direction of the jet (that is along the x axis) now gives:

$$\Sigma F_X = \text{mass flow rate} \times (V_{2X} - V_{1X})$$
$$-F_{RX} = \rho Q[(V_1 - U)\cos\theta - (V_1 - U)]$$
$$-F_{RX} = \rho A V_1(V_1 - U)[\cos\theta - 1] \tag{11.12}$$

| Box 11.4 | **Vanes on a runner** |

1. Assume that the runner is large enough for the jet impact to occur at right angles to the vane, despite the rotation of the runner.

2. Assume that the jet hits only one vane at a time.

3. Assume that the *average* length of the jet to the point of impact with the vane is constant and does not vary (that is the jet is not extending as in Fig. 11.5b). This means that the mass flow rate can be taken as $\rho A V_1$ (or ρQ), as with a stationary vane. The second and third assumptions are interlinked.

Thus the only difference between equations (11.10) and (11.12) is that for the situation involving a runner the mass flow rate is $\rho Q = \rho A V_1$ (and *not* $\rho A (V_1 - U)$). Similarly, the equation for the y direction is $F_{RY} = \rho A V_1 (V_1 - U) \sin \theta$.

A good question is 'at what vane speed, U, is the maximum power obtained from a given jet?' Since power, Pow, is defined as a force multiplied by the distance moved per second in the direction of the force, the power developed by the jet when it moves the vane at U m/s is:

$$Pow = \rho Q (V_1 - U)[\cos \theta - 1] \times U$$

or $\quad Pow = \text{constant} \times (V_1 U - U^2)$

Differentiating Pow with respect to U and equating to zero gives:

$$\mathrm{d}Pow/\mathrm{d}U = \text{constant} \times (V_1 - 2U) = 0$$

so $\quad (V_1 - 2U) = 0$

or $\quad U = V_1/2$ (11.13)

Thus theoretically the maximum power is obtained when the vane velocity is half that of the jet. However, the choice of runner speed is very restricted (see section 11.1.5), so it is the jet velocity that has to be varied to obtain the optimum performance. For this reason a special spear valve is used to optimise efficiency and maintain a high-velocity jet by decreasing the jet diameter as the discharge decreases (see Fig. 11.7 and below).

We now have the tools to analyse impact type turbines. The summary of the equations in Box 11.5 provides a useful memory aid, while Examples 11.1 to 11.5 and Self Test Question 11.1 provide illustrations of their use.

11.2.4 The Pelton wheel

Early impulse turbines with flat vanes only had an efficiency of 40%. The development of curved vanes raised this figure to 65% by avoiding uncontrolled water splash and the associated waste of energy. In 1889 an American, Pelton, devised the curved bucket shown in Fig. 11.3 that raised the efficiency to 80%, which has subsequently been increased to as high as 93% by further refinements. Thus the vane geometry, and the velocity of the jet, are important with respect to the effectiveness of the turbine. The Pelton wheel is at its best when the velocity is high, so it is ideal for sites where there is a large head difference between the supply reservoir and the turbines. Indeed, the Pelton wheel is often the only option when the head exceeds 500 m. One scheme in Austria operates with a 1750 m head. They can also operate at heads below 200 m. Modern machines frequently use two, four or even six jets (Fig. 11.8). The maximum power output is typically about 80 MW, but could be as high as 400 MW. They are capable of operating smoothly and efficiently over a wide range of conditions, a desirable characteristic for a turbine since they cannot always work at full load. Even at 20% of maximum load, Pelton wheels may be able to deliver an efficiency of around 80% (Fig. 11.14).

A typical arrangement for a Pelton wheel is that a nozzle at the end of a pipeline discharges a high-velocity jet of water of up to about 300 mm diameter (d_j) into the atmosphere. A special needle or spear valve near the nozzle outlet controls the discharge by varying the diameter of the jet, so ensuring that the jet velocity is maintained (Fig. 11.7). This enables the turbine to operate with reasonable efficiency (80 to 90%) over a wide range of power outputs.

The jet hits the buckets mounted on the runner, causing it to rotate. The speed of rotation is kept constant automatically by a governor. The diameter of the runner is typically about 10 to 14 times that of the jet (d_j). After impact

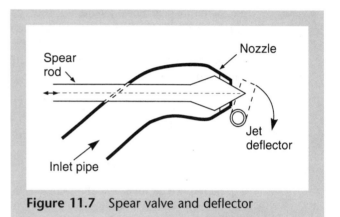

Figure 11.7 Spear valve and deflector

Figure 11.8 A six nozzle Pelton wheel. One of two 260 MW Pelton turbines in the Sellrain-Silz power station, Austria. The operating head is 1233 m, the speed 500 rpm. The runner has a diameter of 2.850 m and weighs 15.3 tonnes. (*Photo courtesy of Sulze-Escher Wyss Ltd*)

Box 11.5 ▶ **Summary of equations**

Vanes not on a runner
The following general equations for force can be adapted to most needs.

$$-F_{RX} = \rho A(V_1 - U)[\eta(V_1 - U)\cos\theta - (V_1 - U)]$$

η = proportion of original velocity

$$F_{RY} = \rho A(V_1 - U)[\eta(V_1 - U)\sin\theta]$$

e.g. hemispherical moving vane, $\theta = 180°$, $\cos\theta = -1$, $\eta = 1.0$ then $F_{RX} = 2\rho A(V_1 - U)^2$
e.g. hemispherical stationary vane, as above but with $U = 0$, then $F_{RX} = 2\rho AV_1^2$
e.g. flat moving vane, $\theta = 90°$, $\cos\theta = 0$, $\eta = 1.0$ then $F_{RX} = \rho A(V_1 - U)^2$
e.g. flat stationary vane, as above but with $U = 0$, then $F_{RX} = \rho AV_1^2$
Similar equations can be developed for the y direction.

Vanes on a runner
The general equation is:

$$-F_{RX} = \rho AV_1[\eta(V_1 - U)\cos\theta - (V_1 - U)]$$
$$F_{RY} = \rho AV_1[\eta(V_1 - U)\sin\theta]$$

e.g. hemispherical vane, $\theta = 180°$, $\cos\theta = -1$, $\eta = 1.0$ then $F_{RX} = 2\rho AV_1(V_1 - U)$
e.g. flat vane with $\theta = 90°$, $\cos\theta = 0$, $\eta = 1.0$ then $F_{RX} = \rho AV_1(V_1 - U)$
Similar equations can be developed for the y direction.

the water leaves the buckets at a relatively low velocity, being directed sideways away from the runner and then falling clear. Since the impact takes place in the atmosphere, the turbine must be located well above the **tailwater** level (that is the level of the discharged water) to ensure that the water leaves the turbine freely. A jet deflector automatically rotates into position in front of the nozzle if the electrical load is rejected, directing the water towards the **tailrace** (discharge channel). The deflector can also be used to govern the runner speed.

EXAMPLE 11.1

A jet of water 50 mm in diameter hits the centre of a single stationary hemispherical cup as in Fig. 11.2d. The deflection angle, θ, is 180°. The velocity of the jet on impact is 11.37 m/s, and it is assumed that there are no energy losses (that is $\eta = 1.0$). What force is exerted on the vane in the original direction of the jet? ($\rho = 1000$ kg/m³)

From above: $F_{RX} = 2\rho AV_1^2 = 2 \times 1000 \times (\pi \times 0.050^2/4) \times 11.37^2 = 508$ N

EXAMPLE 11.2

If the vane in the previous example was moving with a velocity of 5.40 m/s, what would be the force exerted in the direction of the jet now?

From above: $F_{RX} = 2\rho A(V_1 - U)^2$

$$F_{RX} = 2 \times 1000 \times (\pi \times 0.050^2/4) \times (11.37 - 5.40)^2 = 140\,\text{N (note } \textit{decrease})$$

EXAMPLE 11.3

If the conditions are the same as Example 11.2 except that the vane is one of a series on a runner that has a velocity of 5.40 m/s in the direction of the jet, what is F_{RX} now?

From above: $F_{RX} = 2\rho A V_1 (V_1 - U)$

$$F_{RX} = 2 \times 1000 \times (\pi \times 0.050^2/4) \times 11.37(11.37 - 5.40) = 267\,\text{N}$$

$$(\text{note} > 140\,\text{N})$$

EXAMPLE 11.4

A jet of water with a velocity of 20 m/s and a diameter of 75 mm acts on a single moving vane, the water sliding onto the vane tangentially and being turned through an angle of 165°. The velocity of the water leaving the vane is 90% of the original relative velocity of the jet. The velocity of the vane is 9.5 m/s. Calculate the magnitude and direction of the resultant force.

Starting with the general equation for a single moving vane:

$$-F_{RX} = \rho A(V_1 - U)[\eta(V_1 - U)\cos\theta - (V_1 - U)]$$

$\theta = 165°$ and $\cos 165° = -\cos 15°$. $\eta = 0.90$. $A = \pi \times (0.075)^2/4 = 4.418 \times 10^{-3}\,\text{m}^2$.

Thus:

$$F_{RX} = \rho A(V_1 - U)^2[\eta\cos 15° + 1] = 1000 \times 4.418 \times 10^{-3}(20.0 - 9.5)^2[0.90\cos 15° + 1]$$

$$F_{RX} = 911\,\text{N}$$

$$F_{RY} = \rho A(V_1 - U)[\eta(V_1 - U)\sin\theta - 0]$$

$\eta = 0.90$. $A = \pi(0.075)^2/4 = 4.418 \times 10^{-3}\,\text{m}^2$. Thus:

$$F_{RY} = \rho A(V_1 - U)^2[\eta\sin\theta] = 1000 \times 4.418 \times 10^{-3}(20.0 - 9.5)^2[0.90\sin 165°]$$

$$F_{RY} = 113\,\text{N}$$

Resultant force $= \left(F_{RX}^2 + F_{RY}^2\right)^{1/2} = (911^2 + 113^2)^{1/2} = 918\,\text{N}$

Angle of resultant to horizontal $= \tan^{-1}(F_{RY}/F_{RX}) = \tan^{-1}(113/911) = 7.1°$

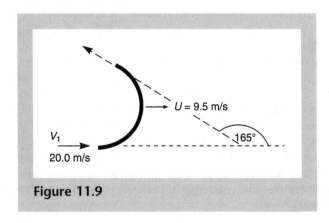

Figure 11.9

EXAMPLE 11.5

Repeat Example 11.4, but this time assuming that the vane is one of many on a runner.

$-F_{RX} = \rho A V_1[\eta(V_1 - U)\cos\theta - (V_1 - U)]$ with $\theta = 165°$

$\begin{aligned}
F_{RX} &= \rho A V_1(V_1 - U)[\eta\cos 15° + 1] \\
&= 1000 \times 4.418 \times 10^{-3} \times 20.0 \times (20.0 - 9.5)[0.90\cos 15° + 1] \\
F_{RX} &= 1734\text{N}
\end{aligned}$

$\begin{aligned}
F_{RY} &= \rho A V_1[\eta(V_1 - U)\sin\theta - 0] \\
&= 1000 \times 4.418 \times 10^{-3} \times 20.0[0.90(20.0 - 9.5)\sin 165°] \\
F_{RY} &= 216\text{N}
\end{aligned}$

Resultant force $= \left(F_{RX}^2 + F_{RY}^2\right)^{1/2} = (1734^2 + 216^2)^{1/2} = 1747\text{N}$

Angle of resultant to horizontal $= \tan^{-1}(F_{RY}/F_{RX}) = \tan^{-1}(216/1734) = 7.1°$

SELF TEST QUESTION 11.1

(i) If $\eta = 1.0$ in Examples 11.4 and 11.5, calculate the percentage change in F_{RX}, F_{RY} and F_R.

(ii) For Example 11.5 with $\eta = 0.9$ and $U = 9.0$, 10.0 and 11.0 m/s, for each U calculate the power developed in the x direction and the percentage change in F_{RX} (from 1734 N).

11.3 Reaction turbines

The **inward flow reaction turbine** was invented by an American, **Francis**, around 1849. In a modern Francis type turbine, water enters from a horizontal pipeline which effectively turns through 360° and reduces in diameter, forming a spiral (Fig. 11.10). This is the volute chamber, which is rather like a snail's shell. The reducing diameter is designed to increase the velocity of the water as it flows through the outer guide blades onto the curved vanes of the central runner, causing it to rotate. The force to drive the runner is obtained from a combination of the velocity of the water, the change in the direction of flow, and the pressure of the water. Having passed through curved slots in the runner, the water falls vertically down the draft tube and then flows away to the tailwater. The draft tube is an integral part of the turbine, generally incorporating a vertical section, a 90° bend of reducing diameter, followed by a gradual expansion to the tailwater. It is designed to 'suck' water through the turbine.

Since the turbine is enclosed in a pressurised casing, these machines are not generally suitable for heads above 500 m because of problems with watertight seals and leakage (in 1993 the world record for a high-pressure Francis turbine was 734 m at a power station in Austria). The shape of the guide blades is an important element in determining the efficiency of reaction turbines. Under optimum conditions a 94% efficiency may be achieved, but under anything other than the optimum load conditions the efficiency falls off rapidly (for example, about 70% efficiency at 20 to 30% of the full load, Fig. 11.14). Thus adjustable guide blades may be used to improve part-load performance. Power output is typically about 60 MW, although the turbine-generators at Dinorwic deliver over 300 MW each (see Table 11.2 and Fig. 11.12a).

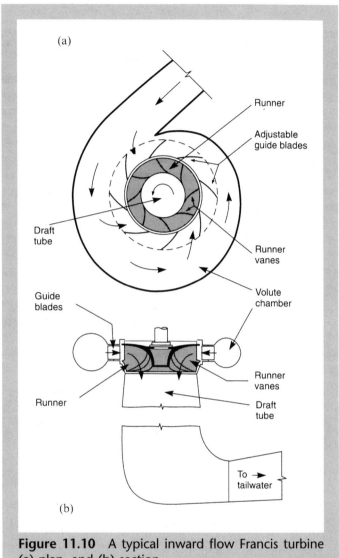

Figure 11.10 A typical inward flow Francis turbine (a) plan, and (b) section

Table 11.2 Turbine types and possible operating range

Turbine type	Head range (m)	Range of output (MW)	Required flow rate (m³/s)
Pelton	50–1700	0.1–400	0.06–80
Francis	20–700	0.1–1000	0.70–1000
Kaplan	10–140	0.3–200	4.00–1000
Bulb generator	2–25	0.1–120	4.00–1000

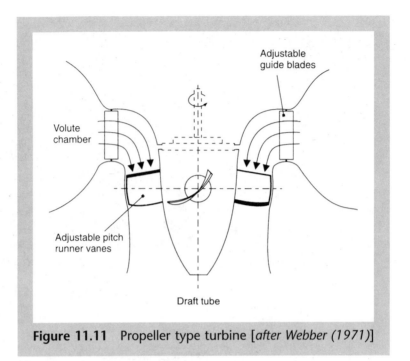

Figure 11.11 Propeller type turbine [*after Webber (1971)*]

The Czech engineer, **Kaplan**, patented the **propeller turbine** around 1913. In this reaction turbine (Fig. 11.11) water flows under pressure parallel to the axis of the machine. It first passes through a set of guide blades, then onto a runner that has a relatively small number of vanes similar to those of a ship's propeller. The propeller is often mounted facing vertically down, so that the propeller shaft runs vertically upwards to the generators mounted above. With this layout, the water enters horizontally from the side, before turning through 90°, passing down through the turbine, being turned through 90° again and discharged. An important part of the design is the ability to adjust the pitch of the runner vanes, like feathering an aircraft's propeller. This enables the turbine vanes to be adjusted automatically to obtain the greatest efficiency over a wide range of loads. The pitch of the guide blades can also be adjusted.

The Kaplan turbine is frequently used with heads of between 10 and 50 m, and a typical power output is around 50 MW. The efficiency of these machines may be up to 94%. One of their principal advantages is flexibility, with over 90% efficiency from 40% to 100% of full load, which is a better part-load performance than other turbines (see Fig. 11.14).

A variation on the Kaplan turbine is the '**bulb generator**'. These units can have runners as large as 7.7 m in diameter (Racine, Ohio). They can be used with heads of less than 8 m provided there is a large flow of water, which makes them ideal for tidal power stations like that on the River Rance, near Saint Malo, France (Fig. 11.13). Completed in 1966, this was the world's first tidal power station. The turbines have a fixed blade propeller housed axially in the short conduit that connects the two sides of the barrage. The generator is located in a watertight bulb within the conduit. The water flows around the bulb as it passes through the barrage. Despite the success of the Rance scheme, tidal power stations are still few in number. If constructed, the Severn Barrage would utilise bulb generators similar to those in Fig. 11.13.

(a)   (b)

Figure 11.12   (a) A large Francis turbine runner for the Tarbela project, Pakistan. Six turbines are designed to deliver a total of 5000 MW. (b) Part of a 6.40 m diameter propeller type runner for the upgraded 1860 MW R.H. Saunders Power Station, Canada. This has 16 units in a 1 km long barrage across the St Lawrence River. Note the size from the man. (*Photos courtesy of Sultzer-Escher Wyss Ltd*)

11.4 ▶ Performance equations and characteristics of turbines

For a turbine the most important relationship is between the head of water and the output power that can be generated with various rates of flow (Table 11.2). Turbine efficiency (equation (11.3)) is another important consideration. Dimensional analysis provides a means of obtaining the equations governing the performance of a turbine. Example 10.2 showed how equation (11.14) can be obtained, while equation (11.15) was derived as equation (10.1).

$$Pow/\rho N^3 D^5 = f[gH/N^2 D^2] \tag{11.14}$$

$$Q/ND^3 = f[gH/N^2 D^2] \tag{11.15}$$

where *Pow* is the output power developed by the turbine, N the speed of runner rotation, D the diameter of the runner, f means a 'function of', and H is the static head of the liquid on entry. Remember that power (*Pow*) is the rate at which energy is produced. For a turbine the speed of the runner may be fixed, since the attached generator must produce electricity at the correct frequency, that is 50 Hz in the UK. The head (H) at a particular site will also be predetermined. Thus equation (11.14) gives the runner diameter needed to obtain the required power output, with the corresponding flow rate being calculated from equation (11.15), or vice versa. Both of the above equations apply to all rotodynamic machines, either pumps or turbines.

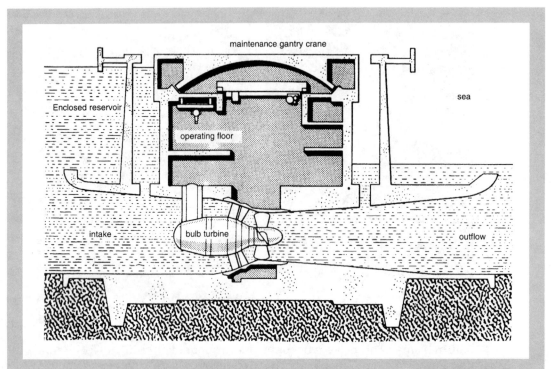

Figure 11.13 Cross-section of La Rance barrage at a bulb generator. The turbine-generator units are surrounded by water inside the passages connecting the two sides of the 700 m long barrage (*diagram courtesy of the New Civil Engineer*)

The term **specific speed**, N_S, is used to denote the performance characteristics of different types of turbine, or turbines of different size. N_S is the speed of a turbine (in rpm) needed to develop 1 kW when operating with a head of 1 m. Under these conditions N_S has the value of N in equation (11.16). By comparing the specific speeds of different types of turbine under similar conditions it is possible to determine which type is best suited to a particular site or duty (for this reason it is also called the **type number**). Specific speed is defined in various ways but a common definition is:

$$N_S = NPow^{1/2}/H^{5/4} \tag{11.16}$$

Note that N_S is not dimensionless because, by custom rather than for any mathematical or engineering reason, gravity and density are omitted from the equation since they are constant. Thus with N in rpm, Pow in kW and H in m, the specific speed range of the turbine types is:

Pelton wheel	N_S from 12 to 60	i.e. high head, low discharge
Francis	N_S from 60 to 500	i.e. moderate head, moderate discharge
Kaplan	N_S from 280 to 800	i.e. low head, large discharge

There is also a specific speed relationship for pumps, which is not the same as equation (11.16). However, equation (11.16) can be derived by following the procedure to obtain the pump specific speed equation in section 11.6.2, but starting with the terms Pow/N^3D^5 and H/N^2D^2. The **affinity laws** derived for pumps can also be applied to turbines.

SELF TEST QUESTION 11.2

With reference to the last paragraph above and section 11.6.2, derive the specific speed expression shown in equation (11.16).

The performance criteria which are generally of importance with respect to a turbine are the efficiency–speed curves, the power–speed curves at different flow rates (expressed as a fraction of the maximum nozzle opening or gate setting), and the efficiency–part load curves (Fig. 11.14).

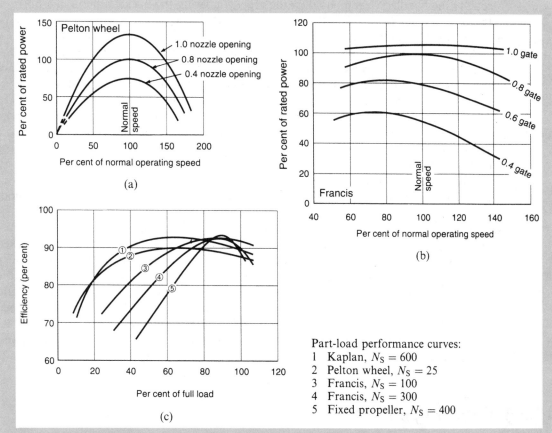

(a)

(b)

(c)

Part-load performance curves:
1 Kaplan, $N_S = 600$
2 Pelton wheel, $N_S = 25$
3 Francis, $N_S = 100$
4 Francis, $N_S = 300$
5 Fixed propeller, $N_S = 400$

Figure 11.14 Turbine performance curves relating to different nozzle openings or gate settings (that is flow rates). (a) Pelton wheel power–speed curve. Note the rapid reduction in power either side of the optimum normal speed (see section 11.2.3). (b) Power–speed curves for a Francis turbine. The maximum efficiency is obtained with a 0.8 gate setting. The efficiency–speed curves have the same general shape as (a) and (b). (c) Part-load performance curves of various turbines. N_S is the specific speed [*after Webber (1971)*]

11.5 Rotodynamic pumps

❝ The centrifugal pump is the most common type, effectively being a Francis turbine operating in reverse. It derives its name from the fact that the pressure head created is largely due to centrifugal action. Axial flow pumps are basically propeller type turbines operating in reverse. There is, of course, no way a Pelton wheel can operate in reverse. ❞

Just a minute, it is all very well saying that centrifugal action causes the increase in pressure. What is centrifugal action, and how does it cause an increase in pressure? I cannot visualise it.

OK, try thinking of it like this. You have a bucket full of water, and you tie a piece of rope to the handle. Now, by holding the rope, you swing the bucket around in a horizontal circle of diameter, D. **Does the water come out of the bucket?**

No, it is held in by the rotation of the bucket. I see, that is the centrifugal action.

Correct. If we put a hole in the bottom of the bucket, the water will be flung out. That is fairly obvious. **Now, if we seal the top of the bucket so that air cannot get in to replace the water lost through the bottom of the bucket, what do you think will happen?**

A vacuum, or a partial vacuum, will form above the water.

Good. Hang on to the idea of the vacuum above the water in the bucket. Now suppose it is possible to have a pipe going from the top cover of the bucket vertically down to a small reservoir just below. You have to imagine that this is being swung around with the bucket. **What will happen?**

Because of the vacuum in the bucket, more water will be sucked up from the reservoir to replace the water leaving through the hole in the bottom, so giving a continuous flow.

Good. Now all we have to imagine is that we have a pipe rising vertically from the hole in the bottom of the bucket. Because of the centrifugal action, the water will be sucked up the suction pipe, then flung out of the bottom of the bucket and forced some distance (H) up the delivery pipe. The faster you swing the bucket around (N), the greater the head, H, and the discharge, Q. The larger the diameter (D) of the swing, the greater H and Q. Thus the bucket and rope are the equivalent of the impeller and casing of a centrifugal pump. Of course, the analogy is a little crude, but it does illustrate some of the principles involved. For instance, several times we have met equations incorporating $Q = f[ND^3]$ and $H = f[N^2D^2]$.

11.5.1 Centrifugal pumps

The key components of this type of pump are the volute casing and the rotating impeller at the centre of the pump (Fig. 11.15). The impeller is the equivalent of the runner in a turbine. The impeller is driven by an electric motor. For a pump mounted with the drive shaft vertical and a motor above, water enters vertically upwards at the centre of the impeller. The water becomes trapped in the passages of the impeller, formed by a number of vanes which curve backwards with respect to the direction of rotation. The angle of the

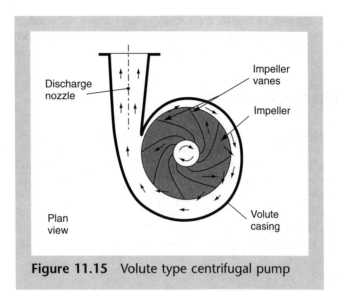

Figure 11.15 Volute type centrifugal pump

vanes on the impeller strongly influences the shape of the *H–Q* and *Pow–Q* curves of the pump. The rotation of the impeller flings the trapped water radially to the outside of the volute casing (the water leaving the impeller tangentially), causing an increase in both the velocity and pressure energy of the water. The water in the volute casing is forced up the delivery pipe as a result of more water being continuously sucked up from the sump or reservoir via the low-pressure supply pipe and being flung outwards by the rotating impeller. Thus the flow through the pump is in the opposite direction to that through a geometrically similar inward flow Francis turbine.

A centrifugal pump will only work satisfactorily if it and the suction pipe are full of water, otherwise overheating will occur. If the pump is mounted above the water level in the sump, then the casing must be filled with water via a tapping, while the air is expelled through another tapping. This is known as priming the pump, and should be carried out before switching on.

One reason why centrifugal pumps are widely used is their versatility, which is due to the availability of a large number of different impeller designs. These enable the head generated to vary between 1 and 120 metres, the discharge to be anything from a trickle to $30\,m^3/s$ or more, and for liquid/solid mixtures like sewage to be handled (provided that the impeller passages are wide enough). With clear water, pump efficiency may be near to 90%, with mixtures somewhat less.

A variation on the centrifugal pump is the multi-stage pump. These tend to be used where the lift required is greater than about 60 m (see Example 11.11). The principle is the same as the centrifugal pump, but in this case a number of identical impellers are mounted in series and driven by the same motor (unlike two identical pumps mounted in series – see section 11.7.3). The water leaving the first impeller is fed back to the centre of the next by an 'S' shaped passage. As many stages as necessary may be used, the total lift obtained being the sum of those generated by each individual stage, so heads as high as 1200 m can be achieved.

Borehole pumps tend to be narrow, vertical multi-stage centrifugal pumps specially designed to lift groundwater from deep wells or drainage pits. Diameters range from 150 to over 350 mm. The larger sizes are capable of lifting as much as $1\,m^3/s$ from a depth of

300 m. The motor can be mounted on the surface, where it is accessible, with the impeller being driven by a drive shaft within the delivery pipe. Alternatively, a combined motor–pump unit can be completely located at the bottom of the well if desired.

11.5.2 Axial flow pumps

This type of pump is basically a propeller turbine in reverse. The impeller is similar to a ship's propeller. If the drive shaft is vertical, as in Fig. 11.16, then water enters the pump axially, in this case vertically upwards. The motor drives the impeller, the blades of which propel or lift the water upwards. The rotational component of the water imparted by the revolution of the impeller is converted into an upward axial flow by the fixed guide blades above the impeller. The vanes also convert kinetic energy to pressure head.

These pumps work best when constantly immersed and ready primed. In any case, the suction lift should **not** exceed one metre, otherwise cavitation may be a problem (see section 11.8).

The axial flow pump is best suited to situations requiring large volumes of liquid to be lifted over a relatively small head (like the axial flow turbine – large volume, small head). The lift of these pumps is very restricted, being of the order of 12 m, although multi-staging is possible. However, they may be ideal for applications like land drainage, irrigation and pumping water or sewage at a treatment works where the required lift is small.

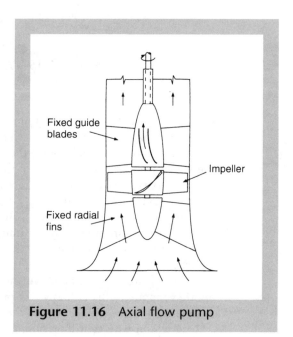

Figure 11.16 Axial flow pump

11.5.3 Mixed flow pumps

These are a cross between a centrifugal pump and an axial flow pump, so the flow is part radial and part axial. As might be surmised, they are suitable for applications which fall between the ideal conditions for a centrifugal or axial flow pump. Thus they are suitable for pumping moderate quantities at moderate heads, say 25 m to 60 m. It is possible to increase the lift by using more than one impeller (a multi-stage unit).

11.6 Pump performance equations, affinity laws and specific speed

The most important relationship for a pump is between the head, *H*, generated (the lift) and the discharge, *Q*. Other important considerations when selecting a pump are its effi-

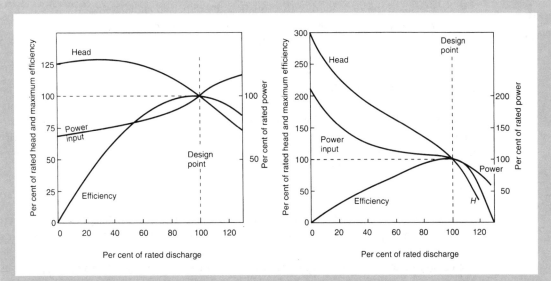

Figure 11.17 Typical performance characteristics of a centrifugal pump (left) and axial flow pump (right)

ciency ε, (equation 11.4)) closely allied to its power requirement, *Pow*. Obviously, the power (frequently electricity) required to drive the pump costs money, so running the pump at maximum efficiency is desirable. Consequently, to assist with the selection of the best pump for a particular duty, it is common practice to combine on one graph the variation of head, discharge, efficiency and power requirement of an individual pump, as shown in Fig. 11.17. Note that the *H–Q* and *Pow–Q* curves of centrifugal and axial flow pumps have different shapes that are characteristic of these types of machine. The mixed flow pump falls somewhere between the two. The design point is usually that at which the pump operates most efficiently.

11.6.1 Performance equations and affinity laws for pumps

The performance equations for centrifugal pumps are the same as for rotodynamic turbines, since they are similar machines only operating in reverse. The only alteration required is to remember that with pumps *Pow* represents the input power (not the output power as with a turbine) and that the head, *H*, with a pump is the useful lift obtained (not the total head at the inlet to a turbine). Apart from these minor differences, the equations are the same as before:

$$Pow/\rho N^3 D^5 = f[gH/N^2 D^2] \tag{11.14}$$

$$Q/ND^3 = f[gH/N^2 D^2] \tag{11.15}$$

where N is the speed of impeller rotation, D the diameter of the impeller, and f means 'a function of'. If a fixed speed motor is used to drive the pump then equation (11.15) gives the diameter needed to lift $Q\,\text{m}^3/\text{s}$ over the required head, *H*. Equation (11.14) gives the corresponding input power.

66 I understand the derivation of the equations by dimensional analysis. However, I do not really understand why they describe the performance of a pump. Can you show me why they work using basic hydraulics? **99**

Suppose we start very simply with the continuity equation, $Q = AV$. Now for a circular impeller of diameter D, then $A \propto D^2$. The peripheral speed (that is the speed on the perimeter) of the impeller, $V \propto ND$. The larger the diameter, the greater the velocity on the circumference. For instance, the outer edge of a record or CD has a greater tangential velocity than the hole at the centre. Substituting for A and V in the continuity equation gives $Q \propto ND^3$, or:

$$Q/ND^3 = c_1 \tag{11.17}$$

where the constant, c_1, takes into account all the numerical constants like π that have been omitted from the equation. Equation (11.17) is, of course, one of the dimensionless terms in equation (11.15).

Let us start once more with the continuity equation $Q = AV$. Many times in previous chapters, when considering weirs and orifices for example, we have assumed that $V = (2gH)^{1/2}$. In other words, $V \propto H^{1/2}$. Taking $A \propto D^2$ again, the continuity equation this time gives $Q \propto D^2H^{1/2}$, or

$$Q/D^2H^{1/2} = c_2 \tag{11.18}$$

where c_2 is another constant. Dividing equation (11.17) by equation (11.18) to eliminate Q gives:

$$\frac{Q}{ND^3} \times \frac{D^2H^{1/2}}{Q} = c_3$$

and cancelling and squaring all of the remaining terms gives:

$$H/N^2D^2 = c_4 \tag{11.19}$$

This is basically the grouping that appears in equations (11.14) and (11.15) with g omitted. The gravity term would be incorporated in the constant c_4.

This time we will start with equation (11.2) which shows that the power, Pow, of a stream of water is ρgQH. Now from equation (11.17) we know that $Q = c_1ND^3$. From equation (11.19) we know that $H = c_4N^2D^2$. Substituting for Q and H in equation (11.2) gives:

$$Pow = \rho gQH \tag{11.2}$$
$$= \rho g(c_1ND^3)(c_4N^2D^2)$$
$$= c_1c_2\rho gN^3D^5$$

If ρ and g and the two constants are incorporated into a new constant, c_5, then:

$$Pow/N^3D^5 = c_5 \tag{11.20}$$

This is basically the left side of equation (11.14) but with ρ included in the constant.

The derivation of equations (11.17) to (11.20) above shows that there is no mystery attached to the form of the pump performance relationships; in fact, the equations are quite logical. It was explained in Chapter 10 that the dimensionless groupings which form the pump equations enable the performance of two geometrically similar pumps, A and B, to be compared. For example, equations (10.4) and (10.5) were:

$$(Q/ND^3)_A = (Q/ND^3)_B \quad \text{and} \quad (H/N^2D^2)_A = (H/N^2D^2)_B$$

Note that the gravity term that was in equation (10.5) has been omitted here since its value is the same in both brackets and therefore cancels. In practice, one of the pumps, say A, could be a model while the other may be a full size prototype. These relationships can be used to calculate the performance of the prototype from the model results. Alternatively, they may be used to determine what changes need to be made to a pump of a particular design to obtain the optimum performance when used in another location with a different head–discharge requirement. The use of the equations tends to be based on the fact that most centrifugal pumps either:

(1) have a variable speed motor, so that the pump speed can be changed to obtain the required head–discharge relationship while retaining the same impeller (D constant), or

(2) have a constant speed motor, so that the pump speed is fixed (N constant) and consequently different diameter impellers have to be used to vary the head–discharge relationship.

If we keep D constant for case 1 and N constant for case 2, then when comparing the performance of two similar pumps (as in equations (10.4) and (10.5)) these terms cancel from the two sides of the expression so that equations (11.17) to (11.20) can be further simplified to those below.

Case 1 – D constant (variable speed)

$$\frac{Q_A}{N_A} = \frac{Q_B}{N_B} \qquad \text{(from eqn (11.17))}$$

$$\frac{H_A}{N_A^2} = \frac{H_B}{N_B^2} \qquad \text{(from eqn (11.19))}$$

$$\frac{Pow_A}{N_A^3} = \frac{Pow_B}{N_B^3} \qquad \text{(from eqn (11.20))}$$

Case 2 – N constant (variable diameter)

$$\frac{Q_A}{D_A^3} = \frac{Q_B}{D_B^3} \qquad \text{(from eqn (11.17))}$$

$$\frac{H_A}{D_A^2} = \frac{H_B}{D_B^2} \qquad \text{(from eqn (11.19))}$$

$$\frac{Pow_A}{D_A^5} = \frac{Pow_B}{D_B^5} \qquad \text{(from eqn (11.20))}$$

The subscripts A and B represent the values of the variables relating to pumps A and B. These relationships are sometimes referred to as the **affinity laws**. In addition to relating the performance of two different pumps, they can be used to investigate the performance of one pump under two different operating conditions, as shown in Examples 11.6 and 11.7.

EXAMPLE 11.6

A pump is fitted with a variable speed motor. At 1200 rpm it delivers 0.12 m³/s of water (subscript A). What speed of rotation would be required to increase the discharge to 0.15 m³/s?

$$\frac{Q_A}{N_A} = \frac{Q_B}{N_B} \quad \text{or} \quad N_B = \frac{Q_B}{Q_A} \times N_A \text{ (with } D \text{ constant)}$$

$$N_B = (0.15/0.12) \times 1200 = 1500 \text{rpm}$$

EXAMPLE 11.7

A pump has a variable speed motor. At 1000 rpm the head over which the pump can lift a given quantity of water is 8 m (subscript A). If the lift has to be increased to 12 m, what speed should the pump now run at?

$$\frac{H_A}{N_A^2} = \frac{H_B}{N_B^2} \quad \text{or} \quad N_B^2 = (H_B/H_A) \times N_A^2 = (12/8) \times (1000)^2$$

$$N_B = 1225 \text{rpm}$$

EXAMPLE 11.8

A pump runs at a constant speed of 1500 rpm with an impeller of 0.9 m diameter. If a similar pump operating at the same speed is fitted with a 1.1 m diameter impeller, what is the percentage increase in power required to drive the larger machine?

$$\frac{Pow_A}{D_A^5} = \frac{Pow_B}{D_B^5} \quad \text{or} \quad Pow_B = \frac{D_B^5}{D_A^5} \times Pow_A = \frac{1.1^5}{0.9^5} \times Pow_A = 2.73 \times Pow_A$$

Thus the power requirement is 2.73 that of the smaller machine, an increase of 173%. Note that because power is proportional to the fifth power of the diameter, a relatively small increase in diameter results in a large additional power demand.

EXAMPLE 11.9

A pump has an impeller diameter of 0.80 m and operates at 1200 rpm. If the speed is increased to 1500 rpm, what impeller diameter would be needed to keep the power requirement the same? How would the change in diameter affect the discharge and head produced by the pump?

From equation (11.20), $\dfrac{Pow_A}{N_A^3 D_A^5} = \dfrac{Pow_B}{N_B^3 D_B^5}$

Since $Pow_A = Pow_B$ then $N_B^3 D_B^5 = N_A^3 D_A^5$ or

$D_B^5 = (N_A/N_B)^3 D_A^5 = (1200/1500)^3 \times 0.8^5$

$D_B^5 = 0.168$ and $D_B = 0.70$ m

From equation (11.17), $\dfrac{Q_A}{N_A D_A^3} = \dfrac{Q_B}{N_B D_B^3}$ or $Q_B = (N_B D_B^3 / N_A D_A^3) \times Q_A$

$Q_B = (1500 \times 0.70^3 / 1200 \times 0.80^3) \times Q_A = 0.84 Q_A$

From equation (11.19), $\dfrac{H_A}{N_A^2 D_A^2} = \dfrac{H_B}{N_B^2 D_B^2}$ or $H_B = (N_B^2 D_B^2 / N_A^2 D_A^2) \times H_A$

$H_B = (1500^2 \times 0.70^2 / 1200^2 \times 0.80^2) \times H_A = 1.20 H_A$

Note that the discharge is *reduced* but the head is *increased*.

11.6.2 Specific speed of a pump

It is possible to calculate the specific speed of a pump and use this as a guide to the appropriate type of pump to use for a particular duty, as with turbines. For a pump, the specific speed equation can be obtained from equations (11.17) and (11.19) by eliminating D:

$$Q/ND^3 = c_1 \qquad (11.17) \qquad\qquad H/N^2D^2 = c_4 \qquad (11.19)$$

$$\text{or} \quad D = Q^{1/3}/c_1^{1/3}N^{1/3} \qquad\qquad\qquad D = H^{1/2}/c_4^{1/2}N$$

$$\text{thus} \quad \frac{Q^{1/3}}{c_1^{1/3}N^{1/3}} = \frac{H^{1/2}}{c_4^{1/2}N}$$

$$\text{or} \quad \frac{N^{2/3}Q^{1/3}}{H^{1/2}} = c_6$$

Multiplying all terms by the power $\frac{3}{2}$ gives the specific speed relationship:

$$\frac{NQ^{1/2}}{H^{3/4}} = c_7 = N_S \qquad (11.21)$$

where c_7 is a constant. N_S is the **specific speed** of the pump, which is the speed (in rpm) needed to discharge $1\,\text{m}^3/\text{s}$ against a $1\,\text{m}$ head. Under these conditions equation (11.21) gives $N = c_7 = N_S$. Generally the specific speed is calculated at the normal operating point of the pump, with all of the variables having the corresponding values (see Example 11.10). It should be noted that the left side of equation (11.21) is not dimensionless, and that the specific speed expression may also be written with a gravity term, g, in front of $H^{3/4}$. Since gravity is a constant, it is often omitted or considered to be included in N_S. Additionally, the specific speeds quoted for various types of pump can be confusingly different since the same units are not always used. However, with N in rpm, Q in m^3/s and H in m the following specific speeds (or **type numbers**) indicate the approximate range of duty of the main types of pump:

Centrifugal	N_S from 10 to 70	i.e. high head, relatively small discharge
Mixed flow	N_S from 70 to 170	i.e. moderate head, moderate discharge
Axial flow	N_S above 110	i.e. low head, large discharge

Mixed flow pumps are essentially a hybrid between centrifugal and axial flow pumps. Note that the specific speed for a pump is defined in terms of discharge and head, since these are the two most important design parameters for a pump. The equivalent expression for a turbine (equation (11.16)) is defined in terms of power and head, because they are the most important parameters. Normally, the higher the specific speed, the smaller the physical size of the unit for a given discharge. It should also be noted that the specific speed equation for a pump (or turbine) does not include a term relating to the size of the unit, such as the diameter. Thus the specific speed is the same for all similar pumps, regardless of size (see section 10.5). With multi-stage pumps, it is assumed that the specific speed is the same for each stage, with H being defined as the total head divided by the number of stages (see Example 11.11).

EXAMPLE 11.10

A pump is needed to operate at 3000 rpm with a lift of 7 m and a discharge of $0.15\,\text{m}^3/\text{s}$. By calculating the specific speed, determine what sort of pump is required.

$$N_S = \frac{NQ^{1/2}}{H^{3/4}} = \frac{3000 \times 0.15^{1/2}}{7^{3/4}} = \frac{3000 \times 0.39}{4.30} = 272$$

$N_S > 110$, so an axial flow pump is required.

EXAMPLE 11.11

At its normal operating point a centrifugal pump with one stage delivers $0.3\,m^3/s$ against a head of 30 m at a speed of 1500 rpm. At another site it is required that $0.4\,m^3/s$ be raised over a height of 105 m by using a similar pump operating at the same speed but with multi-stages in series (one after the other). How many stages are required?

For the first pump, $N_S = NQ^{1/2}/H^{3/4} = 1500 \times 0.3^{1/2}/30^{3/4} = 64$

For the second multi-stage pump N_S also = 64.

For a multi-stage pump with impellers in series the whole discharge passes through each stage, so $Q = 0.4\,m^3/s$ and $N = 1500\,rpm$ as before. Thus:

$N_S = NQ^{1/2}/H^{3/4}$ so $64 = 1500 \times 0.4^{1/2}/H^{3/4}$ where H is the lift per stage.

$H^{3/4} = 1500 \times 0.4^{1/2}/64 = 14.8$

$H = (14.8)^{4/3} = 36.4m$

Therefore, number of stages required = 105/36.4 = 2.9

Thus a three-stage pump is needed.

11.7 ▶ Pump selection for a particular duty

11.7.1 Single pumps

In many Civil Engineering projects it is necessary to select pumps to perform a particular duty. If care is not taken to select the best pump for the particular situation, the result may be operational difficulties and either increased capital or running costs, or possibly all three. Consequently it is worth spending a little time looking at things to avoid.

One of the first steps towards obtaining the best pump for a particular duty is to calculate the specific speed and determine whether a centrifugal, mixed flow or axial flow pump is required. After that, it is often a case of obtaining performance data similar to the graphs in Fig. 11.17. By comparing the head–discharge relationships, efficiency and power requirements of all the available pumps, it should be possible to identify one or two machines that are suitable. It is then a case of finding which one can give the required discharge against the head in question while having the lowest initial capital cost and running cost (in the form of power requirements). Remember, it may be worth paying a little more for a pump initially if this means that the running costs are reduced: the capital cost occurs only once, but running costs are incurred over the entire life of the pump. If the pump operates continuously, or for long periods, the higher running costs may be very significant. However, if the pump is used rarely, then running costs may be secondary to the initial cost of purchase.

Other factors which enter the selection process are the stability of the head–discharge relationship and operational flexibility. Generally, a pump with a relatively steep head–discharge curve should be selected. If the H–Q line is horizontal, or close to it, this indicates that when pumping against this head the discharge could fluctuate significantly in an uncontrolled manner, causing problems with surge and waterhammer (see section 11.8). Flexibility is desirable because at some time it may be necessary to pump rather more than initially calculated, or against a greater head. Depending upon what the pump is being used for, it may be necessary to take into consideration future population increases (increased

Q?), deterioration of the delivery pipework with age (increased frictional resistance), and uncertainty or errors in the original calculations. Some 'reserve' pumping capacity may be a good idea.

Figure 11.18 shows the performance curves of five different centrifugal pumps. A pump is required to lift water 13 m. An acceptable discharge is between 0.4 and 0.6 m³/s. The pump will operate for many hours per day. **Assuming that the initial costs of the pumps are the same, which pump is best suited to the task?** Try to answer the question for yourself before looking at the solution below, and try to justify your choice.

The answer can be obtained by making the following observations. **Pump A** is not running at maximum efficiency in the 0.4 to 0.6 m³/s range. It would deliver about 0.72 m³/s against a 13 m head, much higher than required. **Pump B** operates at near maximum efficiency between 0.4 and 0.6 m³/s. It would deliver about 0.56 m³/s against a 13 m head, which is in the correct range. The power requirement is about 120 kW. A possible option. **Pump C** runs near maximum efficiency in the 0.4 to 0.6 m³/s range, but delivers only 0.36 m³/s against a 13 m head. The discharge is too small, so this pump can be ruled out. **Pump D** operates at maximum efficiency in the required range, but the head–discharge curve is horizontal around 13 m head indicating instability. Thus the discharge could fluctuate causing operational problems, surge and waterhammer, so this pump is rejected. **Pump E** also runs at maximum efficiency between 0.4 and 0.6 m³/s. It discharges approximately 0.5 m³/s against a 13 m head, which is in the centre of the required range. The power demand is 180 kW. Another possible choice.

Thus the final decision is between pumps B and E. Of the two, B is clearly the better option because it has the lower power demand, the larger discharge and a stable head–discharge curve.

11.7.2 Pump selection to suit a rising main; design of a rising main

The design of the pipeline through which a pump will discharge, the **rising main**, cannot be considered separately from the selection of the pump; the two must be considered together, as a unit. The reason for this is quite simple. The head, H_T, against which a pump must discharge is the sum of many components, thus:

$$H_T = H_S + h_{FS} + H_D + h_{FD} \qquad (11.22)$$

where H_S is the static suction lift from the water level in the sump to the datum level of the pump (Fig. 11.19), h_{FS} is head loss due to friction and minor losses in the suction pipe, H_D is the static delivery lift required of the pump to the water level at the discharge point, and h_{FD} is the friction and minor losses in the delivery pipe (see Table 6.4). The friction loss is:

$$h_F = \frac{\lambda L V^2}{2gD} \qquad (6.12)$$

$$\text{or} \quad h_F = \frac{\lambda L Q^2}{12.1D^5} \qquad (11.23)$$

where λ is Darcy's friction factor ($= 4f$ in the UK, $= f$ in the USA), L is the length of the pipe, V the mean velocity and D the pipe diameter. Equation (11.23) is obtained from equation (6.12) by substituting Q/A for V and 9.81 m/s² for g. It is apparent from these equations that h_F increases as the velocity and discharge increase, and as the pipe diameter decreases.

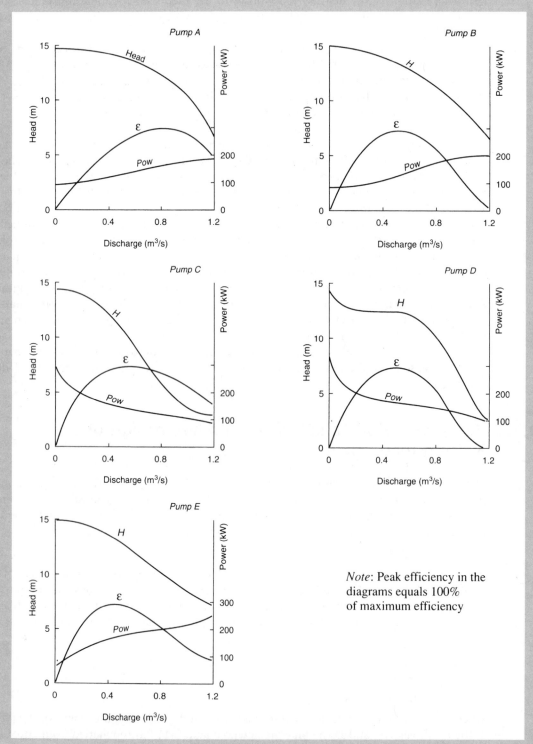

Figure 11.18 Pump performance curves for the question in the text

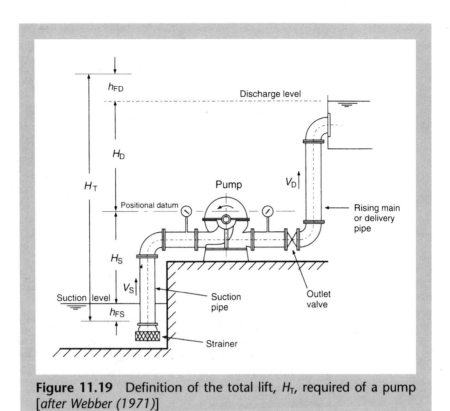

Figure 11.19 Definition of the total lift, H_T, required of a pump [*after Webber (1971)*]

The pump is normally located as close as possible to the sump or suction well to reduce the friction loss in the suction pipe. This and other minor losses can also be minimised by using a suction pipe of generous diameter, since all losses are proportional to V_S^2. As a general guide, V_S should be between 1.5 and 2.5 m/s, and the pipe should be sized accordingly. Thus h_{FS} can be calculated. H_S can be determined from the layout of the pumping station, remembering that the pump should be above the highest liquid level likely to be experienced in the suction well. On the other hand, if a submersible pump is located beneath the liquid level in the suction well then H_S has a negative value. This can result in a reduced efficiency and possibly cavitation.

On the delivery side of the pump, H_D would be known from the design brief. The friction and minor losses are again dependent upon the velocity squared, V_D^2, which in turn depends upon the diameter of the rising main. As a general guide, V_D should be between about 1.2 and 3.0 m/s, but to save energy V_D can be as low as 0.5 m/s provided $V_D > 1.2$ m/s for several hours per day to flush out the system. In any case, the velocity in the rising main should be larger than the settling velocity of any suspended matter, which is about 0.45 m/s for sand up to 2.5 mm diameter and 1.5 m/s for gravel up to 5.0 mm diameter.

In practice two or three possible rising main diameters may be considered (D_1, D_2, and D_3). Since the head loss increases with reducing diameter and increasing discharge, three rising main H–Q (system) curves are obtained diverging from the constant static head (Fig. 11.20). When the H–Q curve of the pump is superimposed three intersection points are obtained (operating points). These show the discharge (for the particular static lift) that will be obtained from the pump when connected to each of the rising mains. Thus the same

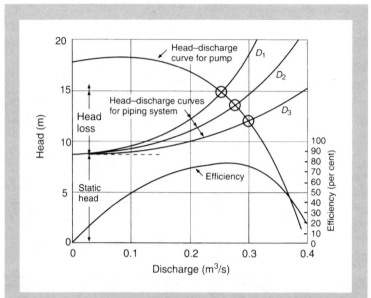

Figure 11.20 *H–Q* or system curves for pipes of diameter D_1, D_2, D_3, with the pump *H–Q* curve superimposed. The intersection of any two *H–Q* curves represents the operating point of the pump (circled) when connected to the rising main [*after Webber (1971)*]

pump gives three different discharges, since the total head (losses included) pumped against is different in each case. The selection of the optimum diameter has to be decided with reference to the efficiency curve and power requirement of the pump, the running cost, and the capital cost of the pipework which increases with diameter (although the total head pumped against and thus the running cost is reduced as the diameter increases).

11.7.3 Pumps in series and in parallel

There are many circumstances where it may be desirable to use two or more pumps (usually identical) operating together instead of one large machine. This may be because the required discharge varies over a wide range and it is preferable to use several small pumps operating at peak efficiency, as needed, instead of one large pump operating much of the time at a low efficiency with the discharge control valve almost shut. Or because either the discharge or the head is out of the range of an individual machine. Or if a pump is to be in continuous use, splitting the duty between several pumps allows one to be on standby to provide emergency backup in the event of a breakdown, and for planned maintenance. In all such cases the choice is between a single machine, pumps in series, or pumps in parallel. Which option is adopted depends upon which best meets the requirements.

When two identical pumps are used in **series (P + P)** this means that each pump has its own motor, but both are fitted to the same suction and delivery pipe so the same water passes though each of the pumps in turn. Thus the discharge is the same as for a single pump, but the head is doubled. *So the H–Q curve may be obtained by doubling H for a given Q.* This arrangement allows a given Q to be pumped

(REMEMBER!!!)

over a wide range of *H*, but there is the problem that if one pump breaks down then the whole system fails. Another option may be to use a multi-stage pump that has one motor driving more than one impeller (see section 11.5.1 and Example 11.11).

When two identical pumps are used in **parallel (P//P)**, this means that each pump has its own suction pipe but delivers into a common delivery pipe. Thus the head obtained from the two pumps is the same as for a single pump, but the discharge is doubled. *So the H–Q curve may be obtained by doubling Q for a particular H.* This may be the best arrangement where *H* is relatively constant but *Q* varies over a wide range, since one or more pumps can be used, as required. (REMEMBER!!!)

Three curves representing a single pump, P, and two identical pumps in series (P + P) and parallel (P//P) are shown superimposed on the system curve in Fig. 11.21. The system curve shows a static lift of 10 m plus the dynamic head loss. The three pump curves intersect the system curve in three different places, indicating the three operating points when the pumps discharge freely into a rising main (that is valves open). From the diagram it is apparent that:

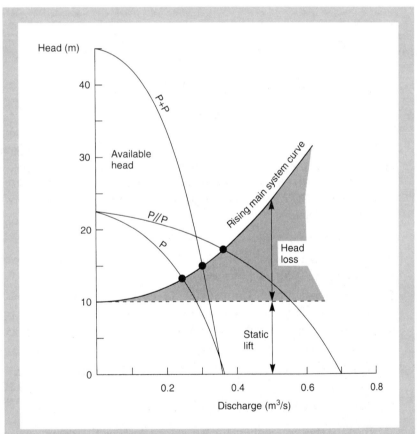

Figure 11.21 Rising main system *H–Q* curve showing a 10 m static lift plus dynamic losses. The *H–Q* curves for a single pump (P) and pumps in parallel (P//P) and series (P + P) are superimposed. Again the intersections of the system curve and the *H–Q* lines of the pumps give the operating points when discharging freely into the rising main

One pump, P, gives:	0.24 m³/s with a 10 m static lift
Two pumps (P + P) give:	0.30 m³/s with a 10 m static lift
Two pumps (P//P) give:	0.37 m³/s with a 10 m static lift

Note that because of the increased system loss, the two pumps in parallel give a discharge which is less than twice that from an individual unit. Note also that if the head produced by the two pump combinations is estimated when the discharge is restricted to that of an individual unit (by partially closing the outlet valve), then the comparative performance becomes:

One pump, P, gives:	10.0 m static lift with a discharge of 0.24 m³/s
Two pumps (P + P) give:	23.8 m static lift with a discharge of 0.24 m³/s
Two pumps (P//P) give:	17.0 m static lift with a discharge of 0.24 m³/s

The relative performance of the pump combinations varies according to the head or discharge where the comparison is made. Study Example 11.12 then try Self Test Question 11.3 below.

EXAMPLE 11.12

The H–Q of a centrifugal pump is shown below:

| H (m) | 22.5 | 22.0 | 20.9 | 19.0 | 16.3 | 12.7 | 7.7 | 0 |
| Q (m³/s) | 0 | 0.05 | 0.10 | 0.15 | 0.20 | 0.25 | 0.30 | 0.35 |

The pump is to be connected to a rising main which has a diameter (D) of 400 mm and a length of 137 m. The entry, exit and minor head losses in the pipeline can be taken as 80D in this particular situation. The friction loss can be approximated by $h_f = \lambda L Q^2 / 12 D^5$ with $\lambda = 0.04$. The static lift is 10 m. By considering a single pump (P), and two of the pumps in series (P + P) and parallel (P//P) determine:

(a) the H–Q curve of the two pump combinations

(b) which combination will be capable of discharging at least 0.3 m³/s, with the ability to pump up to 0.35 m³/s if the need arises.

(a) For two identical pumps in series (P + P), double H for a given Q, thus:

| H (m) | 45.0 | 44.0 | 41.8 | 38.0 | 32.6 | 25.4 | 15.4 | 0 |
| Q (m³/s) | 0 | 0.05 | 0.10 | 0.15 | 0.20 | 0.25 | 0.30 | 0.35 |

For two identical pumps in parallel (P//P), double Q for a given H, thus:

| H (m) | 22.5 | 22.0 | 20.9 | 19.0 | 16.3 | 12.7 | 7.7 | 0 |
| Q (m³/s) | 0 | 0.10 | 0.20 | 0.30 | 0.40 | 0.50 | 0.60 | 0.70 |

(b) First calculate the rising main system curve.

The total effective length of the pipeline = actual length + allowance for minor losses
$$= 137 + 80D = 137 + 80(0.4) = 169\,\text{m}$$

Total head required of pump, H_T = static head + friction loss
$$= 10 + \lambda L Q^2 / 12 D^5$$
$$= 10 + 0.04 \times 169 \times Q^2 / 12 \times 0.4^5$$
$$= 10 + 55.01 Q^2 \text{ m}$$

Q (m³/s)	0	0.10	0.20	0.30	0.40	0.50	0.60
H_T (m)	10.00	10.55	12.20	14.95	18.80	23.75	29.80

The rising main system curve and the P, P + P and P//P curve are those shown in Fig. 11.21. From the intersection of the system curve with the pump curves it is apparent that:

Both P + P and P//P can deliver 0.30 m³/s through the rising main, but P cannot.

Only P//P can pump up to 0.35 m³/s if the need arises.

SELF TEST QUESTION 11.3

Figure 11.21 is plotted from the data in Example 11.12. Re-plot the data, but this time assuming a static lift of zero so that the system curve branches out from the origin. Superimpose the three pump curves: P, (P + P) and (P//P). This time the part of the diagram below the system curve represents the energy loss, while the vertical distance from the system curve to pump curve gives the maximum possible static lift. From this diagram determine:

(a) Which combination can deliver 0.4 m³/s and what is the corresponding maximum static lift?

(b) If the discharge is restricted to 0.15 m³/s, what are the three maximum static lifts?

(c) If the static lift required is 15 m what are the three corresponding discharges?

(d) If the required lift is increased to 20 m, what are the corresponding discharges now?

11.8 Avoiding problems with cavitation and surge

11.8.1 Cavitation

By considering the total energy of a moving fluid it was shown in Chapter 4 that an increase in velocity results in a decrease in pressure, which is why aeroplanes fly. The higher velocity over the top of the wing results in a reduced pressure compared to that underneath. However, this relationship between velocity and pressure produces some unwelcome effects with respect to hydraulic machinery (and sometimes pipes and hydraulic structures). These problems arise when the absolute pressure falls sufficiently for the small quantity of air that is dissolved in water to be released, followed by local vaporisation of the liquid. This combined process is called **cavitation**, and results in small bubbles of vapour being formed that gradually get bigger (like the bubbles in a pan of boiling water). The problem is not just the existence of the bubbles, but the fact that when the pressure increases again they explode violently inwards (implode). This implosion results in very high velocities as the liquid rushes in to fill the void. Bubble growth and implosion only lasts a few milliseconds, but pressures as high as 4000 atmospheres and local temperatures of up to 800°C may be generated. Cavitation can be a very destructive phenomenon, and should be minimised or avoided whenever possible (some cavitation may be unavoidable). The characteristics of cavitation are listed in Box 11.6.

So, cavitation is caused by low pressure, followed by an increase in pressure. The low pressure may be the result of a local increase in velocity, or a general lowering of the static pressure. Before discussing how to design to avoid cavitation, it helps to know where it is likely to occur.

Box 11.6 ▶ **Cavitation**

Cavitation can result in any or all of the following:

(i) Erosion and pitting of the surface on which the bubbles form as a result of the 'hammering' action of the fluid.

(ii) Extremely rapid changes in pressure as a result of the imploding bubbles, causing instability of the flow and thus vibration and noise. The noise may vary in character from an occasional sharp crack to a continuous rattle, or a regular heavy thump accompanied by severe vibration. When submerged, much of the noise from even the quietest of submarines is caused by cavitation of the propeller, which is not unlike the runner of a turbine or the impeller of a pump.

(iii) Constantly changing flow patterns, which reduce efficiency.

With a centrifugal pump, the lowest pressure occurs near the centre of the impeller where the water enters, particularly on the surfaces that are on the downstream or 'sheltered' side of the raised vanes as the impeller rotates. It is here that bubbles of vapour form. The bubbles are then carried with the flow towards the outer part of the impeller. The rotating impeller is designed to cause an increase in pressure (or head) in the volute casing, so as the bubbles move to the outer part of the impeller they implode, causing pitting and damage to the vane tips. There are several things that can be done to minimise this, most of them intended to increase the pressure of the liquid entering the impeller. They are listed below.

(1) Use a generously sized suction pipe to deliver the liquid from the wet well to the pump, to keep the velocity low and the pressure high (and make sure it does not clog or become blocked). This measure can be reinforced by making the suction pipe as short as possible, as there is a limit to the height that a pump can suck up water before it will start to cavitate. Remember that friction and minor losses result in a loss of pressure (the average velocity must remain constant to maintain continuity of flow) so design the pipe with as few bends and constrictions as possible (see Fig. 11.22).

The pressure or head of liquid required to prevent cavitation on entering the impeller is termed the **net positive suction head** (NPSH). The NPSH effectively represents the pressure or head required to force liquid up the suction pipe to the impeller. This varies with the speed of rotation and discharge and has to be determined by the manufacturer from tests performed on a particular type of pump. In such tests the suction lift, H_S, is gradually increased and when there is a marked decrease in efficiency cavitation has started, and this defines the NPSH of the pump. Thus to avoid cavitation, the available suction head should be at least equal to the NPSH, the latter representing the minimum acceptable value and being defined as:

$$\text{NPSH (m)} = H_{\text{ATM}} - H_{\text{VAP}} - H_S - h_{\text{FS}} \qquad (11.24)$$

where H_{ATM} is the atmospheric pressure acting on the free surface of the liquid in the sump (normally about 10 m of water), H_{VAP} is the vapour pressure of the liquid, H_S is the static suction lift, and h_{FS} is the head loss due to friction and minor losses in the suction pipe (see Fig. 11.19). Obviously, all terms in the equation are heads expressed in metres. As an indication of the magnitude of H_{VAP} the height of a column of water equivalent

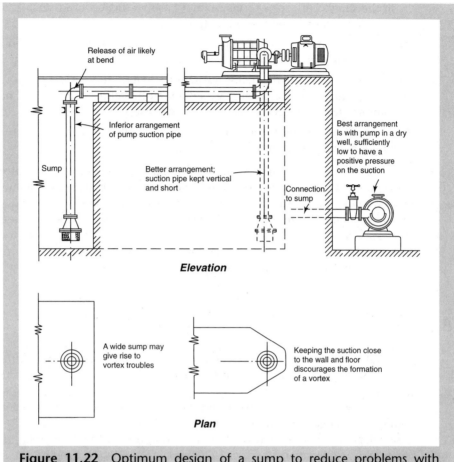

Figure 11.22 Optimum design of a sump to reduce problems with cavitation (*after Twort* et al., *1985*)

to water vapour pressure at various temperatures is roughly as follows: 0 m at 0°C; 0.2 m at 20°C; 0.8 m at 40°C; 2 m at 60°C; 5 m at 80°C; and 10 m at 100°C. Note that H_S is negative (as in equation (11.24)) when it represents a lift from the liquid level to the pump, and that if $(-H_{VAP} - H_S - h_{FS}) > H_{ATM}$ then equation (11.24) yields a negative suction pressure which indicates that cavitation will occur.

It follows from equation (11.24) that cavitation may be triggered by an increase in the static lift (caused by the level in the sump falling), a decrease in atmospheric pressure, or an increase in the temperature of the flowing liquid.

(2) Design the sump so that the liquid does not rotate in the suction pipe before reaching the impeller. This can be achieved by keeping the suction pipe close to the wall and floor of the sump to suppress vortex formation, and by narrowing the sump near the suction pipe (Fig. 11.22).

(3) Locate the pump directly above the sump so as to have a straight suction pipe, or even better, house the pump in a dry well adjacent to the sump and below the water level in the sump. This helps to maintain a positive suction head since H_S is now positive, and eliminates the need to prime the pump.

(4) Increase the pressure at the inlet to the impeller by reducing the velocity. This can be done by enlarging the inlet to the impeller.

(5) Use an impeller material that is resistant to cavitation. In descending order of resistance the best materials are chrome vanadium steel, stainless steel, open hearth steel, aluminium steel, cast steel, nickel bronze and cast iron.

(6) Avoid situations where the pump is trying to deliver a higher flow rate than intended through having overestimated the lift required.

Cavitation can also occur in turbines. Just as cavitation places a limitation on the height that a pump can be set above the water level in a sump, it also places a limit on the height that a reaction turbine can be located above the tailwater level. Remember that a turbine is effectively a pump in reverse. In fact cavitation can happen wherever there is an increase in velocity such as at a restriction in a pipe, at pipe bends where low pressure may occur on the inside of the bend, where the flow passes some projection, and on the surfaces of hydraulic structures like dam spillways.

11.8.2 Surge

The sudden starting and stopping of a pump or turbine, or a change in the flow rate, can cause large pressure variations, and these are significant enough to have to be considered carefully at the design stage.

In a rising main, **surge pressures** (as they are called) occur when the velocity of flow in the pipe is changed, or when valves are opened and closed. The magnitude of the surge largely depends upon the rate of change of flow and the length of pipe. An oversimplified, but graphic, explanation of the cause of surge is as follows. Imagine a rising main with a column of water being pumped through it at a constant rate. Now suppose that the flow of liquid into the pipeline is cut off by stopping a pump or instantaneously shutting a valve. Because the liquid in the pipeline has momentum, it cannot stop suddenly but carries on moving. This leads to separation of the column and the formation of an empty space where the flow was cut off. This results in a negative surge pressure that is less than the static head in this part of the pipeline. Eventually the column comes to rest, then reverses into the empty space with a 'bang' as the empty space fills up and the column is brought to a sudden stop. This causes a high positive surge pressure that is greater than the static head. This happens even in pipes which are not rising towards the outlet: if a vacuum or partial vacuum is formed in the pipe by the separation of the column, the water recoils towards the low pressure.

Surge can never be entirely avoided, since it occurs whenever the pump starts, stops or the flow rate changes. However, steps can be taken to avoid the separation of the water column, and the subsequent positive pressure surge when it re-unites, although some surge will still occur. The problem is much more complex than described above, with surge pressures depending upon the form of the pumping plant, the pipeline profile, pipeline length, diameter, wall thickness and the liquid being pumped. The latter is important because in an unbroken column of liquid pressure changes brought about by variations in flow rate are transmitted at the acoustic velocity (speed of sound), c m/s. Thus in a pipeline of length, L, it will take a time of L/c seconds for the pressure change to reach the end of the pipe, and then the same time for the pressure wave to be reflected back to the source of the disturbance. Thus as a very general rule of thumb, it can be said if $2L/c > 2$ seconds then there is a possibility of significant surge pressures arising, and the system should be designed accordingly. However, this is only a very crude guideline. The acoustic velocity of water, storm water and effluent is 900 to 1250 m/s, raw sewage 500 to 650 m/s and pre-aerated sewage 450 m/s.

Methods of avoiding surge include prolonging the pump run-down period, injecting liquid from another source into the pipeline (see below), and pneumatic loading by means of a compression tank containing air or other non-condensable gas under pressure.

Surge also has to be allowed for in the design of hydro-electric schemes. Dinorwic is designed to go from 0 to 1320 MW in 10 seconds, so there will be surge as the valves open. Similarly, if the load is rejected and the water flow stops suddenly, very large forces will be generated.

Imagine water flowing at high velocity down a large diameter tunnel to a turbine (the tunnels at Dinorwic are 10 m in diameter). Now water has a mass of one tonne per m³, so the water has a lot of momentum. If a valve is suddenly closed stopping the flow, the momentum of the water hitting the closed valve will generate a very large pressure, much larger than the hydrostatic head. This type of surge is called **waterhammer**. The phenomenon is very destructive, and a sudden valve closure is quite capable of bursting a pipe or damaging a tunnel lining. Waterhammer can often be experienced in domestic plumbing, a clonking noise being heard if a downstairs tap is turned off quickly. What happens is that the pressure rebounds from the closed valve to the lower pressure areas upstream, and then back again. At any point near the valve the pressure varies sinusoidally, fluctuating from high to low to high until the waveform is attenuated. With large, high-velocity pipelines special surge chambers are provided so that the water can flow into them when the flow is decelerating, thus converting kinetic energy to potential energy and preventing damage. When the flow starts again the water level in the chamber falls. Generally these chambers are of a modest size, but with hydro-electric schemes where the tripping of a turbine can cause sudden valve closure they are very large, often taking the form of shafts driven through a hill-top and connecting with the water supply tunnels beneath (see Fig. 11.1). At Foyers the shaft is 18.6 m in diameter and 84 m high, at Dinorwic the surge shaft is 30 m in diameter with a depth of 65 m. Both of these are pumped-storage schemes, so when the turbines are acting as pumps and the flow is reversed, the water level in the surge shaft rises when the pumps are switched on, and falls when they are switched off. Example 11.13 provides a simple illustration of the operation of a surge shaft.

EXAMPLE 11.13

A hydro-electric scheme has a 10 m diameter tunnel 1700 m long within which water flows with a mean velocity, V, of 5 m/s. Connected to the tunnel is a 30 m diameter surge shaft. Estimate the rise in water level in the shaft following a sudden closure of the valves leading to the turbines, neglecting the energy losses.

Volume of water in the tunnel $= \pi \times 10^2/4 \times 1700 = 133\,518\,\text{m}^3$

Mass, M_T of water in the tunnel $= 1000 \times 133\,518 = 133.518 \times 10^6\,\text{kg}$

Kinetic energy of the water in the tunnel $= \dfrac{1}{2}M_T V^2 = \dfrac{1}{2}(133.518 \times 10^6 \times 5^2)$

$$= 1669 \times 10^6\,\text{Nm}$$

If the rise in water level is h m, then the increase in volume in the shaft

$$= h \times \pi \times 30^2/4$$
$$= 706.9h \, \text{m}^3$$

Mass, M_S, of water in the shaft $= 1000 \times 706.9h = 706.9h \times 10^3 \, \text{kg}$

The *average* height through which this volume of water is raised is $h/2$

The gain in potential energy $= M_S gh/2$

$$= 706.9h \times 10^3 \times 9.81 \times h/2$$
$$= h^2 \times 3.467 \times 10^6 \, \text{Nm}$$

Assuming no energy loss, the gain in potential energy = loss in kinetic energy

$$h^2 \times 3.467 \times 10^6 = 1669 \times 10^6$$
$$h = 21.9 \, \text{m}$$

11.9 Introduction to the analysis of unsteady pipe flow

The problems associated with surge and waterhammer were described in the previous section. Often it is necessary to evaluate the increase in pressure that arises from closing a valve or varying the flow rate; e.g. to ensure that the pipes can withstand the additional surge pressure. Unfortunately, such problems can be extremely complex to analyse. The fact that unsteady flow is involved means that time becomes a variable, whereas with steady flow time was irrelevant (see section 8.12). Additionally, until now we have assumed that water is incompressible, but in this case the water can be compressed, increasing its density, and the elasticity of the pipe walls must also be taken into account. There are several ways to tackle these complexities. One is simply to assume that the water is incompressible and that the pipe is inelastic. This is the **rigid water column** approach. It is clearly an approximation because it ignores factors which affect the analysis, but provides a relatively quick and cheap means of obtaining a solution. The second approach is to include all of the variables, thereby obtaining a more accurate answer, but a desktop computer and appropriate software are needed to solve the equations. This requires a knowledge of numerical methods for solving partial differential equations, which is beyond the scope of this book (see Roberson *et al.*, 1998). Before computers, approximate solutions were obtained by arithmetical, graphical or algebraic means. Being realistic, anyone who has to analyse a complex pipe–pump/turbine system would need to consult a considerably more detailed text than this and use specialist software. However, he or she may still be well advised to obtain an answer using the rigid water column approach, since it is a simple way of obtaining a check solution (albeit not always very accurate). Thus we will start with the rigid water column approach.

11.9.1 The rigid water column approach

Consider water flowing out of the reservoir in Fig. 11.23 through a long pipeline, at the end of which there is a valve. This could be part of a larger system, such as the first pipe in Fig. 6.4, for example. For convenience, in Fig. 11.23 the pipeline is horizontal. It is assumed that under normal steady flow conditions something downstream controls the pressure at

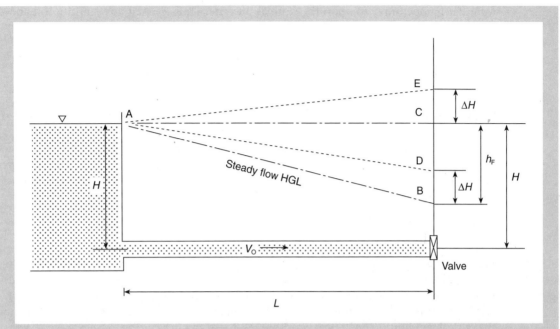

Figure 11.23 The effect of closing the valve is to increase the head by an amount ΔH, the critical condition being as the valve closes making the maximum instantaneous head in the pipeline $H + \Delta H$

the valve, the hydraulic grade line is AB, and the mean velocity $V_0 = Q/A$. With the valve closed and no flow through the pipeline, the horizontal static head line is AC (as in Fig. 6.2). Because the pipeline is long the entrance head loss is ignored, so the difference between AC and AB is the friction loss in the pipeline $h_F = \lambda L V_0^2/2gD$, as in Table 6.1 and equation (6.12). So if AB represents steady flow with the valve open and AC the hydrostatic zero flow condition, what is the maximum head in the pipeline as a result of the surge pressure in the unsteady flow condition as the valve is being closed?

As the name implies, the rigid column approach considers the water in the pipeline to be rigid. For instance, imagine the water in the pipe as a pencil moving at a uniform velocity. If you put your finger in front of the pencil's point and stop it, the other end of the pencil stops at the same time, because it is rigid. With a rigid water column, the elasticity of the pipe walls need not be considered. The mass of water in the pipeline in Fig. 11.23 is ρAL where ρ is the mass density of the water, A the cross-sectional area of the pipe and L its length. With a steady flow at velocity V_0 the water has a momentum of $\rho AL V_0$. Now suppose this steady condition is altered by opening or closing the valve causing the water to accelerate or decelerate by an amount dV/dt, with a corresponding change in momentum. Newton's Second Law tells us that the force (F) required to produce this change equals the rate of change of momentum, or force = mass × acceleration, thus:

$$F = Ma \tag{1.3/4.5}$$

$$F = \rho AL(dV/dt)$$

F exists for only a short period of time while the flow is changing. Now we know that closing the valve causes the water column to decelerate, resulting in an increase in pressure

(ΔP) that is superimposed upon the normal pressure. Thus $F = \Delta PA$ and substitution above gives:

$$\Delta PA = \rho AL(dV/dt)$$

$$\frac{\Delta P}{\rho g} = \frac{L}{g}\frac{dV}{dt}$$

or $$\Delta H = \frac{L}{g}\frac{dV}{dt} \qquad (11.25)$$

where ΔH is the additional head generated by altering the valve setting. The equation shows that ΔH increases with both the pipe length and the speed at which the valve closes. When closing the valve (causing an increase in head) the instantaneous head H' at any time is as shown in Fig. 11.23:

$$H' = H - \frac{\lambda L V^2}{2gD} + \Delta H \qquad (11.26)$$

When the valve starts to shut (i.e. before V_0 and $h_F = \lambda L V_0^2/2gD$ have changed significantly), the diagram shows that at the valve ΔH is added to AB to give the line AD that includes the surge pressure. It is assumed that the value of ΔH varies linearly with distance from the valve, so AD is a straight line. The line AD is below the static head line AC, so this is not the critical condition. The critical condition occurs at the instant the valve closes, i.e. when $V = 0$ (and $h_F = 0$) but ΔH has the same value. This gives the surge head line AE. It is the additional head CE or ΔH above static that may damage the pipe or valve. An instant after closure the pressure will revert to the static head, H. Note that it is assumed above that ΔH remains constant as the valve closes, which will be true only if the valve closes at a uniform rate and produces a uniform reduction of velocity (because of the way valves are constructed this is difficult to achieve and will not result from turning the handle of a valve at a uniform rate).

Equation (11.26) can be written for the maximum instantaneous head condition described above. If in the time required to close the valve (t_C) the velocity changes from the initial value of V_0 to 0 (when $h_F = 0$) then:

$$H'_{MAX} = H + \frac{LV_0}{gt_C} \qquad (11.27)$$

Thus if the valve closure is instantaneous (i.e. $t_C = 0$) then H'_{MAX} is infinite. In reality the head is less than this and instantaneous closure is impossible, although very rapid closures can be achieved. This highlights one of the limitations of the rigid column approach and the error incurred by ignoring the compressibility of the water and the elasticity of the pipe. However, according to Webber (1971) if the deceleration is linear and $t_C > L/60$, the results are reasonably accurate.

The time required for flow to become established (i.e. steady) after opening a valve can also be calculated. In Fig. 11.24 a long pipeline discharges to the atmosphere where $P = 0$. With zero flow AB represents the initial static head. AD shows the final variation in head. This assumes the final, mean steady flow velocity is V_0 (= Q/A) as in Chapter 6, and that with a long pipeline the entrance and exit losses are negligible so the head H is dissipated through friction with $H = h_F = \lambda L V_0^2/2gD$ (if the minor losses are significant, the effective pipeline length from Table 6.4 can be used instead of L). However, when the valve is first opened the hydraulic grade line lies between AB and AD, the difference in head between the reservoir and the atmosphere accelerating the flow so water discharges from the pipeline.

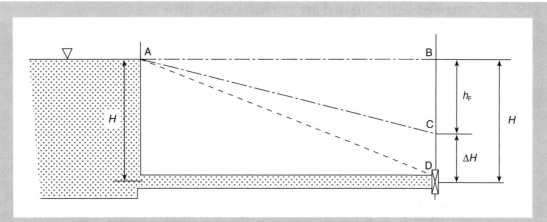

Figure 11.24 The line AB is the static head and AD the final, steady flow hydraulic grade line when the velocity is V_0 and $H = h_F$ and $\Delta H = 0$. AC represents an intermediate condition as the valve opens and the flow accelerates ($V < V_0$ and $h_F < H$). The acceleration causes a reduction in head ΔH (the opposite of closing the valve)

This is the condition represented by AC in Fig. 11.24. While the water is still accelerating, unsteady flow condition prevails with $V < V_0$ and $h_F < H$. Thus the accelerating or surge head ΔH reduces the head at the valve from C to D (i.e. the opposite of what happened in Fig. 11.23 when the valve was shut). If the mean instantaneous velocity during the unsteady, accelerating phase is V then:

$$H = h_F + \Delta H$$

$$\text{or} \quad H - \frac{\lambda L V^2}{2gD} = \frac{L}{g}\frac{dV}{dt}$$

This equation can be rearranged to give an expression for dt and then integrated to find the time t required to accelerate the flow to any given velocity V. The result is:

$$t = \frac{LV_0}{2gH}\ln\left[\frac{V_0 + V}{V_0 - V}\right] \tag{11.28}$$

This suggests that as V approaches V_0 then $t \to \infty$. This problem can be side-stepped by using equation (11.28) to calculate the time ($t_{0.99}$) at which $V = 0.99V_0$ and the flow is essentially steady:

$$t_{0.99} = 2.65\frac{LV_0}{gH} \tag{11.29}$$

This equation conveniently includes the reservoir head H and the final steady state mean velocity V_0. In practice, steady flow is established in a shorter time than the equations above indicate, the assumption of an incompressible liquid resulting in error. So what does happen when the liquid is considered to be compressible and a valve is opened or closed? This is explained below.

11.9.2 Compressible liquid in a rigid pipeline

With the rigid water column above, when the valve was closed it was assumed that the front and back of the water column stopped moving at the same instant. What actually happens is that when the valve closes, a pressure wave is propagated along the pipeline at the speed of sound in the liquid. The celerity (or velocity) of the pressure wave is denoted by c to distinguish it from the liquid velocity V. For an infinite body of clean water $c \approx 1440$–1450 m/s, although the value is much less in real pipelines, as described later. Thus pressure waves are transmitted extremely quickly along a pipeline, although not instantaneously. Figure 11.25a shows the situation one second after the valve has been closed. The pressure wave has propagated a distance $c = 1450$ m upstream (the velocity of the approaching flow V_0 has been ignored since $c \gg V_0$). To the left of the wave front AB, the flow velocity is unaffected and still V_0 (i.e. the steady flow velocity) and the liquid density and pressure are ρ and P respectively. Within the stationary region ABCD, $V = 0$ and the density and pressure are increased to $\rho + \Delta\rho$ and $P + \Delta P$. Imagine this as a series of cars crashing one after another into an immovable concrete block (the valve). The cars become compressed and their density increases. With a liquid this also causes an increase in pressure.

During 1 s the pressure wave travels a distance c upstream and brings to rest a volume of liquid Ac that initally had a mass of ρAc and velocity V_0, where A is the cross-sectional area of the pipe. Considering a control volume (Fig. 11.25b), the force required to reduce the velocity of this body of water from V_0 to 0 is obtained by applying the momentum equation in the direction of motion, as in section 4.5:

$$\Sigma F = \rho Q(V_2 - V_1) \tag{4.9}$$
$$PA - (P + \Delta P)A = \rho Ac(0 - V_0)$$
$$PA - PA - \Delta PA = -\rho AcV_0$$

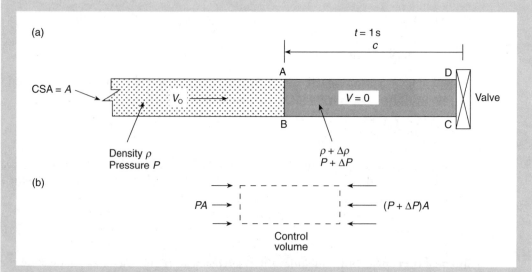

Figure 11.25 At $t = 1$ s the pressure wave AB has travelled a distance c upstream from the valve and has arrested (i.e. $V = 0$) a volume of liquid Ac which has an increased density and pressure, as indicated by the shading

$$\Delta P = \rho c V_0 \tag{11.30}$$

$$\text{or} \quad \Delta H = \frac{c V_0}{g} \tag{11.31}$$

The elasticity of a liquid is defined by its bulk modulus, K, which can be written either in terms of volume or density. This can be used to obtain the sonic velocity c in a large body of still water:

$$c = \sqrt{\frac{K}{\rho}} \tag{11.32}$$

For clean water $K = 2100 \times 10^6 \, \text{N/m}^2$ at 20°C (water is 100 times more compressible than steel) and $\rho = 1000 \, \text{kg/m}^3$ so $c = 1450 \, \text{m/s}$. Substituting these value in equation (11.30) gives:

$$\Delta P = 1450 V_0 \times 10^3 \, \text{N/m}^2 \tag{11.33}$$

In other words, by instantaneously stopping the water column the additional pressure generated increases linearly with the initial velocity V_0 and amounts to $1450 \times 10^3 \, \text{N/m}^2$ or $\Delta H = 148 \, \text{m}$ per unit decrease in velocity. This is independent of the pipe length. In reality, the pipe length is important, and the simple theory above tends to overestimate because the elasticity of the pipe material reduces c to 600–1300 m/s, as shown below.

11.9.3 Compressible liquid in an elastic pipe

An elastic pipe can be deformed by the increase in pressure resulting from surge. Consequently the pipe absorbs energy and reduces the celerity (c) of the pressure wave and hence ΔP or ΔH. If the pipeline is constrained in the longitudinal plane while free to expand circumferentially, it can be assumed that the kinetic energy lost by the water equals the sum of the strain energy gained by the water and the pipeline. This gives the modified celerity of the pressure wave c_P below (Webber, 1971; Daugherty et al., 1985; Roberson et al., 1998):

$$c_P = \frac{c}{\sqrt{1 + \left(\dfrac{KD}{Ed} \right)}} \tag{11.34}$$

where c is the sonic velocity in a large body of stationary liquid, K is the liquid's bulk modulus (N/m²), D is the internal pipe diameter (m), E is Young's modulus (N/m²) for the pipe material (i.e. modulus of elasticity) and d is the pipe's wall thickness. Typical values of K are: water $2100 \times 10^6 \, \text{N/m}^2$, oil (s.a.e. 10) $1670 \times 10^6 \, \text{N/m}^2$ and oil (s.a.e. 30) $1860 \times 10^6 \, \text{N/m}^2$. Typical E values are: steel $2100 \times 10^8 \, \text{N/m}^2$, cast iron $950 \times 10^8 \, \text{N/m}^2$, concrete $250 \times 10^8 \, \text{N/m}^2$ and PVC $26 \times 10^8 \, \text{N/m}^2$. The minimum wall thickness of steel pipes varies from about 4 mm for a 0.16 m diameter pipe to about 14 mm for 2.2 m diameter (see Twort et al., 1994 for details), but if necessary the thickness can be increased to withstand the internal hydraulic pressure or external earth pressure. For the minimum thickness above, steel pipes are tested to pressures of 700–280 m of water (for 0.16 m and 2.2 m diameter respectively) but would have a lower working head. Ductile iron pipes like that in Fig. 6.16a may have a maximum working head of between 600 m (0.08–0.2 m diameter) and 250 m (1.6 m diameter). Unplasticised PVC pipes are available in various classes but can have a maximum working pressure of 150 m when the wall thickness is between 4.5 mm and 20.8 mm (0.05 m and 0.3 m diameter respectively). Thus the choice of pipe material may depend upon the surge pressure and static pressure in addition to cost, availability, maintenance requirements etc.

With water (c = 1450 m/s) in a steel pipe having D = 0.9 m and d = 0.007 m, equation (11.34) gives:

$$c_P = 1450 \Big/ \sqrt{1 + \left(\frac{2100 \times 10^6 \times 0.9}{2100 \times 10^8 \times 0.007} \right)}$$

$$c_P = 959 \, \text{m/s or } 0.66c$$

Sometimes much lower values of c_P are obtained. Thus c_P can be considerably less than the sonic velocity in a large body of water, so when elasticity of the pipeline is taken into consideration and c_P is substituted for c in equations (11.30), (11.31) and (11.33), the value of ΔP or ΔH obtained is also significantly less.

In a pipeline of length L it effectively takes L/c_P seconds for the pressure wave to reach the reservoir at the end of the pipeline (Fig. 11.26). We will see that the ratio L/c_P is used as a counter below. For simplicity it is assumed that the pipeline is horizontal and friction is ignored. Where the pipeline is distended as a result of the surge pressure the water within it has the denser, darker shading in the diagram to indicate an increase in its density, and that it is stationary (V = 0). What happens when the valve closes instantaneously at time t = 0 is this:

t = 0 (Fig. 11.26a). At the instant the valve shuts, water from the reservoir is still entering the pipeline at the steady flow velocity V_0, although the flow has been arrested at the valve itself. The resulting pressure wave at the valve starts to travel upstream at velocity c_P.

t = 0.5L/c_P (Fig. 11.26b). The wave front representing ΔH is half-way to the reservoir, the water behind the wave becoming stationary as it passes.

t = L/c_P (Fig. 11.26c). The wave front reaches the reservoir at t = L/c_P when instantanously all of the water in the pipeline is stationary. This cannot last because the head in the pipe exceeds that in the reservoir, so the pipeline starts to discharge into the reservoir.

t = 1.5L/c_P (Fig. 11.26d). The pipeline is discharging to the reservoir, the pressure falling as it empties causing the wave front to travel back towards the valve at celerity c_P. At this moment in time it is half-way back.

t = 2L/c_P (Fig. 11.26e). The head everywhere in the pipeline is again H, but the entire water column is moving at velocity V_0 towards the reservoir. As the water at the valve moves towards the reservoir, a reduction in head, $-\Delta H$, occurs and a negative pressure wave starts to travel towards the reservoir.

t = 2.5L/c_P (Fig. 11.26f). At this instant the negative pressure wave is half-way to the reservoir.

t = 3L/c_P (Fig. 11.26g). The negative pressure wave is at the reservoir and the entire pipe water column is stationary. The wave front is about to start back towards the valve.

t = 3.5L/c_P (Fig. 11.26h). The wave front is half-way back to the valve. Behind it the water is flowing from the reservoir into the pipeline, regaining the reservoir head H.

t = 4L/c_P (Fig. 11.26i). The wave front has reached the valve and is about to rebound upstream again. All of the flow in the pipeline is towards the valve at velocity V_0. This is the same condition as diagram (a), indicating that one complete cycle has ended. Without friction, the whole cycle repeats indefinitely.

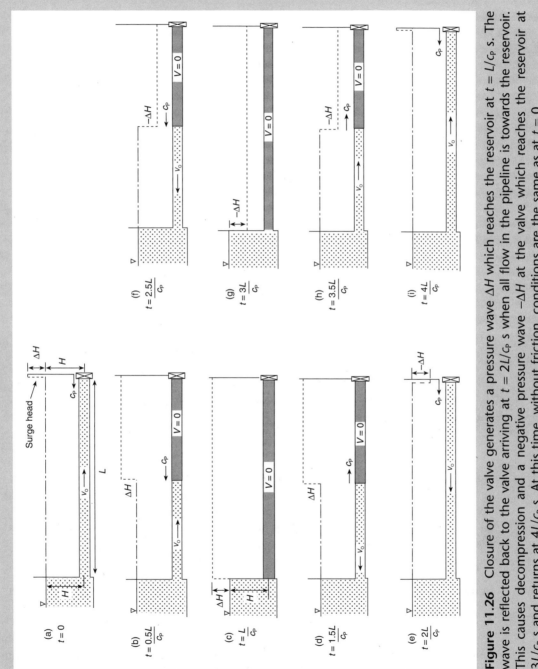

Figure 11.26 Closure of the valve generates a pressure wave ΔH which reaches the valve arriving at $t = L/c_P$ s. The wave is reflected back to the valve arriving at $t = 2L/c_P$ s when all flow in the pipeline is towards the reservoir. This causes decompression and a negative pressure wave $-\Delta H$ at the valve which reaches the reservoir at $3L/c_P$ s and returns at $4L/c_P$ s. At this time, without friction, conditions are the same as at $t = 0$

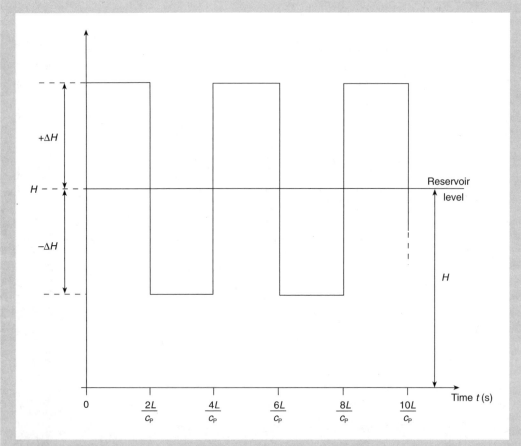

Figure 11.27 The head–time graph for the valve in Fig. 11.26. Without the friction head loss the pattern repeats indefinitely

A head against time graph for the valve can be drawn to summarise the cycle described above (Fig. 11.27). Similar (but different) diagrams can be drawn for the mid-point, reservoir or any other position. In reality, friction damps the pressure oscillations quite quickly, although if there is a high friction loss in the system the initial pressure rise may be higher than that calculated above. This is indicated by the solid line in Fig. 11.28 which is super-imposed on a dashed uniform rectangular-toothed wave.

Instant valve closure is not possible but takes a time t_C. If $t_C < 2L/c_P$ this is referred to as **rapid closure**. In this case the maximum pressure rise is the same but lasts for a shorter period of time; if $t_C = 2L/c_P$ then the rectangular waves of Fig. 11.27 are replaced by a saw-tooth wave pattern. The term **slow closure** indicates that $t_C > 2L/c_P$, which means that the pressure wave can complete one whole cycle before the valve closes. In other words, the head at the valve is reduced by ΔH before the valve is fully closed. This means the pressure rise is smaller than for rapid closure, as would be expected; in one quoted instance, the surge pressure was reduced by about two-thirds.

The celerity of a wave and the surge pressures generated are not affected by pipeline slope, although a slope will vary the hydrostatic head H so the maximum pressure experienced

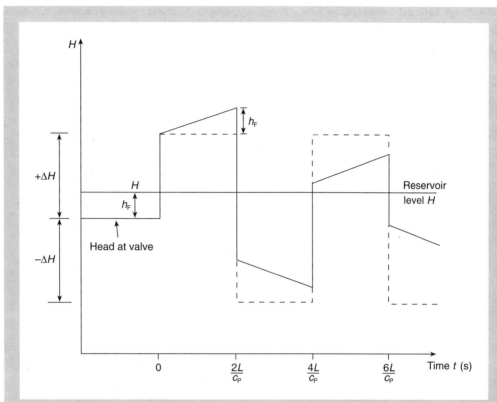

Figure 11.28 Diagrammatic illustration of the damping effect of friction. In long pipelines, friction initially increases the surge head (solid line), but afterwards it causes the amplitude of the uniform surge waves (dashed) to decrease quite rapidly

will also vary according to position. As described in section 11.8.2, surge caused by valve closure is often called waterhammer. The pressure variations are sometimes called hydraulic transients, a transient flow being an intermediate unsteady flow between two steady flow conditions. Whatever the terminology, large +ve surge pressures can burst pipes, while low −ve pressures can collapse them. Protection can be afforded by slowing down valve closure (if possible) or providing some means of surge suppression such as a compression tank or surge chamber.

EXAMPLE 11.14

A reservoir discharges through a 0.6 m diameter, horizontal pipeline 2950 m long that terminates in a valve 37 m below the reservoir surface. (a) Take $\lambda = 0.025$ and assume it is constant throughout. Neglecting minor losses, how long will it take for the velocity to reach 99% of its final steady flow value? (b) Once steady flow is established, calculate the value of c_P in the pipe if it is steel with a wall thickness of 6 mm, and thus determine the head increase resulting from a sudden valve closure. Assume $c = 1440$ m/s. (c) If a slow valve closure is used to reduce the pressure surge, what should be the minimum value of t_C?

(a) First, calculate the final steady flow velocity V_0 using the Darcy equation, without the minor entrance and exit loss.

$$H = \lambda L V_0^2 / 2gD$$
$$37 = 0.025 \times 2950 \times V_0^2 / (19.62 \times 0.6)$$
$$V_0 = 2.430 \, \text{m/s}$$

Now use equation (11.29) to calculate the time at which $V = 0.99V_0$.

$$t_{0.99} = 2.65 \frac{LV_0}{gH} = 2.65\left(\frac{2950 \times 2.430}{9.81 \times 37}\right) = 52 \, \text{s}$$

(b) From equation (11.32), $K = c^2\rho = 1440^2 \times 1000 = 2073 \times 10^6$. From the text, for steel $E = 2100 \times 10^8 \, \text{N/m}^2$. Using equation (11.34):

$$c_p = \frac{c}{\sqrt{1 + \left(\dfrac{KD}{Ed}\right)}} = \frac{1440}{\sqrt{1 + \left(\dfrac{2073 \times 10^6 \times 0.6}{2100 \times 10^8 \times 0.006}\right)}} = 1021.5 \, \text{m/s}$$

$$\Delta H = c_p V_0 / g \tag{11.31}$$
$$= 1021.5 \times 2.430 / 9.81$$
$$= 253 \, \text{m}$$

Thus allowing for the static head, maximum pressure is about $253 + 37 = 290 \, \text{m}$.

(c) With a rapid valve closure of $t_C < 2L/c_p$ the pressure will be as above. If $t_C > 2L/c_p$ the pressure will be reduced. Thus the critical time is $t_C = 2L/c_p = 2 \times 2950/1021.5 = 5.8 \, \text{s}$.

For a slow valve closure and reduced pressure t_C should be larger than $5.8 \, \text{s}$.

Note that the rigid column theory and equation (11.27) indicate a maximum instantaneous head under these conditions of:

$$H'_{MAX} = H + \frac{LV_0}{gt_c} = 37 + \frac{2950 \times 2.430}{9.81 \times 5.8} = 163 \, \text{m}$$

11.10 The ram pump

The ram pump is an interesting application of the waterhammer-surge effect, the high pressure generated as a result of waterhammer being used to lift a small quantity of water over a relatively large head. These pumps are often not regarded as pumps at all, because there is no mechanical input of power. The power to drive the pump is simply the energy of the water in the supply pipe where a relatively large flow Q falls through a distance H (Fig. 11.29). The waterhammer effect enables a smaller quantity of water q to be lifted to a height h.

The ram pump itself has two main components: the pulse valve and the air vessel (Fig. 11.30). The water from the supply pipe flows into the pulse valve. It is the pulse valve that creates the waterhammer. The pulse valve is designed so that it opens and closes as a piston moves up and down. The valve opens as the piston drops as a result of its own weight. With a small flow from the supply pipe, water passes around the piston and spills out through the top of the valve. However, as the flow increases, the friction drag on the piston increases, until at some point it is lifted sharply upwards against its seat, closing the valve. As the valve slams shut, the flow of water in the supply pipe stops instantaneously, its momentum

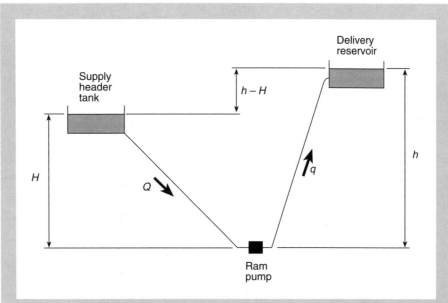

Figure 11.29 A ram pump is supplied with a flow (Q) from a header tank and is capable of raising a part of this (q) to a higher level

being dissipated as a large pressure surge which lasts for a fraction of a second. This large pressure forces water through a non-return valve in the delivery pipe. The non-return valve closes when the delivery flow stops, preventing water flowing back down the delivery pipe. With the flow in the supply pipe stopped, the piston of the pulse valve drops open once more, repeating the cycle.

The air vessel is a compression tank included to prevent the large surge pressures causing any damage, but is also used to smooth the flow in the delivery pipe, which is otherwise spasmodic. Typically the pumping cycle may occur 40–120 times per minute. Only a small quantity of water is pumped on each cycle, but the rapidity of the cycles means that the discharge is significant. For example, Jeffery *et al.* (1992) quoted the performance of the DTU steel ram pump as: supply head range (H) = 2–30 m; supply flow range (Q) = 60–120 l/min, delivery head range (h) = 6–100 m and typical delivery range (q) = 2–20 l/min. The ratio of h/H is typically between 5 and 25 for industrial units. Efficiency (ε) ranges from 50% to 80% for a well-designed system.

The efficiency of the system is $\varepsilon = \dfrac{\text{output}}{\text{input}} = \dfrac{qh}{QH}$ (11.35)

The delivery flow is $q = \dfrac{\varepsilon QH}{h}$ (11.36)

The hydraulic power of the ram pump system is $Pow = \dfrac{9.81qh}{60}$ (11.37)

where *Pow* is in watts or J/s; q (l/min) and h (m) are defined above. Output powers of 10–500 watts are typical.

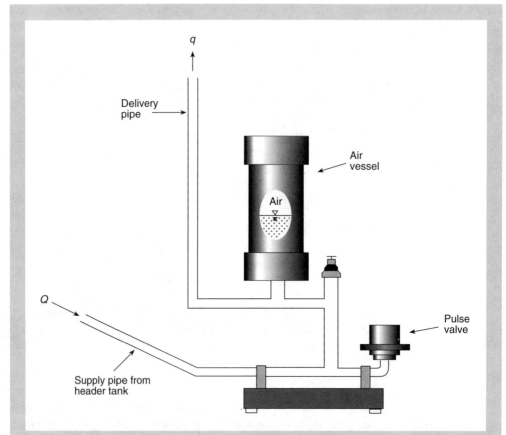

Figure 11.30 The principal components of a ram pump are the pulse valve and the air vessel. The sudden closure of the valve causes a large waterhammer pressure which forces the flow (*q*) up the delivery pipe. The air vessel prevents damage to the pipework and smooths the flow

The ram pump is obviously not suited to pumping large quantities but, because it requires no electrical or mechanical input of energy, has obvious advantages for pumping small quantities in rural areas or developing countries. They can be used to supply drinking water or for small hillside irrigation schemes. They require a clean source of water, such as a stream or spring. A pipe (preferably short) flowing under gravity takes this water to the supply header tank. The vertical height between the water level in this tank and the inlet to the pump is *H*. Ideally the supply flow *Q* should be large and *h*/*H* should be relatively small. The size of a ram pump is usually denoted by the diameter of the supply pipe from the header tank, typically 50 mm or 100 mm. If a larger discharge is required, assuming *Q* is adequate, the best solution is not to employ a larger pump but to use two or more pumps fed by separate pipes from the same header tank. This also allows pumps to be taken out of service if there is a seasonal decrease in *Q*. Most pumps can operate under a range of conditions and can be adjusted to obtain optimum efficiency at a particular site. Their simplicity makes them reliable and, with clean water, well-manufactured pumps have been known to operate continuously for over 10 years (Jeffery *et al.*, 1992).

Summary

1. Positive displacement pumps are like bicycle pumps: they have a piston that moves back and forth by the same amount each time, delivering the same relatively small quantity on each stroke. Alternatively they can use intermeshing gears. This chapter is mostly concerned with larger rotodynamic machines which are characterised by: (a) having a rotating element called a runner (turbines) or impeller (pumps); (b) a continuous flow of liquid through the machine; and (c) a continuous output.

2. Impulse or impact turbines like the Pelton wheel use the impact of a water jet to turn a runner. A reaction turbine uses water pressure to drive the runner; a Francis turbine uses both the velocity and pressure of the water.

3. The efficiency of a turbine ε_T = output power/input power = $Pow/\rho gQH$ (equation (11.3)).
   The efficiency of a pump ε_P = fluid output power/mechanical input power = $\rho gQH/Pow$ (equation (11.4)).

4. For impulse turbines, the force on a single moving or stationary vane can be obtained from:

$$-F_{RX} = \rho A(V_1 - U)[\eta(V_1 - U)\cos\theta - (V_1 - U)]$$
$$+F_{RY} = \rho A(V_1 - U)[\eta(V_1 - U)\sin\theta]$$

When the vane is on a runner, the jet is considered to be of fixed length (instead of extending, as with a single moving vane). Thus its mass flow rate is ρQ or ρAV_1 kg/s, not $\rho A(V_1 - U)$ as above. Hence:

$$-F_{RX} = \rho AV_1[\eta(V_1 - U)\cos\theta - (V_1 - U)]$$
$$+F_{RY} = \rho AV_1[\eta(V_1 - U)\sin\theta]$$

Box 11.5 demonstrates how to adapt the equations to different situations.

5. With turbines the relationship between the head of water available and power output is important. Remember that Pelton wheels require a large head (H) and a relatively small discharge (Q); Francis turbines require moderate H and Q, Kaplan/bulb generators require low H and large Q. With pumps, the head (or lift) and the discharge are important: centrifugal pumps give a large H with relatively small Q; axial flow (propeller) pumps give low H with large Q. These relationships are often summarised by the specific speed or type number below.

6. The performance equations for pumps and turbines were derived in Chapter 10.

$$\frac{Pow}{\rho N^3 D^5} = f\left(\frac{gH}{N^2 D^2}\right) \text{ and } \frac{Q}{ND^3} = f\left[\frac{gH}{N^2 D^2}\right]$$

(11.14) and (11.15)

The specific speed equations are:

$$\text{Turbine: } N_S = \frac{NPow^{1/2}}{H^{5/4}} \qquad (11.16)$$

$$\text{Pump: } N_S = \frac{NQ^{1/2}}{H^{3/4}} \qquad (11.21)$$

Equations (11.14/11.15) allow the performance of one machine under different conditions to be assessed, or the performance of similar pumps of different size. See also the pump affinity laws in section 11.6.1.

7. The lift of a pump can be summarised as $H = \text{static lift} + \dfrac{\lambda LQ^2}{12.1D^5}$. This means that the H–Q line of the rising main that the pump is connected to (the system curve) is higher when D is small and lower when D is large (see Fig. 11.20). The actual discharge from the pump is determined from the intersection point of the pump's H–Q line and the rising main's H–Q line.

8. For two identical pumps in series (P+P): double H for a given Q.
   For two identical pumps in parallel (P//P): double Q for a given H.

9. Cavitation occurs where low pressure enables the air in water to be released to form bubbles of vapour. The implosion of the bubbles when the pressure increases again is called cavitation; this can result in erosion and pitting of the surface, vibration or noise, and reduced efficiency. Surge occurs when pumps start and stop, or there is a change in the flow rate. These changes tend to cause separation of the moving water column and a partial vacuum in the pipe; large surge pressures are generated when water reverses and rushes to fill the empty space. Waterhammer occurs when a valve shuts quickly and the moving water column has to stop suddenly: the momentum of the moving water is dissipated as a sudden, large increase in pressure (often accompanied by a clunking or hammering sound). Waterhammer can burst pipes.

10. Unsteady surge flow can be analysed using the rigid water column approach which assumes that the water in the pipeline is 'solid'

and that all of the water stops at the instant the valve is closed (equations (11.25) and (11.27)). This is unrealistic since water can be compressed, which absorbs energy and reduces the surge head ΔH. Equation (11.31) allows for water compression but assumes a rigid pipe. In reality, if the surge pressure is large the pipe will become distended absorbing yet more energy. The elasticity of the pipeline reduces the sonic velocity from c to c_P (equation (11.34)), so when the smaller value c_P is substituted into equation (11.31) this further reduces the surge head since $\Delta H = c_P V_0 / g$. The value of c_P is also important since the ratio L/c_P determines how long it takes a pressure wave to reach the end of the pipeline. If the time to close the valve $t_C < 2L/c_P$, this is regarded as a rapid closure and the full surge head ΔH may be expected; if $t_C > 2L/c_P$, the closure is slow and a reduced surge head results as a consequence of decompression at the valve (Fig. 11.26).

Revision questions

11.1 Describe the difference between (a) a turbine and a pump; (b) an impulse turbine and a reaction turbine; (c) a reciprocating pump and a rotodynamic pump.

11.2 Define (a) synchronous speed; (b) the overall efficiency of a turbine; (c) the overall efficiency of a pump; (d) relative velocity.

11.3 A 15 mm diameter jet of water hits the centre of a stationary hemispherical cup (as in Fig. 11.2d) with a velocity of 6.0 m/s, divides, and flows smoothly over the cup without loss of velocity. The deflection angle is 180°. (a) What is the force exerted by the jet on the cup? (b) If the velocity of the water leaving the cup (V_2) is 0.97 of the initial velocity, what is the force exerted on the cup now?

[(a) 12.72 N; (b) 12.53 N]

11.4 A jet of water flows tangentially onto a single stationary vane (as in Fig. 11.2e) with a

velocity, V_1, of 16.0 m/s. The jet is turned through 150° and has an exit velocity $V_2 = 0.85 V_1$. The volumetric flow rate of the jet is 0.04 m³/s. What is the magnitude and direction of the resultant force exerted on the vane?

[1144 N at 13.8° to the horizontal]

11.5 (a) A horizontal jet of water hits a flat plate angled at 40° to the jet as in Fig. 11.2a and is deflected smoothly without loss of velocity. The diameter of the jet is 20 mm and V_1 is 7.32 m/s. Calculate the magnitude of the resultant force. (b) If everything is as in part (a) except that the deflection angle has increased to 60°, what is the magnitude of the resultant force now?

[11.52 N; 16.83 N]

11.6 A jet of water with a velocity of 25.0 m/s and a diameter of 200 mm slides tangentially onto a stationary curved vane as in Fig. 11.6a and is turned through an angle of 165°. The velocity of the water leaving the vane is 90% of the original jet velocity. (a) Calculate the magnitude and direction of the resultant force. (b) If everything is as above except that the vane is moving at 12.0 m/s away from the jet and the relative velocity at exit is 90% of the initial relative velocity, what is the new magnitude and direction of the resultant? (c) If the conditions are as in part (b) except that the vane is now mounted on a runner, what is the magnitude and direction of the resultant?

[36 990 N at 7.1°; 10 000 N at 7.1°; 19 230 N at 7.1°]

11.7 Describe and illustrate what the following look like, showing clearly the flow path of the water through the machine: (a) a Pelton wheel; (b) a Francis turbine; (c) a centrifugal pump; (d) an axial flow pump. (e) For the above, describe the most important performance parameters and illustrate how they vary for a particular type of machine.

11.8 Define what is meant by the specific speed of (a) a turbine and (b) a pump. (c) A turbine is required to generate 5.5 MW of electricity from the regulating releases from Kielder Water reservoir. The nominal head of water available is 47.35 m. If the turbine has a rotational speed of 500 rpm, by calculating the specific speed determine what sort of turbine is required. (d) A second turbine is required at Kielder to utilise the compensation flow to generate 500 kW from the 47.35 m head when running at 1000 rpm. What sort of turbine should this be? (e) If the turbine in part (c) has a water requirement of 14.1 m³/s, what would be the required flow rate if the turbine was run at 1000 rpm, and what would be the new power output?

[(c) Kaplan; (d) Francis; (e) 28.2 m³/s, 44 MW]

11.9 Water is to be pumped for several hours per day from a sump into a rising main of either 150 mm or 200 mm diameter. The static lift is 6.0 m and the pipe friction factor (λ) for both pipes is 0.02. The effective length of the pipe is 28 m. Two types of pump, A and B, are under consideration. Both operate at maximum efficiency in the range under consideration. The head–discharge characteristic of each pump is as follows:

Pump A

Hm	9.80	8.68	8.00	7.48	6.88	6.10	4.87
Qm³/s	0	0.01	0.02	0.03	0.04	0.05	0.06

Hm	3.45	1.40
Qm³/s	0.07	0.08

Pump B

Hm	–	–	9.65	7.63	5.88	4.00	1.75
Qm³/s	–	–	0.02	0.03	0.04	0.05	0.06

(a) Plot the rising main system curve for pipes of 150 mm and 200 mm diameter from $Q = 0$ to $Q = 0.10$ m³/s.

(b) Superimpose the two pump curves on the two system curves. What is the discharge obtained from each pump when connected to each of the rising mains?

(c) A discharge of about 0.05 m³/s is required. Which of pumps A and B acting either alone or with another identical pump in parallel or series can deliver this quantity of water with the 6.0 m static lift and either of the rising main diameters?

(d) Select the rising main diameter and the pump or combination of pumps you think most suitable.

(e) For whichever combination of pumps and rising main you have selected, calculate the velocity in the delivery pipe. Is this satisfactory?

(f) Is surge likely to be a serious problem with this installation?

(g) Sketch a possible layout for the pump(s) and the sump.

[(b) A = 0.040 m³/s (150 mm), 0.048 m³/s (200 mm), B = 0.035 m³/s (150 mm), 0.038 m³/s (200 mm); (c) A = 0.48 m³/s (200 mm), A//A = 0.053 m³/s (150 mm), B//B = 0.057 m³/s (150 mm), B + B = 0.051 m³/s (150 mm), B + B = 0.053 m³/s (200 mm); (f) No]

11.10 (a) Describe what is meant by cavitation, net positive suction head, surge and waterhammer. (b) What are the symptoms of cavitation, surge and waterhammer? If you wanted to find out if these phenomena were present in a hydraulic system, what would you look for? (c) If these three phenomena are causing problems, what steps can be taken to minimise their effect?

11.11   In a proposed pipeline a steady flow of water ($K = 2100 \times 10^6\,\text{N/m}^2$) will exist at a velocity of 2.5 m/s. The intention is to use 0.3 m internal diameter PVC pipes ($E = 26 \times 10^8\,\text{N/m}^2$, wall thickness $d = 20\,\text{mm}$). (a) Calculate the wave celerity c_p in the pipe assuming $c = 1450\,\text{m/s}$. (b) Calculate the additional surge head ΔH that will be experienced following an instantaneous valve closure.

(c) If the maximum static head on the pipeline is 9 m and the maximum working pressure of PVC pipes is 150 m of water, determine whether or not PVC is suitable for this application. (d) If the pipeline is to be 100 m long, draw the head–time graph at the valve.

[400 m/s; 101.9 m; yes]

12

Introduction to engineering hydrology

Hydrology involves the movement of water (in all its forms) over, on and through the Earth. Engineering hydrology encompasses subjects such as rainfall, riverflow, groundwater, water supply, flood estimation and forecasting, flood alleviation, the design of storm water sewers, and a host of other things. Everyone needs a continuous supply of fresh water, and expects adequate protection from flooding. These things can literally be a matter of life and death. This became all too apparent in 2000 when the wettest autumn on record resulted in severe and prolonged flooding over large areas of England, the damage to property and agriculture amounting to around £1 billion. It was estimated that in the UK as many as 5 million people may be at risk from flooding. Many properties can no longer be insured because they flood too frequently. Some blamed global warming for the extreme weather. It is far too early to be sure, but recent years have been the warmest on record in England and the warmest in the northern hemisphere for a millennium. Global warming could make extreme events more common.

This chapter provides an introduction to the hydrological cycle, global warming and the main hydrological variables such as rainfall, evapotranspiration and runoff. It provides an outline of the essential knowledge and principles underlying Chapter 13, which covers some applications of engineering hydrology such as flood prediction and water resource evaluation. The questions answered in this chapter include:

What is the hydrological cycle?

How are mankind's activities resulting in global warming and sea level rise?

What causes rainfall, and why do some parts of Britain receive more than others?

Why do many of the heaviest rainfalls occur in summer?

How do we measure rainfall depth and intensity?

What are evapotranspiration and infiltration and why are they important?

How are evapotranspiration and infiltration measured?

What affects the runoff to a river, and how is riverflow measured?

12.1 ▸ The hydrological cycle

It is called a cycle because water evaporates from the oceans, where most of the Earth's water is stored, and is blown as water vapour and cloud over land where it falls as precipitation (Fig. 12.1). Then, over a period of time varying from minutes to millions of years, it makes its way back to the oceans as riverflow or groundwater seepage.

The heat of the Sun, i.e. **solar radiation**, drives the Earth's weather systems and the hydrological cycle. **Evaporation** from the oceans involves turning water into water vapour (gas). When cooled, this vapour condenses around dust particles and falls as **precipitation**, of which rain, drizzle, snow and hail are the most important. On land, **interception** by vegetation means that some precipitation does not reach the ground, the water being evaporated back into the atmosphere instead. However, most precipitation does fall onto the ground where it may evaporate, be used by vegetation, become surface runoff or infiltrate.

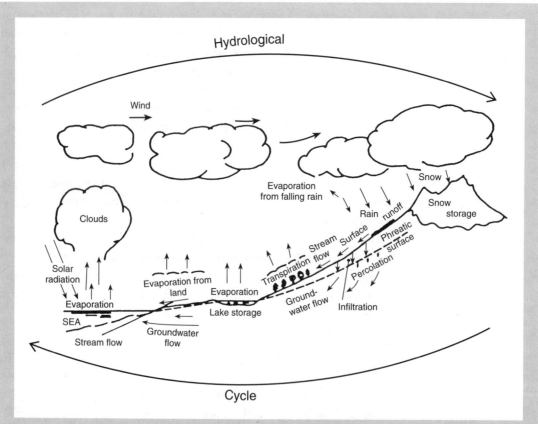

Figure 12.1 The hydrological cycle, so called because water evaporates from the oceans, is blown over land where it falls as precipitation, and then returns to the ocean as surface runoff or groundwater flow [*after Wilson (1990); reproduced by permission of Macmillan, now Palgrave*]

Surface runoff is water that runs over the ground into stream or river channels. **Infiltration** is the water that penetrates the ground surface; as **interflow** it may move horizontally through the earth to a stream channel, or it may **percolate** deeper to the **water table** or **phreatic surface.** Beneath this level the ground is totally saturated and the movement of water is called **groundwater flow. Transpiration** occurs when soil moisture or groundwater is sucked up through the roots of plants and released from the leaves into the atmosphere. On vegetated surfaces it can be difficult to separate evaporation from transpiration, so the two are jointly referred to as **evapotranspiration.**

The estimated quantity of water on the Earth is shown in Table 12.1. Over 97% of the Earth's water is in saline oceans. Given that 2.15% of all water (i.e. 78% of fresh water) is in the form of ice-caps, glaciers and snow it is not surprising that they have been considered as a source of water. One scheme envisaged towing icebergs from Antarctica to Saudi Arabia (Anon, 1976). Not counting ice, only 0.62% of the Earth's water is fresh water. Of this, about 98% is groundwater; half is stored below a depth of 800 m and is too deep to use. The remainder, if spread evenly over the land surface, would be about 30 m deep (Wilson, 1990). Atmospheric moisture accounts for only 0.035% of all fresh water, which is equivalent to about 25 mm of rainfall spread evenly over the world (Smith, 1972). Ironically, on a planet that is mostly covered by water, drinking water can often be in short supply. As populations increase the amount of water available per person falls, so by 2050 one-third of the world's population may go thirsty (Anon, 1999).

The Earth's water cycle is too large to be studied easily, so often hydrological studies are conducted within a catchment. Using a reasonably large scale map, if you mark a point somewhere on a river, the contours on the map can be used to draw the boundary (or **watershed**) of the area that drains to this point. This is its **drainage basin** or **catchment.** Within the catchment, any precipitation which becomes surface runoff will arrive at this point. If a point is selected further upstream the catchment area diminishes, whereas it increases with distance downstream. Because of differences between the land surface and underground topography, the surface and groundwater catchments may be different.

The **water budget** within a catchment (Fig. 12.2) can be studied using equation (12.1), which has many applications in hydrology. Such a study may be undertaken to quantify the major components of the hydrological cycle over a considerable period of time.

Table 12.1 Estimated Earth's water inventory [*after Wilson, 1990*]

Location	Volume (10^3 km^3)	Percentage of total water
Fresh-water lakes	125	0.009
Rivers	1.25	0.00009
Soil moisture	65	0.005
Groundwater	8 250	0.61
Saline lakes and inland seas	105	0.008
Atmosphere	13	0.001
Polar ice-caps, glaciers and snow	29 200	2.15
Seas and oceans	1 320 000	97.22
Total	1 360 000 or 1.36×10^{18} m^3	100.0

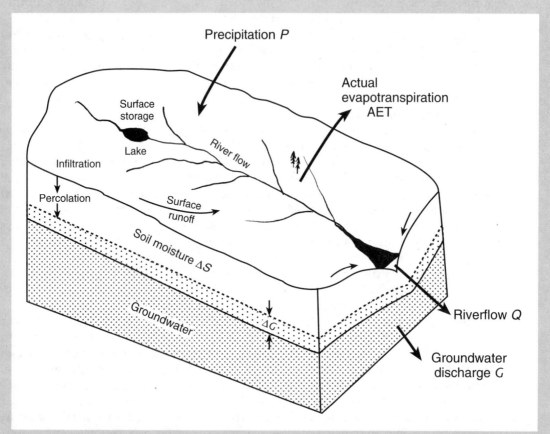

Figure 12.2 Some of the components of the hydrological cycle shown on a catchment scale. In a water budget study inflow = outflow ± change in storage. For example, the inflow is precipitation P, the outflow is river flow Q, actual evapotranspiration AET and groundwater discharge G; change in storage is soil moisture ΔS and groundwater level ΔG

$$\textbf{Inflow} = \textbf{Outflow} \pm \textbf{Change in storage} \tag{12.1}$$

$$\text{or} \quad P = Q + G + AET \pm \Delta S \pm \Delta G \tag{12.2}$$

where P is precipitation, Q is the cumulative riverflow during the study period, G is the cumulative groundwater discharge, AET is the cumulative actual evapotranspiration, ΔS and ΔG are the changes in soil moisture and groundwater storage respectively. Clearly for this equation to work dimensionally and numerically, all of the terms must have the same units. This could be m^3 (e.g. depth of precipitation multiplied by catchment area) but is generally mm per time period as in Tables 12.2 and 13.1. This makes it easy to see the proportion of precipitation that becomes AET and runoff. With respect to runoff, most gauging stations record electronically the river discharge (m^3/s) at 15 minute intervals. Thus it is a simple matter for a computer to calculate the total daily, monthly and annual runoff (m^3). Dividing these values by the catchment area (A m^2) gives the equivalent depth expressed in mm/time period.

Table 12.2 Long-term annual water budget of Great Britain and the continents [*after Barry, in Water, Earth and Man, 1969; reproduced by permission of Routledge*]

Continent	Precipitation $\bar{P}$ (mm/yr)	Total runoff $\bar{Q}$ (mm/yr)	Actual evapo-transpiration $\overline{AET}$ (mm/yr)	$\bar{Q}/\bar{P}$ (%)	$\overline{AET}/\bar{P}$ (%)
Britain	1050	650	400	61.9	38.1
Africa	670	160	510	23.9	76.1
Asia	610	220	390	36.1	63.9
Europe	600	240	360	40.0	60.0
N. America	670	270	400	40.3	59.7
S. America	1350	490	860	36.3	63.7
Australia and New Zealand	470	60	410	12.8	87.2
Mean value derived after weighting according to area	725	243	482	33.5	66.5

There are many variants of equation (12.2), all of them deceptively simple. For example, if the two sides of equation (12.2) are to balance, G must be the total quantity of groundwater discharge resulting from precipitation (*P*) on the catchment. The equation will not balance if G includes underground flow originating from a different surface catchment area. Often it is assumed that there is little or no subsurface flow across the watershed (G = 0), which may be true if the underlying rocks are impermeable, but not if an aquifer is present. In the latter case G should be evaluated, which is not always easy. Studies of less than one year normally mean that ∆S and ∆G must also be evaluated, but ∆G = 0 if the study begins and ends when the the height of the water table is the same. Similarly, ∆S = 0 if the soil moisture content is the same. To facilitate this, in the UK many studies begin on 1 October, which is the start of the **water year** when surface and groundwater reservoirs generally start refilling, and end on 30 September. If the zero values are adopted, over a considerable period of time (represented below by the overbars) equation (12.2) reduces to:

$$\bar{P} = \bar{Q} + \overline{AET} \tag{12.3}$$

Water budget studies are difficult to conduct accurately, as will become more apparent later. Nevertheless, they provide a means of understanding the relative importance of the components of the hydrological cycle within individual catchments. For example, with an impermeable clay covered catchment the surface runoff may be relatively high with little groundwater storage, whereas with a permeable chalk surface the runoff may be small and groundwater storage large. Similarly, equation (12.3) and Table 12.2 illustrate the hydrological differences between the continents. The table shows that actual evapotranspiration is much larger than might be imagined at first, accounting for 87.2% of precipitation in Australia and New Zealand but only 38.1% in Britain. This helps to explain why much of Australia is desert with dry rivers that flood intermittently, whereas Britain is usually a wet, green country with plenty of water.

12.2 Mankind's intervention in the hydrological cycle

The hydrological cycle has evolved over millions of years, but recently the activities of humans have altered it. One obvious intervention is **cloud seeding**. In an attempt to modify the weather and induce rainfall in dry areas, artificial condensation nuclei consisting of dry ice, silver iodide or ammonium nitrate have been sprayed or fired into clouds to simulate the natural dust particles around which rain drops form. This can be a contentious issue, with grievances arising from weather modification having been heard in the US Supreme Court.

Dams have been built to block rivers, with the water in the reservoir being used for hydro-power, drinking water or **irrigation**. There are large irrigation schemes in over 30 countries, particularly China, India, the USA and Pakistan. Irrigation makes relatively dry areas suitable for cultivation, and can increase the yield of many crops by 40%, which is highly significant as populations increase. In 1977 the total global irrigated area was estimated at $2\,230\,000\,km^2$, rising to $2\,730\,000\,km^2$ by 1990 (Cuenca, 1989). This is an area eleven times that of the UK, or equivalent to 29% of the USA. Another estimate is that approximately 15%–20% of global arable land is irrigated, yielding 34%–40% of crop production. Limited irrigation is employed in Britain, mostly in the relatively dry south and east.

Irrigation schemes can be successful if well designed and operated, but in the past irrigated areas have frequently become infertile as a result of waterlogging and/or salinisation. In Pakistan alone, $400\,km^2/yr$ was being lost to production at one time. Early schemes often used distribution channels to flood vast areas of land. Waterlogging occurred because the large quantities of irrigation water raised the water table, effectively turning the cultivated area into an infertile marsh. This can be avoided by using field drains to prevent the groundwater level increasing, but they are expensive to install and operate. Salinisation seriously affects about 10% of the world's irrigated land. Salinisation occurs because (unlike rainwater) surface waters contain a certain amount of salt, some of which is left behind as the irrigation water evaporates or is used by plants. With time, there is a progressive build-up of salt which can render the soil infertile. The salt can be flushed out by adding yet more water, but this increases the risk of waterlogging. It may also leave the problem of disposing of the large quantity of waste salt water that collects in the field drains.

Difficulties with both waterlogging and salinisation can be avoided by using trickle or drip irrigation, which basically involves delivering small quantities of water direct to the plants' roots via a small pipe with regularly spaced holes. However, this system is most likely to be found in developed countries since it requires more technology than simply flooding a field.

Irrigation schemes may also fail if reservoirs fill with sediment and can no longer supply sufficient water. Damming rivers to divert water to irrigation reduces the river flow downstream. This has resulted in the drying up of the Aral Sea in the former USSR. This inland sea was the world's fourth largest lake with a surface area of $62\,000\,km^2$, but it is disappearing, with long-term implications for the local climate.

As populations increase there is a need for more houses, shops, roads and factories. **Urbanisation** is the name given to the spread of built-up areas. It involves building on greenfield sites, turning grass into concrete and tarmac. This can have many significant effects, such as: increasing rainfall, reducing interception and infiltration, increasing both the quantity and speed of runoff from the ground surface, and reducing groundwater recharge. Thus it can never be assumed that the hydrological conditions before and after development will be the same.

Explain how urbanisation can: (a) increase rainfall; (b) reduce interception; (c) reduce infiltration; (d) increase the quantity of runoff; (e) increase the speed of runoff; (f) reduce groundwater recharge. Hint: read sections 12.3, 12.5 and also section 13.5 before answering.

Deforestation is a major problem in many parts of the world. Felling trees for lumber and/or clearing forests to provide land for agriculture significantly alters the hydrology. Trees play a major role in the interception and transpiration of rainfall. Leaf mulch absorbs and stores water. Tree roots break up the ground, increasing infiltration and groundwater recharge, while simultaneuously reducing surface runoff. Thus deforestation tends to increase the speed and quantity of runoff, and on the steep catchments of the Himalayas the consequences are often apparent hundreds of kilometres downstream when floods engulf low-lying areas such as Bangladesh. Bangladesh lies within the flat delta of the Rivers Ganges and Brahmaputra (see Table 12.9). Deforestation exacerbates the devastating monsoon storms that occur; in 1970 half a million people were killed by floods, while a 1988 monsoon flood and cyclone left 30 million homeless and thousands dead (Anon, 1994).

After deforestation, the felled area frequently becomes dry and barren. With no roots to bind the soil together and prevent it blowing away, **soil erosion**, gullying and landslides often occur. It has been estimated that 20% of the world's cultivated top soil was lost between 1950 and 1990. It is because of the soil it carries that the Yellow River in China, for example, got its name. Some of the soil can become trapped behind dams, causing operational problems and reducing the volume of the reservoir. Some reservoirs have virtually filled with sediment; some dams have valves to allow it to be flushed out.

Soil erosion can lead to **desertification**, which is the creation of deserts as a result of changes in climate or human actions. The causes of desertification include human poverty and overpopulation, overcultivation, overgrazing, poor irrigation and (of course) deforestation. The area affected is not easy to distinguish from drought, but the *annual* loss of land to desertification is thought to be about 60000 km² (about half the area of England) distributed among 100 countries. This affects around 135 million people, mainly in Africa, Asia, Australia and North and South America. About 650000 km² of land on the southern edge of the Sahara have become unproductive since 1925; in the USA 400000 km² are damaged beyond practical repair. In India one-third of arable land is threatened with loss of the topsoil. World-wide 600–700 million people live in areas of threatened drylands (Alexander, 1993).

In industrialised and neighbouring countries **acid rain**, in the form of dilute sulphuric and nitric acid, has been a problem for decades. Acid rain forms when water vapour in the atmosphere combines with airborne industrial pollutants and the fumes from car exhausts, particularly the oxides of sulphur and nitrogen. In Scandinavia, Europe and North America this has been responsible for damaging buildings and forests, and for turning the water in some lakes acidic, with the subsequent loss of aquatic flora and fauna, such as insects, fish and birds.

Both industrial and agricultural pollution (e.g. herbicides, pesticides and fertilisers) have significantly altered the quality of the water in the hydrological cycle, some rivers and groundwater sources becoming unsuitable for water supply without expensive treatment. Sometimes minute quantities of chemicals are potentially dangerous (Hamill and Bell, 1986;

1987). However, a more worrying consequence of atmospheric pollution is **global warming**. Short-wavelength radiation from the Sun enters our atmosphere and heats the Earth, but the re-radiated long-wavelength heat is partially prevented from escaping back into space by 'greenhouse' gases, in the same way that a garden greenhouse maintains a higher temperature inside than outside. Without this **greenhouse effect**, the mean temperature of the Earth would be about –40°C (Linsley *et al.*, 1982). Unfortunately, the Earth is 'overheating' because the concentration of the main greenhouse gases is increasing. In 1999 Sir Crispin Tickell, Chairman of the Climate Institute of Washington DC, observed that three of the last eight years had been the hottest in the northern hemisphere for 600 years (Anon, 1999).

The main greenhouse gases are carbon dioxide (CO_2), methane (CH_4), nitrous oxide (N_2O), water vapour and chlorofluorocarbons (CFCs). The relative effect per molecule of gas is $CO_2 = 1$, $CH_4 = 30$, $N_2O = 160$ and CFCs = 17 000. Thus even small amounts of CFCs are very damaging; unfortunately, they have a 100 year life (Anon, 1990).

As a result of burning fossil fuels, such as coal and oil, and slash and burn deforestation, the amount of carbon dioxide in the atmosphere has increased by 25% since the Industrial Revolution, by 10% since 1950, and by 0.4% per year in the 1980s (Watson *et al.*, 1996). This gas is responsible for about 55% of global warming. Methane is a by-product of agricultural activities, such as growing rice and keeping sheep and cattle. In the 1980s its concentration increased by 0.8% per year, while nitrous oxide increased by 0.25% per year as a result of vehicle exhausts and burning fossil fuels. Water vapour arises artificially from cooling towers, and increases naturally as the global temperature rises. CFCs are chemicals used in aerosols, as coolants in refrigerators and air conditioning, and in the plastic foam of fast-food boxes; their level rose at 4% per year in the 1980s. They are responsible for 20% of global warming, and are also partly responsible for the destruction of the **ozone layer**, which allows potentially damaging radiation from the Sun to reach the Earth's surface. In 2000 the ozone hole over Antarctica was reported to be larger than ever before with an area three times that of the USA. There is concern that this could result in increased skin cancer and environmental damage over the whole globe.

Global warming is a complex process, since some of the excess carbon is absorbed by the oceans while the increased cloud (water vapour) can have a cooling effect. Furthermore, warming is likely to be greater at higher latitudes, over the continents, in the winter half of the year. Nevertheless, mean global warming is estimated at 1.5°C to 1.8°C by 2030; for a low estimate this could be decreased by 30%, or for a high estimate increased by 50% (Herschy and Fairbridge, 1998). The accuracy of any prediction is confused by assumptions regarding what steps, if any, will be taken to limit emissions of greenhouse gases in future. In June 1990, 93 nations, including the UK and the USA, agreed to phase out the production of CFCs by the end of the 20th century. However, there is no quick-fix, since many developing countries cannot afford anti-pollution measures or the more expensive, environmentally friendly chemicals and solutions proposed by the richer nations.

Although a mean temperature increase of 1.8°C may sound attractive in cold Britain, it should be remembered that there is only about a 2°C difference between the present and the Earth's warmest ever period, and a 4°C difference between an ice age and our current climate. During the ice ages the amount of water stored in ice sheets was so large it resulted in a mean sea level 100–150 m below today's. Conversely, as a result of melting ice and the expansion of the water in the oceans, global warming could add 0.17 m to present sea levels by 2030, and 0.49 m by 2100 (see Global warming in a glass, Box 12.1). For many UK catchments there could be a 10% increase in winter precipitation and a 5% increase in total annual runoff. In summer, potential evaporation could increase by 10%, and in dry lowland

Box 12.1 **Global warming in a glass**

About half the rise in sea level resulting from global warming will be caused by melting ice, and the remainder by the expansion of the water in the oceans. Because it is easier, this experiment illustrates the reverse of global warming, that is how water reduces in volume when there is a fall in temperature.

Fill a glass (a tall, narrow tumbler is best) or a graduated measuring cylinder with almost boiling water. Make sure that initially the water is level with the top of the rim. Leave it to cool – ideally in the fridge to reduce evaporation – then see by how much the water level has fallen. Now imagine the process in reverse on an ocean scale, and you have half the sea level rise due to global warming. Remember the oceans contain $1.32 \times 10^9 \, km^3$ of water, so there is a lot to expand!

Note that section 4.1.1 pointed out that the mass density of a liquid (= mass/volume) changes with temperature. This demonstration proves it.

areas riverflow may decrease by 4%. Much more work is needed to understand fully global warming and to increase accuracy, but the evidence so far suggests that in the UK both winter floods and summer droughts may be more severe and more frequent. Sea level rise is a concern for all low-lying areas of the world, such as parts of the UK, Bangladesh and many small oceanic islands (see Fig. 12.3). Some 75% of Bangladesh is less than 3 m above sea level. It is protected by around 4000 km of coastal embankments which contain 700 sluices.

From the above it should be apparent that there is a fragile balance between the Earth, the hydrological cycle and humankind. Mankind's intervention could have a significant effect on rainfall patterns, floods, droughts, water supply, irrigation, desertification and soil erosion.

12.3 Precipitation

Precipitation occurs as a result of moist air or water vapour rising, cooling and then condensing around aerosols or **condensation nuclei** (e.g. particles of dust or ice) to form low clouds of water droplets or high clouds of ice crystals. These raindrops and snowflakes fall to Earth when they have sufficient weight to overcome the updrafts of air. In Britain precipitation mainly falls as rain, but snow, hail, sleet, fog and dew are also experienced.

Precipitation can occur as a result of three basic processes. **Cyclonic** or **frontal precipitation** arises from the large scale circulation of air over the Earth's surface. This circulation occurs because air at the equator is heated by the Sun, becomes less dense and rises to be replaced by cooler, denser air moving in from the poles. Thus on a stationary Earth there would be a relatively simple circulation of warm air from equator to poles at high level, and cool air from poles to equator at low level (Linsley *et al.*, 1982). However, this simple pattern is disturbed by the Earth's rotation (1670 km/hr at the equator), which gives rise to the Coriolis force that imparts a sideways deflection to the air masses. Thus as air travels it is deflected to the right in the northern hemisphere, to the left in the southern hemisphere. Friction, oceans and land masses also have an effect on wind speed and direction, so our weather is complex. Nevertheless, air is constantly moving from areas of high pressure (anticyclones)

Figure 12.3 Gold Corner pumping station in the Somer-set Levels. The arrow indicates the height of the high spring tides in the Bristol Channel, so global warming and rising sea levels are of concern in Britain, as in many countries. The station collects floodwater from the River Brue and South Drain and uses four 1.52 m diameter screw-type mixed flow pumps to lift it by 2.4 m so that it can discharge to sea via the man-made River Huntspill. Each pump has a capacity of around 4.3 m³/s. The Huntspill was constructed in 1940–42 to act as a reservoir for an ammunition factory

to low pressure (depressions), although not at 90° to the isobars (lines of equal atmospheric pressure) because of the sideways deflection mentioned above.

The boundary between two masses of air that have different temperature and moisture content is called a **front**, so named because when the boundary is drawn on a map it resembles the fronts or trenches of the First World War. The existence and movement of fronts is usually demonstrated daily on TV weather forecasts. Frontal precipitation occurs when warm air rises over the colder, denser air on the other side of the front. Typically warm-front precipitation occurs when warm air is advancing over the cold air. The rate of ascent is only about 1:100 to 1:300, so continuous light to moderate precipitation may be experienced 300 to 500 km ahead of the front. Cold-front precipitation occurs when it is the cold air that is advancing into a warm air mass, the steep frontal gradient of 1:50 to 1:150 forcing the warm air to rise rapidly, giving intermittent but heavy showers over a relatively small area.

Orographic precipitation occurs when moist air streams have to rise over hills and mountains. Often cloud can be seen forming above the upslopes. Glider pilots use the rising air to gain lift. Most of the precipitation falls on the windward slope that is exposed to the prevailing wind, while the leeward or sheltered side receives less and is in the '**rainshadow**' of the mountains. Thus the location of mountains or hills modifies the pattern of frontal rainfall. Britain experiences westerly winds; in Fig. 12.4 it is easy to see the relationship between ground height and rainfall. Thus the best place for reservoirs is in mountainous regions to the west, such as Wales, the Lake District and Scotland where annual rainfall may exceed 2500 mm. The rainshadow to the east is also apparent; East Anglia is relatively dry (< 600 mm/year) and agriculture sometimes requires irrigation to optimise crop yields. In the USA, westerly winds give San Francisco on the west coast an average of 559 mm of rain per year, but 400 km to the east, in the rainshadow of the mountains, Death Valley records on average only 43.2 mm per annum.

Convective precipitation is caused by localised cells or pockets of warm air rising and cooling to form cloud. These cells occur because the **albedo** or reflectivity of the ground surface varies: light shiny surfaces reflect the heat of the sun, dark mat surfaces (e.g. tarmac, buildings) absorb it. The result is usually isolated patches of cottonwool-like cumulus cloud over the convection cells. Again, glider pilots often head for cumulus cloud where they know they will find updrafts. Convection cells may also be caused by the difference in temperature between land and sea.

Thunderstorms are usually caused by convection. Since this process depends on heat from the Sun, in Britain thunderstorms mostly occur in the south and east on summer afternoons or evenings. Because towns and cities are warmer than the surrounding countryside, there is the possibility of increased thunderstorms and convectional rainfall over built-up areas. For engineers these storms are important: the rainfall experienced is often very heavy, so roads may become flooded as the capacity of storm water sewers is exceeded. There may also be catastrophic flash floods in rivers (see section 13.4).

12.3.1 Measurement of precipitation

From the above it is apparent that both the total depth of rainfall and its intensity (i.e. how heavy it is) are important. The annual depth and monthly distribution of rainfall are often critical factors in water resource studies, whereas **rainfall intensity** is important with respect to flood prediction. Ideally a rain gauge network should be able to measure both. The measurement of snowfall is rather problematic, since wind-blown snow may pass over a rain

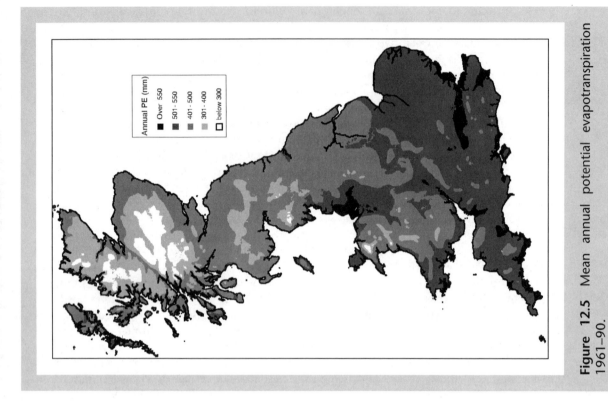

Figure 12.5 Mean annual potential evapotranspiration 1961–90.

Figure 12.4 Mean annual precipitation 1961–90 *[maps prepared by the Centre for Ecology and Hydrology from Meterological Office data; reproduced by permission]*

Figure 12.7 Mean annual surface runoff 1961–90

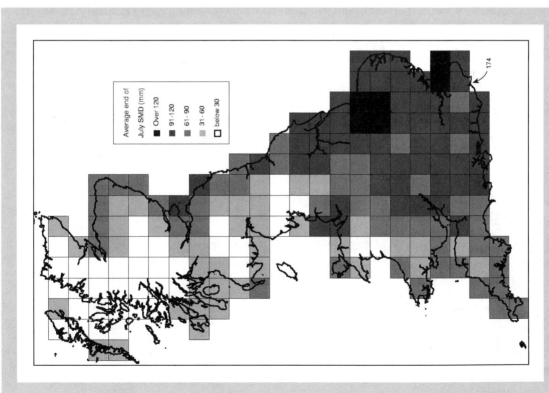

Figure 12.6 Mean soil moisture deficit 1961–90 at the end of July

gauge, while drifts may prevent snow entering. Consequently snowfalls may be unrecorded unless manual measurements are made using a metre rule, when every 12 mm of fresh snow is approximately equivalent to 1 mm of rain. It is important to know how much snow is lying on a catchment, particularly in countries with large snowfalls, since a sudden thaw can release considerable quantities of water (in addition to any rain that may be falling at the time – see Box 9.1). Flooding caused, or exacerbated, by snow melt is a frequent problem, so continuous monitoring and evaluation should be undertaken when the conditions warrant it.

The conventional Meteorological Office Mark II 127 mm (5 inch) copper/brass rain gauge consists of a funnel, which collects the rainfall, and a glass bottle or reservoir that stores it. These are read daily at 0900 h GMT by tipping the contents into a special glass measuring cylinder graduated directly in mm of rainfall. For remote sites monthly-read gauges that have a larger reservoir are used. All gauges, regardless of type, are usually set with their rim horizontal at a height of 300 mm above ground level (to stop rain or hail that hits the ground bouncing or splashing inside). Alternatively, the rim may be set at ground level if an anti-splash grid is placed around the gauge.

A record of rainfall intensity can be obtained using either a tilting syphon or tipping bucket gauge. With the tilting syphon the movement of a float in a collecting chamber causes a pen to draw a record of the rainfall on a rotating drum. When 5 mm of rain has fallen, the chamber tilts and empties. However, the use of hand-wound clocks and paper charts is rather old-fashioned and inconvenient, so often the tipping bucket gauge is preferred. With this gauge, rainfall is collected by a funnel and falls into one of two triangular-shaped buckets located side-by-side on a pivoted beam. When the first bucket is full, its weight causes it to overbalance so that the bucket tips to one side, emptying it, while the second bucket takes its place under the funnel. When it is full it tips to the side, being replaced by the first. The buckets are sized so that, when full, each represents 0.2 mm or 0.5 mm of rainfall. By using a magnetic counting device and timer, the rainfall intensity can be recorded by a datalogger, or obtained in real-time using an electronic telemetry system.

The Meteorological Office has a network of around 6500 conventional rain gauges covering Great Britain and Northern Ireland. There are relatively few tipping bucket gauges since they are more expensive and less reliable, but only a few are needed to monitor the intensity of rain as it falls on key catchments. Computers and telemetry systems have greatly advanced the art of flood forecasting: the flow down rivers can also be measured in real-time using modern instrumentation. Satellites and weather radar provide rainfall maps

---

Box 12.2 See for yourself

There are a number of Internet sites that illustrate nicely some of the topics in this chapter. Photographs, descriptions and specifications of rain gauges can be found at **www.munro-group.co.uk/rain**. Interesting details of weather, climate, forecasting and up-to-date satellite and radar pictures are on the BBC's site **www.bbc.co.uk/weather**. The Meteorological Office's site also gives lots of interesting information about the weather, such as extremes of UK climate, London smog, severe winters and the story of the 1953 flood and storm surge referred to in the solution to Self Test Question 12.2. All of this is at **www.met-office.gov.uk**.

with pixels coloured according to rainfall intensity, so from a Flood Control Room it is now possible to see weather systems arriving, and to monitor rainfall and runoff (see Box 12.2 and Cluckie and Han, 2000).

When siting rain gauges, generally the site selected should be flat and easily accessible, to facilitate data collection and maintenance, but in a quiet location to minimise tampering and vandalism. Exposed, wind-swept locations should be avoided. If this is not possible, to provide some shelter and to prevent wind-blown rain being missed, a 300 mm high turf embankment may be constructed around the gauge at a distance of 1.5 m. Sites which are over-sheltered or in the rainshadow of trees or buildings will also give false readings. A rule of thumb is to site a gauge at least a distance $2H$ from an obstacle of height H, remembering that trees will grow! Similarly, the removal of trees or buildings may affect the catch (see Example 12.1).

Even if a gauge is well sited, its accuracy may be questionable. A conventional gauge set 300 mm above ground may collect 6%–8% less than a ground level gauge. Less rain is caught as wind speed increases and/or drop size decreases. Small amounts of rain may evaporate before the gauge is read, while snow may not be collected effectively. Thus all that is obtained is an inaccurate estimate of the rain that falls into (say) a 0.013 m² collector, which can result in a sizeable discrepancy between the actual and assumed annual volume of water falling on a catchment, perhaps 10^9 times larger.

The depth of precipitation falling at any location varies, of course, from month to month and year to year. In Britain, in any particular year rainfall may be anywhere between 60% and 150% of the long-term mean (Smith, 1972). However, in more than half of all years the annual depth will be within ±10% of the mean, so rainfall is fairly consistent, particularly in the north and west. Rainfall intensity generally varies during a storm, increasing to a peak near the middle. The maximum intensity that may be expected depends upon many factors, as explained below.

EXAMPLE 12.1 DOUBLE MASS CURVE OF PRECIPITATION

A rain gauge installation (A) has been upgraded by replacing a conventional manually read gauge with a tipping bucket type. At the same time a new protective fence has been built around the site. (a) Examine the homogeneity (consistency) of the record of gauge A in column 2 of Table 12.3 compared to the average of five near-by gauges (column 3). Determine whether or not a change in conditions occurred. (b) If a change occurred, adjust the pre-change data to be homogeneous with that currently being recorded. (c) The rain gauge record at A during 1999 has been lost as a result of a computer virus, so estimate the missing annual value.

(a) The principle behind a double mass curve is that any increase or decrease in annual rainfall ought to affect all of the rain gauges in a particular region in a similar way, so if one behaves differently this may be a sign of a change in conditions – such as a change of observer, change of gauge, or alterations to the gauge site or its surrounds. It has been standard procedure to routinely check every gauge against others in the same vicinity. In this example the catch at gauge A (column 2 of Table 12.3) is being checked against the average catch of five adjacent gauges (column 3). Columns 4 and 5 show the cumulative rainfall depths, which are plotted in Fig. 12.8.

From Fig. 12.8 it is apparent that there is a change of slope, the data from 1990 onwards plotting below the extrapolated 1984–90 line (dashed). It would appear the catch at gauge A has been reduced, possibly as a result of the fence making the site more sheltered and/or the replacement of the original gauge.

Table 12.3 Recorded data, cumulative rainfall and corrected catch for Example 12.1

Year	Recorded rainfall (mm)		Cumulative rainfall (mm)		Corrected catch at gauge A
	Gauge A	Five gauge average	Gauge A	Five gauge average	
1984	690	880	690	880	608
1985	852	1040	1542	1920	751
1986	780	979	2322	2899	687
1987	842	1033	3164	3932	742
1988	878	1060	4042	4992	774
1989	801	1008	4843	6000	706
1990	899	1105	5742	7105	792
1991	710	1018	6452	8123	
1992	801	1099	7253	9222	
1993	750	1066	8003	10288	
1994	822	1134	8825	11422	
1995	718	1037	9543	12459	
1996	801	1108	10344	13567	
1997	728	1022	11072	14589	
1998	700	980	11772	15569	
1999	???	1161	—	16730	

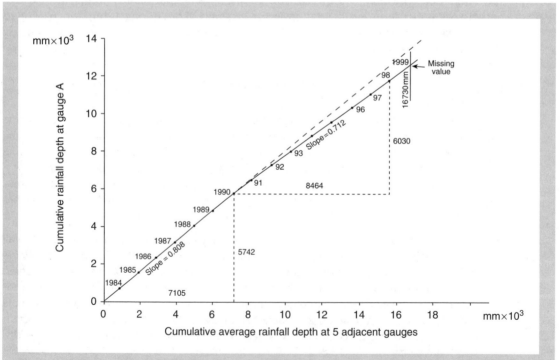

Figure 12.8 Double mass rainfall graph for Example 12.1. By extrapolating the 1984–90 line it is apparent that a change occurred after 1990. The ratio of the slopes of the two lines can be used to adjust the incorrect data

(b) Rather than have to correct all future readings from this site, it is more convenient to make a once-only adjustment to the pre-1991 data. This can be achieved by using the ratio of the slopes of the lines to reduce the readings at gauge A between 1984 and 1990 as follows:

slope of line 1984–90 = 5742/7105 = 0.808
slope of line 1990–98 = 6030/8464 = 0.712
ratio of slopes = adjustment factor = 0.712/0.808 = 0.881

The adjusted gauge A readings in the last column of the table are obtained by multiplying the observed values in column 2 by 0.881. *Note that the adjustment factor in this example had to be 0.712/0.808 (and not 0.808/0.712) because the pre-1991 data is relatively high and has to be reduced. Always think about what you are trying to achieve, rather than memorising some rule for making the adjustment.*

(c) The cumulative five station value for 1999 is shown by the short vertical line corresponding to 16730 on the horizontal axis. This cuts the extrapolated 1990–98 line at a cumulative gauge A value of 12600 on the vertical axis. Subtracting from this the 1998 cumulative value in the table gives the catch at A in 1999 as approximately 12600 – 11772 = 828 mm.

12.3.2 Intensity–duration–frequency relationships

If you have been caught outdoors during a torrential burst of rainfall, you may have sheltered in a doorway or under a tree until it eased off. There is an almost instinctive understanding that really heavy rainfall does not last very long. This understanding is correct. Table 12.4 shows some of the highest point rainfalls recorded in Britain. Dividing the rainfall depth (mm) by the time over which the rain fell (th) gives the **rainfall intensity** ($i$mm/h). The table indicates that i decreases with increasing t. The data can be plotted to form an intensity–duration curve that represents the most intense storms ever recorded. Other curves could be plotted to represent the rainfall intensity experienced on average every year (1 in 1 years) or once in every 10 years (1 in 10 years). The various curves represent a different **return period** or **frequency of occurrence**, that is the average period between storms which exceed this rainfall intensity.

In terms of both rainfall intensity and depth, part (b) of the table acts as a reminder that the British climate is extremely moderate by world standards. Cherrapunji has an altitude of 1310 m and receives rain during the monsoon in late spring or summer. The effect of air rising over the mountains is accentuated by having to squeeze through deep constricting valleys. Even so, 26.5 m of rainfall is extreme even by Cherrapunji's standards, and involved some overlapping of monsoon seasons. The monsoon can result in severe flooding, as described in section 12.2.

Part (a) of the table illustrates that the majority of the most intense British events occurred in eastern and southern England where convectional thunderstorms are most frequent. According to Holford (1977), 34 of the 50 heaviest 2 hour rainfalls in Britain in the 20th century occurred in southern England, 9 in the Midlands and northern England, 4 in Wales, 2 in Scotland and 1 in Northern Ireland. Conversely, the greatest depth of rain in 2 days, 1 month and 1 year predictably occurred in western Britain.

When preparing the table it was tempting to omit the 1906 Ilkley event so that the inten-

Table 12.4 Some extreme rainfall event [*after Wilson, 1990*]

Date	Location	Depth (mm)	Duration *t*	Intensity (*i* mm/h)
(a) British Isles				
1935	Croydon, London	5.1	1 min	306.0
1906	Ilkley, West Yorkshire	12.7	4 min	190.5
1970	Wisbech, Cambridgeshire	50.8	12 min	254.0
1958	Sidcup, Kent	63.5	20 min	190.5
1980	Orra Beg, Northern Ireland	97.0	45 min	129.3
1910	Wheatley, Oxfordshire	110.2	1 h	110.2
1975	Hampstead, London	140	2 h	70.0
1960	Horncastle, Lincolnshire	178	3 h	59.3
1917	Bruton, Somerset	200	8 h	25.0
1955	Martinstown, Dorset	279	24 h	11.6
1974	Sloy, Strathclyde	300	48 h	6.3
1909	Llyn Llydaw, Gwynedd	1436	1 month	—
1954	Sprinkling Tarn, Cumbria	6528	1 year	—
(b) World events				
1970	Guadeloupe	38	1 min	2280
1920	Bavaria	126	8 min	945
1889	Romania	206	20 min	618
1947	Missouri, USA	305	42 min	436
1935	Texas, USA	559	2.75 h	203
1964	Réunion (Indian Ocean island)	1340	12 h	112
1952	Réunion	1870	24 h	78
1952	Réunion	2500	48 h	52
1861	Cherrapunji, India	9300	1 month	—
1861	Cherrapunji	22454	6 months	—
1861	Cherrapunji	26461	1 year	—

sity values decreased consistently, but it was included as a reminder that most rainfall events go unrecorded. More than 12.7 mm of rain in 4 min has probably fallen many times, but not where there was a rain gauge or observer.

The intensity–duration relationship has been analysed and has resulted in many empirical equations. For example, the ones in Table 12.5 appeared in the now obsolete British Code of Practice 2005 'Sewerage'. For a 12 minute storm and return periods of 1, 5 and 10 years, the table gives rainfall intensities of 33 mm/h, 55 mm/h and 65 mm/h respectively. Obviously the intensity increases as the event becomes increasingly rare, but all of these values are much less than the 254.0 mm/h in 1970 at Wisbech (Table 12.4).

Rainfall intensity may now be determined using computer software, the Internet or by analysing individual records. The 1999 Flood Estimation Handbook (FEH) presents a rainfall depth–duration–frequency model which can be used to estimate extreme rainfalls at any UK location (Institute of Hydrology, 1999). It is principally intended for durations between 1 h and 8 days, and return periods between 2 and 2000 years, although cautious extrapolation is possible. The handbook is accompanied by a CD-ROM which provides the parameters for the model and software to do the calculations. The handbook also describes the

Table 12.5 Some examples of empirical British intensity–duration equations (t = duration of storm in minutes)

Return period	$t = 5$ to $20\,min$	$t = 20$ to $120\,min$
1 in 1 year	$i = \dfrac{650}{t+8}\,mm/h$	$i = \dfrac{1000}{t+20}\,mm/h$
1 in 5 years	$i = \dfrac{1200}{t+10}\,mm/h$	$i = \dfrac{1500}{t+18}\,mm/h$
1 in 10 years	$i = \dfrac{1550}{t+12}\,mm/h$	$i = \dfrac{1975}{t+22}\,mm/h$

derivation of rainfall profiles for use in rainfall–runoff modelling. The basis of the new method is the index variable *RMED*, which is defined as the median of annual maximum rainfalls (for a given duration) at a site. This is the rainfall equivalent of *QMED*, and can be derived in a similar way using the procedure in section 13.3.3. The value of *RMED* is potentially susceptible to climate change, and it has been suggested that by 2020 the average intensity of precipitation will have increased modestly in all seasons in all regions of the UK, with an associated shortening of the return period (for the same reason as in Example 13.6).

Note that the FEH handbook replaced the 1975 Flood Studies Report (FSR), which had been the standard reference in the UK when engineers wished to analyse rainfall and predict flood magnitudes (NERC, 1975). Although the FEH and its CD-ROM would be used in practice to calculate rainfall intensity and frequency, for simplicity Table 12.5 will be employed in some examples in this chapter.

SELF TEST QUESTION 12.2

For durations up to 180 min draw the maximum intensity–duration relationship for Britain by plotting on linear graph paper the data in Table 12.4. Then use Table 12.5 to plot the intensity–duration curves corresponding to a 1 in 1 and 1 in 10 year return period. (a) For 10 and 90 minute storms, what are the largest and smallest rainfall intensities on your graph? (b) What is the significance of the answer in (a) for engineers who are using rainfall intensity to design flood alleviation works? (c) What return period would you adopt if you were designing: (i) a dam, (ii) a storm water sewer in an area where houses have basements, and (iii) a normal modern housing estate without basements?

12.3.3 Depth–area–duration relationships

It was explained above how localised convection cells often result in very intense thunderstorms. Although the rainfall intensity is high at the centre of the cell, it usually decreases quickly with distance. For example, the isohyets (i.e. contours of rainfall) for the

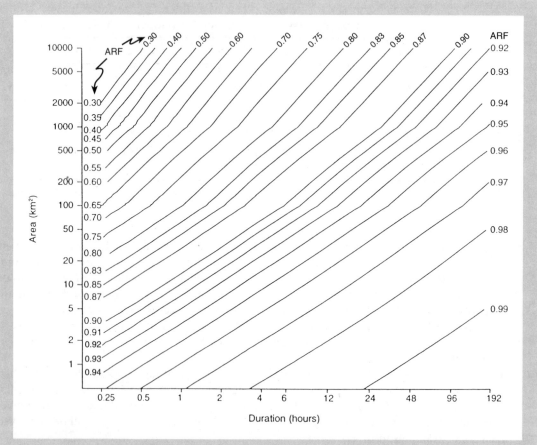

Figure 12.9 The areal reduction factors (ARFs) are the values attached to the lines on the diagram. They are determined from the rainfall duration and catchment area [*reproduced from the Flood Estimation Handbook, courtesy of the Institute of Hydrology*]

Martinstown event in Table 12.4 indicate that while 279 mm of rainfall was recorded near the centre of the storm, about 6.5 km away it was only 127 mm, and 19 km away it was 75 mm. Thus if trying to estimate the potential size of a flood, it cannot be assumed that a very high, localised rainfall depth or intensity occurs over an entire catchment. The **areal reduction factor** (ARF) is the ratio of the rainfall depth over an area to the rainfall depth of the same duration and return period at a representative point in the area. The values are always less than 1.0, as shown in Fig. 12.9, and reduce with increasing catchment size and duration. The ARFs are the values attached to the lines. For example, suppose a rain gauge recorded 41 mm in 1 h and it is required to estimate the equivalent intensity on a catchment of 100 km^2. Moving vertically upwards from 1 h and horizontally from 100 km^2 gives the point of intersection almost on the ARF = 0.80 line. Thus the reduced intensity is 0.80 × 41 = 33 mm in 1 h.

12.3.4 Calculation of areal mean rainfall depth

A large catchment or region of the country may contain many rain gauges which experience different amounts of precipitation, so some method of calculating the mean depth is required. The simplest method is merely to calculate the mathematical mean of all the readings. However, this may be relatively inaccurate. For example, fewer gauges may be located in remote, inaccessible, mountainous regions where the rainfall is highest, so a mathematical mean may give an underestimate. An alternative is to draw contours of rainfall (isohyets) and then work out the mean precipitation from the product of the area between successive contours and the average contour value (see Example 12.2). A third method is to join neighbouring gauges with straight lines, then construct the perpendicular bisectors of these lines so that polygons are created around each gauge. The mean precipitation can be obtained from the product of the area of each polygon and the rainfall depth at its centre.

EXAMPLE 12.2

Isohyets of annual precipitation have been drawn at 100 mm intervals (i.e. 1000 mm, 900 mm etc.). The area enclosed by adjacent contours (A km^2) is shown in Table 12.6, together with the average contour value (P mm). Calculate the mean precipitation falling on the area.
Column 3 of Table 12.6 shows the weighted precipitation $= A \times P$, and the cumulative value $= 213\,800$.

Table 12.6 Areas (A) and precipitation values (P) used in Example 12.2

Area A (km^2)	Average contour value P (mm)	$A \times P$
31	950	29 450
44	850	37 400
59	750	44 250
103	650	66 950
65	550	35 750
302		213 800

Weighted mean precipitation = 213 800/302 = 708 mm.

Note: the calculations would be conducted in a similar fashion using the polygon method.

12.4 Evaporation, transpiration and evapotranspiration

When rainfall stops, wet pavements and roads dry as the water evaporates into the atmosphere. In winter this happens slowly, but in summer it happens quickly, and steam may even be seen rising from the surface. The rate of **evaporation** is greatest when the sun is hottest, the air and water temperature is relatively high, the air is relatively dry (i.e. low humidity) and a drying wind is blowing. Evaporation also takes place from soil and vegetated surfaces, although this is not so visible. Evaporation from the surface of a lake can be considerable, and this must be allowed for when designing or assessing the yield of a reservoir. Linsley *et al.* (1982) indicated that the mean annual evaporation from shallow lakes in the USA varied between 0.5 m and 2.0 m. Over a large surface area, this represents a lot of water.

Potential evaporation (PE) is the maximum rate of evaporation, such as from the free water surface in a lake or an evaporation pan (see below). However, under most conditions the rate of actual evaporation (AE) from the ground surface is limited by the amount of water available in the surface layers of the soil. Thus AE < PE, particularly in summer, as shown in Fig. 13.1.

Plants do not need rainfall every day to survive because, for several days at least, they can draw upon water stored in the soil. Water from the ground moves through the roots, up the stem to the leaves, where water vapour is transpired into the atmosphere via stomata in the leaves. Some large trees are capable of transpiring around $9 m^3$ of water per day. As a result, clay soils which contain a lot of water shrink in dry weather, so near-by buildings may experience subsidence.

The rate of **transpiration** is greatest in daylight, when the temperature is high, the soil contains a plentiful water supply, and the plant has long roots and a large leaf area. Conversely, transpiration is low when the temperature is low (in winter it may be negligible), the soil is dry, the roots are short and the leaves small. Drought tolerant plants like cacti have swollen tough stems and leaves reduced to spines.

Evaporation and transpiration usually occur at the same time from most surfaces, so the combined water loss is called **evapotranspiration** (ET). Potential evapotranspiration (PET) is the theoretical maximum rate of water loss assuming a continuous and plentiful supply of water to plants (Fig. 12.5). Actual evapotranspiration (AET) is the lower, true rate of water loss experienced with whatever water is available naturally from rainfall or groundwater. Only next to rivers, and in irrigated or naturally marshy areas will AET = PET. As shown in Table 12.2, AET is a very significant process that accounts for roughly 38%–87% of rainfall according to location.

12.4.1 Measurement of evapotranspiration (ET)

The measurement of ET is not easy and there is considerable scope for error. Part of the problem is that the rate of ET at any time depends upon many factors, including the amount of water available, the local weather conditions and the type of vegetation. One option is to use a water budget approach to solve equation (12.2) for *AET*, the actual evapotranspiration in a catchment. This assumes that precipitation, runoff and all of the other quantities can be accurately evaluated. Clearly, this would require extensive instrumentation and calculation to obtain accurate answers for a large number of catchments. An alternative approach, which eliminates the question of how much water is available, is to either measure the rate of potential evaporation from an open water surface (E_O), or to measure the rate of PET from an artificially irrigated vegetated surface.

Some relatively simple empirical equations have been devised for the rate of evaporation (E_O) from reservoirs, but their use is often limited: they can be applied only to a certain geographical region; they also require a knowledge of wind speed, and the actual and saturation vapour pressure of the air. Alternatively, the actual evaporative loss from a reservoir can be obtained from equation (12.1) if the inflow, outflow and change in storage are known and an allowance is made for precipitation falling directly onto the water surface. One such measurement at Kempton Park in Britain gave annual evaporation as 663 mm (Shaw, 1994). However, it is easier and more convenient to measure directly the evaporation (E_O) from a pan of water. The standard British evaporation pan is 1.83 m square, contains water to a depth of 550 mm, and is set into the ground. The daily evaporation rate is obtained by measuring the reduction in water level, after which the depth is made up to 550 mm again. Allowance must also be made for any rainfall. This requires very careful and precise measurements if accuracy is to be attained. In the USA a circular 1.21 m diameter pan containing 180 mm of water is used. This is set on a timber grillage above ground level. Both the British and the USA pans suffer from the fact that they contain relatively small quantities of water that heat up more quickly than reservoirs, and so have higher evaporation rates. Thus a correction factor is needed, which is typically about 0.92 and 0.75 respectively for the two types.

The approximate evaporative loss from bare soil (E_B) and turfed soil (E_T) can be obtained from E_O by using a coefficient (Wilson, 1990). For Southern England:

$$E_B = 0.90E_O \tag{12.4}$$

$$E_T = 0.75E_O \quad \text{(whole year)} \tag{12.5}$$

$$E_T = 0.60E_O \quad \text{(November to February)} \tag{12.6}$$

$$E_T = 0.80E_O \quad \text{(May to August)} \tag{12.7}$$

Thus if $E_O = 663$ mm as at Kempton Park earlier, then $E_B = 0.90 \times 663 = 597$ mm and $E_T = 0.75 \times 663 = 497$ mm.

The lysimeter in Fig. 12.10 is basically a research (rather than an operational) device for measuring the rate of PET from an irrigated surface. It consist of a sealed tank that contains

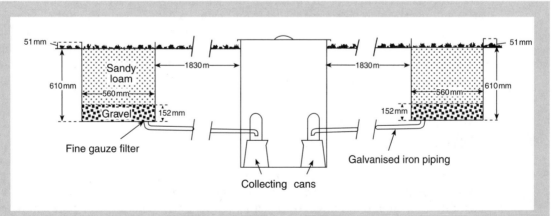

Figure 12.10 A two bucket lysimeter used to measure evapotranspiration. Irrigation water is added to the buckets, some being retained as soil moisture while any surplus water drains through the bottom into the collecting cans. The remaining water (plus any rainfall) must be evapotranspired as indicated by equation (12.8) [*after Smith (1972); reproduced by permission of Macmillan, now Palgrave*]

soil and vegetation representative of its surroundings. Underneath the tank is a device that can measure the change in weight of the tank, and hence the volume of water stored in it. Water is added to the tank continually to ensure that there is always a plentiful supply to the plants, any excess water percolating through the soil and draining from the base of the tank through a pipe into a collecting pit, where it is measured. Thus:

PET = rainfall + irrigated water ± change in stored water − percolation water (12.8)

Enough irrigation water should be added to ensure that there is always water seeping from the drain pipe in the base. If no irrigation water is provided, the measurement will approximate AET. However, the fact that the soil sample is relatively small and has been disturbed may mean that the readings are not truly representative. Nevertheless, the device is relatively simple and can be manufactured from galvanised iron dustbins (Shaw, 1994). Some lysimeters have two or more tanks to improve accuracy.

The water loss from forested areas has been problematical, since trees do not easily fit into lysimeters. However, research (notably at Plynlimon, Wales) suggested that if the annual rainfall is about 600 mm there is little difference in the loss from grassland and forest, but with 2300 mm of rainfall the forest may lose on average about 850 mm per year compared to 405 mm for grassland (Gash and Stewart, 1977; Newson, 1979). Around 22%–38% of rainfall evaporates from the forest canopy, about 50%–55% falls through the canopy while a further 12%–23% reaches the ground by flowing down the stem. Thus, in areas of relatively high rainfall, increased evaporation from the canopy results in a larger evapotranspiration loss from mature forests than grassland, with a corresponding decrease in surface runoff of up to 33% (Wilson, 1990). This has some implications with respect to estimating reservoir yields: generally the loss from forested, upland catchments has been underestimated.

None of the methods of estimating ET described so far are particularly satisfactory for general use. The technique employed by the Meteorological Office for the routine calculation of ET is the Penman method (Penman, 1948; 1950). Penman produced a theory and formula for the estimation of open water evaporation (E_O) from observations of the weather. The basis is a combination of an energy balance and aerodynamic factors, the former providing an indication of the energy available for heating and evaporation, and the latter the efficiency with which atmospheric turbulence removes the water vapour produced. The data required are the standard observations of mean air temperature, relative humidity, wind velocity and hours of sunshine. The derivation of the formula and its application is too long to be presented here, so refer to Wilson (1990) or Shaw (1994) for details. The use of the ET data is described later in section 13.1.

12.5 Infiltration and percolation

Infiltration occurs when water first penetrates the ground surface, whereas percolation is its subsequent movement vertically down through the ground to the water table (Fig. 12.11). Often this distinction is not observed, so either term may be used to describe the combined process.

Because the soil at the ground surface is either broken naturally by roots or cultivated, its capacity to absorb and transmit water (f_1 mm/h) is greater than that of the soil below the surface (i.e. $f_1 > f_2$). As the depth below ground increases the soil becomes more compact, so its permeability decreases ($f_3 < f_2$). Consequently percolation is often a relatively slow process; intergranular seepage vertically in the unsaturated zone above the water table perhaps has

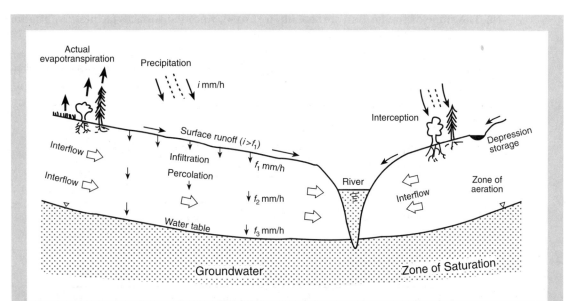

Figure 12.11 Some precipitation may be intercepted by trees and vegetation, the remainder falling on the ground surface. Surface runoff is most likely when $i \gg f_1$, and infiltration is most likely during prolonged, gentle rainfall when $i \leq f_1$. The soil becomes more compact with depth, so the initial infiltration capacity at the ground surface (f_1) reduces as water percolates down to the water table ($f_1 > f_2 > f_3$) so some water in the zone of aeration must become interflow and move horizontally above the water table, discharging to streams and becoming part of the groundwater baseflow

a velocity of about 1 m/year (Smith *et al.*, 1970). Owing to the reduction in permeability with depth, some percolated water has to become interflow that moves horizontally above the water table (Fig. 12.11). Interflow often discharges to streams and becomes part of the groundwater baseflow that enables rivers to continue flowing during periods without rain when there is no direct surface runoff. The conditions which favour infiltration and groundwater recharge are prolonged periods of light rain or drizzle when its intensity $i < f_1$. Surface runoff occurs mostly when $i > f_1$. Very intense rainfalls cannot be absorbed by the ground and result in large amounts of runoff and flooding. This explains why parts of England are sometimes lying under a metre or so of water but, because local water supplies are derived from groundwater, the region is officially experiencing a drought and hosepipe bans!

Factors that affect infiltration are listed briefly below:

(a) The type of soil, namely its permeability and porosity.

(b) The initial moisture content of the soil, since a dry soil can absorb water more readily than a wet or totally saturated one.

(c) Rainfall intensity (which is very important up to a limiting value), since this governs both the amount of water available and the hydraulic head which drives the infiltration process. Intensities greater than the limiting value favour surface runoff.

(d) Time, since a dry soil will become progressively wetter as infiltration continues (see Example 12.3).

(e) **The slope of the ground surface** (up to 1 in 4), since flat surfaces facilitate infiltration, whereas steep slopes promote surface runoff.

(f) **The presence of vegetation**, since both the initial (f_O) and final (f_C) infiltration capacities will be larger for vegetated soil than bare soil. Roots slow runoff and break up the soil, so enhancing infiltration, while dry leaves and organic debris absorb water. Some typical values of f_O and f_C are shown in Table 12.7.

Table 12.7 Typical initial (f_O) and final (f_C) infiltration capacities of some soils [*after Wilson, 1990*]

Soil type		f_O (mm/h)	f_C (mm/h)
Standard agricultural soil	– bare	280	6–220
	– turfed	900	20–290
Peat		325	2–20
Fine sandy clay	– bare	210	2–25
	– turfed	670	10–30

Nothing should be taken for granted with respect to infiltration, since frozen catchments can render the soil type and infiltration capacity irrelevant. Similarly, compaction of the surface by feet, vehicles or the impact of large raindrops can make it relatively impermeable, as can being baked by a very hot sun.

12.5.1 Measurement of infiltration

Logically, one way to measure the infiltration capacity of a soil is to try to simulate the natural infiltration process. This can be achieved using an infiltrometer, which consists of two concentric rings pressed into the ground. Water to a depth of 5 mm is poured or sprinkled into the rings, then the quantity added to the inner ring to maintain this depth over a period of time is recorded. A rain gauge is also needed to compensate for any precipitation. The depth of 5 mm is rather small to achieve accuracy, but larger depths are unrealistic. The outer ring is needed to eliminate the effect of water flowing radially outward into the drier soil surrounding the infiltrometer. Example 12.3 shows how the infiltration rate may be calculated.

Another way to measure infiltration is to simulate rainfall using a sprinkler system. If the runoff from a test area is collected and measured (and any rainfall or natural inflow allowed for) then the difference is assumed to be infiltration. Usually the rate of infiltration is overestimated, naturally vegetated areas tending to have different characteristics to experimental plots.

An approximation of the depth of water 'lost' during a storm on a particular catchment can be obtained by comparing the depth of rainfall to the equivalent depth of surface runoff. The average rainfall depth can be obtained from gauges; the depth of runoff can be found by calculating the volume of surface runoff (VOL m^3), using the techniques described in section 12.6, and then dividing by the catchment area. Thus if $G = 0$ in equation (12.2), the water loss due to evaporation and infiltration is equivalent to $AET + \Delta S + \Delta G$. This requires many doubtful assumptions regarding catchment boundaries, groundwater flow and storage, so the result is unlikely to be accurate.

The **Φ index** is the average rainfall intensity above which the volume of rainfall equals the volume of surface runoff. This is a simple index that varies in value according to soil type and catchment characteristics, but does not vary with time. The index represents all losses, so depression storage, interception, AET and infiltration are all included. Usually the largest loss is infiltration. One of the attractions of the index is its simplicity, particularly when quick estimates of probable runoff are required. The calculation of the Φ index is illustrated in Example 12.4.

If no other data are available, two 'rules of thumb' for infiltration in Britain are 250 mm/year or 40% of rainfall (Hamill and Bell, 1986). However, other investigators have used figures between 150 mm and 300 mm/year. Given the large variation of rainfall over Britain (500 to >2500 mm), using a percentage of precipitation probably makes more sense than adopting a fixed value such as 250 mm/year.

12.5.2 Net rainfall

Gross rainfall is that measured by a rain gauge. Net or effective rainfall is the rainfall that remains after deducting all losses such as interception, ET and (REMEMBER!!!) infiltration. Thus net rainfall can be defined as the part of total rainfall that becomes surface runoff. One simple way to obtain net rainfall is to subtract Φ from gross rainfall (see Example 12.4). If there is a soil moisture deficit (section 13.1.1), this should be subtracted from the first part of the rainfall hyetograph until it is reduced to zero. For instance, if the soil moisture deficit (SMD) was 14 mm then this would cancel out the first three bars of the hyetograph in Fig. 12.13.

EXAMPLE 12.3   CALCULATION OF THE INFILTRATION RATE f USING AN INFILTROMETER

The diameter of the inside ring of a double ring infiltrometer is 0.30 m. The soil under test is a silty clay. The first two columns of Table 12.8 show the volume of water added during each of the time intervals. (a) Determine the infiltration capacity of the soil (f mm/h) during each time interval. (b) Plot a graph of f against time. (c) What is the initial (f_o mm/h) and final (f_c mm/h) infiltration capacity? (d) What is the average infiltration capacity during the first 10 min and 60 min?

(a) The area of the inner ring, $A = (\pi \times 0.30^2)/4 = 0.071\,\text{m}^2$.

Columns 1 and 2 of the table show the amount of water added to the inner ring during each time interval. The other columns show the calculations needed to obtain f mm/h. For example, between 2 and 5 mins (i.e. in 3 min = 0.050 h) the amount added to the ring was $0.455 \times 10^{-3}\,\text{m}^3$, which is equivalent to $(0.455 \times 10^{-3}/0.071) = 0.00641\,\text{m}$ or 6.41 mm. Thus the average infiltration rate during this interval = 6.41/0.050 = 128 mm/h.

(b) The graph in Fig. 12.12 is drawn by plotting the f mm/h values in the table against the mid-point of the time interval, such as 128 mm/h against (2 + 5)/2 = 3.5 min. It shows that infiltration decreases with time, whereas the simple Φ index assumes that losses are constant.

(c) Based on the average values for the time intervals, the table shows that $f_o = 141$ mm/h and $f_c = 11$ mm/h.

Table 12.8 Infiltrometer observations and calculations for Example 12.3

Time since start of test t mins	Volume of water added during time interval VOL m³	Equivalent depth of water added $= (VOL/A) \times 10^3$ D mm	Time interval Δt h	Infiltration rate during interval $= D/\Delta t$ f mm/h
0	—	—	—	—
2	0.330×10^{-3}	4.65	0.033	141
5	0.455×10^{-3}	6.41	0.050	128
10	0.719×10^{-3}	10.13	0.083	122
20	1.232×10^{-3}	17.35	0.167	104
30	0.985×10^{-3}	13.87	0.167	83
60	1.526×10^{-3}	21.49	0.500	43
90	0.639×10^{-3}	9.00	0.500	18
120	0.427×10^{-3}	6.01	0.500	12
180	0.779×10^{-3}	10.97	1.000	11

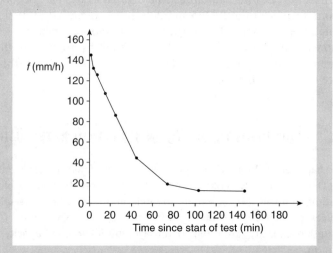

Figure 12.12 Infiltration–time curve for Example 12.3. Based on the results of an infiltrometer test, the graph shows how f declines with time

(d) Over the first 10 min the total volume of water added $= (0.330 + 0.455 + 0.719) \times 10^{-3}$ m³
$$= 1.504 \times 10^{-3} \text{ m}^3.$$

This is equivalent to $(1.504 \times 10^{-3}/0.071) \times 10^3 = 21.18$ mm depth of water over the inner ring. With 10 min $= 0.167$ h, then $f_{10} = 21.18/0.167 = 127$ mm/h.

Repeating this calculation for the first 60 min gives the volume added $= 5.247 \times 10^{-3}$ m³, the equivalent depth as 73.90 mm and $f_{60} = 74$ mm/h.

EXAMPLE 12.4 CALCULATION OF NET RAINFALL AND THE Φ INDEX

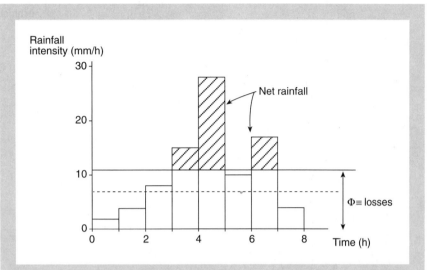

Figure 12.13 A hyetograph or gross rainfall intensity–time graph. The area below the horizontal line represents the combined losses (Φ mm/h). The rainfall above the line becomes surface runoff and is called net rainfall

Figure 12.13 shows the variation of rainfall intensity with time during a storm. There is no soil moisture deficit. (a) Calculate the total depth of gross rainfall. (b) For the type of soil on the catchment, the Φ index is thought to be 11 mm/h. Use this to calculate the total depth of net rainfall. (c) At a nearby gauging station the actual surface runoff from the storm is estimated as equivalent to 43 mm of rainfall. Determine the Φ index for the catchment.

(a) Total depth of gross rainfall = 2 + 4 + 8 + 15 + 28 + 10 + 17 + 4 = 88 mm.

(b) In Fig. 12.13 the solid horizontal line at 11 mm/h represents Φ, and the area below it the losses. The net rainfall above the line = 4 + 17 + 6 = 27 mm. This is the amount of rainfall that would become surface runoff.

(c) If the surface runoff is actually 43 mm then the estimate of Φ above is too large, so the horizontal line needs to be lowered (shown dashed). This requires trial and error until the values above the line = 43 mm. With Φ = 7 mm/h the net rainfall = 1 + 8 + 21 + 3 + 10 = 43 mm.

12.6 Surface runoff

Some proportion of the rain that falls onto the ground will run over it into a stream or river channel, and then down the channel to the sea. The natural movement of water over the surface of the Earth as a result of gravity is called surface runoff. Here hydraulics and hydrology overlap, since Chapter 8 analysed the flow in open channels, such as rivers.

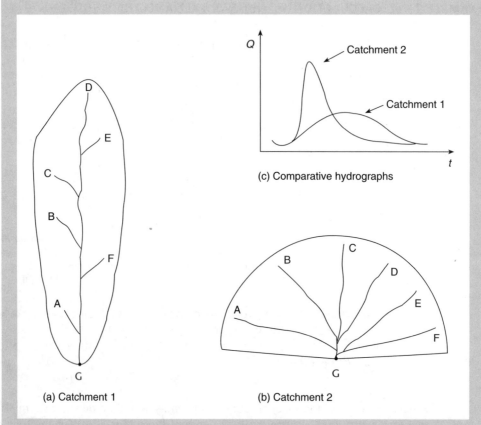

Figure 12.14 The influence of catchment shape on a river hydrograph. (a) With elongated catchment 1 the time of travel from D to G is much greater than A to G, so its hydrograph in diagram (c) is relatively low and elongated. (b) With semicircular catchment 2 the flow distances are much the same length, so runoff arrives simultaneously from all points giving a relatively high, narrow ('flashy') hydrograph. (c) Comparative hydrographs

Surface runoff is an important component of the hydrological cycle: Table 12.2 shows that world-wide 13%–62% of precipitation may become runoff. Figure 12.7 shows Britain's surface runoff (in mm), and these values can be compared to precipitation in Fig. 12.4.

A **hydrograph** is a record of either water level (stage) or discharge against time. An important part of engineering hydrology concerns being able to understand and analyse river flood hydrographs. This is necessary to estimate the magnitude and duration of a flood of specified return period (sections 13.2 and 13.3), and to design suitable flood alleviation works (section 13.4). Floods are mostly the result of surface runoff, so it is important to have an understanding of the factors that affect runoff. These fall into two categories: climatic factors (a to e) and catchment characteristics (f to j), which are summarised below.

(a) **Type of precipitation and intensity.** Snow is stored until it melts, when a sudden thaw can release large amounts of runoff very quicky. Very intense rainfall that significantly exceeds the infiltration capacity of the soil (i.e. $i \gg f_1$ in Fig. 12.11) favours surface runoff; low intensity rain ($i < f_1$) may result in little runoff.

(b) **Duration of precipitation.** Short periods of intense rainfall are capable of producing flash floods, even in summer. However, for a given rainfall intensity, as the catchment becomes increasingly saturated, the longer it rains the greater will be the runoff.

(c) **Areal extent of the storm.** Gentle rain falling uniformly over a large catchment may not exceed the soil's infiltration capacity, resulting in little or no runoff. However, the same volume of water falling as intense rainfall on a small part of the catchment may produce severe local flooding.

(d) **Orientation of the storm and catchment.** For example, consider a very intense rain storm with an area about one-third of the catchment in Fig. 12.14a. If the storm travels the full length of the catchment (i.e. from D to G), a greater volume of water will be deposited on it than if the storm travelled at the same speed across it (say from B to F). If the catchment's long axis has the same orientation as the prevailing winds, the first scenario will occur most frequently, so relatively high runoff may be experienced. Also, with a direction D to G the rain moves in the same direction as the runoff, resulting in a relatively narrow but high hydrograph; from G to D the directions are opposite so the hydrograph will be broader and not so high; B to F may be between the two.

(e) **Weather and antecedent catchment conditions.** High temperatures and long periods of sunshine increase evapotranspiration, making the catchment dry and absorbent (see soil moisture deficits and catchment response in section 13.1). As a result, runoff and river flow are less in summer than in winter. However, one torrential downpour can exceed the ability of the ground to absorb the rain and cause flash flooding, even in summer. Two or more rainfall events in quick succession may also cause severe flooding, especially if the catchment is already saturated and the river level still high when the second event occurs (Fig. 12.15). The diagram shows that the second, smaller rainfall peak produced the largest flood peak. Thus the antecedent catchment condition is also important (i.e. its condition prior to a rainfall event).

(f) **Land use.** Artificial, impermeable surfaces like roads, pavements and house roofs give rapid runoff, so built-up areas and urbanisation can increase the quantity of runoff and the speed with which it reaches a river channel. Similarly, land or field drains may reduce natural storage and result in quicker, increased surface runoff. Deforestation may reduce interception, ET and infiltration and increase runoff.

(g) **Type of soil or rock.** As described in section 12.5 the type of soil affects the infiltration rate, and hence surface runoff. Catchments covered with impermeable clay will have an annual hydrograph with a relatively large amount of surface runoff and a small groundwater baseflow. Typically flood hydrographs in such areas are narrow and high ('peaky' or 'flashy') as a result of very large inputs of intermittent surface runoff. Conversely, where the ground surface is highly permeable (e.g. sand, gravel, chalk, limestone) most rainfall may be absorbed so that the annual hydrograph has a relatively high groundwater baseflow, even in summer. Here, surface runoff results in relatively small fluctuations superimposed on the large baseflow. Some streams (frequently called bournes) are intermittent and only flow seasonally when the water table is high enough to provide the baseflow.

(h) **Catchment shape.** The shape of a catchment governs the rate at which water is supplied to the main channel and the time required to reach a REMEMBER!!!

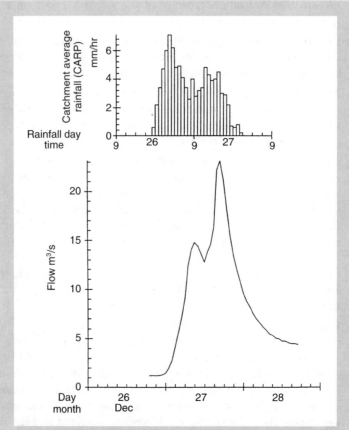

Figure 12.15 Hyetograph of gross rainfall and gauging station hydrograph for the River Warleggan at Trengoffe, Cornwall, 26–28 December 1979. The first period of rainfall has the largest total depth, but the second period occurring after 9.00 am on 27 December produced the highest river discharge since the catchment was already wet and the river in flood [*reproduced by permission of the Centre for Ecology and Hydrology*]

downstream point (i.e. G in Fig. 12.14). The **time of concentration (t_C)** is the time required for water falling on the most remote part of the catchment to reach G. Prior to this only part of the catchment's area is contributing to flow; afterwards the entire area is contributing. The way in which this affects surface runoff and hydrograph shape is as follows. With the elongated catchment in diagram (a), the distance runoff has to travel to reach G varies significantly from A to D. If all other factors are the same (e.g. channel slope and roughness), runoff will arrive at G at different times, in the order A, F, B, C, E, D. This results in a broad, low hydrograph and a relatively long time of concentration (diagram c). Conversely, the tributaries of the semicircular catchment in (b) have almost equal length, so runoff from points A to F will arrive at G almost simultaneously. This gives a relatively high, narrow, flashy hydrograph and a relatively short time of concentration.

(i) **Stream frequency.** If runoff has to travel large distances overland to reach a stream channel, there is every chance it will become obstructed by roots, trapped in puddles, etc. This enhances the opportunity for infiltration. However, if there are numerous or frequent stream channels only a short distance apart, this increases the likelihood of runoff reaching one of the channels quickly, so surface runoff is increased.

(j) **Catchment area.** Runoff can be assumed to increase with catchment area, which is why river discharge increases with distance downstream from the source. However, discharge is not directly proportional to area, since upland catchments are generally steeper, and there is more storage on large catchments than small ones. Many of the factors above also have an effect. Nevertheless, in the early stages of calculation and within a relatively uniform catchment, a rough estimate of the discharge at some point may be obtained using the area ratio. This should be checked subsequently by measurement or more accurate calculations, as necessary.

12.6.1 Measurement of surface runoff

A once-only measurement of the discharge of a stream or river can be obtained using a velocity meter (as described in section 5.7) or by measuring the dilution of a chemical (see BS 3680). However, when a continuous, accurate record of river flow is needed, this is best achieved using a well constructed and calibrated gauging station. Most consist of a weir (Fig. 9.4) and gauging hut containing recording apparatus. The theory of weirs was covered in section 9.5. At a suitable distance upstream of the weir (to avoid drawdown of the water surface), a pipe connects the river to a stilling well located in the gauging hut. The well may contain a float, pressure transducer or other device that is capable of recording the water level. In the past, either pens drawing on paper charts or punched paper tape were used to record the data, but now in the UK electronic data logging would be the norm. In remote sites the equipment can be powered by solar cells or batteries. Frequently the data are transmitted to base using a telemetry system so real time data can be obtained. Pre-programmed 'intelligent' stations are also capable of initiating an alarm when the water level rises above a threshold, or if there is a power failure.

Gauging stations only measure the stage (water level) of the river: the discharge must be obtained from the weir's rating curve, that is the curve that relates stage to discharge. To preserve accuracy, gauging stations must be carefully sited on a relatively straight reach of river, the stage must not be affected seasonally by reed or weed growth in the channel, and normally there must be no bypass flow around the station during flood. If the weir becomes submerged (section 9.5.1) then the discharge may have to be determined by a velocity meter gauging. To facilitate this, or to provide check readings at any stage, many gauging stations have a cableway across the river. The cable is connected to a winch in the gauging hut, so the meter can be positioned where required.

Two alternative but not particularly common types of gauging station are those which rely on the use of ultrasonic and electromagnetic principles. The ultrasonic method works by sending an ultrasonic pulse back and forth between two transducers located diagonally on opposite banks of the river. The pulse travels faster with the flow of water than when travelling against it, so by measuring the difference in the travel time the mean velocity of the river (V) at the level of the transducers can be determined. The discharge $Q = AV$, where A is measured using a water level recorder located in a specially lined, rectangular section of the channel. Advantages of this method are that it is capable of giving an accurate and continuous record of river flow; it can cope with flow reversals; it does not obstruct the river channel so there is no backwater and no barrier to boats; the cost of installation does not increase with the width of the river; and it can be used with channels up to 200 m wide, possibly wider. Disadvantages are that transducers at different depths are required to obtain an accurate mean velocity and to cope with changes in river stage; the pulses can be refracted or blocked by plants, silt, salty water or entrained air; and the method is not suitable for shallow rivers or those with unstable beds or low velocities.

The electromagnetic method works on the same principle as a bicycle dynamo: electrodes in the river bank detect the voltage generated by the water flowing through a vertical magnetic field created by an electromagnetic coil. The voltage can be used to determine the mean velocity, then $Q = AV$ exactly as above. Advantages are that it can be used where there

---

Box 12.3 ▶ ## The world's largest rivers

The rivers of Britain are tiny in world terms. In fact, excluding the former USSR, Europe's rivers are also relatively small. Table 12.9 shows the world's largest and some 'typical' British rivers. At its tidal limit the Amazon carries almost 20% of the earth's total runoff (Nace, 1969; Smith, 1972). The River Rye (Derwent/Ouse) is on the drier, eastern side of Britain. Despite its larger catchment area it has a smaller mean flow than the River Dovey, which is in wetter central Wales (Shaw, 1994). Similarly, the Tay in relatively wet eastern Scotland has a larger flow than the Thames of southern England (see Fig. 12.7).

Table 12.9 Approximate mean discharge of some rivers

River	Location	Mean discharge (m³/s)	Catchment area (km²)
Amazon	South America	778 710	5 775 790
Congo	Africa	39 640	2 977 210
Yangtze	China	21 800	1 941 660
Brahmaputra	East Pakistan	19 820	934 590
Ganges	India	18 690	1 058 850
Yenisei	USSR	17 390	2 588 880
Mississippi	USA/Canada	17 300	3 221 960
Tay	at Ballathie, Scotland	164.8	4 587
Thames	at Kingston, England	66.0	9 948
Dovey	Central Wales	21.7	471
Exe	at Thorverton, England	15.9	600
Rye	Yorkshire, England	9.4	679
Warleggan	at Trengoffe, England	0.83	25.30

Note: the discharge is that at the river's mouth, unless an alternative location is specified. Many UK rivers are affected by reservoirs, abstraction, groundwater recharge, etc.

is a large degree of weed growth, silt or entrained air; it can cope with flow reversals; and some types of installation do not obstruct the channel. Disadvantages are that the method is technically complex; the channel may have to be lined with a heavy insulating membrane; it requires an electricity supply; and background electrical 'noise' can be a problem (e.g. pylons). Additionally, if the coil is buried in the river bed then installation can be expensive when the river is >25 m wide (river diversion may be needed), but if the coil is bridged over the channel it forms an obstruction, and is only practical if the river is relatively narrow.

12.6.2 Hydrograph analysis: separating surface runoff and groundwater baseflow

World-wide, about 70% of riverflow is composed of surface runoff, the other 30% being groundwater discharge and interflow. Surface runoff is intermittent: it is greatest during or just after rainfall, but decreases to zero some time afterwards. Even so, most rivers continue to flow during dry weather because of the groundwater contribution (Fig. 12.16). Where the

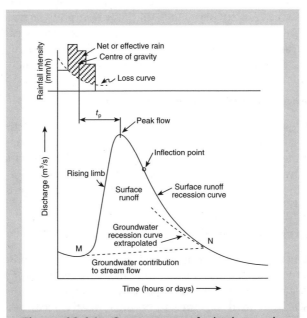

Figure 12.16 Components of a hydrograph. The surface runoff starts at M and ends at N. The master groundwater recession curve can be fitted over the surface water recession curve; N is located where the two converge (see also Fig. 12.17) [*after Wilson (1990); reproduced by permission of Macmillan, now Palgrave*]

geology does not allow a significant groundwater baseflow, rivers may dry-up during pro-longed periods of hot dry weather.

The separation of surface runoff from groundwater baseflow is a necessary step in under-standing and modelling a catchment's response to rainfall. For instance, a rainfall–runoff model such as the Unit Hydrograph (section 13.2) equates net rainfall to surface runoff, so the groundwater baseflow must be removed from the hydrograph during the modelling process. How this is done is described below and illustrated in Example 12.5.

The start of surface runoff can be assumed to occur where the hydrograph first starts rising (i.e. at M in Figs 12.16 and 12.18a). To locate the point N where surface runoff ends, one or more of the four methods listed below can be adopted. Three of these recog-nise that when the river flow is composed entirely of groundwater the recession curve is almost exponential, while the surface water recession curve is not. Thus by locating the change in gradient or curvature it is possible to find N. Although it is a simplification of reality, M and N can be joined by a straight line. Then, if required, the surface runoff at each time interval can be obtained by calculating the vertical difference between MN and the hydrograph. In Example 12.5 these values are used to calculate the total volume of surface runoff.

1 Log Q method

One of the properties of an exponential curve is that when plotted to a log scale it forms a straight line. Thus the data after the inflection point are plotted as a log Q against time graph. The change in slope indicates the end of surface runoff, i.e. N in Fig. 12.18b. To simplify any subsequent calculations, this change is assumed to occur at a time that corresponds to one of the plotted points (e.g. 7.5 days when time interval is 0.5 day).

2 Change in curvature method

Another way to detect the change of curvature is to plot the ratio of $Q_t/Q_{t+\Delta t}$ where Q_t is the discharge at time t and $Q_{t+\Delta t}$ is the discharge at the next time interval (e.g. $Q_{6.5}/Q_{7.0}$ if the time interval is 0.5 day, as in Example 12.5). The ratio is calculated for each point on the hydrograph after the point of inflection, then $Q_t/Q_{t+\Delta t}$ is plotted against t as in Fig. 12.18c. Again, assume the change in curvature to occur at a time corresponding to one of the original data points.

3 Master recession curve

The log Q method above was applied to only one flood hydrograph, but the annual record from a river gauging station will contain many hydrographs that could be analysed. This would yield several straight lower portions that represent individual groundwater recessions. These can be fitted together to form a master recession curve that shows how the groundwater discharge declines over a much larger range of flow.

The master recession curve is produced by first using log Q and t as the axes of a graph, over which the individual log Q flood hydrographs (drawn on transparent film using the same time scale) can be positioned, starting with the lowest Q first (Fig. 12.17a). The second lowest log Q graph is positioned where its straight portion meets the extrapolation of the first. Remember, the individual curves cannot move vertically because they are tied to the log Q scale, but they can be moved anywhere horizontally. This is repeated until the highest and last individual hydrograph has been positioned. A line can now be drawn through all

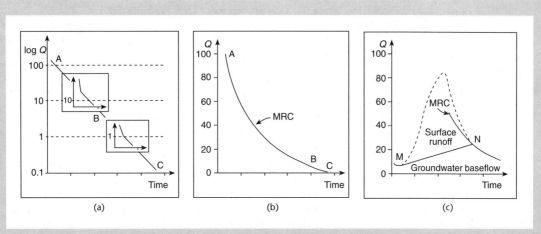

Figure 12.17 (a) The derivation of the groundwater master recession curve starts by plotting individual hydrographs (as log Q against time) on transparent film. These can be placed on the base graph so that the lower sections form a straight line ABC. For clarity only two are shown. (b) When plotted to a natural Q against time scale the master recession curve ABC is obtained. (c) To identify point N, the master recession curve is placed over the gauged hydrograph. Point N is located where the two lines converge and become colinear

of the straight portions, and then redrawn to a natural Q scale (diagram b). This is the master recession curve (MRC). It is used to separate surface and groundwater by positioning the master curve (on transparent film) over the recession limb of a hydrograph. Once again, the master curve is tied to the Q scale, but can be moved horizontally until there is a match between the lower parts of the hydrographs, as in Fig. 12.17c. Where the curves are superimposed, the river flow is entirely groundwater; where they diverge is where the surface runoff ends. Note that each catchment has its own characteristics, so the master recession curve is only valid for one location, and cannot be applied elsewhere.

4 Catchment area (A km²)

Table 12.10 gives an extremely approximate illustration of the time lag between the peak of the hydrograph and N. It shows clearly that on large catchments there is a period of many days between the rainfall, the flood peak and the end of the surface runoff.

Table 12.10 Location of N according to catchment area [*after Wilson, 1990*]

Catchment area, A (km²)	250	1250	5000	12500	25000
Time from peak to N (days)	2	3	4	5	6

Table 12.11 Total hydrograph and calculated values for Example 12.5

Time t days	Q m³/s	log₁₀ Q	$Q_t/Q_{t+\Delta t}$	Baseflow m³/s	Surface runoff q m³/s
3.0	21.9				
3.5	14.2			14.2	0 (M)
4.0	20.3			15.8	4.5
4.5	39.6			17.4	22.2
5.0	80.1			19.0	61.1
5.5	111.1		(1.241*)	20.6	90.5
6.0	89.5		(1.487*)	22.2	67.3
6.5	60.2	1.780	1.580	23.8	36.4
7.0	38.1	1.581	1.411	25.4	12.7
7.5	27.0	1.431	1.268	27.0	0 (N)
8.0	21.3	1.328	1.246	21.3	0
8.5	17.1	1.233	1.230	17.1	0
9.0	13.9	1.143	—	13.9	0

* The increasing values in brackets are not needed; they are included only to show what happens if the data before the point of inflection are analysed.

EXAMPLE 12.5

The first two columns of Table 12.11 show the discharge (Q) in a river at 0.5 day intervals (Δt), starting on 3 March 1998. (a) Use these data to locate the start and end (i.e. M and N) of the surface runoff. (b) Separate the surface runoff from the groundwater baseflow. (c) Calculate the total volume of surface runoff during the flood. (d) If the catchment area is 296 km², calculate the total depth of net rainfall, and comment upon the accuracy of the value obtained.

(a) The first two columns of the table show the recorded flood hydrograph, which is plotted as Fig. 12.18a. It is apparent that M, where surface runoff starts, is at 3.5 days, since this is where the hydrograph starts to rise. Both the log Q and change in curvature method will be used, and columns 3 and 4 show the values needed to plot Fig. 12.18b and c. From these two diagrams, the change in gradient which denotes the end of surface runoff, N, is at 7.5 days. Note that for a catchment of 296 km², Table 12.10 suggests a time from the peak of the hydrograph to N of around 2 days, as in Fig. 12.18a.

(b) In Fig. 12.18a, the surface runoff is separated from the groundwater baseflow by the line MN.

(c) The baseflow is assumed to increase linearly from 14.2 m³/s at 3.5 days to 27.0 m³/s at 7.5 days, which is a change of 12.8 m³/s during 8 half-day time increments. Thus the change every 0.5 day is 12.8/8 = 1.6 m³/s. This gives the baseflow at 4.0 days as 14.2 + 1.6 = 15.8 m³/s, 4.5 days as 15.8 + 1.6 = 17.4 m³/s, etc. The other values are shown in column 5 of the table. By subtracting these from the recorded discharge (Q, in column 2) we obtain the surface runoff, q, in the last column. The q values could be used to re-draw the hydrograph with MN as the horizontal axis, but this is not necessary here.

   The area under the hydrograph and above MN represents the volume of surface runoff (i.e. with time in seconds on both axes of the graph, area = s × m³/s = m³). Assuming the

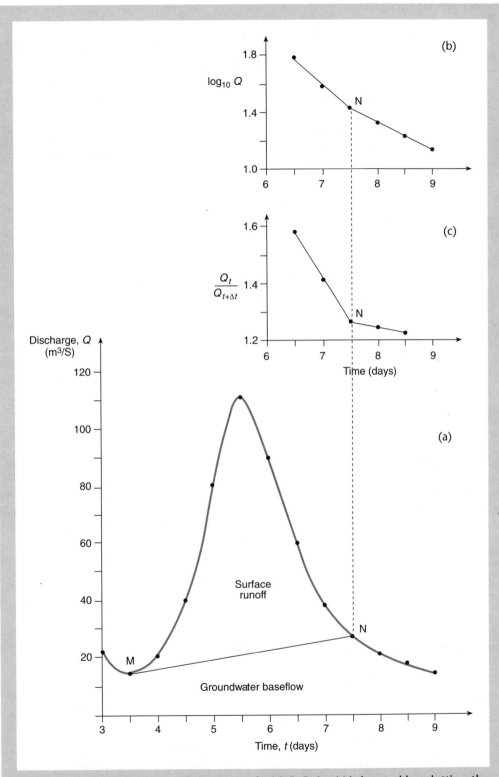

Figure 12.18 (a) Hydrograph for Example 12.5. Point N is located by plotting the data after the point of inflection against time using either (b) $\log_{10} Q$ or the ratio $Q_t / Q_{t+\Delta t}$ as the vertical scale

points of the hydrograph to be joined by straight lines, a general equation for calculating the area (A) above a line consisting of n points spaced at an interval, L, is:

$$A = L\left(\frac{q_1 + q_n}{2} + q_2 + q_3 + q_4 + \ldots q_{n-1}\right)$$

(12.9)

The same result can be obtained by calculating the areas of the triangles at each end of the hydrograph and the trapeziums in-between; this is just a short-hand method of summarising the calculations. Below, A becomes the volume (VOL m³), $L = \Delta t = 0.5 \times 24 \times 60 \times 60 = 43.2 \times 10^3$ s, and q the ordinates of surface runoff (m³/s), so:

$$VOL = 43.2 \times 10^3\left(\frac{0+0}{2} + 4.5 + 22.2 + 61.1 + 90.5 + 67.3 + 36.4 + 12.7\right)$$

$$VOL = 12.731 \times 10^6 \, m^3$$

(d) If the total depth of net rainfall is d (m), then assuming that this is spread evenly over the catchment area of 296 km² or 296×10^6 m²:

$$d \times 296 \times 10^6 = 12.731 \times 10^6$$

$$d = 0.043 \, m$$

Thus the catchment runoff (CR) is equivalent to 43 mm of rainfall (see Table 13.1). This value may not be very accurate owing to errors in measuring Q and in correctly locating the position of N, the assumption that MN is linear, and the approximation involved in calculating VOL.

SELF TEST QUESTION 12.3

The figures below show the discharge (Q) recorded at a gauging station on 26 January between 0600h and 2100h GMT. (a) Separate surface runoff and groundwater baseflow using the log Q and $Q_t/Q_{t+\Delta t}$ methods. (b) Calculate the volume of surface runoff. (c) Comment upon any inconsistency or apparent inaccuracy in the analysis.

Time GMT (h)	Q (m³/s)	Time GMT (h)	Q (m³/s)
0600	4.45	1400	13.52
0700	4.36	1500	10.91
0800	4.97	1600	8.95
0900	5.99	1700	8.16
1000	8.06	1800	7.63
1100	13.68	1900	7.18
1200	16.25	2000	6.84
1300	15.41	2100	6.60

Summary

1. Many hydrological and hydraulic problems can be solved using the basic equation (12.1): inflow = outflow ± change in storage. Substituting the components of the hydrological cycle gives equation (12.2) and, assuming there is no change in storage, the very simple long-term equation (12.3): precipitation = surface runoff + actual evapotranspiration.

2. Precipitation occurs as water vapour rises, cools and condenses to form droplets and then falls to Earth. Causes of precipitation include weather fronts, hills (orographic precipitation) and convection cells. Measurement of both rainfall depth (mm) and intensity (mm/h) is important; intensity decreases as either the duration or rainfall area increases.

3. Evapotranspiration (ET) is the combined rate of water loss from plants and the ground; actual evapotranspiration (AET) is the true rate of loss, potential evapotranspiration (PET) is the maximum rate assuming a plentiful supply of water. ET is high on hot days with lots of sunshine, wind and a low humidity. Consequently ET, SMD and groundwater recharge exhibit a seasonal variation (Fig. 13.1).

4. Infiltration occurs when water penetrates the ground surface, and percolation as it subsequently drains down to the water table. Infiltration rates are high when there is a permeable, porous soil that is initially dry; the rainfall is of moderate intensity and prolonged; the ground surface is flat and the soil broken by plant roots.

5. Net (or effective) rainfall is the part of gross rainfall that becomes surface runoff (the remainder is 'lost' as infiltration, ET or makes up any soil moisture deficit). Surface runoff is the movement of water over the Earth's surface. Runoff is large and flooding most likely when there is prolonged and/or intense rainfall and/or snowmelt; the storm covers a large area and travels in the same direction as the river; the catchment is already wet and/or is relatively impermeable; the catchment is 'semi-circular' and has many streams that collect the runoff efficiently (Fig. 12.14). Surface runoff can be separated from groundwater baseflow using the log Q and change in curvature methods, the master recession curve or catchment area approximation.

Revision questions

12.1 With respect to Britain, answer the following. (a) Why is there a large variation in rainfall from west to east? (b) Why are some areas more likely than others to receive high intensity rainfall? (c) Which are the areas where agriculture is most likely to require irrigation to maximise crop yields? (d) Although the effect of global warming is still not apparent, how might it alter rainfall depth, intensity and irrigation requirements in the future? Is this something that hydrologists and engineers should be aware of?

12.2 In recent years the rainfall recorded at gauge A is thought to be of doubtful accuracy due to an error in the way the catch was measured. Its readings are shown below, along with the average from six near-by gauges (6ga). By drawing a double mass curve, determine whether or not a change occurred at gauge A at some time in the past. If there was a change, correct the most recent data to make them consistent with the earlier part of the record.

Year	Gauge A (mm)	6ga (mm)
1986	1652	1196
1987	1840	1387
1988	1726	1343
1989	1741	1299
1990	1853	1401
1991	1674	1275
1992	1781	1392
1993	1665	1214
1994	1750	950
1995	1795	1020
1996	1882	1136
1997	1843	1089
1998	1756	990
1999	1964	1236

[Change at end of 1993; corrected readings 1994–99: 1337, 1371, 1438, 1408, 1342, 1500 mm]

12.3 A soil consists mostly of silt with some sand and clay. Its infiltration capacity has been determined using an infiltrometer with a 0.350 m diameter inside ring. The total volume of water added to the inside ring since the start of the test is shown below. Calculate the average infiltration rate during each of the time intervals. *Hint*: the total volume added since the start of the test is given, so calculate the amount added during each of the time intervals then proceed as in Example 12.3.

Time since start (min)	Total vol added (10^{-3} m³)
0	0
5	0.865
10	1.722
15	2.459
20	3.004
30	3.679
40	4.016
50	4.193
60	4.354

[108, 107, 92, 68, 42, 21, 11, 10 mm/h]

12.4 The data below show the discharge (Q m³/s) recorded at a gauging station between 26 and 28 December. It is similar to that shown in Fig. 12.15. The catchment area is 25.3 km². (a) Separate the surface runoff and groundwater components of the hydrograph. (b) Calculate the volume of surface runoff. (c) Calculate the total average depth of net rainfall on the catchment. If the answer obtained is significantly different to the rainfall shown in Fig. 12.15, explain why.

Date	Time (h)	Q m³/s
26 December	1800	1.2
	2100	1.1
27 December	0000	1.5
	0300	5.0
	0600	9.7
	0900	14.5
	1200	12.8
	1500	16.3
	1800	23.1
	2100	13.3
28 December	0000	9.5
	0300	7.3
	0600	5.9
	0900	5.2
	1200	4.7
	1500	4.5
	1800	4.4

[(a) N at 0600 on 28 December; (b) 844 560 m³; (c) about 33 mm net rain. Fig. 12.15 shows gross rain, which will be larger]

13

Applications of engineering hydrology

Chapter 12 presented the basic knowledge and principles required to understand the applications outlined in this chapter. Here we consider how to evaluate and solve some of the problems that are commonly encountered in engineering hydrology. This includes a simple, introductory guide to the Flood Estimation Handbook, which can be used to obtain the magnitude and frequency of occurrence of flood events. Given the possibility that global warming may result in damaging floods occurring more often, this is an important and very relevant topic. According to the Centre for Ecology and Hydrology, as a result of climate change the cost of providing the present levels of flood protection may increase fourfold to £1.2 billion over the next 50 years. The measures that can be adopted to try to control or alleviate flooding are also discussed. In urban areas, flooding is prevented by surface water sewers, and a simple design method is introduced.

This chapter also shows how both surface and groundwater resources can be evaluated and quantified with the objective of obtaining a reliable water supply. As the world's population increases, standards of living improve and people become more profligate with water, it will become more difficult to satisfy the demand in future. If climate change occurs it is estimated that the UK may have to invest £5 billion in new water resources over the next 30 years. The questions addressed include:

How can we predict the magnitude of a flood on a catchment?

How can we determine how often flooding will occur?

What can we do to alleviate flooding?

How can we design surface water sewers?

How can we evaluate the potential of a river as a source of water?

How can the storage capacity of a reservoir be determined?

When and how does groundwater recharge occur?

How can we evaluate the potential of groundwater as a water supply?

13.1 ▶ Predicting a catchment's response to rainfall

An important part of engineering hydrology involves predicting how a catchment will respond to rainfall: will there be large amounts of surface runoff and flooding? Or will most of the rain soak harmlessly into the ground? Flood prediction is a skilled and necessary activity (see Cluckie and Han, 2000). Many riverside or low-lying areas flood regularly, and people need sufficient warning to be able to sandbag doors, to move possessions upstairs, or in some cases to evacuate the area. On the other hand, where water supplies are largely derived from groundwater, understanding when and how much recharge will occur is essential, because this may determine the abstraction rate from a well or the yield of an aquifer.

Generally both runoff and groundwater recharge (infiltration) are greatest in winter when ET is low, net rainfall is relatively high and the catchment is wet (Tables 13.1 and 13.2). Conversely, net rainfall, runoff and infiltration are usually small in summer when ET is high and the catchment is dry. Although individual years may vary greatly, this rhythm of the seasons is usually apparent (Fig. 13.1). However, to make accurate predictions a much greater knowledge of the actual soil moisture deficit and water balance at a particular location and time is needed.

Table 13.1 Two examples of the seasonal variation of daily mean flow (DMF m³/s) and catchment runoff (CR mm) [*Data courtesy of the Centre for Ecology and Hydrology*]

	Jan	Feb	Mar	Apr	May	Jun	Jul	Aug	Sep	Oct	Nov	Dec
River Tamar at Gunnislake, 1957–85: Catchment area 916.9 km², average annual rainfall 1227 mm												
DMF (m³/s)	46.0	36.9	25.8	16.3	11.2	6.8	6.0	8.3	11.3	22.0	35.0	45.6
CR (mm)	135	97	75	46	33	19	18	24	32	64	99	133
River Warleggan at Trengoffe, 1970–87: Catchment area 25.3 km², average annual rainfall 1512 mm												
DMF (m³/s)	1.46	1.37	1.02	0.73	0.53	0.43	0.34	0.39	0.46	0.70	1.04	1.41
CR (mm)	154	131	108	75	56	44	36	41	47	74	107	149

Table 13.2 Water balance for Harrogate, Yorkshire (all values in mm) [after Smith, 1972]

		Jan	Feb	Mar	Apr	May	Jun	Jul	Aug	Sep	Oct	Nov	Dec	Year
1	Precipitation	80	62	49	53	62	51	72	74	63	75	79	72	792
2	PET	4	8	29	46	72	86	84	70	41	18	4	3	465
3	Storage change	0	0	0	0	–10	–35	–12	4	22	57	0	0	
4	SMD	0	0	0	0	10	45	57	53	31	0	0	0	
5	Storage balance	75	75	75	75	65	30	18	22	44	75	75	75	
6	Water deficiency	0	0	0	0	0	0	0	0	0	0	0	0	
7	Water surplus	76	54	20	7	0	0	0	0	0	26	75	69	327
8	Total water surplus		157									170		327

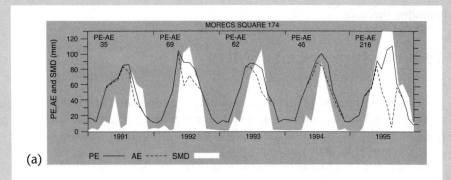

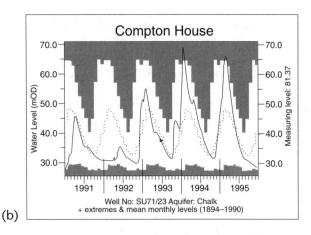

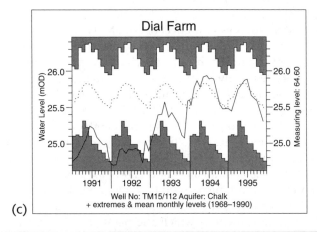

Figure 13.1 The rhythm of the seasons. (a) The variation in potential evaporation (PE), actual evaporation (AE) and soil moisture deficit (SMD) at MORECS square 174 in Kent. Note that AE < PE during part of the summer. (b) and (c) Groundwater hydrographs in the chalk aquifer at Compton House and Dial Farm (see Fig. 13.20 for locations). Note that one displays fluctuations of over 35 m and the other only 1.5 m. The annual variation depends upon factors such as the porosity of the aquifer and the location of the well (e.g. valley bottom near a river, or on top of a hill). The shaded top and bottom of the diagram indicate the historical monthly maxima and minima, the dotted line the monthy mean and the solid line the recorded value [*reproduced by permission of the Centre for Ecology and Hydrology*]

13.1.1 Soil moisture deficit (SMD)

As described in section 12.5, the rate (i mm/h) at which rain falls relative to the infiltration capacity of the soil (f mm/h) is important in determining whether runoff occurs, or whether the ground absorbs much of the rainfall. Another important factor is how dry the soil is, since this affects its ability to soak up water.

Beneath the water table the soil is totally saturated, so any well that penetrates below the water table fills with water. Between the ground surface and the water table there is normally a zone of aeration where the soil contains moisture, but is not saturated. However, if there is a prolonged period of heavy rainfall, a layer of soil at the ground surface may become saturated, which means that it can hold no more water because all of the void spaces between the soil particles are already full. When the rain stops, some of this water is retained as soil moisture in the zone of aeration, while the remainder drains under gravity to the water table. This wet but drained state is sometimes called the soil's **field capacity**. The rainfall that reaches the water table is 'groundwater recharge'. The annual cycle of recharge often results in a large increase in water level (Fig. 13.1b and Example 13.11).

If a soil is at field capacity (or saturated) then plants have an unlimited water supply, so the water loss from the ground surface will be high with evapotranspiration (ET) occurring at the potential rate (PET). Without rain, the soil dries (from the surface downwards) as water is removed progressively by plants and the drying action of the sun and wind. The difference between the soil's field capacity and its actual water content at any time is called the **soil moisture deficit** (SMD), which is usually recorded in mm of water, as explained below. As the SMD gets larger, the plants' water supply becomes increasingly restricted, causing ET to fall to the lower actual rate (AET). ⟨REMEMBER!!!⟩

Plants with short roots feel the effect of any SMD first, since the soil dries from the surface downwards, while trees and other deep-rooted plants continue to survive happily. This gives rise to the concept of a **root constant**, which is basically the amount of soil moisture (mm) that can be extracted easily from a soil by a certain type of plant. This constant is also defined as the maximum SMD that can be accumulated without reducing ET to something less than the potential rate (Ward, 1975). For example, the root constant for grass is 75 mm and woodland 200 mm (note that this notional depth of water is contained in several metres' depth of soil according to its porosity; trees have roots much longer than 200 mm). It may be assumed that PET occurs up to a SMD = root constant + 25 mm, after which ET switches to the actual rate. The additional 25 mm allows for a plant sucking up a little extra water from below root level. As time passes it becomes difficult to extract any water and the plant wilts. The **wilting point** is the plant's maximum survivable SMD or, in other words, the lowest soil moisture content at which the plant can extract water from the soil. At this point, if it rains or if the plant is watered artificially it will recover, but without water it reaches the permanent wilting point and dies.

The terms 'root constant' and 'wilting point' give the impression of accuracy, but in fact this is not the case. A particular area of ground contains many different types of plant so there is a range of root lengths, root constants and wilting points, while the amount of water available depends upon soil depth, soil type and how far it is to the nearest river. Frequently 'average' conditions are assumed where 50% of the area is grass, 30% is long-rooted vegetation and 20% is adjacent to a river where water is plentiful. However, such complexities will be left for more detailed texts, and only a simple example of a water balance will be presented below. More sophisticated models are available (e.g. see Shaw, 1994).

13.1.2 The water balance

A water balance is a method of calculating the 'wetness' or 'dryness' of a place (Thornthwaite, 1948; Smith, 1972). A modified procedure more suited to British conditions was adopted by Penman (1949). This provided a valuable insight to the rhythm of the seasons, when irrigation is required, the periods when flooding is most likely because there is no SMD, when a water surplus occurs, and when groundwater and surface reservoirs will refill. Smith used Harrogate, West Yorkshire, as an example because its water balance is similar to the mean conditions for England and Wales. Example 13.1 shows how the water balance and SMD are determined, and provides a commentary on the calculations.

In Table 13.2, note that precipitation is fairly evenly distributed throughout the year, whereas PET is low during the winter and autumn months but high in summer. This causes a SMD in spring and summer when rainfall *might* be absorbed, resulting in limited surface runoff and a reduced chance of flooding. However, many severe flash floods have occurred in summer during very intense rainfall, so this is not guaranteed. The water surplus between October and April indicates that this is when flooding and groundwater recharge would be most likely to occur.

It is important to realise that this is just one example of what may happen at one location in one particular year. At the same location during a particularly dry year there may be a SMD throughout the winter. In drier locations there may be much larger SMDs so that farmers have to irrigate crops for many months, whereas in wetter areas (e.g. Scottish mountains) there may be a water surplus throughout the year. Nevertheless, the SMD and other data in Table 13.2 are clearly useful, although not very accurate: variations in soil and plant type have already been mentioned; how and when the ET rate switches from the potential to actual rate is complex; the balance is simplistic because ET occurs over a whole month but all the precipitation may occur in one day; the measurement of ET and soil moisture is not easy; and, in practice, information is required not at the end of the month, but on a daily or weekly basis.

In the UK, the Meterological Office Rainfall and Evaporation Calculation System (MORECS) estimates at weekly and monthly intervals the ET and SMD occurring in a grid of 40 km × 40 km squares covering the whole country (Figs 12.5 and 12.6). MORECS tries to take into account many of the factors discussed above. The data are widely used by flood forecasters, and by farmers who wish to assess irrigation requirements. The collection of riverflow and groundwater level data in the UK falls largely under the aegis of the Environment Agency.

EXAMPLE 13.1

The first two rows of Table 13.2 show the monthly precipitation and PET at Harrogate. Assuming that the soil is at field capacity to begin with and that 75 mm of soil moisture is available to plants, calculate the SMD, the monthly change in soil moisture, and any water deficiency or water surplus.

The calculations are conducted on a monthly basis, starting with January. The numbers in the various rows are obtained as follows:

Row 3: the storage change is the reduction in soil moisture storage. It is the numerical difference between precipitation and PET. Initially (and in November and December) these values are positive, but an increase in storage is not possible because the soil is already at field capacity and cannot hold any more water. Thus a 0 is recorded in row 3, with the positive differences between precipitation and PET being recorded as a water surplus in row 7.

Row 4: with the soil at field capacity to begin with, initially SMD = 0. From May through to October the SMD represents the accumulated storage change. From May to July it increases as PET exceeds precipitation, then it declines again as PET diminishes. In October precipitation exceeds PET by 57 mm but only 31 mm are required to return the soil to field capacity (i.e. SMD = 0), so the remaining 57 – 31 = 26 mm are shown as a water surplus in row 7. Since the SMD never exceeds the root constant (75 mm) ET will always occur at the potential rate in this example.

Row 5: the soil remains at field capacity until the end of April, but in May increased PET means that moisture reserves stored in the soil have to be utilised. The storage balance is the initial soil moisture content (75 mm) minus the SMD. It shows the amount of water left in the soil. If negative values were obtained, these would be recorded as a water deficiency in row 6. There is no such deficiency in this example.

Row 6: any water deficiency would indicate that plants could not easily draw upon moisture in the soil, so this amount of water would have to be added to maintain ET at the potential rate and subsequently to prevent permanent wilting.

Rows 7/8: precipitation exceeds PET in spring by a total of 157 mm and in autumn by 170 mm so this is the period when reservoirs will fill, groundwater will be recharged and (since the SMD = 0) runoff and floods are most likely. The water surplus begins in October, which is often taken as the start of the 'water year'. Over the 12 months, the water surplus represents 41% of the precipitation, and PET 59%.

SELF TEST QUESTION 13.1

Below are monthly precipitation and PET data for a location in Cornwall. In January the soil is at field capacity. An arable crop is being grown that has a root constant of 75 mm, but assume that the plants can easily extract another 25 mm of water so that PET is maintained up to a SMD of 100 mm. After this all soil moisture is exhausted and the plants can only use the water that falls as rain. Water deficiency and wilting are considered to occur when the SMD > 100 mm. Use a water balance to determine whether or not a water deficiency does occur and, if so, in which months irrigation would be required to prevent the crop from wilting.

	Jan	Feb	Mar	Apr	May	Jun	Jul	Aug	Sep	Oct	Nov	Dec
Precipitation (mm)	201	246	128	113	53	26	28	86	173	127	109	255
PET (mm)	19	19	36	59	72	83	68	67	48	33	17	16

13.2 The unit hydrograph rainfall–runoff model

In simple terms the unit hydrograph (UH) is a rainfall–runoff model. After allowing for any SMD and other losses, it relates the depth of net rain falling on a catchment to the quantity of surface runoff. This general idea was encountered in Example 12.5 where the product of net rainfall depth (m) and catchment area (m²) was equated to the volume of surface runoff (m³) represented by the area under the hydrograph and above the line MN. The UH extends this basic principle, so that at some point on a particular catchment an observed rainfall event and associated flood hydrograph can be used to obtain the hydrograph that

Table 13.3 An indication of dam design floods [*from Institution of Civil Engineers, Floods and Reservoir Safety, 3rd edn 1996; reproduced by permission of ICE*]

Note: the Reservoirs Act 1975 requires the owners of reservoirs with a capacity of more than 25 000 m³ of water above the level of adjacent ground to provide for their inspection in the interests of public safety.

	Dam category	General standard	Minimum standard if overtopping is tolerable
A	Failure endangers 10 or more lives in a settled community	Probable Maximum Flood (PMF)	1 in 10 000 year flood
B	Failure endangers lives not in a community, or causes extensive damage	1 in 10 000 year flood	1 in 1000 year flood
C	Negligible risk to life and limited flood damage	1 in 1000 year flood	1 in 150 year flood
D	No foreseeable loss of life and very limited flood damage	1 in 150 year flood	Not applicable

would arise from net rainfall of a different intensity and/or duration. Since rainfall records are generally longer than those from gauging stations, it is easier and more reliable to calculate the magnitude of (say) a 1 in 200 year rainfall event and then use the UH to obtain the corresponding flood hydrograph than to conduct a frequency analysis on a short river flow record. The UH is the only method that can be used when the shape of the hydrograph and volume of flood water are required, such as when designing a flood storage reservoir (see section 13.4).

One use of the unit hydrograph is to determine the probable maximum flood (PMF). The PMF is the worst flood imaginable, and is more extreme than a 1 in 10 000 year flood. Often the PMF is required for the design of reservoir spillways in locations where failure would cause the death of 10 or more people in a community (see section 9.1.2, Box 9.1 and Table 13.3). As described in section 13.3.2, the alternative method of calculating the design flood by extrapolating recorded river flow data is usually only reliable up to about a 1 in 250 year event.

The PMF is obtained by first estimating the probable maximum precipitation (PMP), which is the highest rainfall intensity theoretically possible for a given duration, catchment size, location and time of year. This analysis is usually undertaken by a trained meteorologist. The basic idea is that there is an upper limit to the amount of water in the atmosphere available to be precipitated, so by estimating this the PMP can be obtained. Having obtained the PMP, it can be used in a unit hydrograph model to estimate the PMF. For the PMF to be the worst-case scenario, it may be necessary to assume rainfall of summer thunderstorm intensity falling onto a frozen catchment (giving 100% runoff) with snowmelt as well. Sometimes the PMF is so large that for economic or other reasons it is justifiable to use either a proportion of the PMF or a flood of specified return period.

13.2.1 The unit hydrograph: definition, principles and limiting assumptions

The unit hydrograph (UH) was originally based on one inch or one centimetre of rainfall, which is how it got its name. However, rainfall is now usually quoted in mm, so the unit hydrograph is defined as: **the hydrograph of direct surface runoff resulting from 10 mm *depth* of constant intensity net rain falling uniformly over the entire catchment during a specified period of time.**

Note that the 10 mm is the depth of net rainfall, ***not*** the rainfall intensity (see Example 12.4). Also, when we refer to (say) a 2 hour unit hydrograph, the time quoted is the specified period of time (t) during which the net rainfall occurs, ***not*** the duration of the runoff (T) which may last for many days. Assumptions which may limit the accuracy or application of the UH are as follows:

(i) All of the net rainfall must appear as runoff in the UH. Remember: net rainfall (m) × catchment area (m²) = the volume (m³) represented by the area of the surface runoff hydrograph, as in Example 12.5. This can be used to check the validity and accuracy of the estimated net rain and derived UH.

(ii) It is assumed that the rainfall has a constant intensity (e.g. i mm/h) which does not vary during the time t. Variations in rainfall intensity are dealt with later.

(iii) The rain must fall uniformly over the entire catchment (see section 12.3.3). This restricts the UH to catchments of about 500 km² or less. Larger areas have to be split into sub-catchments.

(iv) The UH represents the combined physical characteristics of the catchment, so it is site-specific and cannot be used elsewhere. It may also change with time, as or when the catchment changes.

The way in which the UH rainfall–runoff model is used is generally:

observed event → derivation of UH → calculation of design flood hydrograph

Thus the UH is a stepping-stone that allows an observed rainfall and flood event, perhaps of relatively small return interval, to be used to calculate the hydrograph of a more extreme design flood. It is not possible to go straight from the observed to the design event, since the latter may comprise rainfall of several different intensities and durations, as illustrated in the examples later. However, accuracy is improved if the durations (t) of the observed and design rainfalls are reasonably similar.

13.2.2 Derivation of the unit hydrograph

For a particular gauging station, the UH for the catchment is derived from an observed flood hydrograph by following the procedure shown diagrammatically in Fig. 13.2 and described below.

1. Identify an isolated storm of short uniform intensity net rainfall which covers the whole catchment and which produces a typical hydrograph (diagram a).

2. Separate the surface runoff from the groundwater baseflow; obtain the ordinates of surface runoff (as in Example 12.5); replot the surface runoff hydrograph with 0 on the time scale coinciding with the start of the runoff; and plot the hyetograph (rainfall intensity–duration graph) above the hydrograph as in diagram b.

3. Calculate the volume of surface runoff and check that this equals the product of the total depth of net rainfall multiplied by the catchment area.

4. Use the principle of proportionality in Box 13.1 to adjust the rainfall intensity and surface runoff hydrograph ordinates (q) so that they both represent 10 mm depth of net rain on

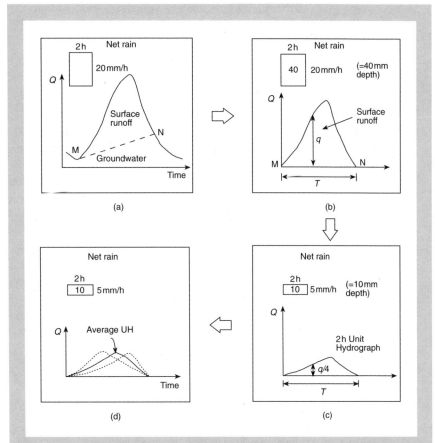

Figure 13.2 Derivation of the Unit Hydrograph (UH). (a) Calculated net rainfall and total hydrograph. (b) Net rainfall and surface runoff hydrograph. (c) Derived individual 2h UH, which is the runoff from 10mm depth of rainfall over the whole catchment in 2h. (d) Two 2h UH (dashed) with the average UH (solid line) determined from the average peak value and average time to peak

the catchment, as required by the definition of the UH. In this example, the observed net rainfall is 20mm/h for 2h so this represents a total depth of 40mm on the catchment. This is four times that required for a UH, so divide both the observed rainfall intensity and surface runoff by 4 to obtain the UH. Thus the intensity becomes 5mm/h for 2h (= 10mm) and the surface runoff is $q/4$ (diagram c).

5. Repeat the above steps for rainfalls of similar duration (t), and plot all the resulting UHs (diagram d). Use the graph to determine the average time to reach a peak, and the average peak value. Superimpose the 'average' hydrograph on the graph. Check that the relationship in step 3 holds true for the UH; if the volumes are not equal, adjust the UH until they are.

6. If the record does not contain enough storms of similar duration for step 5 to be undertaken satisfactorily, derive the UH for constant intensity rainfalls of varying length and then adjust the UHs to a common value (see below).

Box 13.1 ▸ **The basic principles**

REMEMBER!!!

In addition to those mentioned in the main text, the UH rainfall–runoff model is based on these two key principles that are used in most calculations, including the examples below:

1. **The principle of proportionality.** This states that the ordinates of the surface runoff hydrograph are directly proportional to the rainfall intensity. For example, if rain falling at i mm/h results in a hydrograph with a peak value of q m³/s, then rain of the same duration but falling at $4i$ mm/h would result in surface runoff with a peak of $4q$ m³/s. However, because we are simply multiplying the ordinates of the first hydrograph by 4, its duration (T) remains the same despite the much higher peak. This is unlikely: the surface runoff would be expected to last longer, so this is a possible source of error.

2. **The principle of superposition.** This means that hydrographs resulting from contiguous (i.e. touching) and/or isolated periods of uniform intensity net rainfall can be added together to obtain the combined effect. This allows changes in rainfall intensity during a storm to be dealt with. For example, if there are three periods of rainfall with different intensities, the three hydrographs arising from each are calculated separately and then all of the values added together to obtain the combined surface runoff.

Increasing the duration of a UH by multiples of t

This technique is used to increase the duration of the UH by a multiple of the rainfall duration (t). For instance, if we start with a 2h UH then this procedure enables the 4h, 6h, 8h, etc. UH to be obtained. As an illustration, the procedure for increasing the duration of the UH from 2 to 4h is summarised in Fig. 13.3 and described below.

1. Start with the 2h UH, which by definition represents 10mm depth of net rainfall, i.e. 2h at 5mm/h (diagram a).

2. To represent 4 hours of rainfall at 5mm/h, add a second contiguous hyetograph and hydrograph, as shown in diagram b. These are identical to those in diagram a, but offset by t = 2h to the right, since this is when the second period of rainfall starts relative to the first. Using the principle of superposition, the two hydrographs can be added to obtain the total runoff from 4h × 5mm/h = 20mm of net rain.

3. The 4h UH must, by definition, represent the surface runoff from 10mm of net rain, so (using the principle of proportionality) divide both the rainfall intensity and the ordinates of the hydrograph by 2. The result is the 4h UH, i.e. 4h × 2.5mm/h = 10mm (diagram c).

In practice, it is not necessary to draw the diagrams in Fig. 13.3: the calculations are normally conducted using a table, as in Example 13.2. In that table the 2h offset mentioned in step 1 above is represented by 'offset' in the square at the start of row 4.

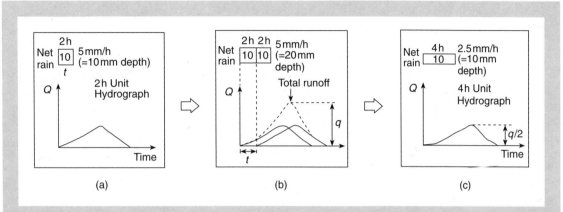

Figure 13.3 Extending the duration of the UH by a multiple of t, i.e. from 2 h to 4 h in this example. (a) Original 2 h UH. (b) Two contiguous periods of net rainfall and corresponding 2 h UHs with the second starting 2 h after the first. The ordinates are added to get the total runoff. (c) The 4 h UH corresponding to 10 mm depth of net rainfall

Decreasing (or increasing) the duration of the UH

The 'S-curve' technique can be used to decrease the duration of the UH (e.g. from 6 h to 4 h) or to increase it by a factor which is not a multiple of t (e.g. 2 h to 3 h). Again, in practice the calculations would be conducted in a table; the diagrams in Fig. 13.4 are only to illustrate the procedure explained below. As an example, suppose we have a 4 h UH (i.e. $t = 4$ h) and want to reduce it to the 3 h UH.

1. Diagram a shows the original 4 h UH and net rainfall, i.e. 4 h at 2.5 mm/h = 10 mm.

2. To produce the hydrograph that represents the runoff arising from continuous rainfall at 2.5 mm/h, draw identical, contiguous hyetographs to the right of the original. For each new hyetograph, draw the equivalent UH, each offset to the right by $t = 4$ h. Add all the hydrographs together to obtain the combined effect; the result is an S-shaped curve, as shown in diagram b. Note that the S-curve reaches a constant runoff value at the catchment's time of concentration (t_C). As a guide, the number of repeated hyetographs needed to obtain this constant value is T/t where T h is the duration of surface runoff in diagram a and t h is the duration of the rainfall. With natural, large river catchments, rainfall never lasts long enough to equal t_C, but this does happen with the small areas that drain to sewers.

3. Draw two identical bands of continuous rainfall (2.5 mm/h) offset by the required duration of the UH, i.e. $t_S = 3$ h in this case. Draw the corresponding identical S-curves, also offset by 3 h (diagram c).

4. Subtract the rainfall diagrams and the S-curves from each other and replot the remainder (diagram d). This is the runoff q arising from net rainfall of 2.5 mm/h × 3 h = 7.5 mm depth.

5. To obtain the surface runoff arising from 10 mm depth of net rainfall, as required by the definition of the UH, multiply the rainfall intensity in diagram d by 10/7.5 = 1.33 so that it becomes 1.33 × 2.5 = 3.33 mm/h. Check: 3.33 mm/h × 3 h = 10 mm. By the principle

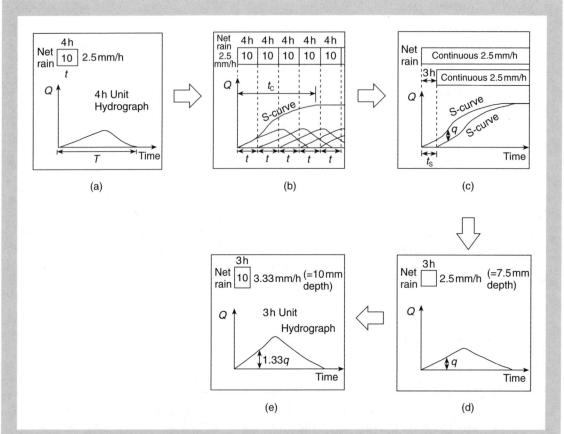

Figure 13.4 Decreasing the duration of the UH using an S-curve, i.e. from 4h to 3h in this example. (a) Original 4h UH. (b) Contiguous periods of rainfall, each offset by 4h, and corresponding hydrographs which are added to get the S-curve. (c) Two periods of continuous rainfall offset by 3h, and corresponding S-curves. (d) The net rain and surface runoff obtained by subtracting the two bands of continuous rainfall and the S-curves. (e) The 3h UH corresponding to 10mm depth of net rainfall

of proportionality, multiply the ordinates of surface runoff (q) by 1.33 also. The result shown in diagram e is the 3h UH.

Obtaining the design storm

The design storm is obtained using the principles of proportionality and superposition, as described in Box 13.1 and illustrated by Fig. 13.5. Here we are starting with the 2h UH (diagram a) and the 4h UH (diagram b). Suppose we want to estimate the combined effect of 20mm/h of net rain for 2h followed by 7.5mm/h for 4h.

1. In the first 2 hours we have $20 \times 2 = 40$mm depth of net rain, which is four times as much as the 2h UH, so multiply the 2h UH by 4.

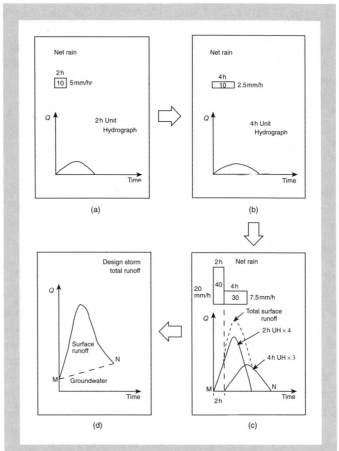

Figure 13.5 Determining the design storm total runoff hydrograph, starting in this example with (a) the 2 h UH and (b) the 4 h UH. (c) The principles of proportionality and superposition are used to obtain the two flood hydrographs, which are added to get the total surface runoff. (d) The groundwater baseflow is added to get the total design storm runoff

2. In the next 4 hours we have $7.5 \times 4 = 30$ mm depth of net rain, which is three times as much as the 4 h UH, so multiply the 4 h UH by 3.

3. Superimpose the hyetographs and hydrographs for the two rainfall events on the same graph, and add the runoff ordinates together (diagram c). Add any groundwater base-flow, and the result is the total design storm discharge (diagram d).

This procedure is usually accomplished numerically using a table, without drawing the diagrams, as shown in the following examples.

EXAMPLE 13.2   INCREASING THE DURATION OF A UH TO OBTAIN THE DESIGN FLOOD

The hydrograph below is the surface runoff (q) recorded at a gauging station on 13/14 December 1997 (i.e. the groundwater baseflow has already been subtracted). The times are hours GMT. The hydrograph is the result of 25 mm/h of net rain falling over a 2 hour period. Calculate the surface runoff that would result from 4 hours of rainfall at 40 mm/h. Assume that the groundwater baseflow is negligible.

Time (h)	0800	1000	1200	1400	1600	1800	2000	2200	2400	0200	0400
q (m³/s)	0	1.4	5.4	12.6	14.6	9.9	7.2	5.0	3.1	1.4	0

The duration of the observed rainfall is 2 h, but the design storm has a duration of 4 hours. Thus we must first derive the 2 h UH, change its duration to 4 h, then calculate the runoff for the design storm. The calculations are shown in Table 13.4 and explained below.

Step 1 – derive the 2 h UH. The observed rainfall represents 25 mm/h × 2 h = 50 mm of net rain on the catchment, which is five times the 10 mm required for the 2 h UH. Therefore, the surface runoff (q) in row 2 must be divided by 5 to obtain the 2 h UH (i.e. 5 mm/h × 2 h = 10 mm). These values are shown in row 3. This step is equivalent to diagrams b and c in Fig. 13.2.

Step 2 – derive the 4 h UH. The 4 h UH is obtained from two identical and contiguous 2 h hyetographs (as in Fig. 13.3b), which means there are also two identical hydrographs with the second offset to the right by t hours. Therefore, row 4 repeats the runoff values in row 3, but offset to the right by $t = 2$ h. Adding rows 3 and 4 gives row 5, i.e. the total surface runoff arising from 4 h at 5 mm/h = 20 mm depth of net rainfall. Thus we must divide by 2 to obtain the 10 mm depth required for the 4 h UH (as Fig. 13.3c), the resulting values being shown in row 6.

Step 3 – calculate the design runoff. The 4 h UH represents 10 mm net rainfall, the design storm 4 h × 40 = 160 mm or 16 times the 4 h UH. Therefore, multiply the values in row 6 by 16 to obtain the design surface runoff in row 7. In this example the groundwater baseflow is assumed to be negligible; had it not been so, it would have been added to the values in row 7.

Table 13.4   Unit hydrograph calculations for Example 13.2

1		Depth of net rainfall	13/14 December 1997: Time (h GMT)											
			0800	1000	1200	1400	1600	1800	2000	2200	2400	0200	0400	0600
2	q (m³/s)	50 mm	0	1.4	5.4	12.6	14.6	9.9	7.2	5.0	3.1	1.4	0	
3	2 h UH = q/5	10 mm	0	0.28	1.08	2.52	2.92	1.98	1.44	1.00	0.62	0.28	0	
4	Repeat 2 h UH	10 mm	offset	0	0.28	1.08	2.52	2.92	1.98	1.44	1.00	0.62	0.28	0
5	Row 3 + 4	20 mm	0	0.28	1.36	3.60	5.44	4.90	3.42	2.44	1.62	0.90	0.28	0
6	4 h UH	10 mm	0	0.14	0.68	1.80	2.72	2.45	1.71	1.22	0.81	0.45	0.14	0
7	Design runoff = 4 h UH × 16	160 mm	0	2.24	10.88	28.80	43.52	39.20	27.36	19.52	12.96	7.20	2.24	0

EXAMPLE 13.3   USING THE S-CURVE TO REDUCE THE UH DURATION AND OBTAIN THE DESIGN FLOOD

A 4 h UH has the values shown below. Determine the runoff that will result from a design storm that consists of the following net rainfall: 20 mm/h for 2 h, no rainfall for 2 h, then 10 mm/h for 2 hours. Assume that the groundwater baseflow during the design storm is constant at $3\,m^3/s$.

Time (h)	0	2	4	6	8	10	12	14
4h UH (m^3/s)	0	2.35	7.00	9.75	7.50	4.20	1.80	0

The calculations are shown in Table 13.5 and explained below. Since we are a given a 4h UH (i.e. net rainfall of 2.5 mm/h) but the design event consists of two 2h periods of rainfall, the first step is to obtain the 2h UH using the S-curve technique. Having done this, the second step is to calculate the surface runoff arising from the design net rainfall. The third step is to add the groundwater baseflow.

Step 1 – derive the 2h UH. Rows 2 to 5 show the 4h UH ordinates, each row being offset 4h to the right. The offset is necessary to achieve contiguous hyetographs as in Fig. 13.4b, with the associated runoff having the equivalent offset. The required number of repetitions is $T/t = 14/4$ ≈ 4. The ordinates in rows 2 to 5 are added to give the S-curve values in row 6. We want the 2h UH, so in row 7 two identical S-curves are offset by 2h and subtracted to obtain the ordinates in row 8. These values represent the runoff from 2h of rain at the original 2.5mm/h = 5mm depth, so they are multiplied by 2 in row 9 to obtain the 2h UH corresponding to 10mm depth of net rain.

Step 2 – calculate the design event surface runoff. The first part of the design rainfall is 20mm/h × 2h = 40mm or 4 × 2h UH, so row 10 shows the ordinates in row 9 multiplied by 4. This rain lasts 2h, then there is no rain for 2h, then at time = 4h the second design rainfall of 10mm/h × 2h = 20mm occurs. Thus starting at time = 4h, row 11 shows the 2h UH multiplied by 2. Row 12 shows the sum of rows 10 + 11, which is the total design surface runoff.

Step 3 – add the groundwater baseflow. It is assumed in row 13 that the average baseflow during the storm is $3.00\,m^3/s$, so this is added to the values in row 12 to get the total design discharge in row 14.

Table 13.5 Unit hydrograph calculations for Example 13.3

1		Depth of net rain	Time (h)								
			0	2	4	6	8	10	12	14	16
2	4h UH	10 mm	0	2.35	7.00	9.75	7.50	4.20	1.80	0	
3	Repeated 4h UH		4h offset		0	2.35	7.00	9.75	7.50	4.20	1.80
4	Repeated 4h UH				4h offset		0	2.35	7.00	9.75	7.50
5	Repeated 4h UH						4h offset		0	2.35	7.00
6	S-curve: add rows 2 to 5	Contin.	0	2.35	7.00	12.10	14.50	16.30	16.30	16.30	16.30
7	S-curve offset 2h	Contin.	offset	0	2.35	7.00	12.10	14.50	16.30	16.30	16.30
8	Row 6 – 7	5 mm	0	2.35	4.65	5.10	2.40	1.80	0	0	0
9	Row 8 × 2 = 2h UH	10 mm	0	4.70	9.30	10.20	4.80	3.60	0		
10	2h UH × 4 (design rain)	40 mm	0	18.80	37.20	40.80	19.20	14.40	0		
11	2h UH × 2 (design rain)	20 mm	offset	no rain	0	9.40	18.60	20.40	9.60	7.20	0
12	Row 10 + 11 = surface runoff	60 mm	0	18.80	37.20	50.20	37.80	34.80	9.60	7.20	0
13	Groundwater baseflow	—	3.00	3.00	3.00	3.00	3.00	3.00	3.00	3.00	3.00
14	Total design flow (12 + 13)	60 mm	3.00	21.80	40.20	53.20	40.80	37.80	12.60	10.20	3.00

SELF TEST QUESTION 13.2

The surface runoff (q) resulting from 3 h of gross rainfall at 16 mm/h is shown below. It is estimated that a 16 mm soil moisture deficit absorbs all of the rainfall during the first hour, after which the catchment losses can be represented by Φ = 6 mm/h. Calculate the total river discharge during a design event that comprises the following net rainfall: 4 mm/h for 6 h, followed immediately by 25 mm/h for 2 h. Assume the average groundwater baseflow is 4.50 m³/s.

Time (h)	0	2	4	6	8	10	12	14	16	18
q (m³/s)	0	1.20	3.28	6.96	9.92	8.80	5.88	3.52	1.56	0

13.3 Statistical analysis of river flow data

Some of the techniques described below can also be used to analyse rainfall depth or intensity, but are employed here specifically to analyse river flow data. With respect to flood estimation, they provide an alternative approach to the unit hydrograph described above.

13.3.1 Full series analysis

Perhaps the easiest introduction to the statistical group of methods is to consider a full series of 365 daily mean flows (DMFs) recorded at a gauging station during one year. Obviously the mean flow on one particular day depends to some extent on the previous day's flow, so these are not independent events, but in this case it does not matter. The objective of the analysis is to determine the percentage of the year that a particular discharge will be equalled or exceeded. How this is accomplished will be explained using the data in Table 13.6, which is part of Example 13.4. The analysis starts by determining how many times (n) during the year the DMF is within a particular range. For illustration purposes only, using equal discharge intervals, this information is plotted as a histogram in Fig. 13.6a. Working from the highest to the lowest discharge, the next step is to calculate Σn for each discharge band, and then divide Σn by 365 to obtain the percentage of the year in which the discharge is equalled or exceeded. This relationship can be plotted as a graph, which is called the **flow–duration curve**. When drawing the graph, plot the percentage value against the lower value of the discharge band, since it is this value that is equalled or exceeded. The plotted points are simply joined together to form the curve (i.e. do not draw the best straight line through them). Using natural scales the result is Fig. 13.6b, but usually a log Q–normal probability scale is adopted (Fig. 13.6c) so that large discharges can be accommodated while also affording greater detail at small discharges. To ensure an even distribution of points on a log Q–probability scale, use smaller increments of discharge during low flows, as in Table 13.6.

With 10 years of data, all 3650 values of DMF can be used to obtain the average flow–duration curve. Alternatively, each year can be analysed separately and ten annual lines plotted on the same graph to obtain some idea of the variation; enveloping curves can be drawn to indicate the upper and lower range of flows corresponding to wet and dry years. Figure 13.6c shows a relatively wet year (1994), a relatively dry year (1984) and the average curve from 1970 to 1995.

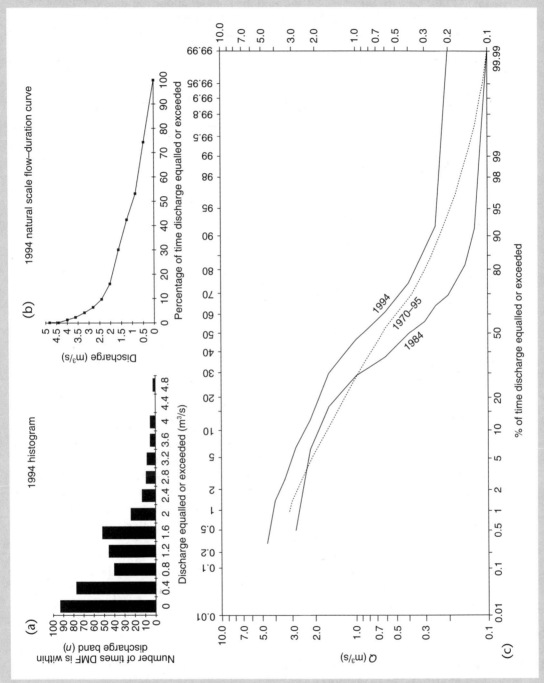

Figure 13.6 Flow–duration curve for Example 13.4. (a) Histogram of number of occurrences. (b) Natural scale flow–duration curve. (c) Log-probability flow–duration curve

Table 13.6 Data for Example 13.4 showing the number of days (n) during 1984 and 1994 that the daily mean flow (DMF) in the River Warleggan was within the discharge intervals, and the calculated percentages [*Data courtesy of the Centre for Ecology and Hydrology*]

Discharge range (m³/s)	1984*			1994		
	n	Σn	$\dfrac{\Sigma n}{366} \times 100$ (%)	n	Σn	$\dfrac{\Sigma n}{365} \times 100$ (%)
>4.600	0	0	0	1	1	0.3
4.000–4.599	0	0	0	4	5	1.4
3.400–3.999	0	0	0	5	10	2.7
2.800–3.399	2	2	0.5	14	24	6.6
2.200–2.799	21	23	6.3	21	45	12.3
1.600–2.199	39	62	16.9	65	110	30.1
1.000–1.599	43	105	28.7	61	171	46.8
0.600–0.999	34	139	38.0	53	224	61.4
0.400–0.599	44	183	50.0	48	272	74.5
0.300–0.399	24	207	56.6	38	310	84.9
0.250–0.299	28	235	64.2	26	336	92.1
0.200–0.249	20	255	69.7	29	365	100.0
0.150–0.199	44	299	81.7	0		
0.125–0.149	37	336	91.8	0		
0.100–0.124	30	366	100.0	0		
0.000–0.099	0			0		

* Leap year with 366 days.

Flow–duration curves have many uses, such as indicating the proportion of time that water can be abstracted from rivers, or the quantity of flow that can be relied upon to dilute sewage effluent. The quantity of flow that is maintained throughout the year may also give some indication of catchment hydrology. A curve with a relatively shallow slope that maintains a large flow throughout the year may suggest a permeable catchment that absorbs winter precipitation and which has a high groundwater component in summer. A relatively steep curve may suggest an impermeable catchment that has a large winter runoff and a low groundwater baseflow in summer.

Note that the daily mean flow (DMF) is used when constructing flow duration–curves. This is the mean flow over a 24 hour period, within which there may be a much higher instantaneous discharge (i.e. a flood peak) which lasts for only a short period of time. For example, at Trengoffe on 27 December 1979 the maximum instantaneous flow was 23.91 m³/s (Fig. 12.15) while the DMF was 12.43 m³/s.

EXAMPLE 13.4 FLOW–DURATION CURVE FOR THE RIVER WARLEGGAN AT TRENGOFFE, CORNWALL

This small, elongated catchment drains south from Bodmin Moor and has an area of 25.30 km². It is about 200–250 m above sea level, entirely rural, with some small villages and woodland in the lower valleys. Average annual rainfall is approximately 1 512 mm and PET is about 475 mm. Geologically, the catchment consists of 70% granite moorland and tors, 30% Devonian slates.

Granite is normally fairly impermeable, but kaolinisation (a hydrothermal alteration caused by the movement of acid solution along joints) has occurred in parts, allowing freer passsage of water and some storage. Consequently the baseflow is relatively high for an upland catchment. This is one of the few gauged catchments on Bodmin Moor with a natural flow regime unaffected by reservoirs or abstraction. The gauging station has a three-bay compound Crump weir. The long-term mean daily flow (i.e. the average over several years) is $0.83\,\mathrm{m^3/s}$. Table 13.6 shows the number of times (n) the daily mean flow (DMF) fell within the discharge intervals during 1984 and 1994. (a) Construct the flow–duration curves for these two years. (b) Determine the flow ($\mathrm{m^3/s}$) which was exceeded for 10%, 50% and 95% of the time in 1984 and 1994. Note that the 10% exceedance value is used as a measure of the variability or 'flashiness' of the flow, the 50% exceedance is the median value, and the 95% exceedance is a significant low flow parameter used in the assessment of river water quality consent conditions (e.g. in specifying the dilution of effluents).

(a) The calculations are in Table 13.6 and the flow–duration curves are plotted in Fig. 13.6c. It can be seen that the two years have different flow regimes: 1984 was relatively dry, while 1994 was relatively wet. For comparison, the average curve for 1970–1995 is also shown. Diagrams a and b show the form of the 1994 histogram and natural scale flow–duration curve (using equal discharge increments).

(b) In 1984 the 10%, 50% and 95% exceedances were 1.90, 0.40 and $0.12\,\mathrm{m^3/s}$, and in 1994 they were 2.40, 0.89 and $0.24\,\mathrm{m^3/s}$.

SELF TEST QUESTION 13.3

To supplement water supplies during drought years there is a proposal to extract $0.20\,\mathrm{m^3/s}$ of water from the River Warleggan at Trengoffe and pump it to nearby Colliford Reservoir. If the flow downstream of the abstraction point must always be at least $0.22\,\mathrm{m^3/s}$, using the 1984 flow duration curve in Fig. 13.6c determine: (a) the number of days when the full amount could be extracted, and hence the total volume of water available during the year; (b) the number of additional days when the flow is between $0.22\,\mathrm{m^3/s}$ and $0.42\,\mathrm{m^3/s}$ and would support some extraction, but not the full amount.

13.3.2 Frequency analysis

A full series analysis is useful for interpolation, but cannot be used for extrapolation. Frequency analysis can be used for either interpolation or extrapolation. It evaluates the relationship between the return period (T) and the magnitude of a particular hydrological event, such as a peak flood discharge, a drought flow, or rainfall depth or intensity. Thus if we want to design a bridge to withstand a 1 in 100 year flood, frequency analysis can provide an estimate of the design discharge. When we say '1 in 100 years' we are quoting an average return period (also commonly called the frequency of occurrence, frequency, or return interval). It is worth spending a few moments considering what this really means.

Almost everyone is familiar with the concept of being paid once per week or once per month, which is a regular, evenly spaced series of similar events. Floods are *not* like this: they are independent, vary in size, and are randomly spaced throughout time. Thus when we speak of a 1 in 100 year flood, we mean the (REMEMBER!!!)

peak flood flow (Q) that **on average** will be exceeded in only one out of every 100 years. However, a 1 in 100 year flood could occur this year, followed by a 1 in 500 year flood next year. Large floods are not evenly spaced throughout time. Similarly, the 1 in 10000 year flood could occur tomorrow, its name does not mean it is due in 10000 years.

The flood with a return period of T years is denoted by Q_T and is called the 1 in T year flood, or often just the T-year flood. The relationship between T and the probability (P) that an event of magnitude Q_T will be exceeded in any one year is:

$$P = \frac{1}{T}$$
(13.1)

It follows that the probability that the event won't be exceeded in any one year, i.e. non-occurrence, is (1 − P).

During the design life (L) of a particular structure, the probability (J) that at least one event will occur which exceeds in magnitude that of a specified return period (T) can be obtained from equation (13.2). Example 13.5 shows that the probability of the design flood occurring is much greater than may be imagined. According to the Flood Estimation Handbook, if L is quite large (say 100 years) then typically the largest flood during this period will have a return period T = 1.45L years (Institute of Hydrology, 1999).

$$J = 1 - \left(1 - \frac{1}{T}\right)^L$$
(13.2)

EXAMPLE 13.5

(a) A dam is to be constructed to withstand a 1 in 10000 year flood. It is anticipated that the dam will have a 100 year life. What is the probability that a flood that exceeds the design flood will occur during the lifetime of the dam?

(b) A bridge has an intended design life of 120 years and is designed for a 1 in 100 year flood. What is the probability that something bigger will be experienced?

(c) A bridge pier is to be constructed inside a temporary cofferdam. If the cofferdam is designed for a 1 in 20 year flood and will be in the river for 2 winters, what is the probability that a larger flood will occur?

(d) What is the significance of the answers to these questions?

The answer to parts (a) to (c) can be obtained from equation (13.2):

(a) $J = 1 - \left(1 - \dfrac{1}{10\,000}\right)^{100} = 0.01$ or 1%

(b) $J = 1 - \left(1 - \dfrac{1}{100}\right)^{120} = 0.70$ or 70%

(c) $J = 1 - \left(1 - \dfrac{1}{20}\right)^{2} = 0.0975$ or 9.75%

(d) The significance is that there is always a risk of failure. Statistically there is often a surprisingly large probability that a flood will occur which is larger than the one you designed for. Thus there is always a very real risk of failure. For engineers, the question is often what degree of risk can be tolerated, and how much money can be spent justifiably to reduce the risk to an acceptable level? For example, see Table 13.3.

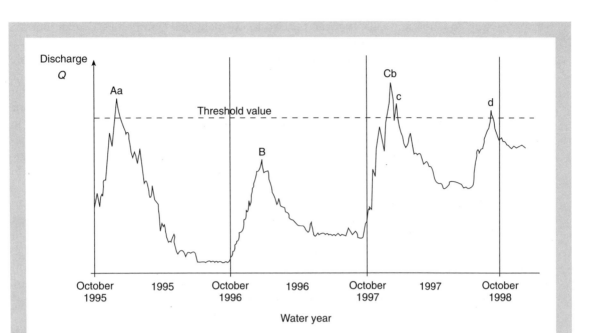

Figure 13.7 Diagrammatic river flow hydrograph based on water years which run from 1 October to 30 September. For an annual maximum series the highest peak in each year is used, i.e. A, B, C. For a peaks over a threshold (POT) series use a, b, c and d (assuming c is judged to be independent)

Using equation (13.2) and Example 13.5 we have now examined the meaning of the return period and understand that even rare events may occur at any time, but we have not yet explored how to obtain the magnitude (Q_T) of the T-year flood. The first step is to select the data to be analysed. There are two alternatives: an annual maximum series or a partial duration series. In both cases it is important that the floods are independent events. One rule for independence is that the peaks should be separated in time by three times the time to peak, and that the flow should decrease between peaks to two-thirds of the first peak. For example, using this guideline, the two peaks b and c in Fig.12.15 are not independent, and they do not look as though they are.

To form an **annual maximum series**, for each year of record select the largest instantaneous gauged value in that water year. Thus in Fig. 13.7 the peak values of floods A, B and C would be used. Note that this figure is purely diagrammatic and of necessity has a compressed horizontal scale. In 1995 a typical seasonal variation is apparent, with high winter and low summer flows. Water years, which run from 1 October to 30 September, are generally used so that the highest winter flow can be selected, regardless of calendar year. Generally an annual series is recommended when more than 14 years of data exist and when the median annual flood is to be calculated (i.e. *QMED*, see section 13.3.3).

A **peaks over a threshold (POT) series** is recommended with 2–13 years of data. In this example, the series would be all of the instantaneous gauged peak discharges above the dashed horizontal line in Fig. 13.7, so they are a, b, c and d. The year in which a peak occurs does not matter. However, peak c may be omitted if it is considered not to be independent of the previous event (i.e. b). The Flood Estimation Handbook (Institute of Hydrology, 1999) gave an equation and table that can be used to calculate *QMED* from a POT series, but the use of an annual mamimum series is preferred.

There are problems with both of the above series. Since the objective is to extrapolate the largest floods on record, with the annual maximum series it is illogical to include peak B while ignoring the much larger peak d (and perhaps c). Leaving aside concerns about independence, with the POT series the fact that three peaks b, c and d all occur in one year raises doubts about the distribution of the data and its extrapolation to find the 1 in *T*-year event. However, there are many possible sources of error when undertaking a frequency analysis, so how the data are selected may not be of critical importance. Only the annual maximum series is considered below, because it provides the best introduction to the subject.

An annual maximum frequency analysis starts with the selection of the data, which is then ranked so that the highest flood discharge (Q_T) has a rank (r) of 1, the second highest 2, and so on down to the lowest value which has a rank of N, where N is the number of values in the series. If two values of Q_T are identical they can be given the same rank (e.g. 5) and the following rank (i.e. 6) omitted. Next, the probability of occurrence (P) of each of the ranked values is calculated using one of the following equations, and multiplied by 100 to get the percentage probability ($P\%$).

Weibull: $\qquad P = \dfrac{r}{N+1}$ $\hfill$ (13.3)

Gringorten: $\quad P = \dfrac{r-0.44}{N+0.12}$ $\hfill$ (13.4)

The Weibull equation is often employed because it is simple, although Gringorten's may be the best, especially for Gumbel distributions (see below). In many cases there is little difference. For example, if there are 20 years of data ($N = 20$) then for $r = 9$ Weibull gives $P = 9/(20 + 1) = 0.429$ or 42.9% and Gringorten $P = (9 - 0.44)/(20 + 0.12) = 0.425$ or 42.5%. Thus the difference is often too small to be plotted, although sometimes it can be significant.

Each observed flood discharge (Q_T) is plotted on a graph against its percentage probability of occurrence ($P\%$). The aim is to obtain a straight-line graph that is easy to extrapolate, and special graph paper is used to achieve this. The type of paper recommended varies according to the length of the record and whether or not it is assumed that the data in the series are normally distributed. For 10 to 25 years of data, either log-normal probability or Gumbel paper may be used; for longer records, other types such as Pearson III may be recommended. Wilson (1990) plotted data for the Thames at Teddington on five types of paper, and it is interesting to compare them to see how linear they are. Here we will only consider log-normal probability (or log-probability for short) and Gumbel graph paper, examples of which are given in Appendix 3. If Gumbel paper is not available, an alternative is to plot Q_T against y on ordinary linear graph paper. The data should lie on a straight line. The Gumbel reduced variate, y, is given by:

$$y = -\ln[-\ln(1 - P)]$$ $\hfill$ (13.5)

This equation can also be used to produce Gumbel graph paper by drawing a linear y-scale, and then marking the values of y that the equation gives as equivalent to $P = 0.1\%$, 0.5%, 1%, 5% etc.

An example of a Q_T against $P\%$ plot on log-probability paper is shown in Fig. 13.8. If the data follow the assumed distribution they should plot as an approximate straight line. In fact, the 'before development' data are reasonably linear. Some scatter should always be expected because of: gauging errors; the method of selecting the data (annual maximum or POT); the method of calculating P; assumptions regarding the distribution of the data; catchment changes (e.g. deforestation, urbanisation); and possibly global warming and climate change. For a simple analysis, draw the best straight line through the data by eye. Initially, using something like a highlighter pen with a wide tip can help locate the optimum line while allowing for inaccuracies. For more sophisticated analyses, the optimum line may be obtained using a least squares technique or method of moments. If the data do not form a straight line then try a different type of graph paper (e.g. Gumbel), check for outliers and investigate the accuracy of the data.

Outliers are points on the graph which do not align with the rest of the data and which are out of position. For example, if a really large event such as the 1000 year flood has been recorded in a short 10 year record, with $r = 1$ Weibull gives $P\% = (1 \times 100)/(10 + 1) = 9\%$. With a longer record of 200 years this outlier could still be first ranked, but now $P\% = (1 \times 100)/(200 + 1) = 0.5\%$ so it plots nearer to its true position, e.g. further to the left when using log-probability paper. One guideline is that anything larger than three times the median value is an outlier, but remember this is only a guideline. For example, a point which is 2.8 times the median value could still be an outlier, but it could also be the result of a gauging error. The accuracy of data can be difficult to check, but sometimes going back to the full gauging station record (instead of just an extracted maximum value) may reveal some sudden jump in the readings. Remember that it is relatively easy for flood debris to become stuck across a weir, raising the upstream water level and *apparently* increasing the discharge.

Having drawn the Q_T–$P\%$ line, it can be used to determine the magnitude of the flood corresponding to any return period T (or $P\%$). The line may be cautiously extrapolated, but for a single station extrapolation should only really be conducted to $T \leq 2N$. The longest gauging station record in Britain started in 1883, so approximately it has $N = 120$ years, while most records started around the 1960s (say $N = 40$ years or less). Thus using the simple technique above we may only be able to estimate floods with a return period of 1 in 50 to 1 in 250 year with relative confidence (i.e. Q_{50} to Q_{250}). Extrapolating beyond this may involve even larger errors than usual. The UK Flood Estimation Handbook (Institute of Hydrology, 1999) contains details of more sophisticated methods that can be employed on these occasions.

When estimating Q_T, it must be appreciated that a range of answers is possible depending upon whether or not possible outliers are included, how the optimum line is drawn, the type of graph paper, etc. For example, in one particular location estimates of Q_{100} were 669 m³/s, 728 m³/s, 742 m³/s, 784 m³/s, 669 m³/s, 725 m³/s, 905 m³/s, 769 m³/s, 705 m³/s and 735 m³/s. If a safe design is essential then possible outliers should be included and the largest figure adopted, otherwise an average figure could be employed. If in doubt, consider the potential consequences of your flood estimate being exceeded.

EXAMPLE 13.6   FREQUENCY ANALYSIS FOR RIVER WARLEGGAN AT TRENGOFFE, CORNWALL

The annual maxima for a series of water years are shown in the first three columns of Table 13.7. (a) Determine the magnitude of a 1 in 5, 1 in 20 and 1 in 100 year flood. (b) This small catchment (25.3 km²) is currently almost entirely rural in character. There is a proposal for a substantial housing/industrial development which, when combined with climate change, may increase

Table 13.7 River Warleggan at Trengoffe – recorded data and annual maximum frequency analysis [*Data from the FEH CD-ROM courtesy of the Institute of Hydrology, Wallingford*]

QMED = (8.629 + 7.525)/2 = 8.077 m³/s.

Recorded data			Prior to development, ranked data and % probability				Post development
Water year	Date	Q m³/s	Rank r	Water year	Q_T m³/s	P% Eqn (13.4)	1.35Q_T m³/s
1981	20 Dec 1981	17.914	1	1981	17.914	3.97	24.184
1982	6 Nov 1982	10.085	2	1992	13.843	11.05	18.688
1983	19 Dec 1983	4.651	3	1985	12.973	18.13	17.514
1984	27 Jan 1985	5.673	4	1982	10.085	25.21	13.615
1985	24 Aug 1986	12.973	5	1986	9.813	32.29	13.248
1986	10 Dec 1986	9.813	6	1987	9.543	39.38	12.883
1987	1 Feb 1988	9.543	7	1988	8.629	46.46	11.649
1988	24 Feb 1989	8.629	8	1994	7.525	53.54	10.159
1989	14 Feb 1990	6.239	9	1993	7.074	60.62	9.550
1990	1 Jan 1991	4.936	10	1989	6.239	67.71	8.423
1991	31 Oct 1991	3.807	11	1984	5.673	74.79	7.659
1992	18 Dec 1992	13.843	12	1990	4.936	81.87	6.664
1993	24 Jan 1994	7.074	13	1983	4.651	88.95	6.279
1994	27 Jan 1995	7.525	14	1991	3.807	96.03	5.139

runoff by up to 35%. Investigate the effect of the development on catchment flood magnitude and frequency.

(a) Columns 4–6 of Table 13.7 show the recorded data ranked in order of decreasing peak flow (Q_T). The percentage probability (P%) is calculated using Gringorten's equation (13.4). The solid line in Fig. 13.8 shows the plot of Q_T against P%. From the line, the values of Q_T corresponding to the following return periods can be obtained:

1 in 5 years or P = 20% gives Q_5 = 11.8 m³/s

1 in 20 years or P = 5% gives Q_{20} = 17.3 m³/s

1 in 100 years or P = 1% gives Q_{100} = 23.8 m³/s

The Q_{100} value is questionable since it involves extrapolation beyond 2N.

(b) The last column of Table 13.7 shows the values of 1.35Q_T, that is the flood peaks after development has increased them by 35%. The rank (r) and number of years in the series (N) are unchanged, so the values of P% are as before. Plotting 1.35Q_T against P% gives the dashed 'after development' line in Fig. 13.8. The effect of the development on flood frequency is indicated by the horizontal distance between the lines. Thus for the floods in part (a), after development:

Q_T = 11.8 m³/s corresponds to P = 42% or T = 2.4 years

Q_T = 17.3 m³/s corresponds to P = 15.5% or T = 6.5 years

Q_T = 23.8 m³/s corresponds to P = 4.7% or T = 21.3 years

Thus the development reduces the return period of the floods so that the existing 23.8 m³/s, 1 in 100 year event will occur in future every 21 years, almost five times as often. This increase

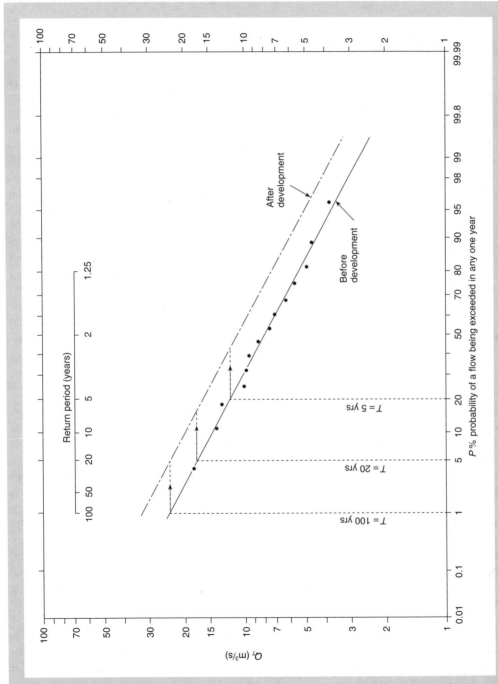

Figure 13.8 The annual maximum flood series (Q_T) at Trengoffe plotted against $P\%$ on log-normal probability paper. The solid line shows the recorded data, the dotted line the relationship if development of the catchment results in a 35% increase in Q_T. Note that after development a flood of any Q_T occurs more often (i.e. a 1 in 100 year flood becomes a 1 in 21 year event)

in flood frequency is rarely understood by developers. More obviously, post development the 1 in 100 year flood would be increased to $1.35 \times 23.8 = 32.13\,\text{m}^3/\text{s}$.

13.3.3 *QMED* and pooled data

The median annual maximum flood (*QMED*) is that which is exceeded on average every other year. In other words, it is the middle value in a ranked series of annual maximum floods. Ideally there should be more than 14 years of data. If there is an even number of annual maxima, *QMED* is the arithmetic mean of the two central values. In either case, half of the values should be larger and half smaller than *QMED*. For example, using the ranked Trengoffe data in Table 13.7 the value of $QMED = (8.629 + 7.525)/2 = 8.077\,\text{m}^3/\text{s}$. The value obtained should be adjusted for climatic variability and catchment change (e.g. urbanisation or the construction of impounding reservoirs). Note that in the 1999 Flood Estimation Handbook (FEH) *QMED* replaced $\bar{Q}$ (*QBAR*), which was used in the earlier Flood Studies Report (FSR; NERC, 1975) and which will still appear in many text books.

The 1975 FSR used a region curve from several gauging stations within a geographical area to form a dimensionless, graphical correlation between flood magnitudes of known return period, Q_T, and the return period, *T*. However, the 1999 FEH discarded regional data sets in favour of pooled data from catchments which are hydrologically similar, regardless of location. Thus river flow records in some parts of Scotland may be used to estimate the return period of a flood on Dartmoor in south-west England. The basis of the idea is that the various catchments can be related using dimensionless flood frequency curves where:

$$Q_T = QMED \times x_T \tag{13.6}$$

$$\text{or} \quad x_T = Q_T/QMED \tag{13.7}$$

where x_T = the *T*-year growth factor or growth curve. For example, Fig. 13.9a shows the data plot for the growth curve of the St Neot River at Craigshill Wood, Cornwall. The vertical scale is $Q_T/QMED$; the horizontal scale is the Gumbel reduced variate, *y*, from equation (13.5) and the return period. The recorded data are shown in Table 13.8, and give $QMED = 8.354\,\text{m}^3/\text{s}$. The graph is produced simply by ranking the annual maximum series, calculating $Q_T/QMED$ and *y*, and then plotting the data.

It is apparent that Craigshill Wood has a short record which ended in 1982. By using pooled data it is possible to produce a more confident and more accurate estimate of the relationship between Q_T and *T*. For example, assuming that the St Neot is hydrologically similar to the River Warleggan 3 km away, the two data sets can be combined to obtain Fig. 13.9b. This diagram uses the full Warleggan data series (from Table 13.7 and Self Test Question 13.4 below) with $QMED = 8.561\,\text{m}^3/\text{s}$. Other hydrologically similar stations can be added, a curve fitted to the data, and the data points removed. Often a seductively thin curve is obtained, apparently indicating great accuracy and certainty. It should always be remembered that this may hide a considerable amount of scatter and uncertainty. Nevertheless, the graph indicates the relationship between Q_T and *T*. For example, moving vertically up from $T = 10$ years in Fig. 13.9b gives $Q_T/QMED = 1.9$ so at Craigshill Wood $Q_{10} = 1.9 \times 8.354 = 15.9\,\text{m}^3/\text{s}$.

When pooling data, to estimate the *T*-year flood a combined record length of at least $5T$ station years is required. The 1999 FEH describes the procedures to be used, and includes a CD-ROM of annual maxima, POT and catchment descriptors for 1000 sites in the UK. This

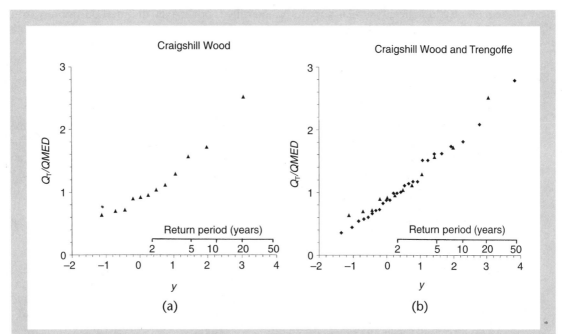

Figure 13.9 (a) Plot of $Q_T/QMED$ against y for the St Neot at Craigshill Wood, Cornwall. (b) Combined Craigshill Wood and Trengoffe data

Table 13.8 St Neot at Craigshill Wood, Cornwall – recorded data and growth curve [*Data courtesy of the Institute of Hydrology, Wallingford*]

$QMED = (8.712 + 7.996)/2 = 8.354\,m^3/s.$

Recorded data			Ranked data for *QMED* and growth curve				
Water year	Date	Q m³/s	Water year	Q_T m³/s	Rank r	$Q_T/QMED$	y Eqn (13.5)
1971	29 Nov 1971	8.712	1979	21.081	1	2.52	3.051
1972	6 Aug 1973	5.319	1981	14.372	2	1.72	1.982
1973	17 Sep 1974	13.094	1973	13.094	3	1.57	1.439
1974	13 Nov 1974	10.823	1974	10.823	4	1.30	1.056
1975	29 Jan 1976	7.996	1982	9.352	5	1.12	0.751
1976	14 Oct 1976	5.832	1971	8.712	6	1.04	0.488
1977	9 Dec 1977	7.699	1975	7.996	7	0.96	0.249
1978	23 Dec 1978	6.009	1977	7.699	8	0.92	0.023
1979	27 Dec 1979	21.081	1980	7.504	9	0.90	−0.203
1980	9 Mar 1981	7.504	1978	6.009	10	0.72	−0.441
1981	19 Dec 1981	14.372	1976	5.832	11	0.70	−0.718
1982	5 Nov 1982	9.352	1972	5.319	12	0.64	−1.123

enables hydrologically similar sites to be identified, but makes the method impractical in books like this, so it will not be taken further.

SELF TEST QUESTION 13.4

(a) The annual maxima for the River Warleggan at Trengoffe during the water years 1969–80 are shown below. Using only these data, the Gringorten equation (13.4) and log-probability paper, determine Q_5, Q_{20} and Q_{100}. Are these values significantly different from those in Example 13.6a? (b) Repeat the above analysis but this time using the full 1969–94 record from Table 13.7 and below. What do you deduce about the importance of a long record? (c) Calculate the values of QMED obtained from the 1969–80 data below and the full 1969–94 record. What are the percentage differences compared to the 8.077 m³/s value in Table 13.7? (d) Using all of the 1969–94 data, calculate the values of Q_T/QMED and y so they can be combined with the data in Table 13.8.

Water year	Date	Q (m³/s)	Water year	Date	Q (m³/s)
1969	17 Jan 1970	14.869	1975	29 Jan 1976	13.907
1970	12 Jan 1971	6.107	1976	3 Feb 1977	3.047
1971	30 Nov 1971	8.493	1977	9 Dec 1977	10.085
1972	6 Aug 1973	8.493	1978	22 Dec 1978	5.167
1973	15 Nov 1973	15.519	1979	27 Dec 1979	23.914
1974	13 Nov 1974	12.973	1980	16 Nov 1980	7.494

13.3.4 Estimating *QMED* from catchment descriptors

For rural catchments > 0.5 km², where no gauging station records exist it is possible to produce a low accuracy estimate of QMED from the catchment descriptors in the 1999 FEH. The basic concept is that the size of a flood depends upon the drainage area of the catchment (*AREA* km²), the 1961–90 standard average annual rainfall (*SAAR* mm), the soil type as represented by the standard percentage runoff (*SPRHOST*), the baseflow index (*BFIHOST*) and any flood attenuation due to storage in reservoirs and lakes (*FARL*, default value = 1.0). Hence:

$$QMED_{rural} = 1.172\, AREA^{AE}\left(\frac{SAAR}{1000}\right)^{1.560} FARL^{2.642}\left(\frac{SPRHOST}{100}\right)^{1.211} 0.0198^{REHOST} \tag{13.8}$$

$$\text{where} \quad AE = 1 - 0.015\ln\left(\frac{AREA}{0.5}\right) \tag{13.9}$$

$$REHOST = BFIHOST + 1.30\left(\frac{SPRHOST}{100}\right) - 0.987 \tag{13.10}$$

To be judged as rural, the value of *URBEXT* for the catchment should be less than 0.025. This value, and those of the other variables, can be found on the CD in the 1999 FEH. Example 13.7 illustrates how the equation can be used to estimate QMED at Trengoffe. You should note that this gives *QMED* = 13.724 m³/s, whereas it was 8.077 m³/s using the short gauged annual maximum series in Example 13.6 and 8.561 m³/s using the complete series in Self Test Question 13.4, so the catchment descriptors method yields an inaccurate value in this instance. This is perhaps to be expected, since some of the variables in section 12.6 (e.g. catchment shape and orientation) are omitted. In fact the FEH indicates that only about 68% of $QMED_{rural}$ estimates will lie in the range (0.65 $QMED_{rural}$, 1.55 $QMED_{rural}$). Consequently this method should never be employed for major engineering work where accuracy is important, and the estimated flows should be quickly verified by actual measurements and/or other methods.

EXAMPLE 13.7

Given the following catchment descriptors for the River Warleggan at Trengoffe, estimate the value of $QMED$: $AREA = 25.30\,km^2$, $BFIHOST = 0.500$, $SPRHOST = 35.7$, $SAAR = 1518\,mm$, $FARL = 0.9728$, $URBEXT = 0.0013$.

From equation (13.9), $\quad AE = 1 - 0.015\ln\left(\dfrac{AREA}{0.5}\right) = 1 - 0.015\ln\left(\dfrac{25.30}{0.5}\right)$

$$AE = 1 - 0.05885$$
$$AE = 0.941$$

From equation (13.10), $\quad REHOST = BFIHOST + 1.30\left(\dfrac{SPRHOST}{100}\right) - 0.987$

$$REHOST = 0.500 + 1.30\left(\dfrac{35.7}{100}\right) - 0.987$$
$$REHOST = -0.0229$$

Now inserting the values into equation (13.8) gives:

$$QMED_{rural} = 1.172\,AREA^{AE}\left(\frac{SAAR}{1000}\right)^{1.560} FARL^{2.642}\left(\frac{SPRHOST}{100}\right)^{1.211} 0.0198^{REHOST}$$

$$= 1.172 \times 25.30^{0.941} \times \left(\frac{1518}{1000}\right)^{1.560} \times 0.9728^{2.642} \times \left(\frac{35.7}{100}\right)^{1.211} \times \left(\frac{1}{0.0198^{0.0229}}\right)$$

$$= 1.172 \times 20.909 \times 1.918 \times 0.930 \times 0.287 \times 1.094$$

$$= 13.724\,m^3/s$$

Actually $QMED = 8.561\,m^3/s$ so the factorial error = 13.724/8.561 = 1.60, which is just outside the 68% confidence limit of 1.55 $QMED_{rural}$. Therefore this is one of the 32% of estimates that fall outside the 0.65–1.55 $QMED_{rural}$ range.

Note that in this example $URBEXT = 0.0013$, which is well below the 0.025 limit, since this is basically an undeveloped catchment. However, $FARL$ is less than the unreservoired default value of 1.0. In fact, this is an area of china clay extraction and there are numerous small lakes in the catchment.

13.3.5 Historical flood marks, regression analysis and channel dimensions

Sometimes historic flood levels marked on buildings can be used to estimate the corresponding flood discharge, or the stage can be treated as a POT series (Fig. 13.10). The rational method described in section 13.5 can also be employed to obtain a very quick estimate

Figure 13.10 Yarm is situated on the tidal River Tees. The arrows show the flood levels of 17 September 1771 and 10 March 1881. Such flood marks can be thought of as peaks over a threshold series [*Photo © L. Hamill*]

of the peak flood at some point on a catchment, as well as for designing sewers. Alternatively, when a gauging station has a record which is too short for analysis or extrapolation, it can be extended using regression analysis. This needs a near-by donor site with a record that overlaps the subject site. Where they overlap, the monthly or annual maxima from the two stations can be plotted against each other on a graph and the regression line obtained (some of the techniques in section 5.6 can be employed). At its simplest, this results in a linear relationship of the form $Q_S = jQ_D$ where j is the gradient of the line. Thus the gauged values from the long donor record (Q_D) can be used to estimate the missing flood values of the short subject record (Q_S). Ideally the donor station should be on the same river, above or below the subject site, although stations on neighbouring catchments may be used if they are hydrologically similar. This technique can work quite well, and it is the preferred method of estimating *QMED* when the subject site has a record of less than two years. Obviously, estimates can be refined as more data are recorded at the subject site.

The dimensions of a natural river channel can be used to estimate *QMED*, the assumption being that the water level in the main channel reaches bankful capacity every year or so. The procedure is to select three typical, natural rectangular to trapezoidal cross-sections spaced at least one channel width apart, measure the average horizontal bankful channel width (*BCW* m), then:

$$QMED = 0.182\ BCW^{1.98} \qquad\qquad (13.11)$$

BCW should correspond to the width at the minimum elevation of the active floodplain, which is often indicated by the height of the lower limit of perennial vegetation such as

trees. About 68% of estimates are expected to lie within the interval (0.58 *QMED*, 1.73 *QMED*). This is less accurate than using catchment descriptors.

13.4 ▶ Riverine flood alleviation

This section is headed flood alleviation because, although it is possible to alleviate or reduce the damage caused by floods in rivers, it is impossible to prevent floods or to totally defend against them. For engineers this raises the question as to what size and frequency of flood to design for?

Suppose there is a village or town that has to be protected by building flood walls on either side of the river. How high should they be? If the walls are designed for a 1000 year, 100 year and 50 year flood level then during a 100 year period, equation (13.2) shows that the chances of something larger being encountered are 10%, 63% and 87% respectively. Thus the risk that any economically designed scheme will be overtopped is quite large. Example 13.5 shows there is still a 1% chance that a 1 in 10 000 year flood will be exceeded during any 100 year period. This is something that the public may not appreciate: tempers can be high when people who thought their homes were totally protected by a flood defence scheme suddenly find themselves inundated and valued possessions ruined.

It would be possible to construct flood walls so tall that they would never be overtopped, but this would be very expensive and may mean protection works for other towns cannot be afforded. They may also be so tall that neighbouring properties completely lose their view of the river, seeing only something that resembles a prison wall instead. The aesthetic and environmental objections to such a scheme may be considerable. Thus even if the design flood can be estimated accurately, finding an acceptable, effective and economical method of alleviation may be difficult. Much depends upon the type of flood and the location.

'Slow' floods often cover huge areas and are the result of widespread, prolonged rainfall. Such floods are characterised by a slow, steady, progressive rise in river stage. Slow floods can be the most deadly, because they cover very large areas for a long time. Following 3 months of rainfall, during July and August 1993 the Missouri/Mississippi rose 1.2 m above its previous highest level, bursting through embankments and causing damage across nine states. Contemporary TV news bulletins reported an area of about 41 400 km^2 inundated (an area twice the size of Wales) with 30 dead, 30 000 people forced to leave their homes and $10 billion of damage. On the other hand, the ingredients for a spectacular 'flash' flood are generally very intense rain, a steep, impermeable catchment, and a relatively narrow river channel. Under these conditions a large proportion of the rainfall becomes runoff, there is a very rapid rise in river stage, and river velocities are high. The results can be catastrophic, with many people killed and houses and bridges destroyed. The flood at Lynmouth in Devon is an example of such a flood.

One of the worst British floods was at Lynmouth on 15 August 1952. Note that this was a summer flood; the worst floods do not necessarily occur in winter. The East and West Lyn rivers fall steeply from Exmoor. In 12 hours, about 229 mm of rain fell onto already sodden ground (Bleasedale and Douglas, 1952; Smith, 1972; Holford, 1977). Despite a catchment area of only 101 km^2 the peak flow was estimated to be about 700 m^3/s. As a result Lynmouth was engulfed by fast flowing flood water carrying thousands of tonnes of large boulders (Fig. 13.11). Around 93 houses were destroyed or rendered unsafe, 3000–4000 people made homeless, 34 lives lost, 28 bridges destroyed, and 66 cars wrecked or swept out to sea.

Figure 13.11   Lynmouth after the flood of August 1952 [*Photo courtesy of Western Morning News, Plymouth; reproduced by permission*]

The estimated $700\,m^3/s$ peak discharge at Lynmouth is the same as the combined capacity of the river channel and flood relief channel at Exeter (Figs 8.6 and 13.14). However, at Exeter the River Exe has a catchment about eleven times that of the Lyn. Lynmouth was something of a shock for British hydrologists. Simple graphs like that in Fig. 13.12, which had been used to establish the normal maximum flood (NMF) for dam spillway design, had to be revised with an added line for a catastrophic flood = 3 × NMF (Institution of Civil Engineers, 1960). In 1975 the Flood Studies Report was published (NERC, 1975); this included a much more sophisticated and comprehensive set of techniques, which were reviewed and updated in the Flood Estimation Handbook (Institute of Hydrology, 1999).

Catastrophic floods are often headline news because they take many lives and cause widespread destruction. Based on reports in British newspapers and journals, one investigator estimated that 2000–3000 people were killed by river and coastal floods in 1976 (Holford, 1977). The true figure may be much higher, because events in some countries are not reported. When the world's major rivers are involved, flooding can spread death and destruction across large areas, so flood alleviation works have to be constructed on a vast scale; in India between 1954 and 1972, some 7250 km of embankments and 9700 km of drains were constructed. The largest scheme affording protection against river floods is on the Lower Mississippi. Siltation is constantly raising the bed of the river, necessitating the heightening of the banks on either side. As a result, cities such as New Orleans are well below the level of the banks, so if overtopping occurs the consequences are serious.

Most flood alleviation schemes only proceed if the benefits of the scheme exceed the cost of construction. The benefits can be evaluated using frequency analysis to assess how often alleviation works of various size would be used during their lifetime, and how much money would be saved by not having to pay for flood damage (see Penning-Rowsell and Chatterton, 1977; Green and Penning-Rowsell, 1986). The damage can be split into two categories: intangible and tangible. **Intangible damage** is real but difficult to evaluate in financial terms, so it is not included as part of the calculated benefits. It includes loss of memorabilia and unique possesions, stress, worry and fear that flooding may happen again. The emergency response to one fatal accident and the subsequent legal consequences can cost around £1 million. The **tangible damage** can be considerable, is easier to evaluate and provides the basis for calculating the benefits of the scheme. It includes not only the cost of repairing homes and replacing furniture, but also business losses when shops are closed, lost output from factories, payments made by insurance companies, and the cost incurred by the public and emergency services. The damage to a house increases up to a flooded depth of about 1.5 m, when it could be 10%–30% of the property's market value. With shops and factories, the cost of ruined stock and loss of production have to be considered, including lost production in other factories dependent upon supplies from the one that is flooded. Disruption to traffic and services, like gas and electricity, has to be counted. Damage arising from the flooding of agricultural land also has to be evaluated, such as lost crops or livestock. However, despite the potentially high cost of flooding, in most cases it is essential to spend no more than necessary on an alleviation scheme, so usually the cheapest options are considered first. In order of increasing cost, the alternatives are usually: channel improvements, flood walls or embankments, flood relief or bypass channels, and flood storage reservoirs.

Channel improvements may involve increasing the cross-sectional area of flow (A) and decreasing the roughness (n). This can be accomplished by dredging the river to remove sand bars and vegetation, and the removal of other obstacles such as narrow bridges. Looking at Example 8.4, it is easy to see how increasing A and reducing Manning's n would

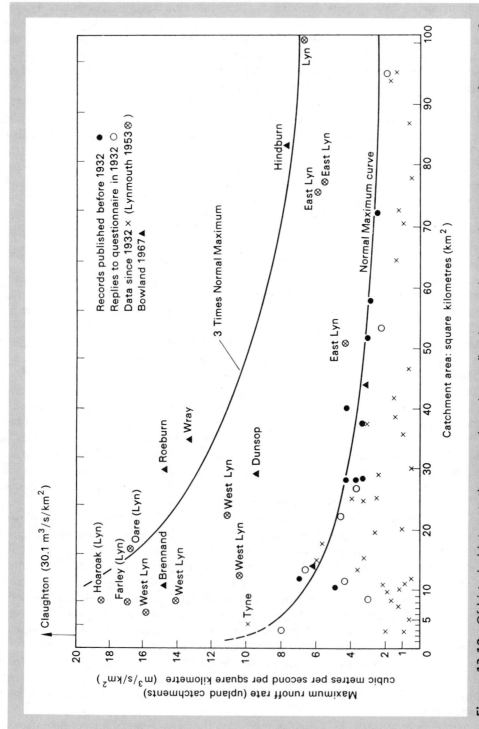

Figure 13.12 Of historical interest, the normal maximum flood curve showing the maximum runoff (m³/s per km²) from upland catchments in the British Isles. The Lynmouth flood, and subsequently others, necessitated a much higher curve = 3 × normal maximum. The 1975 Flood Studies Report and 1999 Flood Estimation Handbook largely replaced such graphs [*after Twort et al., Water Supply, 2nd edn, 1974; reproduced by permission of Butterworth Heineman*]

enable a larger discharge to pass down the river. Similarly, by straightening a meandering channel the flow distance between two points of fixed height is reduced, increasing the bed slope (S_O) and hence the discharge capacity. The Mississippi had so many meanders removed that its length was reduced by some 240 km. However, the river reacted by trying to revert to its natural regime, and it has regained 80 km. Before undertaking such work, all of the environmental repercussions of altering river channels must be considered. Many cities in the USA have vast concrete lined river channels. Although (arguably) necessary, these storm channels are environmentally objectionable and potential death-traps if anyone falls in during a flood. The environment now has a much higher priority than in the past, and current thinking and practice must reflect this (see section 8.5 and Water Space Amenity Commission, 1983; RSPB *et al.*, 1994; Brookes and Shields, 1996).

Flood walls or embankments (levees) are basically designed to contain the flood water (Fig. 13.13). However, their use can create many new problems, such as:

(a) A drainage problem inside the protected area when streams and sewers must pass through the walls.

(b) The need to extend the flood walls up any tributaries until high ground is reached.

(c) High walls can prevent access to the river, or spoil the appearance of the area. Leaving gaps or providing access gates may mean that the scheme is not fail-safe.

Figure 13.13 Canvey Island flood wall. The Thames estuary is to the right. Note that the flood defences are not failsafe: the gates have to be closed manually, which assumes prior warning of an event. In fact, storm surges are closely monitored, since that in 1953 caused the death of 58 people on Canvey Island and over 300 in Britain [*Photo © L. Hamill*]

Figure 13.14 Aerial photograph of part of the Exeter relief channel. A – Control gates on the River Exe force water over the inlet weir. B – Inlet weir to the relief channel. C – Railway bridge over the relief channel (see Fig. 8.6). D – River Exe. E – Outlet weir [*Photo courtsey of the Environment Agency; reproduced by permission*]

(d) Cutting off the floodplain reduces storage so the discharge in the affected reach must be higher to compensate (i.e. the outflow in equation (12.1) increases). Some argue that this was partly responsible for the 1993 Mississippi flood. Thus it may be desirable to set the walls some distance back from the river.

(e) The possibility of increased flooding downstream of the protected area.

(f) Environmental repercussions, such as loss of wildlife habitat and poor aesthetics. An example of a sympathetic approach is illustrated in Fig. 8.14.

If it is not possible to improve the channel or to contain the flood water, it may be possible to construct a new channel to carry some of the flood flow. Figures 8.6 and 13.14 show part of the scheme for Exeter, where the flood relief channel runs parallel to the river. Sometimes, where the topography permits, it may be possible to create a new channel which either cuts across a bend or which bypasses the area to be protected. With small streams and rivers, a tunnel may be a viable option, but with large flows this may be too expensive.

A partially full reservoir can be used to store part of a flood, in which case the outflow will be less than the inflow (equation (12.1)). Water supply reservoirs can be used for this purpose, provided they are located a reasonably short distance above the area to be protected and they are properly managed. To eliminate the possibility of the spillway capacity being exceeded and overtopping of the dam (a common cause of failure), it may be the practice to reduce the reservoir level if a large flood is expected. However, in countries with a water shortage there may be a temptation to keep the reservoir full to maximise supplies,

Figure 13.15 The rolled concrete dam for the flood storage reservoir on the River Lemon above Newton Abbot, Devon. The reservoir is empty, except when storing flood water. The downstream flow is controlled by a sluice gate in the dam. The low part of the dam forms an emergency spillway [*Photo © L. Hamill*]

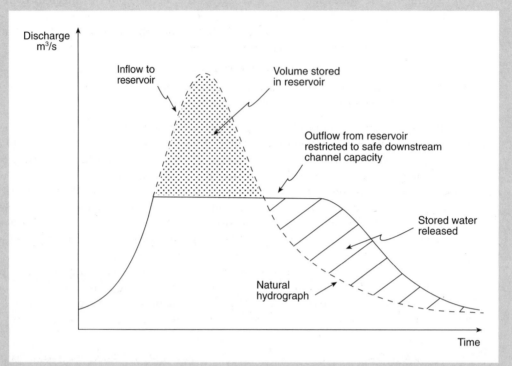

Figure 13.16 A flood storage reservoir restricts the outflow to the safe capacity of the downstream river channel. It effectively cuts the top off a hydrograph with the dotted part representing the volume of water that has to be stored. The cross-hatching shows the stored water being released into the channel after the flood peak has passed

particularly towards the end of the wet season. Consequently there may be insufficient storage available for flood protection should a storm occur suddenly, and this can have tragic results (e.g. Machhu II, Table 9.3). This conflict of interest can be avoided by constructing a dam whose sole purpose is to store flood water, the reservoir normally being empty (Fig. 13.15). This means that very little land is needed and, provided the area is farmed appropriately, little damage should accrue from temporary inundations. It also means that the dam can be located in the ideal position to afford maximum protection. Usually such dams have a sluice gate or similar device to restrict the maximum outflow to the safe capacity of the downstream channel (Fig. 13.16). Flood storage reservoirs effectively cut the top of the flood hydrograph, storing the water until the flood has passed, and then releasing it as quickly as possible (to empty the reservoir, ready for another flood). Thus the design volume equals the area under the hydrograph and above the safe capacity of the downstream channel. This means that the shape of the hydrograph must be known; a unit hydrograph could be used to determine this, but the maximum instantaneous flow derived from frequency analysis would be insufficient by itself.

Flooding can also be a problem in urban areas when the storm water sewers are inadequate.

Table 13.9 Some typical values of the runoff coefficient C in urban/city areas

Type of area	C
Considerable paved areas	0.9–1.0
Average	0.5–0.7
Residential	0.3–0.6
Industrial	0.5–0.9
Playground, parks	0.1–0.4

13.5 Storm water sewer design using the rational method

Intense rain storms can deposit large quantities of water onto streets and pavements, and this often causes flooding of near-by properties (Fig. 13.17). Storm water sewers prevent flooding in built-up areas by carrying away surface runoff. They should not be confused with foul sewers, which carry sewage, and which are assumed below to be separate. Although computer software is available to design storm water sewers, the rational or Lloyd-Davies method is suitable for small schemes. It provides a useful introduction to the subject and illustrates nicely some of the principles discussed earlier.

Several simple equations have been devised to predict the peak flood discharge (Q_P) resulting from rainfall of intensity (i) falling onto a catchment (of area A). In Britain, from 1906 onwards the following became known as the rational or Lloyd-Davies equation:

$$Q_P = CiA \tag{13.12}$$

where C is a coefficient of runoff that represents the characteristics of the catchment, such as the proportion which is impermeable and will give rise to surface runoff (Table 13.9). For large greenfield areas a suitably low value of C would be adopted, while a high value would be used for roads and pavements. The coefficient (C) is dimensionless, so to obtain a numerically and dimensionally correct equation with Q_P in m³/s, A in km² and i in mm/h, equation (13.12) becomes:

$$Q_P = 0.278CiA \tag{13.13}$$

This simple equation can also be used to obtain very quick estimates of peak flows in rivers, although errors of 100% or more might be expected, as indicated by the wide range of C in Table 13.9. With natural catchments, part of the problem lies in knowing what value would be appropriate. However, for the design of surface water sewers in urban areas, some of this uncertainty can be removed by including only the impermeable surfaces (roofs, pavements, roads) for which C will be 1.0, or very close to 1.0. This can be justified on the basis that the most intense convectional storms occur in summer when the unpaved areas have a soil moisture deficit which will absorb most of the rain falling on them. Thus the rational equation illustrates some of the principles discussed earlier.

The area of the impermeable surfaces (A_{IMP}) for which $C = 1.0$ can be determined from maps by direct measurement using a planimeter or digitiser, in which case the rational equation becomes:

$$Q_P = 0.278iA_{IMP} \quad \text{(with } C = 1.0) \tag{13.14}$$

Figure 13.17 Urban flooding at Teignmouth, Devon. Because the storm sewer is overloaded, water is backing up in the manhole and spilling into the street. This can also happen if the sewer cannot discharge freely, perhaps because the outfall is covered by floodwater. Solutions can include larger diameter sewers and providing more storage in the system to delay and reduce the peak flow [*Photo © F. Walters; reproduced by permission*]

The rainfall intensity (i mm/h) is the average value during the time of concentration (t_C, see below), and so varies according to t_C, geographical location and the chosen return period. The intensity can be obtained from actual observations (see section 12.3.2), the Meteorological Office, charts and tables, or it may be included in some computer software. Alternatively, for simple tutorial questions and rough estimates, empirical formulae like those in Table 12.5 can be used with t being taken as t_C, the **time of concentration** at the design point. The time of concentration (t_C) is the time at which runoff from the most remote part of the catchment arrives at the design point, and hence the earliest time at which all of the catchment contributes to flow. The assumption that $t = t_C$ results in the largest value of Q_P. At a design point, t_C (min) consists of two components: the time of entry (t_E min) and the time of flow (t_F min), thus:

$$t_C = t_E + t_F \tag{13.15}$$

where t_E is the time required for a rain falling just inside the catchment boundary to flow over the ground and enter the sewer system, and t_F is the time subsequently required to flow through the sewer system to the design point.

To improve the accuracy of storm sewer design and reduce the tendency to over-estimate Q_P, HR Wallingford produced the modified rational method that can be employed with drainage areas of up to 1.5 km², $t_C < 30$ min and pipes up to 1.0 m diameter. The procedure is largely unchanged, but now C consists of two components:

$$C = C_V C_R \tag{13.16}$$

where the routing coefficient C_R has a recommended constant value of 1.30, and C_V is the volumetric runoff coefficient which (again) is the proportion of rainfall on the catchment that appears as surface runoff in the sewer. This is assumed to be the runoff from impermeable areas, but allowing for some loss of rainfall through cracks, depression storage and some runoff onto absorbent surfaces. Recommended values of C_V are: overall average value about 0.75; catchments with rapidly draining soils about 0.6; and catchments with heavy soils about 0.9. Note that the recommended average values give $C = 1.30 \times 0.75 = 0.98$, almost as assumed in equation (13.14). However, the modified method uses a longer time of entry (see Table 13.10) and so results in lower values of i and Q_P. The larger values of t_E are applicable to large, flat sub-catchments with an area >400 m² and slope <1 in 50, and the smaller values to small, steep sub-catchments with an area <200 m² and a slope >1 in 30.

The design procedure itself is best illustrated by refering to Example 13.8. The comments afterwards explain the calculations in Table 13.11. Note that if the rational method does over-estimate Q_P, this may not altogether be a bad thing since it errs on the safe side and allows for some future increase in the paved area. Under-design does not.

Table 13.10 Recommended time of entry, t_E [after National Water Council, 1981; reproduced by permission of HR Wallingford]

Return period	t_E
5 years	3–6 min
2 years	4–7 min
1 year	4–8 min
1 month	5–10 min

Table 13.11 Sewer design calculations using the rational method in Example 13.8

1 Sewer ref. no.	2 Sewer length L (m)	3 Gradient	4 Guessed diameter (m)	5 Q_{FULL} (m³/s)	6 Mean flow velocity V(m/s)	7 $t_F =$ $L/60V$ (min)	8 t_C (min)	9 i (mm/h)	10 A_{IMP} for each sewer (km²)	11 ΣA_{IMP} (km²)	12 $Q_P =$ $0.278i\Sigma A_{IMP}$ (m³/s)	13 Comments
1.0	40	1 in 250 0.40 m/100 m	0.150	0.013	0.73	0.91	7.91	40.9	0.008	0.008	0.091	Fail
			0.375	0.145	1.30	0.51	7.51	41.9			0.093	OK
2.0	20	1 in 150 0.67 m/100 m	0.150	0.016	0.94	0.35	7.35	42.3	0.001	0.001	0.012	OK
1.1	30	1 in 100 1.00 m/100 m	0.375	0.230	2.1	0.24	7.75	41.3	0.003	0.012	0.138	OK
1.2	60	1 in 300 0.33 m/100 m	0.375	0.130	1.18	0.85	(8.60)		None	(0.012)	0.138	Fail
			0.450	0.210	1.32	0.76	(8.51)		None	(0.012)	0.138	OK

EXAMPLE 13.8

A plan of a sewer network for a housing estate is shown in Fig. 13.18 and additional details are given below. The pipes are concrete with 'O' ring joints and roughness $k = 0.15$ mm. The sewers are to be designed for a return period of 1 in 1 year, and it can be assumed that the rainfall intensity is given by $i = 650/(t + 8)$ mm/h. Using the rational method, determine a suitable diameter for the pipes.

Sewer reference no.	Length (m)	Gradient	Impermeable area (km²)	Comments
1.0	40	1 in 250	0.008	
2.0	20	1 in 150	0.001	Branch
1.1	30	1 in 100	0.003	
1.2	60	1 in 300	None	Carrier sewer

From Table 13.10, take the time of entry as 7 min. Since impermeable areas are given, assume $C = 1.0$. The HR chart in Fig. 6.15 can be used to obtain Q_{FULL} and V. The calculations are shown in Table 13.11 and a commentary is given in the text below.

Column 1. Draw a plan of the sewer network (Fig. 13.18). Working back from the outfall, the longest run of pipes is the primary sewer, the most upstream length of which is designated as 1.0, the next downstream as 1.1, the next as 1.2, and so on. The sewers in the most upstream branch are numbered 2.0, 2.1, 2.2, etc., and the next branch 3.0, 3.1, 3.2, etc. Thus the first number indicates relative position in the network, and the second number after the decimal point the location within the particular branch.

The design process progresses pipe by pipe, from upstream to downstream, picking up the branches like 2.0 and 3.0 as they occur. Each numbered length is designed individually. The design point is always at the downstream end, except for the special case of trunk or carrier sewers (see below). At each design point the value of t_C is calculated and used to obtain i.

Column 2. Determine the length (L) of the sewers. This is normally the plan length measured off a drawing, not the exact slope length (the difference is usually unimportant).

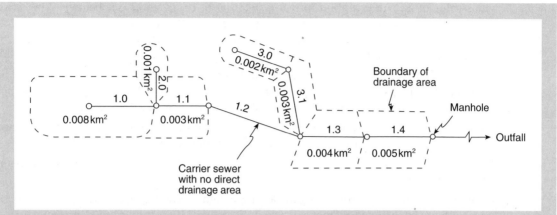

Figure 13.18 Diagrammatic representation of a sewer network and the area draining to individual pipes for Example 13.8. Connections from houses and road gullies are not shown, only the main sewers

Column 3. When possible, make the gradient of the sewer equal to the slope of the ground surface so that the depth of excavation is constant. Pipes should be at least 1.0 m deep to protect them from point loads (e.g. the wheels of vehicles), but generally as shallow as possible since 60% of the cost of construction may arise from excavating and backfilling the trench, and this increases with depth. The gradient is given in Example 13.8, but it must be expressed in m/100 m to be consistent with the right-hand scale of the HR Wallingford pipe design chart in Fig. 6.15.

Column 4. Guess the diameter of pipe required. Unless there is good reason not to, start with the minimum 0.150 m diameter. After that, the standard pipe sizes increase in 0.075 m increments to 1.200 m diameter, and then 0.150 m increment to 2.4 m. Obviously only commercially available pipes in standard sizes would be specified. Within a particular branch, pipe diameters increase in the downstream direction as the flow increases and never reduce in diameter, even if the calculations suggest it as a possibility (e.g. pipe 1.1 is 0.375 m, the same as 1.0, despite the steeper gradient and the possibility that a smaller diameter may suffice).

If the wrong diameter is guessed and the pipe fails, just repeat the calculation on the line below. It should never be necessary to repeat the calculation more than once. For instance, with sewer 1.0 a 0.150 m diameter pipe fails, but we know from column 12 that Q_P is about 0.091 m^3/s, so from the HR chart in Fig. 6.15 it is possible to determine which of the standard pipe sizes could accommodate this. Here it is a 0.375 m pipe.

Column 5. Knowing the pipe gradient and its guessed diameter, the value of Q_{FULL} is obtained from the HR chart (i.e. Q in Fig. 6.15). This is the discharge capacity of the pipe when flowing full, so the value of Q_P estimated from the rational equation in column 12 has to be less than this. If it isn't, then a larger pipe must be adopted and the calculation repeated.

Column 6. Knowing the pipe gradient and its guessed diameter, the mean flow velocity (V) in the full pipe is obtained from the HR chart in Fig. 6.15.

Column 7. The time of flow in a particular sewer length $t_F = (L/60V)$ min.

Column 8. For the first sewer of a branch (i.e. 1.0, 2.0, 3.0, etc.) the time of concentration is $t_C = t_E + t_F$ where in this example $t_E = 7$ min. For the second sewer (i.e. 1.1, 2.1, 3.1, etc.), and all subsequent sewers in that branch, just keep adding the new value of t_F to the previous value of t_C. The exception to this is a carrier or trunk sewer (see below). Where two sewers join, to reduce the tendency to over-design, use the longest upstream value of t_C. For example, for sewer 1.1, use 7.51 min (not 7.35 min) so that t_C = 7.51 + 0.24 = 7.75 min.

REMEMBER!!!

Column 9. In professional practice the rainfall intensity for a given return interval would be obtained from maps and tables or software published with the Wallingford procedure, but for brevity and simplicity the formula in Table 12.5 has been used assuming $t = t_C$.

Column 10. This is the impermeable area (A_{IMP}) drained by each sewer length. It is assumed $C = 1.0$, so C is omitted from the rational equation in column 12.

Column 11. This is the total impermeable area (ΣA_{IMP}) drained by all of the sewers upstream of the design point.

Column 12. With $C = 1.0$ the peak flow at any design point is $Q_P = 0.278\, i\, \Sigma A_{IMP}$ m³/s. If $Q_P < Q_{FULL}$ then the assumed diameter is OK; if $Q_P > Q_{FULL}$ then the assumed diameter is too small and should be increased. Note that increasing the diameter by one increment can increase Q_{FULL} by a relatively large amount. Consequently a marginal failure may sometimes be accepted, but this could result in the sewer's design capacity being exceeded slightly more often than originally intended, perhaps with some slight surface flooding. On a housing estate this may not matter; on a motorway it may be dangerous. It is also possible that a marginally undersize sewer may cause water to rise above the crown of the pipe in the upstream manhole. This would turn the normal gravity flow into pressurised flow, increasing the hydraulic gradient and hence Q_{FULL}. However, if the pipe is too small then water may spill out of the manhole and cause flooding (Fig. 13.17). In the example, a 0.375 m diameter pipe might be acceptable for sewer 1.2 but, since the other pipes have spare capacity and to allow for future development, 0.450 m has been adopted.

Pipe 1.2. This is a carrier or trunk sewer that merely transfers the flow between two points and does not have any area draining directly to it (e.g. a pipe passing under a road embankment). In this special case the design point is assumed to be at the upstream end, so all that is really required is to ensure that the gradient and diameter of the pipe are sufficient to carry the 0.138 m³/s flow from upstream. The time of flow in the carrier sewer is not needed to design it, but t_F is calculated and used to obtain t_C for the pipes downstream. In the example, sewer 1.2 has a time of flow of 0.76 min, giving $t_C = 7.75 + 0.76 = 8.51$ min (shown in brackets because it is not needed on this line). However, when Example 13.8 is continued in Self Test Question 13.5, pipe 1.3 immediately downstream of the carrier sewer has $t_C = 8.51 + 0.47 = 8.98$ min, so the value is used subsequently.

SELF TEST QUESTION 13.5

The data below refer to the sewer network in Fig. 13.18. Continue the calculations in Example 13.8 and hence determine a suitable diameter for the remaining sewers.

Sewer reference no.	Length (m)	Gradient	Impermeable area (km²)	Comments
3.0	35	1 in 160	0.002	Branch
3.1	40	1 in 160	0.003	Branch
1.3	41	1 in 250	0.004	Main sewer
1.4	39	1 in 400	0.005	Main sewer

13.6 ▶ Water supply reservoirs

The discharge of a river is usually high in winter and low in summer (e.g. Table 13.1). At any time the permissible abstraction rate is less than the flow in the river: there always has to be some remaining discharge downstream of the abstraction point (except possibly when the abstraction point is at the tidal limit). The river flow which must never be diminished by abstraction is called the **minimum residual flow** (MRF) until it becomes prescribed by statute, when it is known as the **prescribed flow** (PF). The PF varies widely (e.g. 0–30% of average daily flow) depending upon the importance attached to appearance, fishing, effluent dilution, navigation, etc. (Twort *et al.*, 1994). Thus, only something less than the lowest annual flow can be abstracted straight from a river throughout an entire year. To maximise the yield, a reservoir is needed to balance the seasonal variations in the flow by storing surplus water for later use. They are also needed to smooth fluctuations in the demand for water.

Rainfall and river flow vary from year to year. Usually reservoirs are designed for either the worst drought on record or for a drought of specified return period and duration, e.g. a 1 in 50 year, 12 month drought. In the latter case the frequency analysis technique described in section 13.3.2 is applicable. The minimum observed monthly flows can be ranked, percentage probabilities calculated and plotted, and extrapolation undertaken – as for flood flows. However, with drought flows we are usually interested in the percentage of time that a flow will *not* be exceeded, which is $(1 - P\%)$ when $P\%$ is the probability that a flow will be exceeded (see Self Test Question 13.7).

There are basically three types of impounding reservoir. The water from **direct supply reservoirs** is piped to its destination, with some **compensation flow** being allowed to pass through the dam into the downstream river channel. The level of compensation varies, but typically in the UK it may be equal to the flow which is normally exceeded for 90% of the time (see Example 13.4). **River regulating reservoirs** dispense with the pipeline, thereby reducing costs, and use the river to deliver the water to an abstraction point downstream. Such reservoirs artificially control the river flow, so during dry periods the abstraction rate depends upon the regulating discharge. They result in a slightly higher yield by virtue of the natural increase in river flow between the dam and the abstraction point. However, they are more complex to analyse and operate, since releases take several days to reach the abstraction point. Of course, the river must flow towards the water's final destination, and the use of the river to transport the water does greatly increase the risk of water supplies being badly polluted through agricultural accidents or spillages. **Pumped storage reservoirs** are generally not located on the river which supplies the water. Surplus river flow is abstracted and pumped to the reservoir for later use. The stored water can either be released back to the river or transferred via a pipeline to where it is needed.

Only direct supply reservoirs will be considered below. Many simple reservoir problems can be solved using equation (12.1). Over a specified period of time:

Inflow – Outflow = ±Change in storage   (12.1)

If the calculations are performed on a monthly basis as in Example 13.9, the inflow is the total river flow into the reservoir during a particular month. Outflow is the total monthly demand plus the compensation flow that must be allowed (for environmental and legal reasons) through the dam into the river downstream. Note that the examples assume a constant monthly demand, but it could be varied throughout the year to reflect actual water use. The difference between the inflow and outflow indicates by how much the content of the reservoir increases or decreases during that month. The cumulative monthly change gives the overall storage deficit or surplus.

Reservoir storage problems can be solved using either a table or a graph. In Example 13.9, the problem is first solved using a table, then in Example 13.10 the data are plotted as a mass flow curve (Fig. 13.19). The same answers are obtained either way, although the mass flow curve is more versatile. It consists of a plot of the cumulative inflow (ΣI) and cumulative outflow (ΣO) volume against time. Some points to remember are as follows.

■ Where the two curves touch $\Sigma I = \Sigma O$, so if the reservoir is initially full, it will again be full when the curves cross. If the reservoir is partially full initially, it will be at the same level again where the curves cross.

■ A steep gradient represents a large inflow or outflow, a flat gradient a small inflow or outflow.

■ Where the ΣI curve drops below the ΣO curve, the cumulative inflow is less than the cumulative outlow, so the deficit has to be supplied from storage.

■ Where the ΣI curve rises above the ΣO curve, the cumulative inflow exceeds the cumulative outlow so, if the reservoir was full initially, this indicates the inflow surplus that will discharge over the dam's spillway. To stop this happening, either the reservoir's storage capacity must be increased or the demand (outflow) increased.

It should also be appreciated that the reservoir may have some **dead storage**, that is water that cannot be released or used for water supply. This may occur because the outlet is set above the valley bottom to avoid entraining bed sediment. Some reservoirs experience severe sedimentation problems as a result of the river-borne silt carried into the reservoir. Dams in areas of severe soil erosion have been known to fill with sediment in only a few years (Twort et al., 1994). Depending upon the size of the catchment, the yearly suspended sediment load can be anywhere between 10^4 tonnes (from a catchment area of $10^4 \, km^2$) and 10^9 tonnes (from 10^6 to $10^8 \, km^2$). In hot countries, allowance for evaporation may have to be made when calculating the effective storage and yield.

EXAMPLE 13.9

The figures below show the river flow in units of $10^6 \, m^3$ at a potential reservoir site during a drought year. To be viable, the reservoir must provide at least $5.00 \times 10^6 \, m^3$ per month throughout the year. A constant compensation flow of $0.81 \times 10^6 \, m^3$ per month must pass through the dam (including the months when flow passes over the spillway). Assume that the reservoir is initially full, that it must refill by the following April, and that all months are of equal length. Using a tabular method, determine (a) whether or not the reservoir could sustain the demand during this drought year; (b) the volume of storage required; (c) whether any flow passes over the dam's spillway.

Month	Apr	May	Jun	Jul	Aug	Sep	Oct	Nov	Dec	Jan	Feb	Mar
Inflow	3.15	2.53	2.05	1.88	1.50	1.85	8.70	9.13	13.85	12.65	10.98	6.83

The total demand per month $= (5.00 + 0.81) \times 10^6 = 5.81 \times 10^6 \, m^3$. Assume that this is an average value that can be considered constant throughout the year. The calculations are shown in Table 13.12. They proceed month by month with the change in storage S = Inflow − Outflow.

Table 13.12 Inflow, outflow and storage calculations for Example 13.9

Month	Inflow $(10^6\,\text{m}^3)$	Outflow $(10^6\,\text{m}^3)$	Change in storage, S $(10^6\,\text{m}^3)$	ΣS $(10^6\,\text{m}^3)$
Apr	3.15	5.81	−2.66	−2.66
May	2.53	5.81	−3.28	−5.94
Jun	2.05	5.81	−3.76	−9.70
Jul	1.88	5.81	−3.93	−13.63
Aug	1.50	5.81	−4.31	−17.94
Sep	1.85	5.81	−3.96	−21.90
Oct	8.70	5.81	+2.89	−19.01
Nov	9.13	5.81	+3.32	−15.69
Dec	13.85	5.81	+8.04	−7.65
Jan	12.65	5.81	+6.84	−0.81
Feb	10.98	5.81	+5.17	+4.36
Mar	6.83	5.81	+1.02	+5.38

(a) The yield can be sustained during the dought year since the total storage change over the 12 month period is $+ 5.38 \times 10^6\,\text{m}^3$, indicating that inflow exceeds outflow by this amount.

(b) The volume of storage required is $21.90 \times 10^6\,\text{m}^3$.

(c) Since the reservoir is full initially, the surplus inflow ($5.38 \times 10^6\,\text{m}^3$) discharges over the spillway.

EXAMPLE 13.10

Using the same data as in the previous example, draw the mass flow curve and (a) determine the storage required to meet the given demand; (b) investigate the maximum possible demand that could be obtained, and any additional storage requirement. For convenience, assume each month has 30 days.

The cumulative mass inflow and outflow curves must be drawn. Assuming a uniform demand throughout the year, the cumulative outflow line (ΣO) will be straight with a value of zero initially and $12 \times 5.81 \times 10^6 = 69.72 \times 10^6\,\text{m}^3$ at the end of the year. With units of $10^6\,\text{m}^3$, the monthly inflow (I) and cumulative inflow (ΣI) is:

Month	Apr	May	Jun	Jul	Aug	Sep	Oct	Nov	Dec	Jan	Feb	Mar
I	3.15	2.53	2.05	1.88	1.50	1.85	8.70	9.13	13.85	12.65	10.98	6.83
ΣI	3.15	5.68	7.73	9.61	11.11	12.96	21.66	30.79	44.64	57.29	68.27	75.10

The cumulative inflow and outflow curves are shown in Fig. 13.19. The reservoir is initially full. From April to the end of September, ΣI is less than ΣO so the reservoir empties, the difference AB representing the volume of reservoir storage required if the demand is to be met. From October onwards, inflow exceeds outflow and the reservoir refills. The reservoir is full again where the two lines (ΣO and ΣI) cross. In February and March, ΣI exceeds ΣO so the difference CD represents the total amount of water that passes over the dam's spillway. This could be prevented by providing additional reservoir capacity equivalent to CD, i.e. by raising the height of the dam.

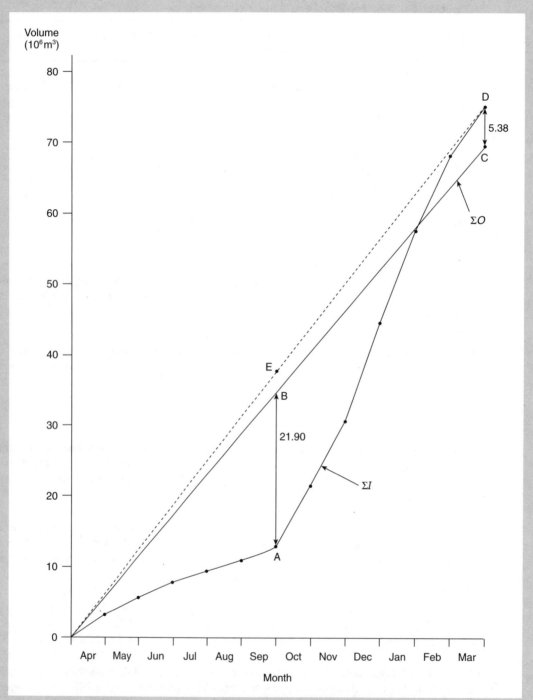

Figure 13.19   Mass flow curve for Example 13.10 showing the cumulative inflow (ΣI) and cumulative outflow (ΣO) volumes. The reservoir is full initially, and full again when the two lines cross. Thus AB represents the storage volume needed to meet the demand, and CD the volume discharged over the spillway

(a) For the specified total demand of $5.81 \times 10^6 \, \text{m}^3$/month, the required storage is given by the ordinate AB = $21.90 \times 10^6 \, \text{m}^3$.

(b) As indicated above, a volume CD = $5.38 \times 10^6 \, \text{m}^3$ passes over the spillway and is 'wasted'. Thus the reservoir could support a slightly higher demand as indicated by the dashed line OD. This represents a total demand of $6.26 \times 10^6 \, \text{m}^3$/month, or $5.45 \times 10^6 \, \text{m}^3$/month with the compensation flow deducted. However, to achieve this yield the reservoir storage must be increased to AE = $24.60 \times 10^6 \, \text{m}^3$. The reservoir would still refill at D, at the end of the 12 month period.

SELF TEST QUESTION 13.6

Continue Example 13.10 by assuming that the reservoir is still full initially, but the maximum reservoir storage is now limited to $15.00 \times 10^6 \, \text{m}^3$. (a) What is the maximum uniform demand that can be maintained throughout the year, and (b) how much water will discharge over the spillway?

SELF TEST QUESTION 13.7

The data below are the minimum instantaneous discharges recorded at a gauging station during a period of 10 consecutive years. There is a proposal to abstract water from just above the gauging station. Determine in how many years on average the flow is likely to be less than $1.50 \, \text{m}^3$/s. Comment upon the accuracy of the result.

Year	1	2	3	4	5	6	7	8	9	10
Min. flow ($Q\,\text{m}^3$/s)	3.97	2.77	2.42	2.90	4.37	1.99	2.11	2.69	2.73	2.54

Hint: rank the data with the highest flow = 1; calculate $P\% = 100r/(N + 1)$ and then plot $\log Q$ against $P\%$ on log-normal probability paper. Extrapolate the line to $1.50 \, \text{m}^3$/s and determine the corresponding probability (P) that this flow will be exceeded. The probability that the flow will be less than this is $(1 - P)$ and the return period is $1/(1 - P)$ years.

13.7 ▶ Groundwater

As shown in Table 12.1, groundwater comprises about 98% of the Earth's fresh water, although half of this is too deep to be usable. Nevertheless, the remainder means that groundwater is a very important source of water. In Britain the principal aquifers are the chalk (a fine grained, white limestone) of southern and eastern England, and the Bunter and Keuper Sandstones of the West Midlands (Fig. 13.20). As much as 40% of the statutory undertakings' water demands for England and Wales have been met using groundwater (Smith, 1972). In the USA, again groundwater supplies about 40% of the water used, with 96% of rural homes being supplied by wells (Todd, 1980). During a drought in California some 11 300 wells were drilled or deepened in 1976, and 19 950 in 1977. As part of a United Nations programme, 55 200 public water wells were constructed in Korea over a five year period (Anon, 1983).

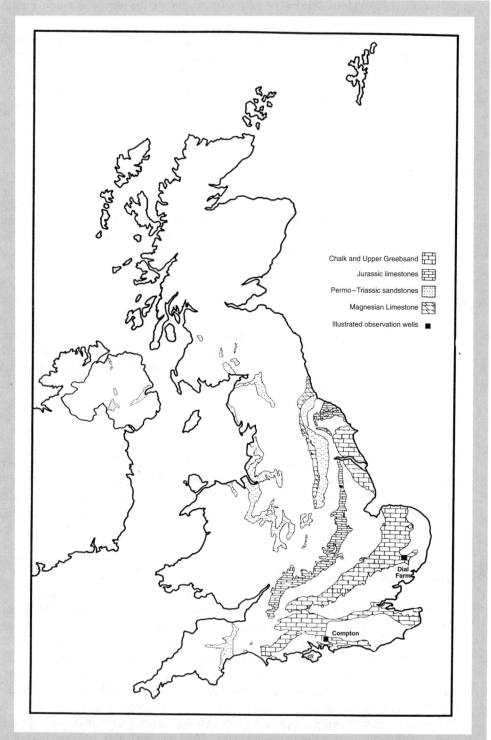

Figure 13.20 Principal aquifers in the UK. Groundwater is an important source of water in the south and east [*reproduced by permission of the Centre for Ecology and Hydrology*]

Where populations are widely dispersed or the infrastructure poorly developed, utilising groundwater locally is an attractive option; connecting every home to a water main supplied from a surface reservoir is not. Some of the other advantages of developing groundwater resources are:

■ a small visual environmental impact and a very modest land take, since the water is stored underground in aquifers, whereas Kielder Reservoir (one of the largest in Europe) has a surface area of $10.9 \, km^2$;

■ with proper planning, additional wells can be phased-in easily as demand increases;

■ groundwater recharge and availability often operate on a different time cycle to surface water, so groundwater may provide an alternative source during some dry periods;

■ if a well is suitably located, of adequate depth and effectively operated and managed, the groundwater obtained is generally of very good quality: often only minimal treatment is needed before being put into supply.

Some of the disadvantages are:

■ groundwater is stored below ground, making it 'invisible', so locating, developing and managing aquifers requires some skill and experience if the venture is to be successful;

■ groundwater is usually pumped to the surface, so there may be additional running costs;

■ wells have a limited life owing to blockage, encrustation, corrosion or collapse, after which extensive maintenance, repair or rehabilitation may be required;

■ some aquifers are now becoming polluted as a result of over-abstraction or agricultural practices (e.g. nitrate from fertilisers, chemicals from herbicides and pesticides), so careful management of groundwater resources is needed.

13.7.1 Aquifers – types and properties

An **aquifer** is a saturated formation which yields water in sufficient quantities to be of significance as a source of supply. This means, of course, that the aquifer has to be relatively permeable so that water can flow easily through it to replace the water that is abstracted. Thus aquifers tend to consist of loose deposits like sand and gravel, or cemented material such as chalk, limestone and coarse sandstone.

Semi-permeable or impermeable formations are called **aquicludes** and **aquifuges** respectively. These may contain water, but are of low permeability so that it cannot be abstracted or replaced easily. Examples may be clay, shale and slate. Impermeable material is often to be found lying under an unconfined aquifer or sandwiching a confined aquifer.

An **unconfined aquifer** exists where the permeable material extends to the ground surface and the water table is open to the atmosphere (Fig. 13.21a). In this case, groundwater flow is the result of gravity, like water flowing down an open channel or a part-full pipe (as described in Chapter 8). On the other hand, a **confined aquifer** is covered by a layer of impermeable material (diagram b). In this case the water is under pressure, just like the pipe in Fig. 6.2. Thus a confined aquifer has a **piezometric level** which is above the top of the aquifer (called the **piezometric surface** when viewed two-dimensionally in plan). A **confined artesian aquifer** occurs where the piezometric level is above the ground surface, such as in a river valley or where the aquifer forms a U-shaped basin (Fig. 13.21c). If a well is drilled in the valley, the pressurised water will gush from it without the need to use a pump.

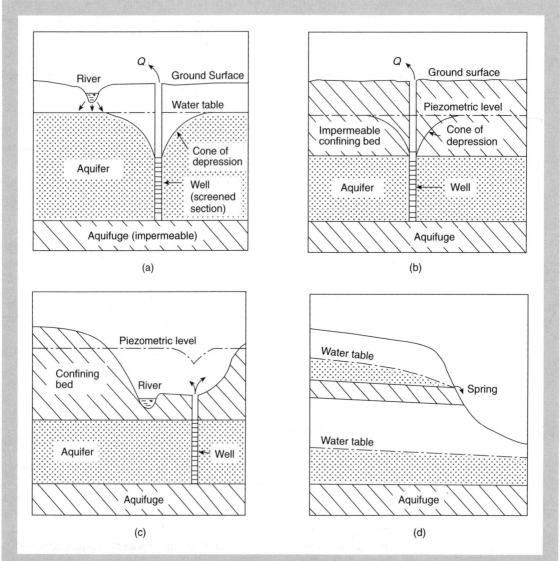

Figure 13.21 (a) Unconfined aquifer with a water table. The aquifer material extends to the ground surface. The stream may be effluent, that is it recharges the aquifer by seepage through the bed. Because of the drawdown of the water table caused by abstraction (Q), only the bottom 50% of the well casing needs slots to allow water to enter. (b) Confined aquifer sandwiched between impermeable layers. The piezometric level is above the aquifer. Drawdown is usually limited to the top of the aquifer, so all of the casing in the aquifer is slotted. (c) A confined artesian aquifer where the piezometric level is above the ground surface, so the well would initially overflow without having to use a pump. (d) A situation where there are two aquifers, one giving rise to a spring

Sometimes there may be more than one aquifer, and more than one water table or piezometric surface. Figure 13.21d shows an example. Here the impermeable material underlying the upper aquifer gives rise to a **spring**. In the past houses or villages were often constructed to take advantage of the springs, giving rise to spring-line settlements along the side of valleys.

It is not always obvious which rock formations can be classed as aquifers and will yield significant quantities of water. This is because the material's permeability can be derived in two ways: its **primary permeability** reflects the ability of a uniformly solid piece of rock to transmit water, while its **secondary permeability** arises from cracks, joints, fissures and solution openings (i.e. modification processes after the rock was formed). Thus the secondary permeability is extremely important, because it is much easier for water to flow through a wide crack in the rock than through the solid material. The success or failure of a water well can depend upon the number and location of the cracks it intersects below the water table. If the flow through the cracks is sufficient to replace the abstracted water, this may allow a modestly successful well to be constructed even in material with a low primary permeability.

The **coefficient of permeability** or hydraulic conductivity (k) is defined as the flow in m^3/d through a $1\,m^2$ cross-section of the aquifer (denoted by A in Fig. 13.22). Thus the units of permeability are m^3/d per m^2, or m/d (the unit of time is days, not seconds, simply because groundwater velocities are very small). Some typical permeabilities are shown in Table 13.13. Normally it is easier to estimate or measure the flow through the full thickness of an aquifer

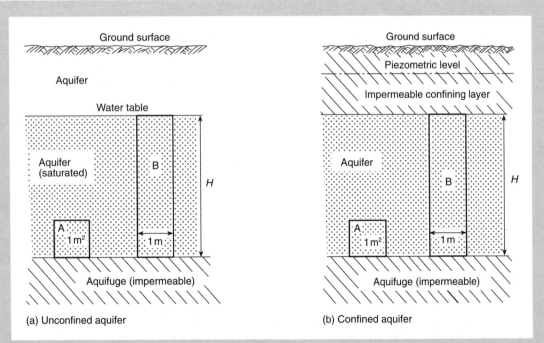

(a) Unconfined aquifer

(b) Confined aquifer

Figure 13.22 For both (a) unconfined and (b) confined aquifers permeability, k, is the flow in m^3/d through a $1\,m^2$ section (i.e. A). Transmissivity, T, is the flow in m^3/d through a $1\,m$ wide section (i.e. B) that has the full thickness H of the aquifer. Thus $T = kH$

Table 13.13 Approximate properties of some common materials

Note the high specific retention (i.e. porosity minus specific yield) of fine grained material. If clay has a porosity of 50% and specific yield of 5%, its specific retention is 45%. Clay is relatively impermeable but contains a lot of water (high porosity), so it shrinks during a drought and can cause subsidence problems with buildings.

Material	Permeability (m/d)	Porosity (%)	Specific yield (%)	Classification
Sand – fine	10^0–10^1	40–45	10–25	Aquifer
– coarse	10^1–10^2	35–40	30–35	Aquifer
Gravel – fine	10^2–10^3	30–35	25–30	Aquifer
– coarse	10^3–10^5	18–25	15–20	Aquifer
Sandstone	10^{-2}–10^3	10–20	5–15	Aquifer/aquiclude
Limestone	10^{-3}–10^2	1–20	0.5–10	Aquifer/aquiclude
Clay	10^{-5}–10^{-3}	45–55	1–10	Aquiclude/aquifuge
Shale	10^{-6}–10^{-4}	1–10	0.5–5	Aquifuge/aquiclude

(i.e. B in Fig. 13.22). Thus the **coefficient of transmissivity** (T) is defined as the flow in m^3/d through a 1 m wide strip of the aquifer, and its units are m^3/d per m, or m^2/d. If the aquifer thickness is H, then clearly $T = kH$.

The quantity of water that can be abstracted from a well also depends upon how much water is actually stored in the pore spaces of the material, such as in the spaces between sand or gravel particles or between the grains of a rock. Thus considering $1\,m^3$ of material, its **porosity** (%) is the proportion occupied by voids that can be filled with water (or air or oil). However, it is not possible to abstract all of this water: water can be wrung out of a wet bath sponge, but some moisture will be retained within it. The **specific yield** (%) is the proportion of the $1\,m^3$ occupied by water that can be released, while the **specific retention** (%) represents the proportion that is retained, so:

porosity = specific yield + specific retention (13.17)

Some typical values of porosity and specific yield are shown in Table 13.13. Another measure of the quantity of water available is the **coefficient of storage**, S, of an aquifer. This is the volume of water released when there is a 1 m decrease in head over a $1\,m^2$ surface area of the aquifer (Fig. 13.23). With an unconfined aquifer there is no difference between the coefficient of storage and the specific yield (other than the latter is expressed as a percentage) since both represent the volume of water obtained by 'de-saturating' $1\,m^3$ of aquifer material. Typical values of S would be between 0.02 and 0.30. However, with a confined aquifer it is the piezometric level (or pressure head) that is being reduced by 1 m, and this may not involve de-saturating the aquifer. Consequently the coefficient of storage is not the same as the specific yield, and S has values in the range 0.00001–0.001, much smaller than the values for unconfined aquifers.

The transmissivity (T) and storativity (S) are two of the most important properties of an aquifer. Often they are determined from pumping tests where water is pumped from a well and the **cone of depression** is monitored. The cone is shown in cross-section in Fig. 13.21a and b; the amount by which the normal (static) water level is reduced at any point is called the drawdown. In a uniform aquifer, when seen in plan the affected area would be circular, so in three dimensions it resembles an inverted cone that represents the part of the

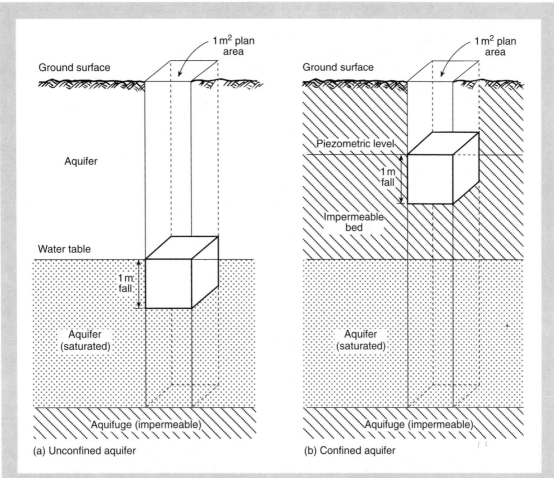

Figure 13.23 The coefficient of storage (S) represents the volume of water released from an aquifer when the water level falls by 1 m over a plan area of 1 m². (a) With an unconfined aquifer, S is the same as the specific yield. (b) With a confined aquifer, the piezometric level or pressure head is reduced by 1 m but the aquifer itself is not dewatered. Thus S is not the same as specific yield

aquifer that is no longer saturated. The size of the cone depends upon T, S, the pumping rate (Q) and how long the well is pumped (t). This relationship is explored later when pumping tests are considered. In practice, the cone is often distorted, since modification processes, faults and variations in permeability, porosity and aquifer thickness result in aquifers that are heterogeneous and anisotropic (i.e. not uniform with properties that depend upon direction).

The discharge from a well increases with the drawdown. Usually the drawdown in a confined aquifer is restricted to the top of the confining layer, otherwise the aquifer itself will be dewatered so that part of it is unconfined and part confined. This condition is not easy to analyse, and also results in a reduction in the discharge achieved per metre of drawdown (i.e. a reduction in **specific capacity**). With unconfined aquifers, the drawdown is limited to 50%–67% of the aquifer thickness. However, there is a paradox: with 100% drawdown in a totally penetrating well, there is no contact between the saturated part of an unconfined aquifer and the well, and theoretically no flow into it!

It is desirable that a well should completely penetrate the aquifer, that is it should extend through the full thickness as in Fig. 13.21. A partially penetrating well results in a smaller yield. For example, a well that penetrates only 50% of a confined aquifer has a yield which is 60%–90% of a well that is the full depth (Hamill and Bell, 1986). One reason for this is that the permeability of an aquifer is much greater horizontally than vertically, so anything which causes vertical flow results in inefficiency. However, for reasons of economy, convenience or practicality, partially penetrating wells are often constructed; in an unconfined aquifer only the bottom 33%–50% is usually screened, since the cone of depression renders the upper part dry. In any case, the open or screened part of the hole through which the water enters should be long enough to allow the well to operate with its maximum planned drawdown and abstraction rate, allowing for fluctuations in water level (see Fig. 13.1) and drought years when levels are especially low.

13.7.2 Safe yield, groundwater recharge and Darcy's law

With any groundwater scheme there is a limit to the quantity of water that can be abstracted. The **safe yield** has been defined as the long-term rate at which water can be abstracted from an aquifer without a progressive decline of its water level or any other adverse effect (such as a reduction in water quality). One sign of over-abstraction may be that streams and rivers have a significantly decreased flow or become dry.

Unfortunately, it is difficult or impossible to determine accurately the safe yield until a groundwater scheme has been operational for some time, and even then the safe limit may not be apparent. Kazmann (1956) likened the safe yield concept to the former speed limit in the State of Tennessee, where no limit was specified in law only a request to 'please drive carefully'. However, the speed limit was deemed to have been broken if an accident occurred, since someone obviously had not been driving carefully or slow enough! Similarly, it may take an incident before the safe yield of an aquifer is known.

A more modern term for safe yield is 'sustainable yield', that is the annual yield that can be sustained well into the future. Whatever it is called, the yield can never exceed the rate at which groundwater is replenished. Thus one method of estimating the long-term yield is to establish the volume of infiltration over the aquifer's recharge area (section 12.5.1). It may then be assumed that some or all of this infiltration percolates to the water table, becomes groundwater flow and can be abstracted by wells. As described in section 13.1, groundwater recharge (via infiltration) occurs mostly in autumn and winter, so another option may be to evaluate the annual rise in groundwater level (see Fig. 13.1). If the average rise, areal extent and specific yield of an aquifer are known, then its yield can be estimated (Example 13.11).

The flow through an aquifer can be investigated by drawing a flow net (Fig. 13.24b) and by using Darcy's law. This describes the steady flow of water through a uniformly permeable material:

$$Q = khA \tag{13.18}$$

where Q is the flow rate (m^3/d), k is the permeability of the material (m/d), h is the hydraulic gradient (fraction) and A is the cross-sectional area of flow (m^2). The permeability can be determined from samples of the material, pumping tests or, much less satisfactorily, from

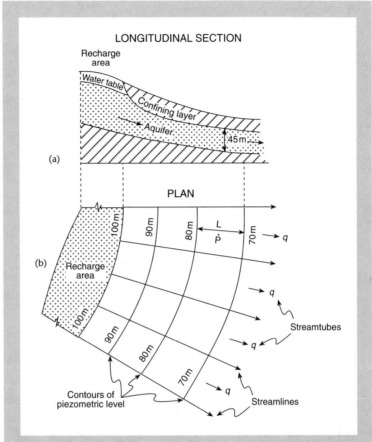

Figure 13.24 (a) Longitudinal section through a confined aquifer showing the recharge area. (b) Part of a flow net for the aquifer, consisting of streamlines and contours of piezometric level (equipotential lines). This is used in Example 13.12

tables (e.g. Table 13.13). The hydraulic gradient can be determined from the spacing of the groundwater contours (i.e. $h = \Delta h/L$ where Δh is the change in height of the water table or piezometric surface in a distance L). The area A is the width of the streamtube multiplied by the aquifer thickness (see Example 13.12).

With an aquifer of uniform thickness, if a flow net with a constant contour interval has been drawn and if the permeabilty at some point is known (k_1), then the permeability (k_2) at any other point in the aquifer can be determined from the relative distance (L) between successive groundwater contours:

$$k_2 = \frac{k_1 L_2}{L_1}$$

(13.19)

Thus widely spaced contours indicate a high permeability, and closely spaced contours indicate a relatively low permeability where the large resistance to flow results in a large head loss. Flow nets can also be used to investigate the area of diversion to a discharging well,

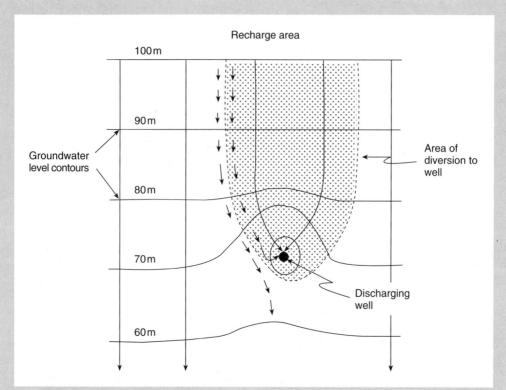

Figure 13.25 Part of a flow net in the vicinity of a well that is being pumped. The area within which water is diverted to the well is dotted. This can be used when estimating the potential yield of a well and aquifer

and hence the potential yield of the well or the aquifer (Fig. 13.25). Note that the small arrows outside and to the left of the area of diversion miss the well, whereas those just inside lead to it.

EXAMPLE 13.11

It is observed that the water table in an unconfined aquifer that extends over an area of $250\,km^2$ has an average annual rise of 5 m. The porosity of the material is estimated to be 12% and its specific yield 7%. (a) Determine the annual volume of recharge. (b) Estimate the maximum volume of water that potentially could be abstracted annually from the aquifer.

(a) The recharge volume = annual rise × area × porosity

$$= 5 \times 250 \times 10^6 \times 0.12$$

$$= 150 \times 10^6\,m^3 \text{ per annum}$$

(b) The maximum volume that could be abstracted = annual rise × area × specific yield

$$= 5 \times 250 \times 10^6 \times 0.07$$

$$= 87.5 \times 10^6\,m^3 \text{ per annum}$$

Note: it is unlikely (and undesirable) that the wells would be able to abstract all of this amount. Some flow through the aquifer downstream of the wells should be preserved, as should some groundwater discharge to rivers to maintain the baseflow.

EXAMPLE 13.12

A flow net for part of a confined aquifer is shown in Fig. 13.24. At point P the permeability is estimated as 2.5 m/d and the corresponding distance (L) between successive contours is 3 km. The aquifer is approximately 45 m thick. Estimate the flow though the aquifer.

The contour interval $\Delta h = 10$ m, so $h = \Delta h/L = 10/3000 = 0.00333$.
Assuming that the 'square' of the flow net has a width of 3 km, then the cross-sectional area of flow $A = 3000 \times 45 = 135\,000$ m^2. Thus the flow through one streamtube of the flow net is:

$$q = khA = 2.5 \times 0.00333 \times 135\,000\,\text{m}^2$$
$$= 1125\,\text{m}^3/\text{d}$$

Since there are four streamtubes in Fig. 13.24, the total flow through this part of the aquifer = $1125 \times 4 = 4500$ m^3/d. Note that this flow will vary seasonally. As in the previous example, it is unlikely that wells could abstract all of this amount, nor would it be desirable to do so.

13.7.3 Well hydraulics and pumping tests

Essentially a water well is created by drilling a hole into the aquifer. In competent rock this may be left unlined, but normally a casing will be used to provide structural support to the side of the well. The casing consists of solid steel pipe down to around the anticipated pumped water level, when it is replaced by a well screen that contains openings through which the water can enter (Fig. 13.26). Sometimes a gravel pack may be placed around the screen to increase its effective radius and to prevent particles of aquifer material entering the well. Water is abstracted from the well using a pump; this can be located on the surface with a drive shaft running down the well to the impellers, but usually a submersible bore-hole pump is located some distance below the water table. It is essential that the impellers or pump are set deep enough to remain submersed even during droughts when ground-water levels are low and the well is discharging at its maximum rate. Lifts of up to 400 m are possible, but well construction and pumping costs increase with depth, so shallow wells are usually desirable.

Under natural conditions, groundwater has a very small velocity (V) so the flow is laminar. However, when water is abstracted from a well, the flow has to pass through a progressively smaller cylindrical cross-section of the aquifer, and ultimately through the well screen itself (say 0.05–1.00 m diameter). Here the velocity is large, and the flow is turbulent. As described in section 6.5.1, in laminar flow the head loss (h_F) is proportional to V, whereas in turbulent flow h_F is proportional to V^2. Consequently the cone of depression is relatively shallow initially but steepens as the well is approached and the head loss increases (Fig. 13.26). There is also an entrance loss as the water passes through the screen into the well. This loss can be reduced by using a commercially designed well screen, which can have an open area of up to 60%. The head loss is relatively large if slots are simply cut into an otherwise solid steel pipe, when the maximum open area is only about 12%. If the entrance loss

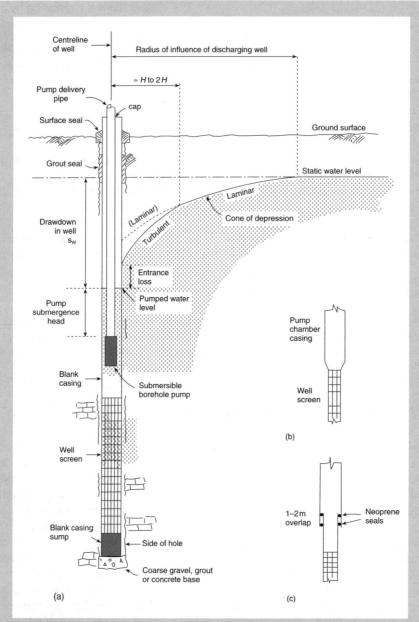

Figure 13.26 (a) Some of the key features of a water supply well with respect to construction and the geometry of the cone of depression. The drawing is not to scale: the drawdown in the laminar part of the cone of depression exceeds that in the turbulent part. The pump submergence head should be large enough to allow for seasonal fluctuations in water level and periods of overpumping. (b) Some wells reduce in diameter with depth, typically below the pump, which may determine the diameter required to that level. Construction can be a welded single string or (c) a telescopic design where the diameter gradually reduces, the smaller casings passing through the larger. The overlapping sections have neoprene seals

and the well loss caused by turbulent flow are not allowed for, it is easy to underestimate the drawdown and set the pump too high. Similarly, to avoid the complications of the turbulent flow area when conducting a pumping test, usually the nearest observation hole is located at least 1.0–1.5 times the aquifer thickness from the well. With a partially penetrating well this distance increases to 1.5–2.0 times the aquifer thickness.

The diameter of a well must be determined from a consideration of the amount of water to be discharged ($Q \text{m}^3/\text{s}$), screen length, the flow velocity through the screens, the size of the pump or impellers, and the type of construction (e.g. constant diameter or reducing with depth). The entrance velocity as the water enters the screens ($V_s = Q/a_0$) should be 0.03 m/s or less for optimum efficiency; if it exceeds 0.045 m/s then it must be reduced. This can be achieved by using a screen type with a larger open area ($a_0 \text{ m}^2$), or by increasing the screen diameter and/or length. Some indicative surface well casing diameters are: 250 mm for up to 500 m^3/d; 300 mm for 270 to 820 m^3/d; 350 mm for 550 to 2700 m^3/d, and 450 mm for 1600 to 8000 m^3/d. However, increasing the diameter does not automatically result in a larger yield. For example, increasing the diameter of a well from 150 mm to 600 mm may increase the discharge by only about 17%–25%. It is virtually impossible to double the discharge by increasing the diameter; increasing the depth of the well, increasing the screen length and increasing the percentage open area are all better options (Hamill and Bell, 1986).

The likely radius of the cone of depression needs to be known when planning a pumping test, so that the observation holes can be suitably located. Some approximate guidelines for confined aquifers are: 250–500 m in fine grained material, 750–1500 m in coarse grained material and 1000–1500 m in fissured rock. For unconfined aquifers the equivalent values are: 100–200 m, 300–500 m and 500–1000 m respectively (Brown *et al.*, 1972). Thus a cone of depression can have a diameter of up to 3 km and, if circular, a plan area of 7 km^2.

Equilibrium pumping tests

An equilibrium pumping test assumes that the well is pumped at a constant rate until the cone of depression stabilises and a constant drawdown is achieved in two observation holes located some distance away from the well. It also assumes that the aquifer is homogeneous, of uniform thickness, the water table or piezometric surface is horizontal initially, and that the well penetrates at least 85% of the aquifer. These assumptions are unlikely to be met in full. It should also be realised that because the groundwater level is always changing (Fig. 13.1), a constant drawdown is not necessarily the same as a constant water level in the observation holes.

The equilibrium Dupuit–Thiem equation is a century or more old and provides the easiest method of analysing pumping test data. It is simply based on the continuity equation $Q = AV$ where the cylindrical cross-sectional area of flow in an aquifer of thickness H(m) at a radius r (m) from the well is $2\pi rH$, and the velocity (m/d) is obtained from Darcy's law $V = k(\text{d}h/\text{d}r)$ where k is the permeability of the aquifer (m/d) and $\text{d}h/\text{d}r$ is the dimensionless slope of the groundwater surface – such as between the two observation holes. If this equation is integrated between limits representing the conditions at the two observation holes which are at distances r_1 and r_2 from the pumped well where the drawdown is s_1 and s_2 (m) respectively, then:

$$Q = \frac{2\pi kH(s_1 - s_2)}{\ln(r_2/r_1)} \qquad (13.20)$$

One disadvantage of this equation is that ideally two observation holes are required. The pumped well itself can be used as one of the observation holes if the drawdown data are modified to allow for the well losses. Even less satisfactorily, by assuming r_2 is the maximum radius of the cone of depression (see above) where $s_2 = 0$, the following very approximate expressions can be obtained:

$$\textbf{unconfined aquifer} \quad T = 1.2Q/s_w \tag{13.21}$$

$$\textbf{confined aquifer} \quad T = 1.6Q/s_w \tag{13.22}$$

where $T (= kH)$ is the transmissivity of the aquifer (m²/d) and s_w (m) is the drawdown in the well. The discharge per metre of drawdown (Q/s_w) is the **specific capacity** of the well, and is often quoted as a means of comparing yields or efficiencies.

The duration of continuous pumping required to obtain a constant drawdown can be as little as 24 h in a confined aquifer and 72 h in an unconfined aquifer when the discharge is less than 500 m³/d. Otherwise durations should be roughly 2 days (500–1000 m³/d), 4 days (1000–3000 m³/d), 7 days (3000–5000 m³/d) and 10 days (>5000 m³/d). In Britain, about 2 weeks has been an average length (see BS 6316, 1983).

Non-equilibrium pumping tests

Equilibrium pumping tests take quite a while to conduct and give no information regarding the storativity of the aquifer. More useful information can be obtained by analysing the rate of drawdown in the aquifer, when both the coefficient of transmissivity (T) and storage (S) can be evaluated. Unfortunately this is a much more difficult problem to solve mathematically. The first solution proposed by Theis in 1935 was based on a heat conduction analogy: a cold metal rod placed on a hot metal plate causes a reduction in temperature which resembles the cone of depression around a discharging water well. However, the Theis method of analysis is not easy to undertake. In 1946 Jacob proposed the simplified method described below.

Most of the Dupuit–Thiem assumptions also apply here, except that now the drawdown increases progressively during the test. There is an additional assumption that the water obtained from storage is released instantaneously with the fall in head. The simplifications made by Jacob mean that when drawing the drawdown–log time graph that forms the basis of the method, the data obtained during approximately the first hour of a test in a confined aquifer will not plot on a straight line and should be ignored; with an unconfined aquifer, the first 12 hours of data may have to be ignored. The transmissivity (T m²/d) is given by:

$$T = \frac{2.30Q}{4\pi\Delta s} \tag{13.23}$$

where Q is the pumping rate (m³/d) and Δs is the drawdown per log cycle obtained from the drawdown–log time graph (Fig. 13.27). The coefficient of storage (S) is:

$$S = \frac{2.25Tt_0}{r^2} \tag{13.24}$$

where t_0 (days) is the intercept on the time axis when the drawdown $s = 0$, and r is the radius or distance (m) from the well to the observation hole. Example 13.13 illustrates the technique. See also Box 13.2.

EXAMPLE 13.13

A non-equilibrium pumping test has been conducted in a confined aquifer. The time since pumping started and the corresponding drawdown in an observation hole 91 m from the pumped well are listed below. The pumping rate was 1800 m³/d. Using a Jacob analysis, determine the value of T and S.

Time ($t$min)	Drawdown (sm)	Time ($t$min)	Drawdown (sm) (continued)
2	0.16	60	2.39
4	0.44	80	2.72
6	0.60	100	2.87
8	0.85	120	3.13
10	0.91	180	3.54
15	1.15	260	3.78
20	1.30	360	4.20
30	1.68	600	4.71
40	1.98	840	5.05
50	2.25		

The data are plotted as the drawdown–log time graph shown in Fig. 13.27. From the graph, $\Delta s = 2.33$ m and $t_0 = 5.6$ min = $5.6/(60 \times 24)$ days. Thus:

$$T = \frac{2.30Q}{4\pi\Delta s} = \frac{2.30 \times 1800}{4 \times \pi \times 2.33} = 141 \text{m}^2/\text{d} \quad \text{and} \quad S = \frac{2.25Tt_0}{r^2} = \frac{2.25 \times 141 \times 5.6}{91^2 \times 60 \times 24} = 0.000\,149$$

13.7.4 Aquifer location and management

It should be apparent from the above that constructing and testing a large-diameter water well is a relatively expensive venture, so it should not be undertaken without a good deal of prior investigation and planning. One of the first steps may be to establish the quantity and quality of the water to be supplied, and the location of the demand. The quantity gives an indication of the required well and aquifer yield. The quality expected varies according to whether the water will be used for drinking or industrial purposes. At a later date, water samples obtained from the aquifers under investigation will have to be compared with this requirement to assess the amount and type of treatment that will be needed, and hence acceptability. Ideally any groundwater development should be located close to the centre of demand and/or a water treatment plant, to reduce the cost of distribution. These initial guidelines may indicate where (ideally) the wells should be located, assuming there is an aquifer present.

The types of formation that make a good aquifer are known (e.g. Table 13.13), and they can be located by a variety of means including geological maps and memoirs, commercial borehole logs and site investigations, aerial photographs and topographic maps. These may be used to indentify one or more possible aquifers, and to form an initial assessment of the distance from source to demand, the depth to water, transmissivity and storage, and the approximate annual yield. Geophysical surveying, such as resistivity, seismic, magnetic and gravity techniques, may be employed to provide additional information. Any aquifer which is likely to be affected by domestic, industrial or agricultural pollution, or which does not

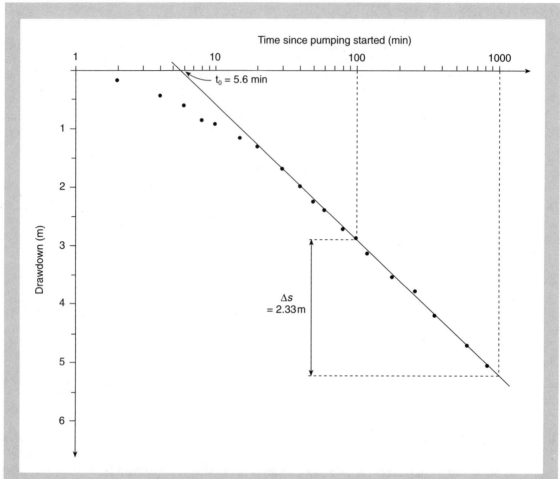

Figure 13.27 Drawdown–time graph for the Jacob pumping test analysis in Example 13.13. The drawdown per log cycle is ΔS, and t_0 is the intercept on the time axis

meet the initial guidelines, must be rejected. After this, what appears to be the best site for development should be identified, and then one small-diameter exploration or pilot hole sunk to confirm that the hydrogeology is expected, and that the water quality is acceptable. This being so, plans can be made to sink a larger well which can be pump tested, together with suitable observation holes. If the hydrogeology or water quality is unsuitable, then another location must be identified.

Once a suitable aquifer and well site have been established, the next step is to determine the long-term yield of the aquifer, the number of wells needed to meet the demand, and the design details of the wells. Factors to be considered include annual and long-term variations in groundwater level, the depth of setting of the pump, and the effect of abstraction on other wells, surface watercourses and the environment. A well or aquifer may have to be rejected if abstraction will reduce unacceptably the yield from another well, or cause surface streams and rivers to dry up. There are many other problems that may arise.

Induced infiltration occurs when abstraction lowers the water table beneath a surface stream, so that either groundwater that would have discharged to the stream is diverted to the well, or water from the stream flows into the aquifer. In the latter case, if the stream is polluted then the aquifer will also become polluted and unusable. In both cases the flow in the stream is reduced. **Saline intrusion** can occur in coastal aquifers that discharge to the sea: if the flow through the aquifer is stopped or reduced, salt water may encroach into the aquifer. Wells located close to the coast may draw sea water to them through the aquifer and have to be abandoned. There are also many examples of groundwater abstraction causing subsidence or structural damage to buildings. Best known is Mexico City, where from 1928 to 1970, subsidence of up to 8.6 m was recorded over an area of 128 km², but similar problems exist in Japan, Taiwan and several areas of the USA (Hamill and Bell, 1986). All of these potential problems must be evaluated.

Once an aquifer has been developed, there is still the task of managing the resource to ensure that it continues to provide water of the required quantity and quality for many years. This may include establishing the sustainable yield, and identifying sensitive areas where commercial or other activities have to be limited or controlled. For example, the aquifer's recharge area would be an unsuitable location for any form of domestic or industrial waste disposal activities. Old quarries are often used for this purpose, but this would be unacceptable if it provided a means for highly polluted leachate to enter the aquifer and flow to the wells. With shallow unconfined aquifers, some consideration may need to be given to farming practices. For example, the liberal application of nitrate fertilisers has caused nitrate pollution in some aquifers, with the result that the groundwater has to be treated or mixed with water from another source. The potential for accidents, such as traffic accidents involving spillage from tankers carrying toxic chemicals, must also be considered, and contingency plans developed.

Good management also involves operating and maintaining the wells properly. This includes not overpumping a well or well group for long periods, which can result in highly mineralised water from deep in the aquifer being drawn into the well, or in a much higher velocity through the well screen than intended. The latter may erode the finer fraction of the aquifer material so that the well becomes a 'sand-pumper'. Another consequence of overpumping may be that the much reduced groundwater level leaves some borehole pumps above the water table, or just below it so that the well has to operate with a much reduced drawdown and yield.

Good maintenance is required to safeguard water quality and operational efficiency. One of the easiest and most common ways for pollutants to enter an aquifer (even a confined aquifer) is down the inside or outside of the well casing. This can occur because of: the surface seal around the casing becoming cracked due to age; cracking or subsidence of the well structure; or a leaking cap (Fig. 13.26). This may also apply to any seals between the layers of a stratified aquifer. Consequently the quality and integrity of the seals should be inspected regularly. There may also be problems with well screens becoming clogged by sand grains bridging across the openings or the build-up of mineral deposits on the screens. Solutions vary according to circumstances, but can include deliberately overpumping for short periods, surging using a plunger, using compressed air to dislodge particles, hydraulic jetting, using a dispersant chemical (or acid in limestone aquifers) and even the use of explosives. Obviously, the pump also needs regular maintenance to ensure it works efficiently.

As the brief description above indicates, there is much that needs to be done after an aquifer is developed and wells constructed to ensure that they continue to be productive and efficient over a period of many years. Sometimes a computer model of the aquifer may be created, so that alternative pumping regimes can be studied, the effect of introducing new wells evaluated, and the dispersion of pollutants investigated. Such models are extremely valuable management tools but, if they are to be reliable, much hydrological and hydrogeological data are required to construct, calibrate and verify the model.

Box 13.2 **See for yourself**

If you have access to the hydrology apparatus pictured in Fig. 13.28, it can be used to illustrate many of the topics in Chapters 12 and 13. The apparatus uses a sprinkler system to generate rainfall, the intensity of which can be varied. The bed of the apparatus is sand. Runoff is collected and measured by a weir. You can use the apparatus to generate identical period of rainfall so you can compare the runoff from dry and saturated permeable catchments, and from impermeable catchments. It can also simulate multiple and moving storms. Because piezometers are connected to the base of the tank, you can also compare the observed and theoretical drawdown of the cone of depression surrounding a well (Fig. 13.29).

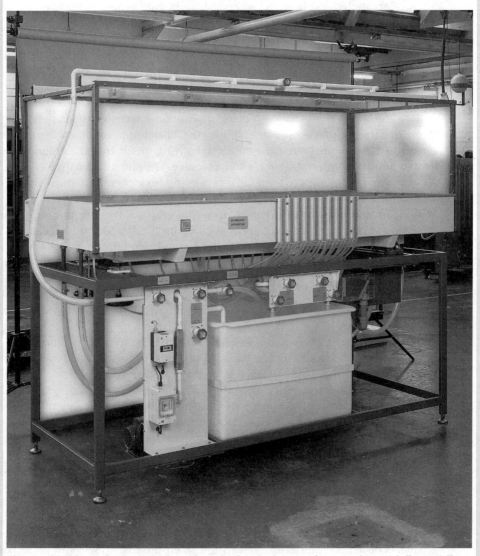

Figure 13.28 A hydrology bench, which can be used to study rainfall–runoff on a catchment, or to investigate the drawdown around a well (Fig. 13.29) [*Photo courtesy of TecQuipment Ltd; reproduced by permission*]

Box 13.2  *Continued:*

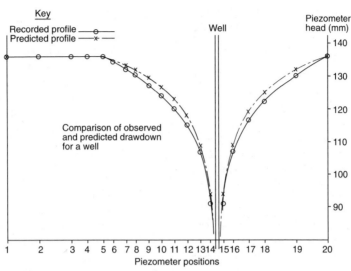

Figure 13.29 Theoretical and recorded drawdown in a well compared using the hydrology bench [*courtesy of Tec-Quipment Ltd; reproduced by permission*]

Summary

1. The soil moisture deficit (SMD) is the accumulated amount by which PET exceeds precipitation. When the SMD exceeds an assumed value (the root constant) ET occurs at the actual rate, a water deficiency occurs and plants may wilt and die. The monthly SMD is calculated using a water balance (e.g. Table 13.2).

2. The unit hydrograph (UH) is a rainfall–runoff model. The unit hydrograph is defined as: the hydrograph of direct surface runoff resulting from 10 mm depth of constant intensity net rain falling uniformly over the entire catchment during a specified period of time. The UH can be manipulated to obtain the surface runoff hydrograph resulting from a different rainfall intensity or duration from that originally observed.

3. Statistical frequency analysis provides a means of estimating the return period (T) of a particular flood magnitude (Q_T). The probability (P) that an event of magnitude Q_T will be exceeded in any one year is $P = 1/T$ and equation (13.4) gives $P = (r - 0.44)/(N + 0.12)$ where r is the rank of an observed flood and N the number of annual maxima in the data series. A graph of Q_T against $P\%$ can be drawn (e.g. on log-probablity paper) and used to obtain the return period of any flood. Data from different gauging stations can be pooled and analysed by using the ratio $Q_T/QMED$ on the graph's y-axis, where $QMED$ is the median annual maximum flood (i.e. the middle ranked value). Alternatively $QMED$ can be estimated from catchment descriptors.

4. The rational method of sewer design is based on equation (13.13): $Q_P = 0.278CiA$ where Q_P (m³/s) is the peak flood at the design point for a particular sewer, C is the coefficient of runoff or proportion of the catchment that is impermeable ($= 1.0$ if only impermeable surfaces are counted in A), i is the rainfall intensity (mm/h) corresponding to the time of concentration (t_C min) at the design point, and A is the total drainage area (km²) at the design point. At the design point, equation (13.15) gives $t_C = t_E + t_F$ where t_E is the time to enter the sewer system (therefore included once only) and t_F is the time to flow along all the sewers above the design point.

5. Problems involving water storage reservoirs can be solved using equation (12.1). A table can be used to record the difference between inflow and outflow, which is the change in storage. The cumulative change gives the reservoir storage capacity needed to meet the demand. Alternatively the cumulative inflow and outflow volumes can be plotted on a graph, the vertical distance between the lines enabling the required reservoir storage to be determined.

6. Useful amounts of groundwater are found in aquifers: unconfined aquifers have a water table open to the atmosphere, confined aquifers are covered by impermeable material so they contain water under pressure. A successful water well requires an aquifer with a high porosity/specific yield/storativity and high permeability. Often the permeability depends upon the presence of cracks and fissures in the aquifer. The amount of water abstracted from the well–aquifer system must not exceed the sustainable limit or the recharge to the aquifer. Often pumping tests are used to estimate permeability, storativity and yield. This involves analysing the size and rate of growth of the cone of depression that results from pumping a well. Once the wells are operational, good management is needed to ensure continuing supplies of good-quality water and to avoid problems with overpumping, pollution, induced infiltration, saline intrusion or well deterioration.

Revision questions

13.1 The data below are the 6h unit hydrograph for a particular catchment. Use this to determine the total discharge hydrograph of a design event that comprises 6h of net rainfall at 5mm/h followed immediately by 4h of net rainfall at 10mm/h. Assume that the average groundwater baseflow is 5.50 m³/s.

Time (h)	0	2	4	6	8	10
6h UH (m³/s)	0	0.98	3.89	6.70	9.20	9.70

Time (h)	12	14	16	18	20	22
6h UH (m³/s)	8.47	6.40	4.41	2.83	1.42	0

[5.50, 8.44, 17.17, 25.60, 38.98, 57.96, 65.23, 62.46, 60.09, 43.95, 27.72, 22.50, 14.02, 5.50 m³/s]

13.2 (a) The annual maxima for Craigshill Wood are listed in Table 13.8. Use the data to perform a frequency analysis and thus determine the magnitude of the 10 and 20 year floods (i.e. Q_{10} and Q_{20}). Try using both the Weibull and Gringorten equations (13.3 and 13.4), and log-probability and Gumbel paper, to see what effect this has. Do your answers confirm the value of Q_{10} derived in the text using the pooled data? Which answer do you think is the most reliable?

13.3 The figures below are the annual maxima for the River Tamar at Gunnislake in the water years 1956–93. (a) Determine how many of the maxima occurred in each month of the year. What does this tell you? (b) Determine the probable size of the 50 and 100 year floods. *Hint*: since there is quite a lot of data, if you can, use a spreadsheet to calculate the values of the Gumbel reduced variate, y, using equations (13.4) and (13.5), and to plot on linear scales a graph of Q_T against y. Your answer will depend upon how you treat the largest flood peak.

Water year	Date	Q (m³/s)
1956	07 Feb 1957	263.333
1957	25 Aug 1958	223.764
1958	05 Jan 1959	314.335
1959	26 Nov 1959	515.004
1960	26 Oct 1960	395.039
1961	16 Jan 1962	214.587
1962	06 Feb 1963	244.002
1963	17 Nov 1963	218.575
1964	13 Jan 1965	244.002
1965	18 Dec 1965	326.455
1966	29 Dec 1966	258.894
1967	04 Nov 1967	384.669
1968	25 Dec 1968	446.687
1969	15 Nov 1969	218.520
1970	21 Jan 1971	341.207
1971	30 Nov 1971	277.902
1972	06 Dec 1972	227.866
1973	27 Sep 1974	398.550
1974	13 Nov 1974	326.567
1975	12 Feb 1976	246.015
1976	15 Oct 1976	280.182
1977	23 Jan 1978	190.561
1978	01 Feb 1979	168.336
1979	28 Dec 1979	703.561
1980	09 Mar 1981	447.408
1981	20 Dec 1981	382.518
1982	04 Jan 1983	308.812
1983	26 Jan 1984	266.472
1984	22 Nov 1984	208.090
1985	24 Dec 1985	248.879
1986	18 Nov 1986	365.900
1987	18 Oct 1987	257.583
1988	09 Oct 1988	319.502
1989	02 Feb 1990	255.632
1990	02 Jan 1991	258.786
1991	31 Oct 1991	100.300
1992	12 Jun 1993	363.687
1993	04 Apr 1994	225.032

[(b) Q_{50} about 570–650 m³/s; Q_{100} about 630–730 m³/s]

13.4 Use the modified rational method to design the surface water sewer system detailed below. Assume that there is no flow in the sewers other than that arising from the design rainfall. Use a return interval of 1 in 5 years, and obtain the rainfall intensities from Table 12.5. Assume that the pipes flow full and have a roughness $k = 0.15$ mm as in Fig. 6.15. Assume $C_V = 0.75$, $C_R = 1.30$ and that the time of entry $t_E = 5$ min. The impermeable area drained by each individual sewer length is given below.

Sewer ref. no.	Length (m)	Gradient	Impermeable area (km²)
1.0	75	1 in 50	0.0020
2.0	50	1 in 75	0.0012
1.1	40	1 in 100	none
1.2	65	1 in 80	0.0018

[1.0 use 0.225 m; 2.0 use 0.150 m; 1.1 use 0.300 m; 1.2 use 0.300 m]

13.5 The inflow to a reservoir during a design drought is shown below in units of 10^6 m³. Every month is assumed to be 30 days long for convenience. The reservoir is partially full to begin with. Compensation flow of 1.80×10^6 m³ must be allowed through the dam at all times. Draw the mass flow curve and then determine: (a) the maximum uniform yield that can be maintained over the 12 month period if no flow passes over the spillway; (b) the initial storage capacity required below the starting level; (c) the additional storage required above the starting level; (d) the total storage required, including an additional 10% for dead storage; (e) the maximum uniform yield that could be obtained over the 12 months if the initial reservoir capacity is limited to 30×10^6 m³.

Month	Jun	Jul	Aug	Sep	Oct	Nov
Inflow (10^6 m³)	5.3	5.5	5.9	7.5	11.8	28.1

Month	Dec	Jan	Feb	Mar	Apr	May
Inflow (10^6 m³)	47.8	42.5	17.9	11.2	6.5	2.9

[(a) 14.3×10^6 m³/month; (b) 44.4×10^6 m³; (c) 27.6×10^6 m³; (d) 79.2×10^6 m³; (e) 11.4×10^6 m³/month]

13.6 (a) The following diameters represent a typical particle size for some types of soil: fine gravel 10 mm, coarse sand 1.0 mm, fine sand 0.1 mm and clay 0.001 mm. Use Table 13.13 to calculate their specific retention. Is there a relationship between particle size and specific retention? (b) A very short pumping test in an unconfined aquifer reduced the level of the water table by an average of 4.5 m over a plan area of $700 \times 10^3 m^2$. The aquifer material is a coarse sand with a porosity of about 35%. How much water was removed during the test? (c) In another part of the unconfined aquifer a flow net has been used to calculate that 1700 m³/d passes through a 1 km wide section of the aquifer. The aquifer is 25 m thick on average and the observed slope of the water table is 0.006. Calculate the permeability and transmissivity of the aquifer.

[(a) Specific retention increases with decreasing particle size; (b) $945 \times 10^3 m^3$; (d) 11.3 m/d and 283 m²/d]

13.7 (a) When searching for an aquifer suitable for water supply and when subsequently designing an abstraction borehole, what are the principal requirements needed to make the well successful? List briefly the most important things needed to obtain the optimum yield and efficiency. (b) The pumping test data below show the drawdown in an observation hole located 91 m from a well that is being pumped at 3900 m³/d. (i) Calculate the coefficients of transmissivity and storativity using Jacob's equation. (ii) Is the aquifer confined or unconfined? Explain your answer. (iii) If the satu-

rated thickness of the aquifer (H) is 30 m, what is the permeability of the aquifer material? (iv) What would be the approximate drawdown in the pumped well during the test? (v) What is the approximate specific capacity of the pumped well?

Time since pumping started (days)	Drawdown (m)
0.4	0.025
0.7	0.090
1.0	0.201
1.5	0.450
2.0	0.555
2.5	0.634
3.0	0.926
3.5	1.122
4.0	1.398
5.0	1.474
6.0	1.837
7.0	2.034
8.0	2.251
9.0	2.270
10.0	2.549
12.0	2.625
14.0	2.920
16.0	3.127
18.0	3.199
20.0	3.373

[(b)(i) 240 m²/d, 0.10; (ii) large S = unconfined; (iii) 8.0 m/d; (iv) 19.5 m; (v) 200 m³/d per m]

Bibliography and references

General

Ackers P (1992). Hydraulic design of two stage channels. *Proceedings of the Institution of Civil Engineers, Water Maritime and Energy*, 96, December, 247–57.

Ackers P (1993). Stage–discharge functions for two stage channels: the impact of new research. *Journal of the Institution of Water and Environmental Management*, 7, February, 52–61.

Alexander D (1993). *Natural Disasters*. UCL Press Ltd, London.

American Iron and Steel Institute (1984). *Handbook of Steel Drainage & Highway Construction Products*, 1st Canadian edn. American Iron and Steel Institute, Washington DC.

Anon (1976). Scientist floats Saudi iceberg plan. *New Civil Engineer*, 18 Nov, Thomas Telford, London.

Anon (1983). Korean village water supply. *Environmentalist*, 3, No. 1, 55–6.

Anon (1990). *Engineers and the Environment*. Engineering Council, London.

Anon (1994). *The Hutchinson Encyclopedia*, 1995 Edition. Helicon Publishing, Oxford.

Anon (1999). Sustainable Optimism (interview with Sir Crispin Tickell). *Water & Environment Manager*, 4(3), May, 23–4.

ARMCO (undated). *Design manual*. Newport, Wales.

Arnell N (1996). *Global Warming, River Flows and Water Resources*. John Wiley, Chichester, UK.

Barry R G (1969). The world hydrological cycle. In *Water, Earth and Man* (ed. R J Chorley), Chapter 1, 11–29. Methuen, London.

Barry R G and Chorley R J (1998). *Atmosphere, Weather and Climate*, 7th edn. Routledge, London.

Binnie G M (1981). *Early Victorian Water Engineers*. Thomas Telford, London.

Bleasedale A and Douglas C K M (1952) Storm over Exmoor on August 15, 1952. *The Meteorological Magazine*, Vol. 81, No. 966, December, 353–67.

Bradley J N (1978). *Hydraulics of Bridge Waterways*, 2nd edn. US Department of Transportation/Federal Highways Administration, Washington DC.

Brater E F, King H W, Lindell J E and Wei C Y (1996). *Handbook of Hydraulics*, 7th edn. McGraw-Hill, Boston, Massachusetts.

British Standards – see Flow measurement section below.

Brookes A and Shields F D (Eds) (1996). *River Channel Restoration: Guiding Principles for Sustainable Projects*. John Wiley, Chichester, UK.

Brown R H, Konoplyantsev A A, Ineson J and Kovalevsky V S (Eds) (1972). *Ground-water Studies – An International Guide for Research and Practice.* UNESCO, Paris.

Central Electricity Generating Board (1979). *Dinorwic pumped storage power station.* CEGB.

Chadwick A J and Morfett J C (1998). *Hydraulics in Civil Engineering*, 3rd edn. E & F N Spon, London.

Chow V T (1981). *Open-channel Hydraulics*, International Student Edition. McGraw-Hill, Tokyo.

CIRIA (1997). *Culvert Design Guide*, report 168. Construction Industry Research and Information Association (CIRIA), London.

Cluckie I D and Han D (2000). Fluvial flood forecasting. *J. CIWEM*, 14, August, 270–6.

Cuenca R H (1989). *Irrigation System Design: An Engineering Approach.* Prentice-Hall, Englewood Cliffs, New Jersey.

Daugherty R L, Franzini J B and Finnemore (1985). *Fluid Mechanics with Engineering Applications*, 8th edn, International Student Edition. McGraw-Hill, Singapore.

Douglas J F, Gasiorek J M and Swaffield J A (1995). *Fluid Mechanics*, 3rd edn. Longman, Harlow, Essex, UK.

Featherstone R E and Nalluri C (1982). *Civil Engineering Hydraulics: Essential Theory with Worked Examples.* Granada, St Albans, Herts, UK.

French R H (1986). *Open-channel Hydraulics*, International Student Edition. McGraw-Hill, Singapore.

Gash J H C and Stewart J B (1977). The evaporation from Thetford Forest during 1975. *J. Hydrol.*, 35, 385–96.

Gilbert S and Horner R (1984). *The Thames Barrier.* Thomas Telford, London.

Green C H and Penning-Rowsell E C (1986). Evaluating the intangible benefits of a flood alleviation proposal. *Journal of the Institution of Water Engineers and Scientists*, 40 (3), June, 229–48.

Hamill L (1993). A guide to the hydraulic analysis of single span arch bridges. *Proc. Instn Civ. Engrs, Mun. Engr.*, 98, March, 1–11.

Hamill L (1997). Improved flow through bridge waterways by entrance rounding. *Proc. Instn Civ. Engrs, Mun. Engr.*, 121, March, 7–21.

Hamill L (1999). *Bridge Hydraulics.* E & F N Spon, London.

Hamill L and Bell F G (1986). *Groundwater Resource Development.* Butterworths, London.

Hamill L and Bell F G (1987). Groundwater pollution and public health. *Bulletin of the International Association of Engineering Geology*, 35, 72–8.

Henderson F M (1966). *Open Channel Flow.* Macmillan, New York.

Herschy R W and Fairbridge R W (Eds) (1998). *Encyclopedia of Hydrology and Water Resources.* Kluwer Academic Publishers, Dordrecht, The Netherlands.

Highways Agency (1994). *Design Manual for Road Bridges*, Vol. 1, Section 3, Part 6, BA 59/94, *The Design of Highway Bridges for Hydraulic Action.* HMSO, London.

Holford I (1977). *The Guinness Book of Weather Facts and Feats.* Guinness Superlatives, London.

Hunsaker J C and Rightmire B G (1947). *Engineering Applications of Fluid Mechanics.* McGraw-Hill, New York.

Hydraulics Research (1988). *Assessing the Hydraulic Performance of Environmentally Acceptable Channels*, Report EX 1799. Hydraulics Research, Wallingford, UK.

Hydraulics Research (1990). *Charts for the Hydraulic Design of Pipelines.* Hydraulics Research, Wallingford, UK.

Institute of Hydrology (1999). *Flood Estimation Handbook.* Institute of Hydrology, Wallingford, UK.

Volume 1: Overview.

Volume 2: Rainfall frequency estimation.

Volume 3: Statistical procedures for flood frequency estimation.

Volume 4: Restatement and application of the Flood Studies Report rainfall–runoff method.

Volume 5: Catchment descriptors.

Institution of Civil Engineers (1960). Floods in the British Isles. Report by Subcommittee on Rainfall and Run-off: Allard W, Glasspoole J and Wolf P O. *Proceedings of the ICE*, Vol. 15, Feb, 119–44.

Institution of Civil Engineers (1996). *Floods and Reservoir Safety*, 3rd edn. Thomas Telford, London.

Jeffery T D, Thomas T H, Smith A V, Glover P B and Fountain P D (1992). *Hydraulic Ram Pumps: A Guide to Ram Pump Water Supply Systems*. Intermediate Technology Publications, London.

Jones J S (1984). Comparison of prediction equations for bridge pier and abutment scour. In *Transportation Research Record 950, Second Bridge Engineering Conference, Vol. 2*, pp 202–9. Transportation Research Board/National Research Council, Washington DC.

Kazmann R G (1956). 'Safe yield' in groundwater development, reality or illusion? *Proc ASCE, J. Irrig. Drain. Div.*, 82, IR3, 1–12.

Kiersch G A (1964). Vaiont reservoir disaster. *Civil Engineering (American Society of Civil Engineers)*, March, pp 32–9.

Laursen E M (1962). Scour at bridge crossings. *Transactions of the American Society of Civil Engineers*, 127, Part 1, 166–80.

Lewin J (1995). *Hydraulic Gates and Valves*. Thomas Telford, London.

Linsley R K, Franzini J B, Freyberg D L and Tchobanoglous G (1992). *Water Resources Engineering*, 4th edn. McGraw-Hill, New York.

Linsley R K, Kohler M A and Paulhus J H L (1982). *Hydrology for Engineers*, 3rd edn. McGraw-Hill, New York.

Markland E (1994). *A First Course in Hydraulics*. TecQuipment Ltd, Nottingham, UK.

Matthai H F (1967). Measurement of peak discharge at width constrictions by indirect methods. *Techniques of Water Resource Investigations of the United States Geological Survey*, Chapter A4, Book 3, Applications of Hydraulics. US Government Printing Office, Washington DC.

Melville B W (1988). Scour at bridge sites. In *Civil Engineering Practice, 2* (Hydraulics/Mechanics) (eds P N Cheremisinoff, N P Cheremisinoff and S L Cheng), pp 337–62. Technomic Publishing Company, Lancaster, Pennsylvania.

Nace R L (1969). World water inventory and control. In *Water, Earth and Man* (ed. R J Chorley), Chapter 2, 31–42. Methuen, London.

National Water Council, Standing Technical Committee on Sewers and Water Mains (1981). *Design and Analysis of Urban Storm Drainage, The Wallingford Procedure, Volume 4, The Modified Rational Method*. Department of the Environment, London.

Neill C R (Ed.) (1973) *Guide to Bridge Hydraulics*. Roads and Transportation Association of Canada, University of Toronto Press, Toronto.

NERC (1975). *Flood Studies Report* (in 5 volumes). Natural Environment Research Council, London.

New Civil Engineer (1979a). Thousands die in Indian dam breach, 16 Aug.

New Civil Engineer (1979b). Three gates failed in India's death dam, 23 Aug.

New Civil Engineer (1979c). Dam collapse enquiry, 13 Sept.

Newson M D (1979). The results of ten years' experimental study on Plynlimon,

Mid-Wales, and their importance for the water industry. *J. Inst. Water Eng. Sci.*, 33, 321–33.

North of Scotland Hydro-Electric Board (1971). *The Foyers Scheme*.

Novak P and Cabelka J (1981). *Models in Hydraulic Engineering: Physical Principles and Design Applications*. Pitman, London.

Novak P, Moffat A I B, Nalluri C and Narayanan R (1990). *Hydraulic Structures*. Unwin Hyman, London.

Olivier H (1972). *Irrigation and Water Resources Engineering*. Edward Arnold, London.

Penman H L (1948). Evaporation from open water, bare soil and grass. *Proc. Roy. Soc. Lond.*, 193, April, 120–45.

Penman H L (1949). The dependence of transpiration on weather and soil conditions. *J. Soil Sci.*, 1, 74–89.

Penman H L (1950). Evaporation over the British Isles. *Quart. J. Roy. Met. Soc.*, 76, 372–83.

Penning-Rowsell E C and Chatterton J B (1977). *The Benefits of Flood Alleviation: A Manual of Assessment Techniques*. Gower Press, Farnborough, UK.

Richardson E V and Davis S R (1995). *Evaluating Scour at Bridges*. Hydraulic Engineering Circular (HEC) 18, US Department of Transportation/Federal Highway Administration, Washington DC.

Roberson J A, Cassidy J J and Chaudhry H M (1998). *Hydraulic Engineering*, 2nd edn. John Wiley, New York.

RSPB, National Rivers Authority and Royal Society for Nature Conservation (1994). *The New Rivers and Wildlife Handbook*. Royal Society for the Protection of Birds, Sandy, Bedfordshire, UK.

Schnitter N J (1993) Dam failures due to overtopping. *Proc. Workshop Dam Safety Evaluation, Grindelwald*, 1, 13–19.

Sharp B B (1981). *Water Hammer: Problems and Solutions*. Edward Arnold, London.

Sharp J J (1981). *Hydraulic Modelling*. Butterworths, London.

Shaw E M (1994). *Hydrology in Practice*, 3rd edn. Chapman & Hall, London.

Singh V P (1996). *Dam Break Modelling Technology*. Kluwer Academic Publishers, Dordrecht, The Netherlands.

Smith D B, Wearn P L, Richards H J and Rowe P C (1970). Water movement in the unsaturated zone of high and low permeability strata by measuring natural tritium. *Proc. Symp.*, 'Isotope Hydrology', IAEA, Vienna, 73–87.

Smith K (1972). *Water in Britain – a study in applied hydrology and resource geography*. Macmillan (now Palgrave), Basingstoke, UK.

Streeter V L and Wylie E B (1983). *Fluid Mechanics*, International Student Edition. McGraw-Hill, Japan.

Thornthwaite C W (1948). An approach towards a rational classification of climate. *Geogrl. Rev.*, 38, 55–94.

Todd D K (1980). *Groundwater Hydrology*, 2nd edn. John Wiley, New York.

Twort A C, Law F M, Crowley F M and Ratnayaka D D (1994). *Water Supply*, 4th edn. Edward Arnold, London.

US Department of the Interior, Bureau of Reclamation (1987). *Design of Small Dams*, 3rd ed. US Government Printing Office, Denver, Colorado.

Vardy A E (1990). *Fluid Principles*. McGraw-Hill, Maidenhead, UK.

Vischer D L and Hager W H (1998). *Dam Hydraulics*. John Wiley, Chichester, UK.

Ward R C (1975). *Principles of Hydrology*. McGraw-Hill, London.

Water Space Amenity Commission (1983). *Conservation and Land Drainage Guidelines*, 2nd edn. Water Space Amenity Commission, London.

Watson R T, Zinyowera M C, Moss R H and Dokken D J (1996). *Climate Change, 1995, Impacts, Adaptions and Mitigations.* Cambridge University Press, Cambridge.

Webber N B (1971). *Fluid Mechanics for Civil Engineers.* Chapman & Hall, London.

Widmann R (1984). Possibilities of improving the safety of large dams, in *Safety of Dams, Proceedings of the International Conference on Safety of Dams, Coimbra/23–28 April 1984*, J Laginha Serafim (ed.), 291–5.

Wilson E M (1990). *Engineering Hydrology*, 4th edn. Macmillan (now Palgrave), Basingstoke, UK.

Yallin M S (1971). *Theory of Hydraulic Models.* Macmillan (now Palgrave), Basingstoke, UK.

Yarnell D L (1934). Bridge piers as channel obstructions. *Technical Bulletin no. 442*, US Department of Agriculture, Washington DC.

Flow measurement in pipelines and channels

BS 1042 Measurement of fluid flow in closed conduits. British Standards Institution, London.

1997 BS EN ISO 5167–1: Orifice plates, nozzles and Venturi tubes inserted in circular cross-section conduits running full. (Replaced BS 1042 Part 1, Section 1.1, 1981/1992.)

1983 Section 2.1 (1998): Methods using Pitot static tubes (= ISO 3966: 1977).

BS 3680 Measurement of liquid flow in open channels. British Standards Institution, London. (First date below is publication date; date in brackets indicates when currency of the standard was confirmed without full revision).

1995 Part 2A: Dilution methods: General (= ISO 9555-1: 1994).

1980 Part 3A: Velocity-area methods (replaced by BS ISO 748: 1997).

1997 Part 3B: Guide for the establishment and operation of a gauging station (= ISO 1100-1: 1996).

1983 Part 3C: Methods for the determination of the stage-discharge relation (replaced by BS ISO 1100-2: 1998).

1993 Part 3E (1999): Measurement of discharge by the ultrasonic (acoustic) method (= ISO 6416: 1992).

1990 Part 3G: General guidelines for the selection of methods (replaced by BS ISO TR 8363: 1997).

1993 Part 3H: Electromagnetic method using a full channel-width coil (= ISO 9213: 1992).

1981 Part 4A (1998): Methods using thin-plate weirs (= ISO 1438/1: 1980).

1986 Part 4B (1998): Triangular profile weirs (= ISO 4360: 1984).

1981 Part 4C (1998): Flumes.

1989 Part 4D: Compound gauging structures.

1990 Part 4E (1998): Rectangular broad-crested weirs (= ISO 3846: 1989).

1990 Part 4G (1999): Flat-V weirs (= ISO 4377: 1990).

1986 Part 4H (1992): Guide to the selection of flow gauging structures (= ISO 8368: 1985).

1986 Part 4I (1998): V-shaped broad crested weirs (= ISO 8333: 1985).

1989 Part 8A (1998): Current meters incorporating a rotating element (= ISO 2537: 1998).

1980 Part 8D (1986): Cableway systems for stream gauging (= ISO 4375: 1979).

1992 Part 11A: Methods of measurement of free surface flow in closed conduits (= ISO/TR 9824-1: 1990).

BS 6316 Code of practice for test pumping water wells. British standards Institution, London.

Useful websites

Beginners guide to propulsion: **http://www.grc.nasa.gov/WWW/K-12/airplane/shortp. html**

Cracking dams: **http://simscience.org**

Dam statistics, World Commission on Large Dams: **http://www.dams.org**

Dam types: **http://www.usbr.gov/cdams**

Flood defence, climate, MAFF: **http://www.maff.gov.uk/environ/envindx.htm**

Floods, Environment Agency: **http://www.environment-agency.gov.uk/subjects/flood/**

Fluid flow, why golf balls have dimples: **http://simscience.org**

Hydrological extremes, global change, Centre for Ecology and Hydrology: **http://www. ceh.ac.uk**

Meteorological Office, world weather, historic events: **http://www.met-office.gov.uk**

Rain gauges: **http://www.munro-group.co.uk**

Rocket motors: **http://www.im.lcs.mit.edu/rocket/intro.html**

UK Climate Change Impacts Programme: **http://www.ukeip.org.uk**

Water use and conservation: **http://www.environment-agency.gov.uk/subjects/waterres/**

Weather basics, extremes, world weather: **http://www.bbc.co.uk/weather**

Derivation of equations

Proof 1.1 – Force on a plane, immersed surface

A plane surface of area A is immersed in a liquid of weight density ρg. The surface is set so that all of it is at an angle α to the surface of the liquid, as shown in Fig. A1.1.

Consider an elemental strip of area δA which forms part of the total surface.

The pressure on one side of the element $P = \rho g h$ (equation (1.8)).

The force on the element is $\delta F = \rho g h \delta A$ (equation (1.2)).

If the distance from O to the element is L, then by geometry the vertical depth of the element below the surface is $h = L \sin \alpha$, so $\boldsymbol{\delta F = \rho g L \sin \alpha \delta A}$ (1)

The total force, F, on the surface is obtained by integrating over the whole area A, so $F = \rho g \sin \alpha \int_0^A L \mathrm{d} A$.

Now $\int_0^A L \mathrm{d} A = A L_G$, where L_G is the inclined distance from O to the centroid of the surface, G. The integral is the first moment of area of A about O, and essentially is the summation of all the elements δA over the whole surface area multiplied by the distance from O to each elemental area. Thus: $\boldsymbol{F = \rho g \sin \alpha A L_G}$ (2)

But by geometry $h_G = L_G \sin \alpha$ so:

$$F = \rho g h_G A \qquad (1.11)$$

The position of the resultant force, F, can be obtained by taking moments. The sum of the moments (M) about O of all the elemental forces, δF, acting on all the elemental areas, δA, will be equal to the total force, F, multiplied by L_P. In other words $M = \Sigma \delta M = F L_P$.

Now for the single element shown in the diagram, $\delta M = \delta F L$. Substituting for δF from equation (1) above gives: $\delta M = \rho g L \sin \alpha \delta A L$.

Integrating this expression between the limits 0 and A to get the total moment acting on all of the elemental strips that comprise the surface: $M = \rho g \sin \alpha \int_0^A L^2 \mathrm{d} A$.

The integral in this equation is the second moment of area of the surface calculated about an axis through O. This can be denoted by I_O. Thus the equation for M becomes:

$$M = \rho g \sin \alpha I_O \qquad (3)$$

However, M is also equal to the resultant force, F, acting at the inclined depth, L_P, so $M = F L_P$ and combining this expression with equation (3) gives $F L_P = \rho g \sin \alpha I_O$.

From equation (2) above, $F = \rho g \sin \alpha A L_G$, so: $\rho g \sin \alpha A L_G L_P = \rho g \sin \alpha I_O$. Cancelling $\rho g \sin \alpha$ from both sides of this expression and rearranging gives:

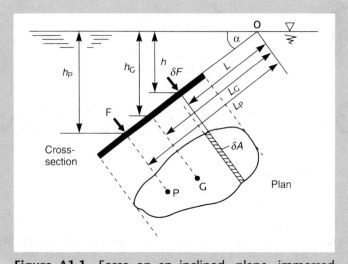

Figure A1.1 Force on an inclined, plane, immersed surface

$$L_p = I_O/AL_G \tag{4}$$

This defines the position of the centre of pressure, P, where the resultant force acts, but is not a very convenient expression to use since the second moment of area of the body has to be calculated about an axis through O. The use of the parallel axis theorem makes the expression more usable by shifting the axis about which I is calculated from O to G, so I_G appears in the equation instead of I_O, where I_G is the second moment of area calculated about an axis through the centroid of the surface, G. This is the 'normal' value used in mechanics. The parallel axis theorem is: $I_O = I_G + AL_G^2$. Dividing by AL_G and combining with equation (4), gives: $L_P = (I_O/AL_G) = (I_G/AL_G) + (AL_G^2/AL_G)$ or:

$$L_P = (I_G / AL_G) + L_G \tag{1.13}$$

Since $h = L\sin\alpha$ and $\sin 90° = 1$, for a vertical surface the above equation becomes:

$$h_P = (I_G / Ah_G) + h_G \tag{1.12}$$

Proof 1.2 – The pressure intensity at a point is the same in any direction

Figure A1.2 shows a small, infinitely thin triangle of liquid which is at rest within a larger body of stationary liquid. The *average* pressure intensities acting at 90° to the sides are P_1, P_2 and P_3, while the weight of the liquid in the triangle is W.

From equation (1.2) ($F = P_{AV}A$) the forces on the sides of the triangle are P_1BC, P_2AC and P_3AB (the dimension into the page can be ignored when calculating the area over which the pressure acts, since the triangle is infinitely thin). The only other force is the weight, which is obtained by multiplying the weight density of the liquid, ρg, by the area of the triangle, so $W = \rho g \frac{1}{2}$BC AC. Again the third dimension is ignored since the triangle is infinitely thin.

Since the triangle is at rest in a body of static liquid, the sum of the forces in any direction must balance (otherwise there would be movement). So the sum of the forces in the horizontal direction must be equal to zero: $P_2 AC - P_3 \sin\theta AB = 0$. By geometry $AC = AB\sin\theta$, so the above equation becomes $P_2 AB\sin\theta - P_3\sin\theta AB = 0$. Cancelling the product $AB\sin\theta$ from both sides of the equation gives $P_2 = P_3$.

Now equating the forces in the vertical direction gives:

$$P_1 BC - P_3\cos\theta AB - \frac{1}{2}\rho g BC\, AC = 0$$

From geometry, $AB\cos\theta = BC$, which can be substituted in the last equation to give:

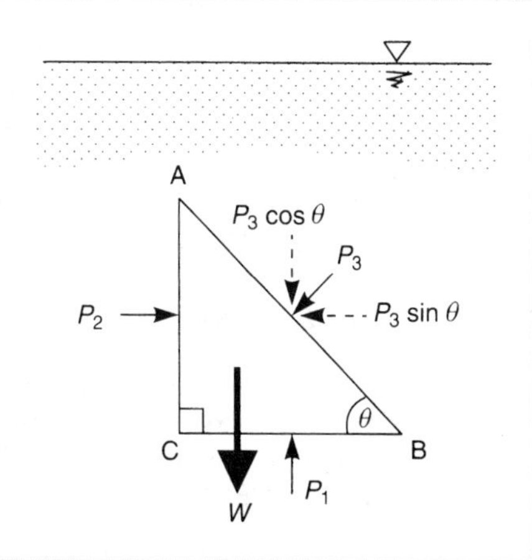

$$P_1 BC - P_3 BC - \frac{1}{2}\rho g BC\, AC = 0$$

Figure A1.2 Triangle of static liquid

The final part of the proof considers what would happen if the triangle was made progressively smaller. As it becomes infinitely small $AC \rightarrow 0$, so that $\frac{1}{2}\rho g BC\, AC \rightarrow 0$. Therefore, eliminating the weight term and cancelling BC in the previous equation gives $P_1 = P_3$.

Thus if the triangle is infinitely small so that A, B, and C are effectively the same point, then $P_1 = P_2 = P_3$ so the pressure intensity at a point is the same in all directions.

Proof 1.3 – The hydrostatic equation

Figure A1.3 shows a small, vertical cylinder of liquid of cross-sectional area A which is at rest in a larger body of static liquid of density ρ. The bottom of the cylinder is at a height z above an arbitrary datum. In this case, since z is increasing in an upward direction, the other variables will also be assumed to increase in an upward direction.

The pressure intensity on the bottom of the cylinder is P while the pressure on the top is $P + \delta P$. The forces acting on the cylinder are:

(a) on the bottom acting vertically up, $F_B = PA$;

(b) on the top acting vertically down, $F_T = (P + \delta P)A$;

(c) acting vertically down, the weight of the cylinder, $W = \rho g A\delta z$.

Now since the liquid is static, the sum of the forces in the vertical direction must be equal to zero (the forces on the sides act horizontally and have no component in the vertical direction, and would balance anyway). If the positive direction is vertically up then:

$$PA - (P + \delta P)\,A - \rho g A\delta z = 0$$
$$-\delta PA - \rho g A\delta z = 0$$
$$\delta P/\delta z = -\rho g$$

If δz is progressively decreased in size, at the limit the pressure gradient becomes $dP/dz = -\rho g$. This is the hydrostatic equation for the pressure gradient at a point. It is of more use when integrated to find P, thus:

$$\int_{P_1}^{P_2} dP = -\rho g \int_{z_1}^{z_2} dz$$

$$P_2 - P_1 = -\rho g(z_2 - z_1) \qquad (1.21)$$

In other words, the change in pressure intensity between two points is proportional to the vertical distance between them (see Fig. 1.34).

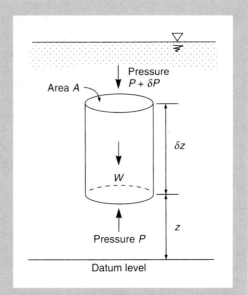

Figure A1.3 Cylinder of static liquid

Proof 3.1 – Theoretical height of the metacentre

Consider the tilted body in Fig. A1.4a. The restoring moment is due to the additional buoyancy arising from the wedge of liquid ACD to one side of the centreline and the reduced buoyancy represented by CEF on the other side. Now take an infinitely small element of ACD (shown hatched) at a distance x from the centreline. In plan this element has an area, δA (Fig. A1.4b). The height, h, of this element measured at 90° to the line CD is:

$$\tan \delta\theta = h/x$$
or $\quad h = x \tan \delta\theta$

When $\delta\theta$ is small $\tan \delta\theta = \delta\theta$ radians so:

$$h = x\delta\theta$$

The volume of the element $= h\delta A$ or $x\delta\theta\delta A$. If the body floats in a liquid of weight density ρg, then the weight of the hatched element is:

$$\rho g x \delta\theta \delta A$$

This represents the additional buoyancy force due to the element (the buoyancy force equals the weight of water displaced).

The additional moment calculated about C due to the element is the product of the elemental force and the distance from C, that is:

$$\rho g x^2 \delta\theta \delta A$$

To obtain the total restoring moment, RM, about C we have to integrate over the whole area, A, of the body in the plane of the water surface, thus:

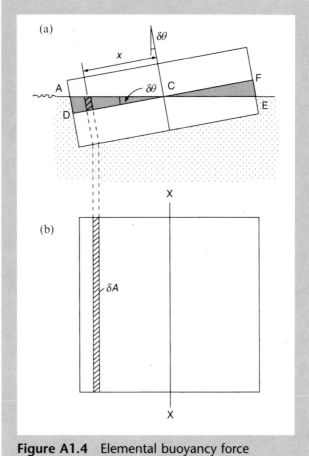

Figure A1.4 Elemental buoyancy force

$$RM = \rho g \delta\theta \int_0^A x^2 \delta A$$

The integral is denoted by I_{ws} and this is the second moment of area calculated about the longitudinal axis X–X in the plane of the water surface, so:

$$RM = \rho g \delta\theta I_{ws} \qquad (1)$$

Now the total restoring moment, RM, can also be calculated by multiplying the total bouyancy force, F, by the distance BB* (Fig. A1.5) so $RM = F \times$ BB*. If V is the volume of water displaced by the floating body then $F = \rho g V$ so $RM = \rho g V$BB*. The distance BB* is obtained from:

$$\sin \delta\theta = \text{BB*}/\text{BM}$$

$$\text{BB*} = \text{BM} \sin \delta\theta$$

and if $\delta\theta$ is small so $\sin \delta\theta = \delta\theta$ radians then:

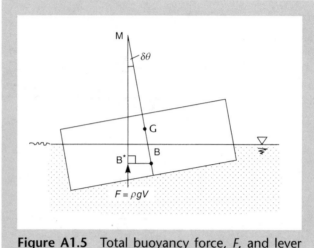

Figure A1.5 Total buoyancy force, F, and lever arm BB*

$$BB^* = BM\delta\theta$$

Consequently

$$RM = \rho g V BM \delta\theta \qquad (2)$$

Equations (1) and (2) above are both expressions for the total restoring moment, RM, and must equal each other:

$$\rho g \delta\theta I_{ws} = \rho g V BM \delta\theta$$

and cancelling ρ, g and $\delta\theta$ gives:

$$\mathbf{BM} = I_{ws}/V \qquad (3.1)$$

Proof 4.1 – The continuity equation

Consider fluid flowing through one streamtube of a conduit whose total cross-sectional area consists of a large number of streamtubes. Suffix 1 refers to the entrance to the tube and 2 to the exit, so that the areas of flow are δA_1 and δA_2 respectively (Fig. A1.6). It is assumed that the fluid is incompressible and that its density, ρ, is constant. The derivation of the continuity equation is based on the assumption that the mass per second entering the streamtube must equal the mass per second leaving. Remember that flow can enter and leave the tube only at the ends.

Since mass per second = density × flow rate: $\rho \delta Q_1 = \rho \delta Q_2$ where δQ is the flow rate in the streamtube.

Now $\delta Q_1 = \delta A_1 v_1$ and $\delta Q_2 = \delta A_2 v_2$ where v_1 and v_2 are the velocities at the ends of the streamtube. The only proof of this step is that m³/s = m² × m/s. Thus: $\rho \delta A_1 v_1 = \rho \delta A_2 v_2$. If the equations for all of the streamtubes that comprise the total cross-sectional area, A, of the conduit are summed:

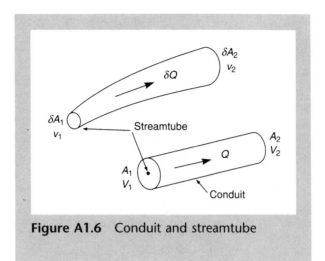

Figure A1.6 Conduit and streamtube

$$\Sigma(\rho\delta A_1 v_1) = \Sigma(\rho\delta A_2 v_2)$$

Now ρ is constant and can be cancelled in the above equation. If $\Sigma\delta A_1 = A_1$ and $\Sigma\delta A_2 = A_2$, and if V_1 and V_2 are the mean velocities at entry and exit to the conduit:

$$A_1V_1 = A_2V_2 \tag{4.4}$$

Proof 4.2 – The momentum equation

Consider fluid flowing through one streamtube of a conduit whose total cross-sectional area consists of a large number of streamtubes. Let suffix 1 refer to the entrance conditions and suffix 2 to those at the exit. The flow through the streamtube is δQ, the areas of the ends of the tube are δA_1 and δA_2, and the corresponding velocities are v_1 and v_2 (Fig. A1.7). The mass of fluid in the streamtube is δM, and δt is the time taken for an element of fluid to flow between 1 and 2.

The derivation of the momentum equation for the fluid system is based on Newton's Second Law, which is:

Force = mass × acceleration (1)

or Force = mass × rate of change of velocity (2)

For the streamtube this can be expressed as: $\delta F = \delta M \times (\delta v/\delta t)$ (3)

where δF is the force on the individual streamtube and δv is the change in velocity between 1 and 2. Now the mass of the streamtube is:

δM = density × volume of streamtube

δM = density × volume per second × time to flow from 1 to 2

$\delta M = \rho\delta Q\delta t$ [check: kg = (kg/m³) × (m³/s) × s = kg]

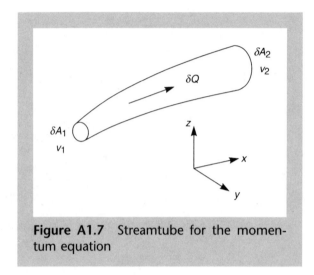

Figure A1.7 Streamtube for the momentum equation

Substituting for δM in equation (3) and applying the Second Law of Motion to the x direction (velocity is a vector quantity) gives:

$$\delta F_X = (\rho \delta Q \delta t) \times (\delta v_X / \delta t)$$
$$\delta F_X = \rho \delta Q \delta v_X$$

The change in velocity between 1 and 2 in the x direction is $\delta v_X = (v_{2X} - v_{1X})$, thus:

$$\delta F_X = \rho \delta Q (v_{2X} - v_{1X})$$

Summing together all the equations for all of the streamtubes comprising the total cross-sectional area of flow of the conduit:

$$F_X = \rho Q (V_{2X} - V_{1X}) \tag{4.9}$$

where V_{1X} and V_{2X} are the mean velocities in the x direction at the entrance and exit of the conduit, and Q is the total volumetric flow rate. This is the equivalent of equation (2) above, and is the momentum equation in the x direction for a fluid system. Similar equations may be derived for the y and z directions. The equation assumes that the velocity distribution is uniform, the mass flow rate is constant, and that there is a continuous change of momentum.

Proof 4.3 – The Bernoulli equation

The derivation of the Bernoulli (or energy) equation is achieved by applying the momentum equation to a streamtube. Flow can only enter and leave the streamtube through the two ends, which are small enough for the velocity to be considered uniform over the cross-sectional area of flow. An ideal fluid flows through the streamtube. The total flow consists of a large number of streamtubes similar to the one shown in Fig. A1.8.

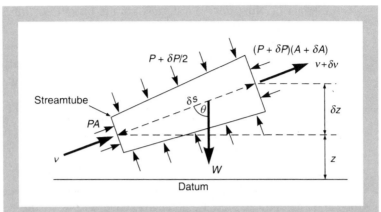

Figure A1.8 Streamtube for the derivation of the Bernoulli equation (shown two-dimensionally for clarity). All variables are shown as increasing in the direction of motion, which is from left to right

The length of the streamtube is δs. The cross-sectional area at the entrance of the streamtube is A, and that at the exit $A + \delta A$. Thus the increase in area in the direction of flow is δA. The corresponding velocities are v and $v + \delta v$. Note that all of the variables are assumed to increase in the direction of flow, even though the continuity equation would indicate that the velocity at the exit should be less than that at the entrance as a result of the increased area of flow. The pressure acting externally on the control volume increases from P at the entrance to $P + \delta P$ at the exit. The elevation of the centreline of the streamtube above an arbitrary datum increases from z at the entrance to $z + \delta z$ at the exit.

Newton's Second Law of Motion for a hydraulic system can be written as:

$$\Sigma F = \rho Q(V_2 - V_1) \tag{4.6}$$

In other words, the sum of the external forces acting on the control volume equals the rate of change of momentum. In this case the momentum equation will be applied ***in the direction of flow***, that is along the centreline of the streamtube so there are no suffixes x or y. First, consider the forces acting ***externally*** on the control volume. Forces will be assumed to be positive in the direction of motion, and negative if they act in the opposite direction.

1. Force due to P acting in the direction of motion, $F_1 = +PA$.

2. Force due to $P + \delta P$ opposing motion, $F_2 = -(P + \delta P)(A + \delta A)$.

3. Force due to the pressure on the sides of the tube which has a component in the direction of motion. Because δs is small, the force in the direction of motion can be approximated by the product of the mean pressure intensity on the sides and the net projected increase in the area in the direction perpendicular to the flow, which is δA. The mean pressure intensity is $(P + [P + \delta P])/2$ or $(P + \delta P/2)$ so the force in the direction of motion is $F_3 = (P + \delta P/2)\delta A$. This can be rationalised by arguing that if the streamtube had parallel sides then the pressure on the two sides would cancel each other out or would have no component in the direction of flow. This is indicated by the equation for F_3, because if the sides are parallel $\delta A = 0$. Similarly, the greater the divergence of the sides, the larger

δA, the larger F_3. Now since δ indicates a finite but almost infinitely small quantity, it follows that the product of two δ's results in something too small to be significant. Hence the equation for F_3 can be simplified to $F_3 = +P\delta A$.

4. The weight of the fluid in the control volume which acts vertically downwards but which has a component that opposes motion when resolved into the direction of flow, thus $F_4 = -W\cos\theta$.

Now: W = mass of fluid enclosed in control volume × gravity

$\quad\quad$ = mass density × volume enclosed × gravity

$\quad\quad$ = $\rho \times \delta s(A + [A + \delta A])/2 \times g$

$\quad\quad$ = $\rho g(A + \delta A/2)\delta s$

Thus $F_4 = -\rho g(A + \delta A/2)\delta s \cos\theta$, where θ is the angle between the centreline of the stream-tube and the vertical. Now by geometry, $\cos\theta = \delta z/\delta s$ so the equation for F_4 becomes $F_4 = -\rho g(A + \delta A/2)\delta z$. If the product of two δ's is again ignored the final expression, remembering that the force opposes motion, is $F_4 = -\rho g A \delta z$.

Therefore $\Sigma F = F_1 + F_2 + F_3 + F_4$

$\quad\quad = PA - (P + \delta P)(A + \delta A) + P\delta A - \rho g A \delta z$

$\quad\quad = PA - (PA + P\delta A + \delta PA + \delta P\delta A) + P\delta A - \rho g A \delta z$

$\quad\quad = PA - PA - P\delta A - \delta PA - \delta P\delta A + P\delta A - \rho g A \delta z$

Cancelling terms and ignoring the product of two δ's:

$\Sigma F = -\delta PA - \rho g A \delta z$

$\Sigma F = -A\delta P - \rho g A \delta z$ $\quad\quad$ (1)

But from equation (4.6), ΣF equals the rate of change of momentum of the fluid.

ΣF = rate of change of momentum

$\quad\quad$ = mass flow rate × change in velocity in direction of motion

$\quad\quad$ = $\rho Q \delta v$ where $Q = Av$ from the continuity equation, so:

$\Sigma F = \rho A v \delta v$ $\quad\quad$ (2)

Equating equations (1) and (2) gives:

$-A\delta P - \rho g A \delta z = \rho A v \delta v$

Dividing through by $\rho g A$:

$-\delta P/\rho g - \delta z = v\delta v/g$

Rearranging and replacing the finite increments (δ's) by infinitesimals (d's) gives:

$dz + (v/g)dv + dP/\rho g = 0$

This is Euler's equation which is applicable to an ideal fluid. For incompressible fluids of constant density, integration of both sides of Euler's equation gives:

$$z + v^2/2g + P/\rho g = \text{constant} = H \quad\quad (4.22)$$

This is the Bernoulli equation which shows the relationship between velocity, pressure and elevation for the steady flow of an ideal fluid of constant density. The terms represent the energy per unit weight, with the constant, H, being the total energy or head. The equation applies to a single streamline (or infinitely small streamtube). Different streamlines would have different values of velocity, pressure and elevation. When applying the Bernoulli equation the streamline used is often the centreline of the conduit, which is taken as being representative of the whole (see section 4.7). The mean velocity ($V = Q/A$) may be used instead of v on the assumption that the velocity is constant over the total cross-sectional area of flow, A.

Proof 5.1 – Inclined Venturi meter

A Venturi meter that has its axis inclined at an angle to the horizontal is shown in Fig. A1.9. Assuming no loss of energy, applying the Bernoulli equation to two points (1 and 2) on the centreline at the entrance and at the throat of the meter respectively:

$$z_1 + V_1^2/2g + P_1/\rho g = z_2 + V_2^2/2g + P_2/\rho g$$

thus $$\left(\frac{V_2^2}{2g} - \frac{V_1^2}{2g}\right) = \left(\frac{P_1}{\rho g} - \frac{P_2}{\rho g}\right) + (z_1 - z_2) \tag{1}$$

For continuity of flow $A_1V_1 = A_2V_2$ or $V_2 = (A_1/A_2)V_1$. Substituting for V_2 in equation (1):

$$\frac{V_1^2}{2g}\left[\left(\frac{A_1}{A_2}\right)^2 - 1\right] = \left(\frac{P_1}{\rho g} - \frac{P_2}{\rho g}\right) + (z_1 - z_2)$$

so $$V_1 = \sqrt{\frac{2g}{\left[(A_1/A_2)^2 - 1\right]} \times \left[\left(\frac{P_1}{\rho g} - \frac{P_2}{\rho g}\right) + (z_1 - z_2)\right]} \tag{2}$$

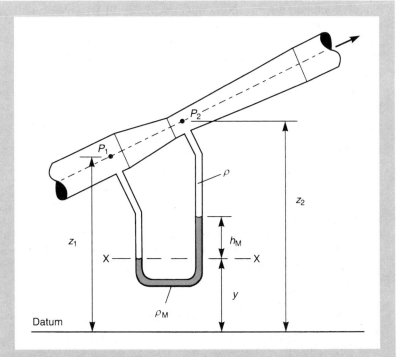

Figure A1.9 An inclined Venturi meter with manometer

This can be used to calculate the discharge ($Q = A_1V_1$) but it is more convenient to eliminate the 'z' terms from the equation. This can be done by considering the manometer readings. Let the pipe liquid in the upper part of the limbs have a mass density ρ and the other liquid in the lower part of the manometer have a mass density ρ_M. The differential head between the limbs is h_M.

Pressure in left limb at level $X - X = P_1 + \rho g(z_1 - y)$

Pressure in right limb at level $X - X = P_2 + \rho g(z_2 - y - h_M) + \rho_M g h_M$

By the equal level, equal pressure principle the pressure at level X–X is the same in both limbs, so equating gives: $P_1 + \rho g(z_1 - y) = P_2 + \rho g(z_2 - y - h_M) + \rho_M g h_M$

Dividing by ρg and rearranging: $\left(\dfrac{P_1}{\rho g} - \dfrac{P_2}{\rho g}\right) + (z_1 - z_2) = h_M[(\rho_M/\rho) - 1]$ (3)

The left-hand side of equation (3) also appears in equation (2) and can be eliminated. Also recognising that $Q = A_1V_1$ and introducing the coefficient of discharge, C_D, then:

$$Q_A = C_D A_1 \sqrt{\frac{2gh_M[(\rho_M/\rho) - 1]}{[(A_1/A_2)^2 - 1]}}$$

This equation does not contain the elevation terms and so is independent of the inclination of the meter.

Proof 6.1 – The Darcy equation

Consider a liquid flowing along the pipe in Fig. A1.10. Over the distance L between points 1 and 2 on the centreline of the pipe the head loss due to friction is h_F. The pressures at the two points are P_1 and P_2 respectively. The pipe has a diameter, D, and flows completely full. The cross-sectional area of flow is A, P is the wetted perimeter and V is the mean velocity of flow.

As the water flows along the pipe the resistance to the flow results in a reduction of water pressure from P_1 to P_2, that is a fall of $(P_1 - P_2)$ N/m². Remembering that a force is equal to a pressure (N/m²) multiplied by an area (m²), it follows that:

Resistance force $= (P_1 - P_2)A$ (1)

Now the general expression for total frictional resistance on a plane surface is:

Resistance force $= KA_P V^N$ (2)

where K is a constant dependent upon surface roughness, A_P is the area of contact between the water and the wetted perimeter, and V is the mean velocity raised to the power of N, which usually has a value around 2. Since equations (1) and (2) above are both expressions for the resistance force they can be equated thus:

$(P_1 - P_2)A = KA_P V^N$ (3)

To obtain the fall in pressure as a head loss, divide both sides of equation (3) by ρg:

$$\frac{(P_1 - P_2)A}{\rho g} = \frac{KA_P V^N}{\rho g}$$

Now $h_F = (P_1 - P_2)/\rho g$ so substituting and rearranging gives:

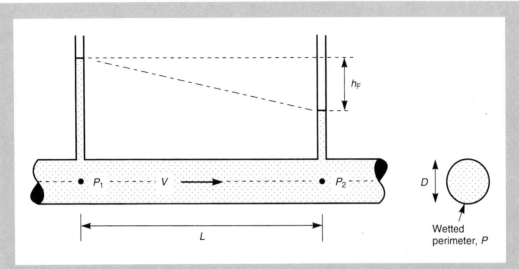

Figure A1.10 Flow along a pipe showing the head loss, h_F, caused by frictional resistance

$$h_F = \frac{KA_P V^N}{A\rho g}$$

But the area of the wetted surface A_P = wetted perimeter × length so $A_P = PL$. Making this substitution gives:

$$h_F = \frac{KPLV^N}{A\rho g}$$

Recognising that the hydraulic radius $R = A/P$ then the equation can be rewritten as:

$$h_F = \frac{KLV^N}{R\rho g} \qquad \text{or} \qquad h_F = \left(\frac{K2g}{\rho g}\right)\left(\frac{LV^N}{2gR}\right) \qquad (4)$$

Darcy's friction factor $f = (K2g/\rho g)$ and for a pipe running full $R = A/P$ where $A = \pi D^2/4$ and $P = \pi D$ so that $R = D/4$. If the power N is assumed to be exactly 2 then:

$$h_F = \frac{4fLV^2}{2gD}$$

and if $\lambda = 4f$:

$$h_F = \frac{\lambda LV^2}{2gD} \qquad (6.12)$$

Proof 8.1 – Chezy equation

Water flows down a straight rectangular channel of constant cross-sectional area, A, as shown in plan in Fig. A1.11a and in longitudinal section in Fig. A1.11b. Consider a rectangular element of the water of length L.

The force producing motion is the component of the weight of the rectangular element acting down the channel, which has a slope S_O (Fig. A1.11b).

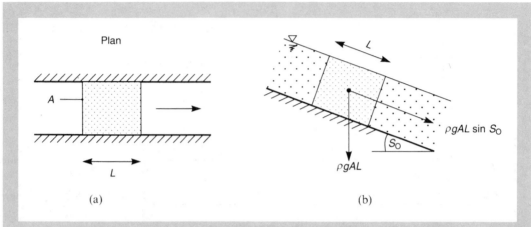

Figure A1.11 (a) Plan of channel showing element of water of length *L*. (b) Longitudinal section

Volume of element = $A \times L$

Therefore weight of element = $\rho g AL$

Component of weight down the channel = $\rho g AL \sin S_O$

If S_O is small then $\sin S_O = \tan S_O$ so the gradient can be represented by the vertical distance over the horizontal distance, such as 1/800 (that is 1 in 800). Thus:

Force producing motion = $\rho g ALS_O$ (1)

The general expression for total frictional resistance on a plane surface is:

Resistance force = $KA_P V^N$ (2)

where K is a constant dependent upon surface roughness, A_P is the surface area of the channel perimeter in contact with the water and V is the mean velocity of flow raised to the power N, which is usually taken as 2 for turbulent flow. Note that equation (2) above is exactly the same as equation (2) in Proof 6.1 for the derivation of the Darcy equation. Now the wetted area $A_P = PL$ where P is the wetted perimeter, so:

Force resisting motion = $KPLV^2$ (3)

For uniform flow the depth of water and the velocity in the channel are constant along its length, so the flow is not accelerating or decelerating. Therefore, all of the forces must balance, so that:

Force producing motion = Force resisting motion

$\rho g ALS_O = KPLV^2$

Cancelling the L's and rearranging gives:

$$V = \sqrt{\frac{\rho g}{K} \times \frac{A}{P} \times S_O}$$

If the Chezy roughness coefficient, $C = \sqrt{\rho g / K}$ and the hydraulic radius, $R = A/P$ then:

$$V = C\sqrt{RS_O}$$ (8.6)

Proof 8.2 – Gradually varying flow equation

As shown in Fig. 8.2b, at any cross-section of an open channel the total energy head is:

$$H = z + D + V^2/2g$$

Differentiating this equation with respect to L, the distance along the channel in the direction of flow gives:

$$\frac{dH}{dL} = \frac{dz}{dL} + \frac{dD}{dL} + \frac{d}{dL}\left(\frac{V^2}{2g}\right) \qquad (1)$$

where $\dfrac{dH}{dL} = -S_F$, the slope of the total energy line or friction gradient (–ve because the energy head decreases in the direction of flow)

$$\frac{dz}{dL} = -S_O,\text{ the bed slope (–ve because the bed falls in the direction of flow)}$$

$$\frac{dD}{dL} = \text{ the slope of the water surface relative to the bed}$$

Thus equation (1) can be written as:

$$-S_F = -S_O + \frac{dD}{dL} + \frac{d}{dL}\left(\frac{V^2}{2g}\right)$$

Rearranging to obtain an expression for the water surface slope gives:

$$\frac{dD}{dL} = S_O - S_F - \frac{d}{dL}\left(\frac{V^2}{2g}\right) \qquad (2)$$

Now $\dfrac{V^2}{2g} = \dfrac{Q^2}{2gA^2}$ where A is the cross-sectional area of the channel and Q the discharge.

Using this relationship and differentiating with respect to A, applying the chain rule gives:

$$\frac{d}{dL}\left(\frac{V^2}{2g}\right) = \frac{dA}{dL} \times \frac{d}{dA}\left(\frac{Q^2}{2gA^2}\right)$$

$$= \frac{dA}{dL}\left(\frac{-Q^2}{gA^3}\right)$$

Applying the chain rule again gives:

$$\frac{dA}{dL}\left(\frac{-Q^2}{gA^3}\right) = \frac{dA}{dD} \times \frac{dD}{dL}\left(\frac{-Q^2}{gA^3}\right) \qquad (3)$$

In deriving equation (8.25) it was assumed that $dA/dD = B_S$, the surface width of the channel at the water surface. With $dA/dD = B$, equation (3) becomes:

$$\frac{dA}{dL}\left(\frac{-Q^2}{gA^3}\right) = \frac{dD}{dL}\left(\frac{-Q^2 B}{gA^3}\right)$$

Thus we finally have an expression to replace the velocity term:

$$\frac{d}{dL}\left(\frac{V^2}{2g}\right) = \frac{dD}{dL}\left(\frac{-Q^2 B}{gA^3}\right)$$

so equation (2) can be written as:

$$\frac{dD}{dL} = S_O - S_F + \frac{dD}{dL}\left(\frac{Q^2 B}{gA^3}\right)$$

$$\frac{dD}{dL} - \frac{dD}{dL}\left(\frac{Q^2 B}{gA^3}\right) = S_O - S_F$$

$$\frac{dD}{dL}\left(1 - \frac{Q^2 B}{gA^3}\right) = S_O - S_F$$

$$\frac{dD}{dL} = \frac{S_O - S_F}{(1 - Q^2 B / gA^3)} \tag{4}$$

Now the Froude number, $F = \dfrac{V}{(gD)^{1/2}}$ so $F^2 = \dfrac{V^2}{gD}$

With $V = Q/A$ and $D = A/B$, we get $F^2 = \dfrac{Q^2 B}{gA^3}$ so equation (4) becomes:

$$\frac{dD}{dL} = \frac{S_O - S_F}{1 - F^2} \tag{8.41}$$

This is the gradually varied flow equation.

2

Solutions to self test questions

STQ1.1

(a) $P = \rho_Q g h = 7850 \times 0.6 = 4710\,\mathrm{N/m^2}$.
(b) The weight of the liquid $W = 7850 \times 2.0 \times 1.0 \times 0.6 = 9420\,\mathrm{N}$.
(c) Total area of timber in contact with the tank $= 2 \times 1.0 \times 0.1 = 0.2\,\mathrm{m^2}$.
   $P_{AV} = W/A = 9420/0.2 = 47\,100\,\mathrm{N/m^2}$.

STQ1.2

$h_G = 1.9 + 1.0/2 = 2.4\,\mathrm{m}$ and $A = 1.8 \times 1.0 = 1.8\,\mathrm{m^2}$
so $F = \rho g h_G A = 9810 \times 2.4 \times 1.8 = 42\,379\,\mathrm{N}$.
$h_P = (I_G/A h_G) + h_G$ where $I_G = LD^3/12 = (1.8 \times 1.0^3)/12 = 0.15\,\mathrm{m^4}$.
$h_P = (0.15/1.8 \times 2.4) + 2.4 = 2.435\,\mathrm{m}$.

STQ1.3

Relative density of oil $= 0.8$ so its mass density $= 0.8 \times 1000 = 800\,\mathrm{kg/m^3}$.
To obtain F_H project the curved surface onto a vertical plane.
Area of projected surface, $A = 1.5 \times 1.0 = 1.5\,\mathrm{m^2}$ (per metre length).
$h_G = (3.4 - 1.5) + 1.5/2 = 1.90 + 0.75 = 2.65\,\mathrm{m}$.
$F_H = \rho_Q g h_G A = 800 \times 9.81 \times 2.65 \times 1.5 = 31\,196\,\mathrm{N/m}$ length.
F_V is obtained by calculating the weight of the oil above the curved surface, so $F_V = \rho_Q g V$
$V = (1/4 \times \pi \times 1.5^2 \times 1.0) + (1.9 \times 1.5 \times 1.0) = 1.767 + 2.85 = 4.617\,\mathrm{m^3}$ (per metre length).
$F_V = 800 \times 9.81 \times 4.617 = 36\,234\,\mathrm{N/m}$ length.
The resultant force $F_R = (F_H^2 + F_V^2)^{1/2} = 10^3(31.196^2 + 36.234^2)^{1/2}$
$$= 47.81 \times 10^3\,\mathrm{N/m\ length}.$$
Angle of resultant to horizontal $\phi = \tan^{-1}(F_V/F_H) = \tan^{-1}(36\,234/31\,196) = 49.3°$

STQ2.1

Draw X–X through lower surface of separation and calculate pressures in left and right limbs at level X–X, then equate and solve to find absolute pressure, P_{ABS}, at the centre of the pipe.

Left: $P_X = \rho_M g h_M + \rho g z + P_{ABS}$ (1)

Right: $P_X = $ atmospheric pressure (P_{ATM}) (2)

        $\rho_M g h_M + \rho g z + P_{ABS} = P_{ATM}$

        $13\,600 \times 9.81 \times 0.1 + 1000 \times 9.81 \times 0.4 + P_{ABS} = 1000 \times 9.81 \times 10.3$

        $P_{ABS} = 101\,043 - 13\,342 - 3924 = 83\,777\,\text{N/m}^2$

Note that this is less than atmospheric. If gauge pressure was required then $P_{ATM} = 0$ in (2).

STQ2.2

Draw X–X through lower surface of separation and calculate pressures in left and right limbs at level X–X, then equate and solve to find pressure difference $(P_1 - P_2)$.

Left: $P_X = 13\,600 \times 9.81 \times 0.4 + 1000 \times 9.81 \times 0.8 + P_1$

       $= 61\,214 + P_1$

Right: $P_X = 1000 \times 9.81 \times (0.4 + 1.3) + P_2$

       $P_X = 16\,677 + P_2$

       $61\,214 + P_1 = 16\,677 + P_2$ so $(P_1 - P_2) = -44\,537\,\text{N/m}^2$

STQ2.3

Draw X–X through upper surface of separation. With same notation as in Fig. 2.9:

$P_1 = \rho g z_1 + \rho g h + P_X$ and $P_2 = \rho g z_2 + \rho_0 g h + P_X$ so:

$(P_1 - P_2) = \rho g z_1 + \rho g h + P_X - [\rho g z_2 + \rho_0 g h + P_X]$

               $= \rho g [z_1 + h - z_2] - \rho_0 g h$ but $a = z_1 - z_2$ so:

               $= \rho g [h + a] - \rho_0 g h$ (So, yes equation changes).

Now $z_1 = a + z_2 = 0.15 + 0.45 = 0.60\,\text{m}$. $h = 0.78 - 0.60 = 0.18\,\text{m}$.

$(P_1 - P_2) = 1.0 \times 1000 \times 9.81[0.18 + 0.15] - 0.8 \times 1000 \times 9.81 \times 0.18 = 1824\,\text{N/m}^2$

STQ3.1

The calculations follow the same procedure as in Example 3.3. Thus:

(a) With load $= 1500 \times 10^3\,\text{N}$, total weight $= 2200 \times 10^3\,\text{N}$, $V = 218.79\,\text{m}^3$, $h = 2.08\,\text{m}$, OB $= 1.04\,\text{m}$, OG $= 2.86\,\text{m}$, BG $= 1.82\,\text{m}$, BM $= 1.96\,\text{m}$, GM $= 0.14\,\text{m}$. Improved stability but GM is still small, remembering that the calculations may be slightly inaccurate.

    With load $= 1400 \times 10^3\,\text{N}$, total weight $= 2100 \times 10^3\,\text{N}$, $V = 208.85\,\text{m}^3$, $h = 1.99\,\text{m}$, OB $= 1.00\,\text{m}$, OG $= 2.83\,\text{m}$, BG $= 1.83\,\text{m}$, BM $= 2.05\,\text{m}$, GM $= 0.22\,\text{m}$. Better, but GM still relatively small.

With load $= 1100 \times 10^3$ N, total weight $= 1800 \times 10^3$ N, $V = 179.01\,\text{m}^3$, $h = 1.70\,\text{m}$, OB $= 0.85\,\text{m}$, OG $= 2.72\,\text{m}$, BG $= 1.87\,\text{m}$, BM $= 2.40\,\text{m}$, GM $= 0.53\,\text{m}$. Relatively stable.

(b) With $b = 7.5\,\text{m}$, total weight $= 2300 \times 10^3$ N, $V = 228.74\,\text{m}^3$, $h = 2.03\,\text{m}$, OB $= 1.02\,\text{m}$, OG $= 2.89\,\text{m}$, BG $= 1.87\,\text{m}$, BM $= 2.31\,\text{m}$, GM $= 0.44\,\text{m}$. Significantly improved stability.

With $b = 8.0\,\text{m}$, total weight $= 2300 \times 10^3$ N, $V = 228.74\,\text{m}^3$, $h = 1.91\,\text{m}$, OB $= 0.95\,\text{m}$, OG $= 2.89\,\text{m}$, BG $= 1.94\,\text{m}$, BM $= 2.80\,\text{m}$, GM $= 0.86\,\text{m}$.

With $b = 9.0\,\text{m}$, total weight $= 2300 \times 10^3$ N, $V = 228.74\,\text{m}^3$, $h = 1.69\,\text{m}$, OB $= 0.85\,\text{m}$, OG $= 2.89\,\text{m}$, BG $= 2.04\,\text{m}$, BM $= 3.98\,\text{m}$, GM $= 1.94\,\text{m}$. Very stable, but is period of roll too short?

STQ3.2

GM $= (300 \times 10^3/50 \times 10^6) \times (8 \times 57.3)/10 = 0.28\,\text{m}$. Reduced stability, small GM, just OK?
GM $= (300 \times 10^3/50 \times 10^6) \times (8 \times 57.3)/40 = 0.07\,\text{m}$. Much too small to be acceptable. Note that by this method of calculation GM cannot become negative (indicating instability), it can only approach zero.

STQ4.1

(a) Steady, uniform. (b) Unsteady, non-uniform. (c) Unsteady, non-uniform. (d) Steady, non-uniform. (e) Steady, uniform.

STQ4.2

$Q_1 = A_1V_1 = (\pi D_1^2/4) \times V_1 = (\pi \times 0.70^2/4) \times 1.60 = 0.616\,\text{m}^3/\text{s}$
$Q_3 = A_3V_3 = (\pi D_3^2/4) \times V_3 = (\pi \times 0.25^2/4) \times 2.30 = 0.113\,\text{m}^3/\text{s}$
$Q_2 = Q_1 - Q_3 = 0.616 - 0.113 = 0.503\,\text{m}^3/\text{s}$
$Q_2 = A_2V_2 = (\pi D_2^2/4) \times 1.10 = 0.503$ giving $D_2 = 0.76\,\text{m}$

STQ4.3

$F_R = P_1A_1 - \rho Q(V_2 - V_1)$
$V_1 = (0.65 \times 4)/(\pi \times 0.30^2) = 9.196\,\text{m/s}$. $V_2 = 9.196(0.30/0.20)^2 = 20.691\,\text{m/s}$
$F_R = 129.40 \times 10^3 \times (\pi \times 0.30^2/4) - 10^3 \times 0.65(20.691 - 9.196)$
$\quad = 9.147 \times 10^3 - 7.471 \times 10^3 = 1.676 \times 10^3$ N
The water exerts a force of 1.68×10^3 N on the nozzle from left to right.

STQ4.4

The energy equation becomes: $3.4 + (0.353V_2)^2/2g = 1.2 + V_2^2/2g$
$0.875V_2^2/2g = 2.2$ giving $V_2 = 7.02\,\text{m/s}$
$Q = A_2V_2 = 1.2 \times 4.0 \times 7.02 = 33.70\,\text{m}^3/\text{s}$
% difference $= 100(33.70 - 26.54)/26.54 = 27\%$

STQ4.5

$V_1 = Q/A_1 = (0.5 \times 4)/(\pi \times 0.6^2) = 1.768\,\text{m/s}$. $V_2 = 1.768(0.6/0.4)^2 = 3.978\,\text{m/s}$

Apply the energy equation to the centreline before and after the bend to obtain P_2.

$V_1^2/2g + P_1/\rho g = V_2^2/2g + P_2/\rho g$

$(1.768^2/19.62) + (150 \times 10^3/10^3 \times 9.81) = (3.978^2/19.62) + P_2/\rho g$

$P_2/\rho g = 14.643\,\text{m}$ so $P_2 = 14.643 \times 9.81 \times 10^3 = 143.648 \times 10^3\,\text{N/m}^2$

$P_1A_1 = 150 \times 10^3 \times (\pi \times 0.6^2/4) = 42.412 \times 10^3\,\text{N}$

$P_2A_2 = 143.648 \times 10^3 \times (\pi \times 0.4^2/4) = 18.051 \times 10^3\,\text{N}$

Apply the momentum equation assuming that on the control volume F_{RX} acts externally from right to left and F_{RY} acts vertically upwards. In the x direction:

$P_1A_1 - P_2A_2 \cos\theta - F_{RX} = \rho Q(V_2 \cos\theta - V_1)$

$42.412 \times 10^3 - 18.051 \times 10^3 \cos 45° - F_{RX} = 10^3 \times 0.5(3.978 \cos 45° - 1.768)$

$29.648 \times 10^3 - F_{RX} = 0.523 \times 10^3$ giving $F_{RX} = 29.125 \times 10^3\,\text{N}$ (right to left)

In the y direction: $F_{RY} - P_2A_2 \sin\theta = \rho Q(V_2 \sin\theta)$

$F_{RY} - 18.051 \times 10^3 \sin 45° = 10^3 \times 0.5(3.978 \sin 45°)$

$F_{RY} - 12.764 \times 10^3 = 1.406 \times 10^3$ giving $F_{RY} = 14.170 \times 10^3\,\text{N}$ (upwards)

$F_R = 10^3(29.125^2 + 14.170^2)^{1/2} = 32.389 \times 10^3\,\text{N}$

$\theta = \tan^{-1}(14.170/29.125) = 25.9°$ to the horizontal (bottom right to top left).

The water exerts an internal force on the bend of $32.39 \times 10^3\,\text{N}$ in the opposite direction, i.e. top left to bottom right.

STQ5.1

$$V_1 = \sqrt{\frac{2gH}{\left[(A_1/A_2)^2 - 1\right]}} \quad \text{so} \quad V_1^2 = \frac{2gH}{\left[(A_1/A_2)^2 - 1\right]}$$

where $V_1 = 2.3\,\text{m/s}$, $V_1^2 = 5.290$, $A_1 = \pi \times 0.15^2/4 = 0.0177\,\text{m}^2$, and $H = 1.2\,\text{m}$. Thus:

$$5.290 = \frac{19.62 \times 1.2}{\left[(0.0177/A_2)^2 - 1\right]} \quad \text{and hence} \quad \left[\frac{0.0177^2}{A_2^2} - 1\right] = \frac{19.62 \times 1.2}{5.290}$$

$0.000313/A_2^2 = 4.451 + 1$ which gives $A_2^2 = 0.000313/5.451$ and $A_2 = 0.00758\,\text{m}^2$.

$$D_2 = \sqrt{\frac{0.00758 \times 4}{\pi}} = 0.098\,\text{m (say 100 mm)}$$

STQ5.2

Apply the Bernoulli equation to a streamline joining point 1 on the water surface to point 2 in the vena contracta: $z_1 + V_1^2/2g + P_1/\rho g = z_2 + V_2^2/2g + P_2/\rho g$. On the water surface $V_1 = 0$, at the vena contracta $P_2 = 0$. Take the datum level through point 2 so $z_2 = 0$ and let $z_1 = H$. Thus:

$H + P_1/\rho g = V_2^2/2g$ so $V_2 = [2g(H + P_1/\rho g)]^{1/2}$. But $Q = A_2V_2$ so

$Q_A = C_D A_2[2g(H + P_1/\rho g)]^{1/2}$.

Now $P_1 = 113.4 \times 10^3\,\text{N/m}^2$ so $P_1/\rho g = 113.4 \times 10^3/(9.81 \times 10^3) = 11.560\,\text{m}$ water. Thus:

$Q_A = 0.62 \times 0.00196[19.62(2.50 + 11.560)]^{1/2} = 0.0202\,\text{m}^3/\text{s}$.

STQ5.3

Large orifice: $Q_A = \frac{2}{3} C_D b (2g)^{1/2} [H_2^{3/2} - H_1^{3/2}]$ where $H_2 = 1.35 + 3.0 = 4.35\,\text{m}$ and $H_1 = 1.35\,\text{m}$.

$61.7 = \frac{2}{3} C_D \times 6.0 \times (19.62)^{1/2} [4.35^{3/2} - 1.35^{3/2}]$ thus $61.7 = C_D \times 132.96$ giving $C_D = 0.46$.

Small orifice: $Q_A = C_D A (2gH)^{1/2}$ where H = height of water surface above centre of opening = $1.5 + 1.35 = 2.85\,\text{m}$. Thus $61.7 = C_D \times 6 \times 3 (19.62 \times 2.85)^{1/2}$ giving $C_D = 0.46$.

STQ5.4

The total depth of water above the crest is H. Consider a horizontal strip of thickness, δh, at a depth, h, from the water surface (as in Fig. 5.14). The discharge through the strip is $\delta Q = V\delta A$ where V is the velocity through the strip = $(2gh)^{1/2}$ and δA is the area of the strip = $y\delta h$. Thus $\delta Q = (2gh)^{1/2} y\delta h$ where y is the width of the strip at the depth h. Thus $y = L + 2b$ where b is the width of the strip resulting from the outward sloping sides of the weir (see Fig. 5.14). Now $b = \tan\theta (H - h)$, so $y = L + 2\tan\theta (H - h)$ and $\delta Q = (2gh)^{1/2}[L + 2\tan\theta (H - h)]\delta h$. Rearranging:

$$\delta Q = (2g)^{1/2} [L h^{1/2} \delta h + 2\tan\theta (H - h) h^{1/2} \delta h].$$

Integrating between $h = 0$ and $h = H$ to obtain the total theoretical discharge:

$$Q_T = (2g)^{1/2} \left[L \int_0^H h^{1/2} dh + 2\tan\theta \int_0^H (H h^{1/2} - h^{3/2}) dh \right]$$

$$Q_T = (2g)^{1/2} L \left[\frac{2h^{3/2}}{3} \right]_0^H + (2g)^{1/2} 2\tan\theta \left[\frac{2Hh^{3/2}}{3} - \frac{2h^{5/2}}{5} \right]_0^H$$

$$Q_T = \frac{2}{3} L (2g)^{1/2} H^{3/2} + \frac{8}{15} (2g)^{1/2} \tan\theta H^{5/2} \qquad (1)$$

$$Q_A = \frac{2}{3} C_D (2g)^{1/2} H^{3/2} \left[L + \frac{4}{5} H \tan\theta \right] \qquad (2)$$

Note that equation (1) is the equivalent of the theoretical discharge equations for a rectangular and triangular weir added together (which effectively is what a trapezoidal weir is) allowing for the slightly different notation ($L = b$ and $\theta = \theta/2$).

(b) From equation (2): $Q_A = \frac{2}{3} \times 0.60 \times (19.62)^{1/2} \times 0.17^{3/2} \left[0.20 + \frac{4}{5} \times 0.17 \tan 20° \right]$

$$= 0.124[0.249] = 0.031\,\text{m}^3/\text{s}$$

STQ6.1

(a) Apply the Bernoulli equation between the water surface in the two reservoirs ignoring minor losses: $z_1 + V_1^2/2g + P_1/\rho g = z_2 + V_2^2/2g + P_2/\rho g + \lambda L V^2/2gD$. On the surface of a large reservoir, $V_1 = 0$, $V_2 = 0$, $P_1 = P_2 = 0$ and $(z_1 - z_2) = 50$ thus: $50 = \lambda L V^2/2gD$.

$50 = 0.04 \times 15\,000 \times V^2/19.62 \times 0.9$. This gives $50 = 33.979 V^2$ and $V = 1.213\,\text{m/s}$.

$Q = AV = (\pi \times 0.9^2/4) \times 1.213 = 0.772\,\text{m}^3/\text{s}$.

(b) Including minor losses at the entrance and exit:

$50 = 0.5V^2/2g + 33.979V^2 + V^2/2g$ which gives $50 = 34.055V^2$ and $V = 1.212\,\text{m/s}$.

$Q = (\pi \times 0.9^2/4) \times 1.212 = 0.771\,\text{m}^3/\text{s}$. The difference between answers (a) and (b) is $0.001\,\text{m}^3/\text{s}$.

STQ6.2

Apply Bernoulli from A to C:

$Z_{AC} = \lambda_1 L_1 V_1^2/2gD_1 + \lambda_2 L_2 V_2^2/2gD_2$. $Z_{AC} = 250 - 220 = 30\,\text{m}$.

$30 = [0.05 \times 17\,500 \times V_1^2/19.62 \times 1.2] + [0.04 \times 5300 \times V_2^2/19.62 \times 0.9]$

$30 = 37.164V_1^2 + 12.006V_2^2$ which gives: $V_2 = (2.499 - 3.095V_1^2)^{1/2}$ (1)

Apply Bernoulli from A to D:

$Z_{AD} = \lambda_1 L_1 V_1^2/2gD_1 + \lambda_3 L_3 V_3^2/2gD_3$. $Z_{AD} = 250 - 190 = 60\,\text{m}$.

$60 = 37.164V_1^2 + [0.04 \times 6400 \times V_3^2/19.62 \times 0.9]$ thus $60 = 37.164V_1^2 + 14.498V_3^2$. This gives: $V_3 = (4.139 - 2.563V_1^2)^{1/2}$ (2)

Apply the continuity equation: $Q_1 = Q_2 + Q_3$ so $D_1^2V_1 = D_2^2V_2 + D_3^2V_3$.

Thus $1.2^2V_1 = 0.9^2V_2 + 0.9^2V_3$ and hence $1.440V_1 = 0.810V_2 + 0.810V_3$.

Thus: $V_1 = 0.563V_2 + 0.563V_3$ (3)

Substituting for V_2 and V_3 in equation (3) from equations (1) and (2):

$V_1 = 0.563(2.499 - 3.095V_1^2)^{1/2} + 0.563(4.139 - 2.563V_1^2)^{1/2}$ (4)

For a real solution $3.095V_1^2 < 2.499$ and $2.563V_1^2 < 4.193$ giving $V_1 < 0.9\,\text{m/s}$ and $1.3\,\text{m/s}$. Solving equation (4) by trial and error gives $V_1 = 0.895\,\text{m/s}$. Substituting in equation (1) gives $V_2 = 0.141\,\text{m/s}$ and $V_3 = 1.444\,\text{m/s}$ from equation (2). Thus

$Q_1 = (\pi \times 1.2^2/4) \times 0.895 = 1.012\,\text{m}^3/\text{s}$. $Q_2 = (\pi \times 0.9^2/4) \times 0.141 = 0.090\,\text{m}^3/\text{s}$.

$Q_3 = (\pi \times 0.9^2/4) \times 1.444 = 0.919\,\text{m}^3/\text{s}$.

STQ6.3

Ignoring minor losses, applying the Bernoulli equation to a streamline joining the water surface of the two reservoirs via pipeline 1 gives: $23 = \lambda_1 L_1 Q_1^2/12.1D_1^5$

$23 = 0.06 \times 5400 \times 0.4^2/12.1 \times D_1^5$ so $D_1^5 = 0.186$ and $D_1 = 0.71\,\text{m}$.

Similarly for pipeline 2: $23 = \lambda_2 L_2 Q_2^2/12.1D_2^5 = 0.06 \times 5400 \times 0.6^2/12.1 \times D_2^5$ so $D_2 = 0.84\,\text{m}$.

For pipeline 3: $23 = \lambda_3 L_3 Q_3^2/12.1D_3^5 = 0.06 \times 5400 \times 1.1^2/12.1 \times D_3^5$ giving $D_3 = 1.07\,\text{m}$.

STQ6.4

Applying the Bernoulli equation to a streamline joining the surface of the reservoirs via pipes 1 and 2, ignoring minor losses: $Z = \lambda_1 L_1 V_1^2/2gD_1 + \lambda_2 L_2 V_2^2/2gD_2$ where $Z = 27\,\text{m}$. Thus:

$27 = [0.04 \times 10\,000 \times V_1^2/19.62 \times 0.9] + [0.07 \times 21\,000 \times V_2^2/19.62 \times 0.75]$

$27 = 22.653V_1^2 + 99.898V_2^2$ (1)

Applying the Bernoulli equation to a streamline joining the surface of the reservoirs via pipes 1 and 3, ignoring minor losses: $Z = \lambda_1 L_1 V_1^2/2gD_1 + \lambda_3 L_3 V_3^2/2gD_3$ where $Z = 27\,\text{m}$ again.

Thus:

$27 = 22.653V_1^2 + [0.05 \times 23\,000 \times V_3^2/19.62 \times 0.60]$ so:

$27 = 22.653V_1^2 + 97.689V_3^2$ (2)

Subtracting equation (2) from equation (1) gives: $0 = 99.898V_2^2 - 97.689V_3^2$.

Thus $V_3^2 = (99.898/97.689)V_2^2$ giving $V_3 = 1.011V_2$ (3)

Applying the continuity equation: $Q_1 = Q_2 + Q_3$ or $D_1^2V_1 = D_2^2V_2 + D_3^2V_3$.

Thus $0.9^2V_1 = 0.75^2V_2 + 0.6^2V_3$ giving $0.81V_1 = 0.563V_2 + 0.36V_3$. Now substituting for V_3

from equation (3): $0.81V_1 = 0.563V_2 + 0.36(1.011V_2)$ thus

$0.81V_1 = 0.927V_2$ or $V_1 = 1.144V_2$ (4)

Substituting for V_1 in equation (1): $27 = 22.653(1.144V_2)^2 + 99.898V_2^2$

$27 = 129.545V_2^2$ thus $V_2 = 0.457\,\text{m/s}$.

Substituting for V_2 in equation (4) gives $V_1 = 0.523\,\text{m/s}$.

Substituting for V_2 in equation (3) gives $V_3 = 0.462\,\text{m/s}$.

Thus $Q_1 = (\pi \times 0.9^2/4) \times 0.523 = 0.333\,\text{m}^3/\text{s}$. $Q_2 = (\pi \times 0.75^2/4) \times 0.457 = 0.202\,\text{m}^3/\text{s}$.

$Q_3 = (\pi \times 0.6^2/4) \times 0.462 = 0.131\,\text{m}^3/\text{s}$.

STQ7.1

(a) See Example 7.5. Thus: $T = \dfrac{30A_{\text{WS}}}{24C_D(2g)^{1/2}\tan(\theta/2)}\left[\dfrac{1}{H_2^{3/2}} - \dfrac{1}{H_1^{3/2}}\right]$

(b) Putting $A_{\text{WS}} = 6 \times 3 = 18\,\text{m}^2$, $C_D = 0.6$, $\theta/2 = 30°$, $H_1 = 0.3\,\text{m}$, and $H_2 = 0.1\,\text{m}$ then:

$T = \dfrac{30 \times 18}{24 \times 0.6 \times (19.62)^{1/2}\tan 30°}\left[\dfrac{1}{0.1^{3/2}} - \dfrac{1}{0.3^{3/2}}\right] = 14.664[31.623 - 6.086] = 374.5\,\text{s}$.

STQ7.2

Water surface is 6.0 m above apex of inverted pyramid. The orifice is 1.5 m above apex. Thus there are three horizontal slices each 1.5 m thick with base areas of 4.0 m × 4.0 m, 3.0 m × 3.0 m, 2.0 m × 2.0 m and 1.0 m × 1.0 m (see Fig. 7.2) at heights of 6.0 m, 4.5 m, 3.0 m and 1.5 m above the apex respectively. The volumes of the three slices are:

$Vol_1 = 1/3 \times 4.0^2 \times 6.0 - 1/3 \times 3.0^2 \times 4.5 = 32.0 - 13.5 = 18.5\,\text{m}^3$.

$Vol_2 = 1/3 \times 3.0^2 \times 4.5 - 1/3 \times 2.0^2 \times 3.0 = 13.5 - 4.0 = 9.5\,\text{m}^3$.

$Vol_3 = 1/3 \times 2.0^2 \times 3.0 - 1/3 \times 1.0^2 \times 1.5 = 4.0 - 0.5 = 3.5\,\text{m}^3$.

$Q_A = C_D A(2gh)^{1/2}$ so with $C_D = 0.60$ and $A = \pi \times 0.1^2/4 = 0.00785\,\text{m}^2$ then $Q_A = 0.0209h^{1/2}$. Thus with water surface 4.5 m above orifice $Q_A = 0.0209 \times 4.5^{1/2} = 0.0443\,\text{m}^3/\text{s}$. At 3.0 m, $Q_A = 0.0362\,\text{m}^3/\text{s}$. Thus average discharge for first slice, $Q_1 = (0.0443 + 0.0362)/2 = 0.0403\,\text{m}^3/\text{s}$.

Time to empty slice, $t_1 = Vol_1/Q_1 = 18.5/0.0403 = 459.1\,\text{s}$.

With water surface 1.5 m above orifice, $Q_A = 0.0209 \times 1.5^{1/2} = 0.0256\,\text{m}^3/\text{s}$. Average discharge for second slice,

$Q_2 = (0.0362 + 0.0256)/2 = 0.0309\,\text{m}^3/\text{s}$ so $t_2 = 9.5/0.0309 = 307.4\,\text{s}$.

With water surface at level of orifice $Q_A = 0$, so average for third slice = $(0.0256 + 0)/2 = 0.0128\,\text{m}^3/\text{s}$. Thus $t_3 = 3.5/0.0128 = 273.4$.

Time to empty tank = $\Sigma t = 459.1 + 307.4 + 273.4 = 1039.9\,\text{s}$ or 17.33 mins.

Percentage difference to previous answer = $100 \times (1040 - 977)/977 = +6.4\%$.

STQ7.3

If water level in tank 1 falls by δx then increase in tank 2, δy, is given by the ratio of the areas:

$\delta y = \delta x(2^2/8^2)$ giving $\delta y = 0.0625\delta x$. The change in differential head producing flow is:
$\delta h_D = \delta x + \delta y = \delta x + 0.0625\delta x$. Thus $\delta h_D = 1.0625\delta x$ so $\delta x = \delta h_D/1.0625$ (1)
Now change in volume of tank 1 = amount discharged through orifice, so
$-A_{WS} \times \delta x = Q_A \times \delta t$. Thus $-(2 \times 2)\delta x = Q_A\delta t$. But from (1) $\delta x = \delta h_D/1.0625$, so
$-4\delta h_D/1.0625 = Q_A\delta t$.
Also $Q_A = C_D A(2gh_D)^{1/2}$. Thus $-4\delta h_D/1.0625 = C_D A(2g)^{1/2}h_D^{1/2}\delta t$. With $C_D = 0.80$ and $A =$
0.00785 m², rearranging gives: $\delta t = -\dfrac{4}{1.0625 \times 0.80 \times 0.00785 (19.62)^{1/2}} h_D^{-1/2}\delta h_D$.

Thus $\delta t = -135.5h_D^{-1/2}\delta h_D$. Integrating this expression to find the time for the head to fall from $H_1 = 19$ m to $H_2 = 0$: $T = -135.3\int_{19}^{0} h_D^{-1/2}\mathrm{d}h_D$. Thus $T = -135.3[+2h_D^{1/2}]_{19}^{0}$ so
$T = -135.3[2 \times 0 - 2 \times 19^{1/2}] = 1179.5$ s or 19.66 mins.

STQ8.1

Sides slope at 1:2 so with a depth of 2.7 m the surface width increases by $2 \times 2.7 = 5.4$ m per side. The bottom width $B = 8.3$ m. This makes the surface width, $B_S = 5.4 + 8.3 + 5.4 =$ 19.1 m.

$$A = \frac{1}{2}(8.3+19.1) \times 2.7 = 36.990 \text{ m}^2.$$

The length of the wetted side slopes are $(2.7^2 + 5.4^2)^{1/2} = 6.037$ m. Thus the length of the wetted perimeter, $P = 6.037 + 8.3 + 6.037 = 20.374$ m.
The hydraulic radius, $R = A/P = 36.990/20.374 = 1.816$ m.
Mean velocity, $V = (1/n)R^{2/3}S_0^{1/2} = (1/0.035) \times 1.816^{2/3} \times 0.001^{1/2} = 1.345$ m/s
Discharge, $Q = AV = 36.990 \times 1.345 = 49.75$ m³/s.

STQ8.2

For $D = 1.0$ m, $A = \left[\dfrac{1}{4}(\pi \times 1.0^2) \times 2\right] + 1.0 \times 1.0 = 2.571$ m².

$P = \left[\dfrac{1}{4}(2 \times \pi \times 1.0) \times 2\right] + 1.0 = 4.142$ m and $R = A/P = 2.571/4.142 = 0.621$ m.
$Q = (A/n)R^{2/3}S_0^{1/2} = (2.571/0.017)0.621^{2/3}(1/600)^{1/2} = 4.49$ m³/s.
For $D = 2.0$ m, $A = 2.571 + 3.0 \times 1.0 = 5.571$ m² and $P = 4.142 + 2 \times 1.0 = 6.142$ m.
$R = A/P = 5.571/6.142 = 0.907$ m.
$Q = (A/n)R^{2/3}S_0^{1/2} = (5.571/0.017)0.907^{2/3}(1/600)^{1/2} = 12.53$ m³/s.
For $D = 3.0$ m, $A = 5.571 + 3.0 \times 1.0 = 8.571$ m² and $P = 6.142 + 2 \times 1.0 = 8.142$ m.
$R = 8.571/8.142 = 1.053$ m. $Q = (8.571/0.017)1.053^{2/3}(1/600)^{1/2} = 21.29$ m³/s.

STQ8.3

Most of the calculations are conducted in Table STQ8.3.

(a) Using equation (8.18), the total discharge is the sum of the discharges in the subsections, which are calculated using the Manning equation. These are shown in column 9, and $Q = 178.719\,\text{m}^3/\text{s}$.

Using equation (8.19) and the values in columns 2, 4, 5 and 6 of the table, the average Manning n value of the entire channel cross-section (n_{AV}) is:

$$n_{AV} = \frac{PR^{5/3}}{\sum_{i=1}^{N}\left(\dfrac{P_i R_i^{5/3}}{n_i}\right)} = \frac{41.426 \times 1.967^{5/3}}{3996.430} = 0.032\ \text{s/m}^{1/3}$$

and $Q = \dfrac{81.500}{0.032} \times 1.967^{2/3} \times \left(\dfrac{1}{500}\right)^{1/2} = 178.797\ \text{m}^3/\text{s}$

This is the same answer as that obtained above since the inherent assumptions are the same (the slight difference is due to rounding errors).

Using equation (8.20) and the values in columns 2, 4 and 7, n_{AV} is:

$$n_{AV} = \left[\frac{\sum_{i=1}^{N}\left(P_i n_i^{3/2}\right)}{P}\right]^{2/3} = \left[\frac{0.3722}{41.426}\right]^{2/3} = 0.043\ \text{s/m}^{1/3}$$

and $Q = \dfrac{81.500}{0.043} \times 1.967^{2/3} \times \left(\dfrac{1}{500}\right)^{1/2} = 133.058\ \text{m}^3/\text{s}$

Using equation (8.21) and the values in columns 2, 4 and 8, n_{AV} is:

$$n_{AV} = \left[\frac{\sum_{i=1}^{N}\left(P_i n_i^{2}\right)}{P}\right]^{1/2} = \left[\frac{0.0793}{41.426}\right]^{1/2} = 0.044\ \text{s/m}^{1/3}$$

and $Q = \dfrac{81.500}{0.044} \times 1.967^{2/3} \times \left(\dfrac{1}{500}\right)^{1/2} = 130.034\ \text{m}^3/\text{s}$

Table STQ8.3

Col. 1 Sub-section	2 n_i s/m$^{1/3}$	3 A_i m^2	4 P_i m	5 $R_i = A_i/P_i$ m	6 $P_i R_i^{5/3}/n_i$ Eqn (8.19)	7 $P_i n_i^{3/2}$ Eqn (8.20)	8 $P_i n_i^{2}$ Eqn (8.21)	9 Q_i m^3/s	10 V_i m/s	11 $V_i^3 A_i$
1	0.055	8.000	8.246	0.970	142.506	0.1064	0.0249	6.374	0.797	4.050
2	0.050	18.000	9.000	2.000	571.438	0.1006	0.0225	25.555	1.420	51.539
3	0.035	8.250	3.354	2.460	429.564	0.0220	0.0041	19.208	2.328	104.089
4	0.030	21.000	6.000	3.500	1613.518	0.0312	0.0054	72.157	3.436	851.881
5	0.025	8.250	3.354	2.460	601.389	0.0133	0.0021	26.891	3.260	285.829
6	0.040	14.000	7.000	2.000	555.565	0.0560	0.0112	24.845	1.775	78.293
7	0.045	4.000	4.472	0.894	82.450	0.0427	0.0091	3.689	0.922	3.135
Total		81.500	41.426	1.967 $= \dfrac{81.500}{41.426}$	3996.430	0.3722	0.0793	178.719	2.193 $= \dfrac{178.719}{81.500}$	1378.816

(b) From the above, Q is between $130\,\text{m}^3/\text{s}$ and $179\,\text{m}^3/\text{s}$, a difference of 38% based on the smaller value (27% based on the larger value). This is not surprising: as explained in the text, the assumptions inherent in the equations are different so they give different answers. It may be logical to expect that the discharges based on the average hydraulic radius and average roughness would be in error when the depth and roughness vary considerably over the cross-section. Indeed, in some problems where n is relatively constant around the perimeter, all of the above equations yield similar answers; the discrepancies increase where n varies considerably. Which of the Q values to use depends upon the circumstances. For example, if the cross-section in the question is part of a flood relief channel, then the safe option is to use the lower value so that there is less danger of under-designing the channel. The failure of the channel to carry the required flood discharge would undoubtedly attract unwelcome publicity and legal problems.

(c) From equation (4.26) and columns 3, 10 and 11:

$$\alpha = \frac{\sum_{i=1}^{N}(V_i^3 A_i)}{V^3 A} = \frac{1378.816}{2.193^3 \times 81.500} = 1.60$$

STQ8.4

$[(B + SD_C)D_C]^3 = (Q^2/g)(B + 2SD_C)$ so $[(9 + 4D_C)D_C]^3 = (35^2/9.81)(9 + 2 \times 4 \times D_C)$
$[(9 + 4D_C)D_C]^3 = 124.873(9 + 8D_C)$

Try $D_C = 1\,\text{m}$	Left side = 2197	Right side = 3122
$D_C = 0.95\,\text{m}$	= 1798	= 2073
$D_C = 0.995\,\text{m}$	= 2154	= 2118
$D_C = 0.990\,\text{m}$	= 2112	= 2113 OK

(a) Thus $D_C = 0.990\,\text{m}$.

(b) $A_C = (B + SD_C)D_C = (9 + 4 \times 0.990)0.990 = 12.830\,\text{m}^2$.
$B_{SC} = (B + 2SD_C) = (9 + 2 \times 4 \times 0.990) = 16.920\,\text{m}$.
$D_{MC} = A_C/B_{SC} = 12.830/16.920 = 0.758\,\text{m}$.

(c) $V_C = (gD_{MC})^{1/2} = (9.81 \times 0.758)^{1/2} = 2.727\,\text{m/s}$.

STQ8.5

The problem is illustrated in Fig. 8.35.
From equation (8.32), critical depth $D_C = (Q^2/gB^2)^{1/3} = (35^2/9.81 \times 6^2)^{1/3} = 1.51\,\text{m}$.
From equation (8.28), critical velocity $V_C = (gD_C)^{1/2} = (9.81 \times 1.51)^{1/2} = 3.85\,\text{m/s}$.
At critical depth, $P_C = 6 + (2 \times 1.51) = 9.02\,\text{m}$, $A_C = 9.06\,\text{m}^2$, and $R_C = A_C/P_C = 9.06/9.02 = 1.00\,\text{m}$.
From equation (8.30), critical slope $S_C = V_C^2 n^2/R_C^{4/3} = 3.85^2 \times 0.017^2/1.00^{4/3} = 0.0043$ or 1 in 233.
Thus $S_O < S_C$ and $D_N > D > D_C$ so there is a M2 drawdown curve (Fig. 8.29 and Table 8.6). The drop acts as the control point, with $D = D_C = 1.51\,\text{m}$ when $Q = 35.00\,\text{m}^3/\text{s}$.

   The calculations are summarised in Table STQ8.5. Since the flow is subcritical ($D > D_C$), we start at the control point where the flow passes through critical depth and work back upstream. Looking at the Depth–E diagram in Fig. 8.35, we are on the part of the curve

Table STQ8.5

Note that the values in the table were based on a computer spreadsheet that used more significant figures than shown below, so there may be some trivial discrepancies arising from rounding errors etc. In column 11, E_2 is the value for the current row; E_1 is the value in column 6 of the previous row.

Col. 1 Chainage m	2 ΔL m	3 D m	4 A m²	5 V m/s	6 $E = D + V^2/2g$ Eqn (8.49) m	7 n s/m$^{1/3}$	8 R m	9 S_F Eqn (8.40)	10 $\overline{S_F}$ Eqn (8.43)	11 $E_2 = E_1 - (S_o - \overline{S_F})\Delta L$ Eqn (8.50) m
(CP) 0	0	1.510	9.060	3.863	2.271	0.017	1.004	0.00429	—	—
100	100	2.032	12.192	2.871	2.452	0.017	1.211	0.00184	0.00307	2.452
200	100	2.114	12.684	2.759	2.502	0.017	1.240	0.00165	0.00175	2.502
300	100	2.168	13.008	2.691	2.537	0.017	1.259	0.00154	0.00160	2.537
400	100	2.205	13.230	2.646	2.562	0.017	1.271	0.00147	0.00151	2.562
500	100	2.233	13.398	2.612	2.581	0.017	1.280	0.00142	0.00144	2.581
600	100	2.255	13.530	2.587	2.596	0.017	1.287	0.00138	0.00140	2.596
700	100	2.272	13.632	2.567	2.608	0.017	1.293	0.00135	0.00137	2.608
800	100	2.285	13.710	2.553	2.617	0.017	1.297	0.00133	0.00134	2.617
900	100	2.295	13.770	2.542	2.624	0.017	1.300	0.00132	0.00132	2.625
1000	100	2.303	13.818	2.533	2.630	0.017	1.303	0.00130	0.00131	2.630
1100	100	2.310	13.860	2.525	2.635	0.017	1.305	0.00129	0.00130	2.635
1200	100	2.316	13.896	2.519	2.639	0.017	1.307	0.00128	0.00129	2.639
1300	100	2.320	13.920	2.514	2.642	0.017	1.308	0.00128	0.00128	2.642
1400	100	2.324	13.944	2.510	2.645	0.017	1.310	0.00127	0.00127	2.645
1500	100	2.327	13.962	2.507	2.647	0.017	1.310	0.00127	0.00127	2.647
1600	100	2.330	13.980	2.504	2.649	0.017	1.311	0.00126	0.00126	2.649

where E increases as D increases. With $S_o = 1/800 = 0.00125$, the value of $(S_o - \overline{S_F})\Delta L$ has a negative value so to obtain the desired increase in E we must write equation (8.50) in column 11 as: $E_2 = E_1 - (S_o - \overline{S_F})\Delta L$.

(a) The values of D in column 3 show that the water surface is within 10 mm of normal depth (i.e. $2.34 - 0.01 = 2.33$ m) at a distance of 1600 m from the drop.

(b) In the first 100 m, the change is $(2.032 - 1.510) = 0.522$ m.

(c) Between 1500 m and 1600 m, the change is $(2.330 - 2.327) = 0.003$ m. Thus with an M2 curve the curvature is greatest near the drop and very shallow at the upstream end.

See also Revision Question 8.10 which relates to this problem.

STQ9.1

As in Example 9.1, the hydrostatic force due to the water $F = 4414.50 \times 10^3$ N.

The weight of the dam = weight density of masonry × volume of dam per m length

$$= 2700 \times 9.81 \times (30 \times B \times 1)$$

$$= 794.61B \times 10^3$$

By similar triangles: $\dfrac{794.61B \times 10^3}{4414.50 \times 10^3} = \dfrac{10}{(B/6)}$

$$794.61B^2 = 264\,870$$
$$B = 18.26\,\text{m}$$

Therefore compared to concrete, the width is reduced by $(19.57 - 18.26) = 1.31\,\text{m}$.

STQ9.2

$B = 67.5\,\text{m}$. Say the piers are 1.5 m wide, then $b = 67.5 - (5 \times 1.5) = 60.0\,\text{m}$.
Thus $m = (1 - b/B) = (1 - 60.0/67.5) = 0.11$ (just less than Yarnell's minimum value in Table 9.6).
Use piers with a semicircular nose and tail, and assume $K_Y = 0.90$.
As in Example 9.3, $D_N = 4.40\,\text{m}$ and $F_N = 0.44$.

$$H_1^* = K_Y D_N F_N^2 \left(K_Y + 5F_N^2 - 0.6\right)(m + 15m^4)$$
$$= 0.90 \times 4.40 \times 0.44^2 (0.90 + 5 \times 0.44^2 - 0.6)(0.11 + 15 \times 0.11^4)$$
$$= 0.77 \times 1.27 \times 0.11$$
$$H_1^* = 0.11\,\text{m}$$

Thus the afflux can be reduced from 0.27 m to 0.11 m by using narrower 1.5 m piers with rounded noses and tails.

STQ10.1

The fundamental dimensions of the quantities involved are: discharge (L^3T^{-1}), length (L), gravity (LT^{-2}), and head (L). C_D is assumed to be dimensionless. Putting these dimensions into the equation in the question gives: $L^3T^{-1} = L \times (LT^{-2})^{1/2} \times L^{3/2}$. Using the laws of indices to determine the powers of the right-hand side of the equation gives: $L^3T^{-1} = L^3T^{-1}$. Thus dimensional homogeneity is satisfied without the inclusion of C_D, so C_D must be dimensionless.

STQ10.2

Let $F = K\rho^a A^b V^c$ where K is a constant. Inserting the fundamental dimensions gives: $MLT^{-2} = [ML^{-3}]^a[L^2]^b[LT^{-1}]^c$. Equating powers of M: $1 = a$. Equating powers of T: $-2 = -c$, so $c = 2$. Equating powers of L: $1 = -3a + 2b + c$. Substituting for a and c gives: $1 = -3 + 2b + 2$ so $b = 1$. Thus $F = K\rho A V^2$.

STQ10.3

$\Pi_1 = \Delta H/l$ as before. $\Pi_2 = \rho^{a_2}\mu^{b_2}g^{c_2}V$ and $\Pi_3 = \rho^{a_3}\mu^{b_3}g^{c_3}D$. Substituting the fundamental dimensions into the equation for Π_2 gives: $0 = [ML^{-3}]^{a_2}[ML^{-1}T^{-1}]^{b_2}[LT^{-2}]^{c_2}LT^{-1}$. Equating powers of M: $0 = a_2 + b_2$ so $a_2 = -b_2$. Equating powers of L: $0 = -3a_2 - b_2 + c_2 + 1$. Substituting $a_2 = -b_2$ and multiplying by 2 gives $0 = 4b_2 + 2c_2 + 2 \ldots$ (1). Equating powers of T: $0 = -b_2 - 2c_2 - 1 \ldots$ (2). Adding equations (1) and (2) gives: $0 = 3b_2 + 1$ so $b_2 = -1/3$ which makes $a_2 = +1/3$. Substituting these values in (2) gives $c_2 = -1/3$. Thus $\Pi_2 = \rho^{1/3}\mu^{-1/3}g^{-1/3}V$ or $(\rho V^3/\mu g)^{1/3}$.

$\Pi_3 = \rho^{a_3}\mu^{b_3}g^{c_3}D$. Substituting MLT gives $0 = [ML^{-3}]^{a_3}[ML^{-1}T^{-1}]^{b_3}[LT^{-2}]^{c_3}L$. Equating powers of M: $0 = a_3 + b_3$ so $a_3 = -b_3$. Equating powers of T: $0 = -b_3 - 2c_3$ so $c_3 = -b_3/2$. Equating powers of L: $0 = -3a_3 - b_3 + c_3 + 1$ and substituting for a_3 and c_3 gives: $0 = -3(-b_3) - b_3 + (-b_3/2) + 1$ or $b_3 = -2/3$. Thus $a_3 = +2/3$ and $c_3 = +1/3$ and $\Pi_3 = \rho^{2/3}\mu^{-2/3}g^{1/3}D$ or $(\rho^2 gD^3/\mu^2)^{1/3}$. Hence $\Delta H/l = f[(\rho V^3/\mu g)^{1/3}, (\rho^2 gD^3/\mu^2)^{1/3}]$ or $\Delta H/l = f'[(\rho V^3/\mu g), (\rho^2 gD^3/\mu^2)]$.

STQ11.1

(i) Example 11.4. The equations are as before but with $\eta = 1.0$ giving $F_{RX} = 958\,N$ (+5.2%), $F_{RY} = 126\,N$ (+11.5%) and $F_R = 966\,N$ (+5.2%). Example 11.5, with $\eta = 1.0$, $F_{RX} = 1824\,N$ (+5.2%), $F_{RY} = 240\,N$ (+11.1%) and $F_R = 1840\,N$ (+5.3%).

(ii) The equations are as in Example 11.5. With $U = 9.0\,m/s$, $F_{RX} = 1817\,N$ so $Pow = 1817 \times 9.0 = 16\,353\,W$ and change in $F_{RX} = +4.8\%$. With $U = 10.0\,m/s$, $F_{RX} = 1652\,N$ so $Pow = 1652 \times 10.0 = 16\,520\,W$ and change in $F_{RX} = -4.7\%$. With $U = 11.0\,m/s$, $F_{RX} = 1487\,N$ so $Pow = 1487 \times 11.0 = 16\,357\,W$ and change in $F_{RX} = -14.2\%$. Note that F_{RX} decreases as U increases but maximum power is obtained when $U = V_1/2 = 10.0\,m/s$.

STQ11.2

$Pow/N^3 D^5 = c_5$ so $D = (Pow/N^3 c_5)^{1/5}$ or $Pow^{1/5}/N^{3/5}c_5^{1/5}$. Also $H/N^2 D^2 = c_4$ so $D = (H/N^2 c_4)^{1/2}$ or $H^{1/2}/N c_4^{1/2}$. Equating the two expressions for D gives: $Pow^{1/5}/N^{3/5}c_5^{1/5} = H^{1/2}/N c_4^{1/2}$. Rearranging gives: $NPow^{1/5}/N^{3/5}H^{1/2} = c_8$ where c_8 is a new constant incorporating c_4 and c_5. Thus $N^{2/5}Pow^{1/5}/H^{1/2} = c_8$. Multiplying all exponents by $\frac{5}{2}$ gives $NPow^{1/2}/H^{5/4} = c_9 = N_S$.

STQ11.3

Note that the answers are subjective since they involve drawing graphs and measuring values, so slight variations are to be expected. (a) Only P//P can deliver $0.4\,m^3/s$ with a maximum static lift of $7.5\,m$ (measure the vertical distance from the system curve to pump curve). (b) With $Q = 0.15\,m^3/s$ the maximum lifts are: P = $17.8\,m$, (P + P) = $36.9\,m$ and (P//P) = $20.4\,m$ (measure vertical distance between system and pump curves). (c) With a static lift of $15\,m$, the corresponding discharges are: P = $0.190\,m^3/s$, (P + P) = $0.285\,m^3/s$ and (P//P) = $0.285\,m^3/s$ (find the discharge at which the system and pump curves are apart by $15\,m$). (d) With a static lift of $20\,m$: P = $0.105\,m^3/s$, (P + P) = $0.260\,m^3/s$ and (P//P) = $0.165\,m^3/s$ (find the discharge where the curves are $20\,m$ apart).

STQ12.1

(a) Urban areas tend to be warmer than the surrounding countryside; the mean temperature of central London is $10.9°C$ compared to $9.6°C$ outside the urban area (Shaw, 1994). Generally, the annual mean temperature of urbanised areas is 0.5–$1.0°C$ higher with 10 times more condensation nuclei and particles. Thus the likelihood of convection cells, cloud formation and precipitation is increased; typically urban areas have 5%–10% more cloud and 5%–10% more total precipitation.

(b) If trees are felled as part of the urbanisation process, then less rainfall will be intercepted by the forest canopy and more rain will fall directly onto the ground surface.

(c) Clearly, if absorbent soil and grass is replaced by impermeable roads, pavements, car parks and house roofs, then the area over which water can infiltrate into the ground is much reduced.

(d) Following on from (c), if rainfall cannot infiltrate, then the amount that runs off the surface will be increased (see Table 13.9). This alters the hydrograph, as described below.

(e) With natural, vegetated surfaces, runoff is likely to be slowed as a result of depression storage in puddles, obstruction by roots or other surface irregularities, and perhaps by temporary absorption by leaf mulch etc. On the other hand, smoothly sloping, impermeable, man-made surfaces are usually designed to promote rapid runoff and to minimise flooding. Thus after urbanisation, the hydrograph peaks earlier and has a more rapid rise than for similar rainfall prior to development. Improvements to river channels (e.g. smoother, straighter alignment) and the loss of floodplain storage may also result in rivers having a higher velocity and increased discharge, so the floods are larger and travel downstream more quickly (see Example 13.6 and section 13.4).

(f) If urbanisation results in large impermeable areas and reduced infiltration, then the amount of rainfall that can percolate to the water table and become groundwater will also be reduced.

STQ12.2

(a) The data are plotted in Fig. STQ12.2. For a 10 min storm, the largest and smallest intensities are about 250 mm/h and 36 mm/h; for 90 min they are about 90 mm/h and 9 mm/h.

(b) The significance is that there is no single rainfall intensity that can be adopted in design calculations, but a whole range of possible values. Any of them may be correct in appropriate circumstances, but wrong in others. All of them have an associated risk of failure, i.e. something larger may occur causing the alleviation works to fail (e.g. see Example 13.5). Even if the British maximum observed intensity in Table 12.4 is used, there is still a small risk that this will be exceeded. For example, there is a controversial observation that 193 mm of rain fell in 2 h at Calderdale in Yorkshire on 19 May 1989. If valid, this is higher than the equivalent value in the table (Institute of Hydrology, 1999). See also probable maximum precipitation in section 13.2.

(c) Normally the return interval adopted must reflect the consequences of failure and the potential loss of life or damage that would result. Consequently any general guidelines must be applied in such a way as to take into consideration local conditions and any special circumstances.

(i) Between a 1 in 150 year and a 1 in 10000 year or probable maximum flood according to the category of dam (see Table 13.3).

(ii) A return interval of about 1 in 100 years may be appropriate. Basements can be flooded as a result of rivers bursting their banks or by large quantities of runoff from streets flowing into the rooms which are below ground level. They may also be flooded indirectly by floodwater backing up along drains (Fig. 13.17). If water suddenly bursts into a basement, it may fill very quickly, giving little time to escape. Many people have been trapped and drowned

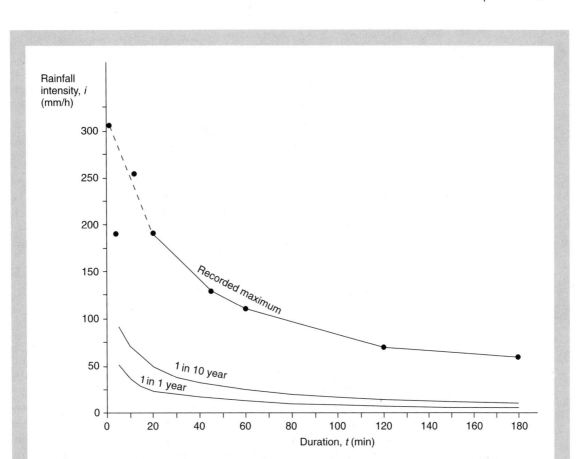

STQ 12.2

by in-rushing flood water. Sometimes escape has been prevented by security bars fixed over the windows, or doors that could not be opened against the force of the water. An illustration of how bad this problem can be is provided by the 1953 surge tide that rose 1.13 m above the high-water level of a high spring tide at London Bridge. This resulted in over-topping of the flood defences along much of the east coast of England, the flooding of 650 km² of farmland and 24 000 houses. Over 300 people drowned, including invalids in bed and engineers inspecting flood defences. Around 11 000 cattle, 9000 sheep, 2400 pigs, 34 000 poultry and 70 horses also drowned. When London's flood defences were assessed after the event, the Thames Technical Panel considered the possibility of a surge 1.83 m above the 1953 flood tide. This would have caused widespread flooding of base-ments with some properties having up to 3.05 m of water above ground floor level. Not sur-prisingly, London is now protected by the Thames Barrier, while the estuary downstream has been provided with higher flood banks (Gilbert and Horner, 1984). Figure 13.13 shows part of the defences on Canvey Island, where 58 people drowned in 1953 when the old wall collapsed (see also Box 12.2).

(iii) Return periods of 1 in 1, 1 in 2 or 1 in 5 years are common for storm water sewers.

STQ12.3

Columns 1 and 2 of Table STQ12.3 show the original discharge data. Columns 3 and 4 show the calculated ratios that should be plotted to locate N.

(a) Point M is at 0700h where the discharge starts to increase. The total discharge hydrograph is shown in Fig. STQ12.3 with the $\log_{10} Q$ and $Q_t/Q_{t+\Delta t}$ ratio plotted above. From these graphs it is apparent that N is at 1600h. M and N are joined by a straight line.

(b) The groundwater baseflow increases from $4.36\,\text{m}^3/\text{s}$ at M to $8.95\,\text{m}^3/\text{s}$ at N, a change of $4.59\,\text{m}^3/\text{s}$ in 9h, or $0.51\,\text{m}^3/\text{s}$ per hour. Thus the baseflow at 0800h is $4.36 + 0.51 = 4.87\,\text{m}^3/\text{s}$, at 0900h it is $4.87 + 0.51 = 5.38\,\text{m}^3/\text{s}$ etc., as shown in column 5. The difference between the values in columns 2 and 5 is the surface runoff ($q\,\text{m}^3/\text{s}$) shown in the last column.

As in Example 12.5, the total volume of surface runoff is equivalent to the area of the hydrograph above MN. The time interval is $1\text{h} = 3600\text{s}$. Thus using equation (12.9):

$$VOL = 3600\left(\frac{0+0}{2}+0.10+0.61+2.17+7.28+9.34+7.99+5.59+2.47\right)$$
$$VOL = 3600\times35.55$$
$$= 127980\,\text{m}^3$$

(c) Here the practice of joining M to N with a straight line looks wrong. Between 0700 and 0800h the total discharge increases by $0.61\,\text{m}^3/\text{s}$ (from 4.36 to $4.97\,\text{m}^3/\text{s}$) but only $0.10\,\text{m}^3/\text{s}$ of surface runoff is recorded in the last column of the table at 0800h. This would indicate that the initial rise of the hydrograph is mostly the result of an increase in groundwater baseflow, which is unlikely since groundwater responds to rainfall much more slowly than surface runoff. It is more likely that the baseflow would have continued to decline for a while before starting to rise (dotted line), and that the surface runoff is responsible for the entire initial rise of the hydrograph. The straight line is too simple to reflect reality in this example.

Table STQ12.3

Time GMT t (h)	Q (m³/s)	log₁₀ Q	Qt/Qt+Δt	Baseflow (m³/s)	Surface runoff q (m³/s)
0600	4.45			—	—
0700	4.36			4.36	0 (M)
0800	4.97			4.87	0.10
0900	5.99			5.38	0.61
1000	8.06			5.89	2.17
1100	13.68			6.40	7.28
1200	16.25			6.91	9.34
1300	15.41			7.42	7.99
1400	13.52	1.131	1.239	7.93	5.59
1500	10.91	1.038	1.219	8.44	2.47
1600	8.95	0.952	1.097	8.95	0 (N)
1700	8.16	0.912	1.069	8.16	
1800	7.63	0.883	1.063	7.63	
1900	7.18	0.856	1.050	7.18	
2000	6.84	0.835	1.036	6.84	
2100	6.60	0.820	—	6.60	

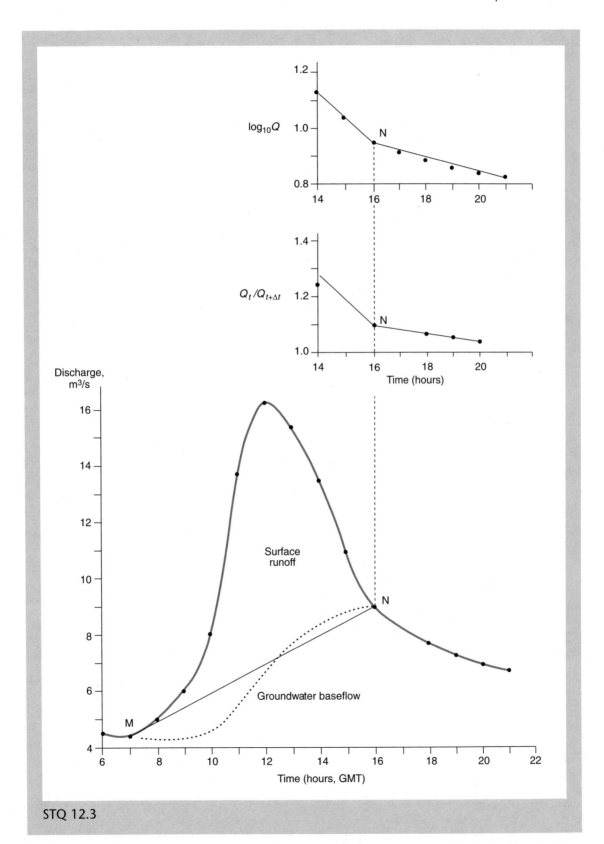

STQ 12.3

STQ13.1

The calculations are shown in Table STQ13.1. Precipitation exceeds PET until May when a SMD starts to build. This is similar to Example 13.1. However, in this question the SMD is allowed to increase up to 75 + 25 = 100 mm, which is reached in July. In this month there is a storage change of –40 mm, but only 24 mm is needed to take the SMD to its 100 mm limit, so the remaining 16 mm represents the water deficiency recorded in row 6. In August, precipitation exceeds PET by 19 mm so the SMD is reduced by this amount and becomes 81 mm. In September, 125 mm of excess precipitation eliminates the 81 mm SMD with the remaining 44 mm being a water surplus. During this month the soil returns to its field capacity (75 mm), which is maintained to the end of the year.

Despite the high precipitation in many months, it is apparent from the table that the low summer rainfall results in a water deficiency in July, when there is insufficient rainfall and remaining soil moisture to meet the requirements of the crop. Thus wilting is likely unless irrigation is used. Whether or not the permanent wilting point is reached depends upon the distribution of rainfall during July and August.

STQ13.2

The first step is to identify the net rainfall and derive the UH. The design storm consists of two periods of constant intensity net rainfall of 6 h and 2 h duration, so steps 2 and 3 are to calculate the 2 h and 6 h UHs respectively. The last step is to obtain the final design hydrograph. The calculations are in Table STQ13.2, which has the given time–discharge data in rows 1 and 2. Where necessary, numbers have been rounded up to two decimal places.

Step 1 – observed net rainfall. The gross rainfall is given in the question. The first hour of rainfall is used eliminating the 16 mm soil moisture deficit, so that leaves 2 h of gross rainfall at 16 mm/h. Subtracting Φ gives 2 h of net rainfall at 10 mm/h, which is a total depth of 20 mm as shown in Fig. STQ13.2 and row 2 of the table.

Step 2 – derive the 2 h UH. The observed total depth of net rainfall is 20 mm, so dividing the ordinates in row 2 by 2 gives the 2 h UH (= 10 mm depth) recorded in row 3.

Step 3 – derive the 6 h UH. The 6 h UH is created from three contiguous 2 h periods of rainfall, so the surface runoff ordinates must also be repeated three times, the second and third repetitions each having a 2 h offset as in rows 4 and 5. Adding rows 3 to 5 gives the runoff

		Jan	Feb	Mar	Apr	May	Jun	Jul	Aug	Sep	Oct	Nov	Dec	Year
1	Precipitation	201	246	128	113	53	26	28	86	173	127	109	255	1545
2	PET	19	19	36	59	72	83	68	67	48	33	17	16	537
3	Storage change	0	0	0	0	–19	–57	–40	+19	+125	0	0	0	
4	SMD	0	0	0	0	19	76	100	81	0	0	0	0	
5	Storage balance	75	75	75	75	56	0	0	0	75	75	75	75	
6	Water deficiency	0	0	0	0	0	0	16	0	0	0	0	0	16
7	Water surplus	182	227	92	54	0	0	0	0	44	94	92	239	1024
8	Total water surplus			555								469		1024

Table STQ13.1

Table STQ13.2

1	Depth of net rain	0	2	4	6	8	10	12	14	16	18	20	22	24
2 q (m³/s)	20 mm	0	1.20	3.28	6.96	9.92	8.80	5.88	3.52	1.56	0			
3 $q/2$ = 2 h UH	10 mm	0	0.60	1.64	3.48	4.96	4.40	2.94	1.76	0.78	0			
4 Repeated 2 h UH	10 mm	offset	0	0.60	1.64	3.48	4.96	4.40	2.94	1.76	0.78	0		
5 Repeated 2 h UH	10 mm		offset	0	0.60	1.64	3.48	4.96	4.40	2.94	1.76	0.78	0	
6 Add rows 3 to 5	30 mm	0	0.60	2.24	5.72	10.08	12.84	12.30	9.10	5.48	2.54	0.78	0	
7 Row 6/3 = 6 h UH	10 mm	0	0.20	0.75	1.91	3.36	4.28	4.10	3.03	1.83	0.85	0.26	0	
8 Rain 1: 6 h UH × 2.4	24 mm	0	0.48	1.80	4.58	8.06	10.27	9.84	7.27	4.39	2.04	0.62	0	
9 Rain 2: 2 h UH × 5	50 mm		offset		0	3.00	8.20	17.40	24.80	22.00	14.70	8.80	3.90	0
10 Ground'r baseflow		4.50	4.50	4.50	4.50	4.50	4.50	4.50	4.50	4.50	4.50	4.50	4.50	4.50
11 Total design flow	74 mm	4.50	4.98	6.30	9.08	15.56	22.97	31.74	36.57	30.89	21.24	13.92	8.40	4.50

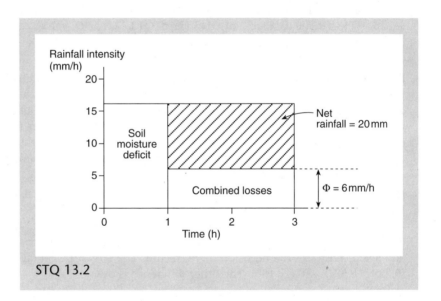

STQ 13.2

(row 6) from 30 mm depth of net rain. Divide row 6 by 3 to obtain the 6 h UH (in row 7) equivalent to 10 mm net depth of rainfall.

Step 4 – calculate the design hydrograph. The first period of rainfall is 6 h at 4 mm/h = 24 mm depth, so multiply the 6 h UH by 2.4 to get the values in row 8. The second period of rainfall starts after 6 h, so there is a 6 h offset at the start of row 9. This rain is 2 h at 25 mm/h = 50 mm depth, so multiply the 2 h UH by 5 to get the values in row 9. Row 10 shows the groundwater baseflow (assumed constant). Adding rows 8 to 10 gives the total design discharge in row 11, which can be plotted if desired.

STQ13.3

(a) To abstract 0.20 m³/s while leaving 0.22 m³/s downstream requires a minimum river flow of 0.42 m³/s. From the 1984 flow–duration curve a flow of 0.42 m³/s is equalled or exceeded for around 48% of the year or $0.48 \times 365 = 175.20$ days. This means the total volume of water available $= 0.20 \times 175.20 \times 24 \times 60 \times 60 = 3.03 \times 10^6 \, \mathrm{m}^3$. The amount that could be abstracted in an average or wetter year (1994) would be larger than this. The potential volume of abstraction could also increase by an amount calculated from (b).

(b) A flow of $0.22 \, \mathrm{m}^3/\mathrm{s}$ is equalled or exceeded for 67% of the time or $0.67 \times 365 = 244.55$ days. Thus there are $(244.55 - 175.20) = 69.35$ days when some abstraction is possible. If this was $0.10 \, \mathrm{m}^3/\mathrm{s}$ on average, then a further $0.10 \times 69.35 \times 24 \times 60 \times 60 = 0.60 \times 10^6 \, \mathrm{m}^3$ could be obtained. However, when the river's discharge is near the $0.22 \, \mathrm{m}^3/\mathrm{s}$ limit it may not be efficient to run the pumps with a much reduced discharge or for short periods, so not all of this may be extracted.

STQ13.4

(a) The ranked data are shown in Table STQ13.4a, along with the $P \, \%$ values calculated from equation (13.4). The data are plotted as Fig. STQ13.4a. The Q_T values equivalent to $P = 20\%$, 5% and 1% give Q_5, Q_{20} and $Q_{100} = 15.2 \, \mathrm{m}^3/\mathrm{s}$, $24.0 \, \mathrm{m}^3/\mathrm{s}$ and $36.0 \, \mathrm{m}^3/\mathrm{s}$ respectively. Example 13.6 gave much lower values ($11.8 \, \mathrm{m}^3/\mathrm{s}$, $17.3 \, \mathrm{m}^3/\mathrm{s}$, $23.8 \, \mathrm{m}^3/\mathrm{s}$).

(b) Repeating the above analysis but using all 26 years of record results in Table STQ13.4b and Fig. STQ13.4b. The data show less scatter and plot easier. This time Q_5, Q_{20} and $Q_{100} = 13.0 \, \mathrm{m}^3/\mathrm{s}$, $19.6 \, \mathrm{m}^3/\mathrm{s}$ and $28.0 \, \mathrm{m}^3/\mathrm{s}$. These are between the two values obtained from the short records. It may be surmised that the 1969–80 record contains larger flood peaks than the other (see the values of $QMED$ below). This is not unexpected: weather often fluctuates in cycles, such as a sequence of wetter than average years. A long record averages these fluctuations out and is more reliable.

(c) Using the 1969–80 record in Table STQ13.4a, $QMED = (10.085 + 8.493)/2 = 9.289 \, \mathrm{m}^3/\mathrm{s}$. From the 1981–94 record in Table 13.7, $QMED = 8.077 \, \mathrm{m}^3/\mathrm{s}$.
Using the full 1969–94 record in Table STQ13.4b, $QMED = (8.629 + 8.493)/2 = 8.561 \, \mathrm{m}^3/\mathrm{s}$. This confirms the statement above that the first part of the record contained larger floods, and again illustrates the need for a long record. The largest percentage difference is $100 \times (9.289 - 8.077)/8.077 = 15\%$. The other difference is 6%.

(d) The values are listed in Table STQ13.4b. They are included in Fig. 13.9b in the text.

STQ13.5

This question continues the calculations started in Example 13.8, which stopped at sewer 1.2. The remaining calculations are shown in Table STQ13.5. Note the following.

(a) Sewer 1.2 is a carrier sewer. The t_F and t_C of 0.76 and 8.51 min repectively are not used for the design of this sewer, but are used to design sewer 1.3 where the $t_C = 8.51 + 0.47 = 8.98 \, \mathrm{min}$.

(b) Sewer 3.0 is a branch, so the calculations effectively start afresh.

(c) Sewer 1.3 continues the calculations for the main branch. Since it is the first pipe downstream of a branch, the longest of the upstream t_C values is used (i.e. 8.51, not 7.96 min). Its cumulative impermeable area is the total of everything upstream ($= 0.021 \, \mathrm{km}^2$).

Table STQ13.4a

Water year	Q_T	Rank r	P %
1979	23.914	1	4.6
1973	15.519	2	12.9
1969	14.869	3	21.1
1975	13.907	4	29.4
1974	12.973	5	37.6
1977	**10.085**	6	45.9
1972	**8.493**	7	54.1
1971	8.493	—	—
1980	7.494	9	70.6
1970	6.107	10	78.9
1978	5.167	11	87.1
1976	3.047	12	95.4

Table STQ13.4b

Water year	Q_T	Rank r	P %	$Q_T/QMED$	y
1979	23.914	1	2.14	2.793	3.832
1981	17.914	2	5.97	2.093	2.787
1973	15.519	3	9.80	1.813	2.272
1969	14.869	4	13.63	1.737	1.921
1975	13.907	5	17.46	1.624	1.651
1992	13.843	6	21.29	1.617	1.430
1974	12.973	7	25.11	1.515	1.241
1985	12.973	—	—	—	—
1977	10.085	9	32.77	1.178	0.924
1982	10.085	—	—	—	—
1986	9.813	11	40.43	1.146	0.658
1987	9.543	12	44.26	1.115	0.537
1988	**8.629**	13	48.09	1.008	0.422
1971	**8.493**	14	51.91	0.992	0.312
1972	8.493	—	—	—	—
1994	7.525	16	59.57	0.879	0.099
1980	7.494	17	63.40	0.875	−0.005
1993	7.074	18	67.23	0.826	−0.109
1989	6.239	19	71.06	0.729	−0.215
1970	6.107	20	74.89	0.713	−0.323
1984	5.673	21	78.71	0.663	−0.436
1978	5.167	22	82.54	0.604	−0.557
1990	4.936	23	86.37	0.577	−0.690
1983	4.561	24	90.20	0.543	−0.843
1991	3.807	25	94.03	0.445	−1.036
1976	3.047	26	97.86	0.356	−1.346

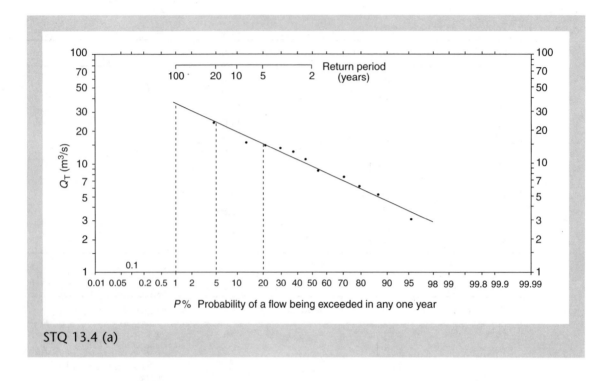

STQ 13.4 (a)

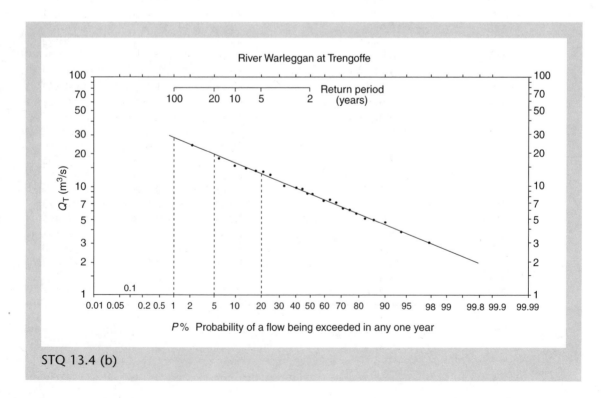

STQ 13.4 (b)

Table STQ13.5

Sewer ref. no.	Sewer length L (m)	Gradient	Guessed diameter (m)	Q_{FULL} (m³/s)	Mean flow velocity V(m/s)	$t_f =$ $L/60V$ (min)	t_c (min)	i (mm/h)	A_{IMP} for each sewer (km²)	ΣA_{IMP} (km²)	$Q_P =$ $0.278i\Sigma A_{IMP}$ (m³/s)	Comments
1.2	60	1 in 300 0.33 m/100 m	0.375	0.130	1.18	0.85	(8.60)		None	(0.012)	0.138	Fail
			0.450	0.210	1.32	0.76	(8.51)		None	(0.012)	0.138	OK
3.0	35	1 in 160 0.63 m/100 m	0.150	0.016	0.92	0.63	7.63	41.6	0.002	0.002	0.023	Fail
			0.225	0.048	1.20	0.49	7.49	42.0			0.023	OK
3.1	40	1 in 160 0.63 m/100 m	0.225	0.048	1.20	0.56	8.05	40.5	0.003	0.005	0.056	Fail
			0.300	0.101	1.42	0.47	7.96	40.7			0.057	OK
1.3	41	1 in 250 0.40 m/100 m	0.450	0.230	1.45	0.47	8.98	38.3	0.004	0.021	0.224	OK
1.4	39	1 in 400 0.25 m/100 m	0.525	0.275	1.25	0.52	9.50	37.1	0.005	0.026	0.268	OK

STQ13.6

The cumulative inflow, ΣI, is as in Example 13.10, and is shown below and plotted in Fig. 13.19.

Month	Apr	May	Jun	Jul	Aug	Sep	Oct	Nov	Dec	Jan	Feb	Mar
ΣI	3.15	5.68	7.73	9.61	11.11	12.96	21.66	30.79	44.64	57.29	68.27	75.10

(a) With the maximum storage capacity limited to $15.00 \times 10^6 \, m^3$, the graph must be replotted so that AB = $15.00 \times 10^6 \, m^3$, as in Fig. STQ13.6. The new uniform demand that can be supported is given by a line passing through 0 and B, which can be extrapolated to C. At C it has a value of about $55.90 \times 10^6 \, m^3$. Thus the gross uniform monthly demand that can be supported is $55.90 \times 10^6/12 = 4.66 \times 10^6 \, m^3$, which becomes $(4.66 - 0.81) = 3.85 \times 10^6 \, m^3$ with the compensation flow deducted. (Numerical check: B has a value of $(12.96 + 15.00) = 27.96 \times 10^6 \, m^3$, so the gross yield is $27.96 \times 10^6 \, m^3/6 = 4.66 \times 10^6 \, m^3$, as above.)

(b) The reservoir is initially full, and is full again when the demand line ΣO and ΣI lines cross in December. From December to the end of March the inflow exceeds the demand so the surplus inflow is passing over the spillway. The cumulative volume discharged over the spillway at the end of March is given by the ordinate CD in Fig. STQ13.6, which is about $19.20 \times 10^6 \, m^3$. (Numerical check: from above the total demand at C is $55.90 \times 10^6 \, m^3$ and ΣI at D = $75.10 \times 10^6 \, m^3$, so the difference CD is $19.20 \times 10^6 \, m^3$.)

STQ13.7

Table STQ13.7 shows the ranked data and calculated percentage probabilities, which are plotted in Fig. STQ13.7. It is apparent from the bottom scale that the probability of a flow of $1.5 \, m^3/s$ being exceeded is 98%, which means that the probability that the flow will be less than $1.5 \, m^3/s$ is 2% (this value can be read directly off the top scale). Thus on average

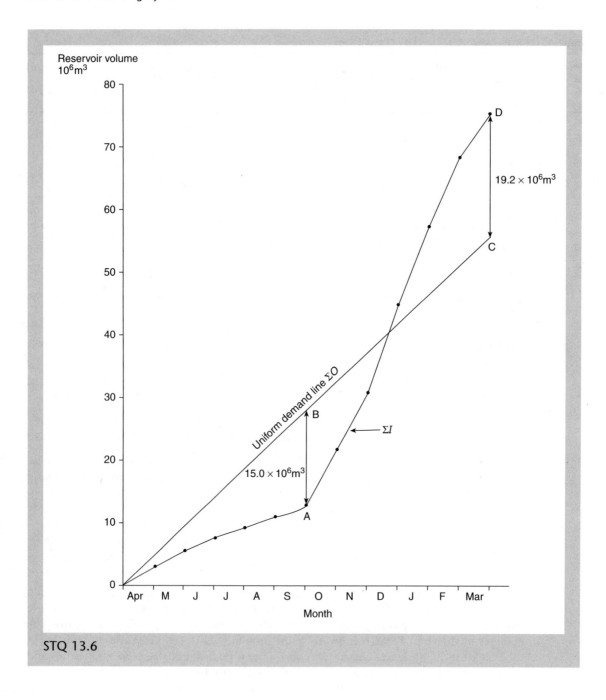

STQ 13.6

the flow will be less than required once every 50 years. With only 10 years of data, extrapolation to 50 years is not recommended. Additionally in this range the answer is affected significantly by only a small change in gradient, e.g. the line could be drawn to give a probability of 1% or 1 in 100 years, so doubling the calculated return period.

Table STQ13.7

Year	Min. flow (m³/s)	Rank (r)	P % = 100r/(N + 1)
5	4.37	1	9.1%
1	3.97	2	18.2%
4	2.90	3	27.3%
2	2.77	4	36.4%
9	2.73	5	45.5%
8	2.69	6	54.5%
10	2.54	7	63.6%
3	2.42	8	72.7%
7	2.11	9	81.8%
6	1.99	10	90.9%

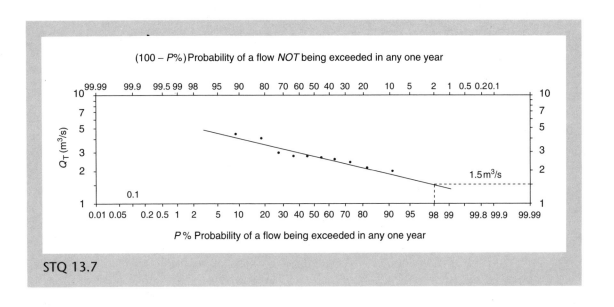

STQ 13.7

3

Graph paper

Log 3 Cycles × Probability

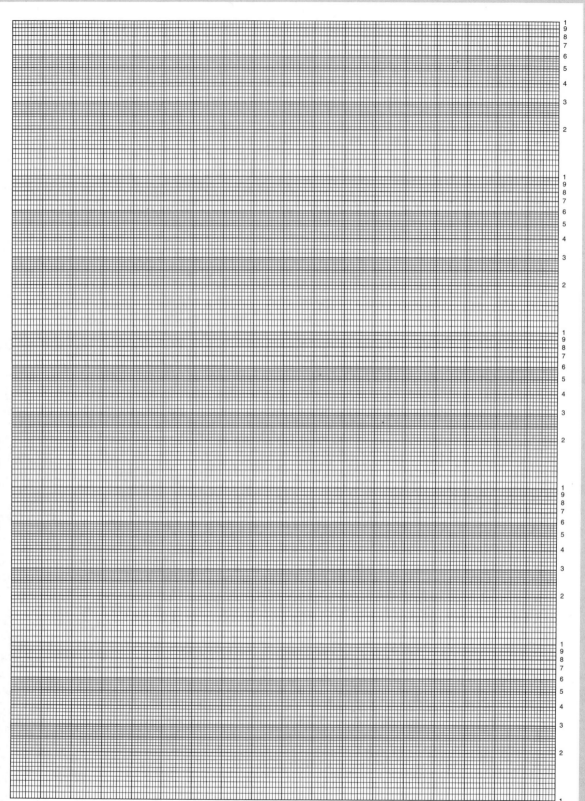

Log 5 Cycles × equal divisions

Gumbel probability

Index